Atomic and Molecular Physics of Controlled Thermonuclear Fusion

NATO Advanced Science Institutes Series

A series of edited volumes comprising multifaceted studies of contemporary scientific issues by some of the best scientific minds in the world, assembled in cooperation with NATO Scientific Affairs Division.

This series is published by an international board of publishers in conjunction with NATO Scientific Affairs Division

A Life Sciences **B Physics**	Plenum Publishing Corporation New York and London
C Mathematical and Physical Sciences	D. Reidel Publishing Company Dordrecht, Boston, and London
D Behavioral and Social Sciences **E Applied Sciences**	Martinus Nijhoff Publishers The Hague, Boston, and London
F Computer and Systems Sciences **G Ecological Sciences**	Springer Verlag Heidelberg, Berlin, and New York

Recent Volumes in Series B: Physics

Volume 96 —Atomic Physics of Highly Ionized Atoms
edited by Richard Marrus

Volume 97 —Mass Transport in Solids
edited by F. Bénière and C. R. A. Catlow

Volume 98 —Quantum Metrology and Fundamental Physical Constants
edited by Paul H. Cutler and Amand A. Lucas

Volume 99 —Techniques and Concepts in High-Energy Physics II
edited by Thomas Ferbel

Volume 100—Advances in Superconductivity
edited by B. Deaver and John Ruvalds

Volume 101—Atomic and Molecular Physics of Controlled Thermonuclear Fusion
edited by Charles J. Joachain and Douglass E. Post

Volume 102—Magnetic Monopoles
edited by Richard A. Carrigan, Jr., and W. Peter Trower

Volume 103—Fundamental Processes in Energetic Atomic Collisions
edited by H. O. Lutz, J. S. Briggs, and H. Kleinpoppen

Atomic and Molecular Physics of Controlled Thermonuclear Fusion

Edited by

Charles J. Joachain

University of Brussels
Brussels, Belgium

and

Douglass E. Post

Princeton University
Princeton, New Jersey

Plenum Press
New York and London
Published in cooperation with NATO Scientific Affairs Division

Proceedings of a NATO Advanced Study Institute,
held July 19-30, 1982,
at the Hotel Zagarella, Santa Flavia, Italy

Library of Congress Cataloging in Publication Data

Main entry under title:

Atomic and molecular physics of controlled thermonuclear fusion.

(NATO advanced science institutes series. Series B, Physics; v. 101)
"Published in cooperation with NATO Scientific Affairs Division."
"Proceedings of a NATO Advanced Study Institute, held July 19-30, 1982, at the Hotel Zagarella, Santa Flavia, Italy"—Verso of t.p.
Includes index.
1. Controlled fusion—Congresses. I. Joachain, C. J. (Charles Jean) II. Post, D. E. (Douglass Edmund), 1945- III. Series.
QC791.7.A86 1983 539.7′64 83-11128
ISBN-13: 978-1-4613-3765-2 e-ISBN-13: 978-1-4613-3763-8
DOI: 10.1007/978-1-4613-3763-8

Softcover reprint of the hardcover 1st edition 1983
A Division of Plenum Publishing Corporation
233 Spring Street, New York, N.Y. 10013

PREFACE

The need for long-term energy sources, in particular for our highly technological society, has become increasingly apparent during the last decade. One of these sources, of tremendous potential importance, is controlled thermonuclear fusion.

The goal of controlled thermonuclear fusion research is to produce a high-temperature, completely ionized plasma in which the nuclei of two hydrogen isotopes, deuterium and tritium, undergo enough fusion reactions so that the nuclear energy released by these fusion reactions can be transformed into heat and electricity with an overall gain in energy. This requires average kinetic energies for the nuclei of the order of 10 keV, corresponding to temperatures of about 100 million degrees. Moreover, the plasma must remain confined for a certain time interval, during which sufficient energy must be produced to heat the plasma, overcome the energy losses and supply heat to the power station.

At present, two main approaches are being investigated to achieve these objectives: magnetic confinement and inertial confinement. In magnetic confinement research, a low-density plasma is heated by electric currents, assisted by additional heating methods such as radio-frequency heating or neutral beam injection, and the confinement is achieved by using various magnetic field configurations. Examples of these are the plasmas produced in stellarator and tokamak devices. In the inertial confinement approach, a small pellet containing deuterium and tritium is compressed by pulses of intense laser radiation or beams of high-energy particles, to produce a high-density, hot plasma in such a short time that the nuclei can fuse and release energy before the plasma expands appreciably. The recent favorable results obtained with tokamaks and the steady progress on other fusion experiments throughout the world demonstrate that fusion power shows a great deal of promise. Large experiments such as the Toroidal Fusion Test Reactor (TFTR) at Princeton, the Joint European Torus (JET) at Culham, JT-60 in Japan and NOVA at Livermore are under way, and even larger experiments such as INTOR are being planned.

Atomic and molecular physics questions are proving to be key issues in the operation of these experiments. There are four main areas where these processes are particularly important : (1) the hot central plasma in which the fusion reactions are to occur, (2) the plasma edge where the plasma comes into contact with the external environment, (3) plasma heating methods and (4) diagnostic techniques used to measure the physical characteristics of the plasma.

In magnetic confinement research, atomic processes affect the hot center of the plasma mainly because of the fueling through charge exchange of protons and hydrogen atoms, and of energy losses due to line radiation from impurity ions. The greatest hope for reducing the impurity radiation is through control of the plasma-wall interface which is the source of these impurities; atomic and molecular processes are of fundamental importance in determining the plasma edge parameters. Neutral beam heating has been crucial to the success of recent tokamak and mirror experiments, and many of the techniques used for the diagnostics of magnetic fusion experiments rely on atomic processes.

In research on inertially confined plasmas, atomic processes play a very important role in the central plasma, where they influence the pellet implosion hydrodynamics; the atomic physics required is that of dense matter, and constitutes a particularly interesting new field. A second area in inertial confinement research which involves atomic physics is that of the interaction of the pellet debris and the containment vessel. Both the construction of drivers based on laser or ion beams, and the interaction of drivers with pellets involve atomic physics problems. Finally, as in the case of magnetic confinement, many diagnostic methods rely heavily on atomic physics.

The aim of the Advanced Study Institute from which this volume originated was to bring together senior researchers and students in both atomic physics and fusion research, to survey the role of atomic and molecular processes in fusion plasmas, and to review recent developments in theoretical and experimental research dealing with these processes. The Advanced Study Institute was held at the Hotel Zagarella, Santa Flavia, near Palermo, from 19th to 30th July 1982. In addition to the invited lectures printed in this volume, forty research seminars were given by participants.

Following the scientific program of the meeting, the lectures published in this book are divided into three parts : (1) Overview of fusion energy research (2) The calculation and measurement of atomic and molecular processes relevant to fusion and (3) The atomic and molecular physics of controlled thermonuclear research devices. We hope that these lectures will provide to scientists from various disciplines a comprehensive introduction to the atomic and molecular physics of controlled thermonuclear fusion, and also a self-contained source from which to start a systematic study of the field.

We would like to express our gratitude to the NATO Scientific Affairs Division for its support, which allowed this Advanced Study Institute to take place, and for continuous advice during the preparation of the meeting. It is also a pleasure to thank all the lecturers who accepted to teach at this Institute. We would particularly like to thank the others members of the Scientific Organizing Committee : Professor M.R.C. McDowell, who directed a previous NATO Advanced Study Institute on this subject in 1979 and whose help was invaluable, Dr. H. Drawin, Dr. M. Harrison, Dr. F.J. de Heer, Dr. J. Hogan and Professor G. Ferrante, the Chairman of the Local Committee. We also wish to thank Professor P. Cavaliere, Dr. R. Daniele and all the people of the University of Palermo who worked very hard to make our stay at Zagarella an interesting and enjoyable one. Dr. E.H. Mund was a perfect treasurer, while excellent secretarial assistance was provided by Miss C. Carbone and Miss N. Coisman. It is also a pleasure to thank Mme E. Péan for her help in preparing the manuscript of this volume.

Charles J. Joachain

Douglass E. Post

Directors of the Advanced
Study Institute
December 1982

CONTENTS

PART III. THE ATOMIC AND MOLECULAR PHYSICS OF CONTROLLED THERMONUCLEAR RESEARCH DEVICES

PART I

FUSION ENERGY RESEARCH

OVERVIEW OF FUSION ENERGY RESEARCH

Günter Grieger

Max-Planck-Institut für Plasmaphysik
EURATOM Association
D-8046 Garching, FRG

This paper consists of two parts, a general introduction on how to use efficiently fusion processes for large scale power production. This part will go through the general principles, will extract the essentials and limitations and define the available working space set by the constraints of nature. The second part will use INTOR as an example of a reactor-like device designed on the basis of the knowledge worked out in the first part.

1. INTRODUCTION

Fusion energy is generated if atomic nuclei fuse together to produce heavier nuclei and if the total mass of the reaction products is smaller than that of the initial atoms. Thus fusion power can be extracted only by the fusion of light nuclei.

There are many fusion processes known but the main candidates also representing the various basic principles are the following:

$$\begin{array}{llll} D + T & \rightarrow & {}^{4}He + n & + 17.6\ MeV \\ D + D & \rightarrow & T + p & + 4.0\ MeV^{+)} \\ & \rightarrow & {}^{3}He + n & + 3.3\ MeV^{+)} \\ D + {}^{3}He & \rightarrow & {}^{4}He + p & + 18.3\ MeV \\ {}^{11}B + p & \rightarrow & 3\ {}^{4}He & + 8.7\ MeV \end{array} \quad (1)$$

+) Both processes occur with about equal probability.

The reaction energy occurs as kinetic energy of the reaction products and is distributed among them according to their inverse mass ratio. From the figures quoted it is immediately apparent that the process energy is less by an order of magnitude than for typical fission reactions.

The fusion power density of a reacting plasma can be expressed in very simple terms by the following equation:

$$P = n_A \cdot n_B \cdot \langle \sigma v \rangle_{A,B} \cdot e\, U_{A,B} \left[\frac{W}{m^3}\right] \quad (2)$$

where n_A and n_B are the densities of the two constituents, and eU_{AB} the corresponding energy generation per process. If one keeps $n_A + n_B = 2\,n = \text{const.}$, this equation exhibits a broad maximum for $n_A = n_B$ which therefore is the intended operation point:

$$P = n^2 \langle \sigma v \rangle_{AB} \cdot e\, U_{AB} \left[\frac{W}{m^3}\right] \quad (3)$$

$\langle \sigma v \rangle$ is a function of the relative velocity of the two particles and of the collision parameter. Fig. 1[1] gives the average values for a Maxwell-Boltzmann distribution of the ions as function of their temperature.

For the most effective reaction, which is D-T, the reaction cross-section is of the order of 10^{-28} m^2 which is by many orders of magnitude smaller than typical atomic cross-sections. This means that Coulomb-forces will come into play long before any reaction can occur and it is for this reason that fusion by directed beams is not possible. Any ordered motion will be destroyed by frequent Coulomb collisions before sufficient fusion reactions are possible.

This means that one has to create conditions which, on the average, conserve the particle energies over a large number of collisions which is the case in a plasma of sufficient temperature. Since the fusion power density is proportional to n^2 one also has to confine the plasma in a limited volume and to prevent it from too fast expansion.

Such a plasma exhibits a pressure, n k T, and it is reasonable to ask what the optimum plasma temperature is for constant plasma pressure assuming that the plasma pressure is a good measure for the difficulties arising from confinement. Fig. 1 shows steep rises of $\langle \sigma v \rangle$ with increasing temperature but at certain values this rise flattens and does no longer exceed the effect of the accompanied reduction in density needed for keeping n k T constant. This is better displayed in fig. 2 where the normalized fusion power density is shown for constant plasma pressure. All curves show maxima in the range from 10 to 100 keV. Once the temperature is selected the fusion power density is proportional to the square

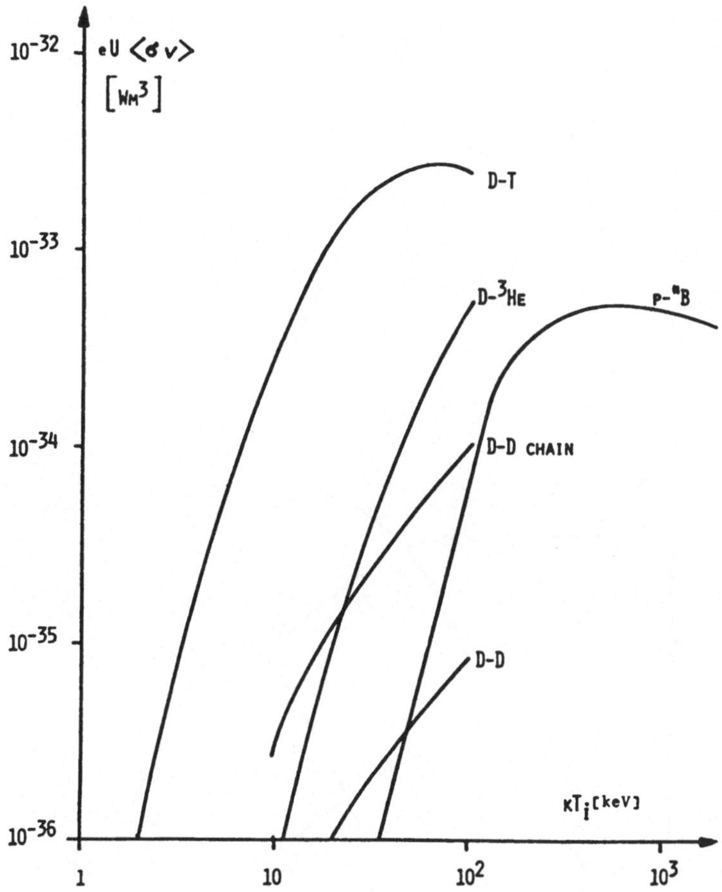

Fig. 1. Fusion power density, normalized to the density of the fuel constituents (see eq. (2)), the alphas containing 1/5.

of the pressure. Thus, the D-D reaction requires 10 times higher pressure to yield a similar fusion power density as the D-T-reaction whereas this factor even approaches 100 for the p $-^{11}$B reaction.

It is a bit intriguing that the extremely high temperature and thus the high pressure is needed solely for fighting against the repulsive Coulomb forces and to allow enough tunneling processes to occur. One is led to look for simpler methods and, in principle, there are possibilities though not feasible at present.

One way would be to shield the positive charge of one of the partners by forming an atom. Unfortunately the electron mass is so low and thus the atomic diameter so large that the ionization energy is only a few volts. If one, however, replaces the electron by a μ^- (2), which has 207 times the electron mass, then the atom would

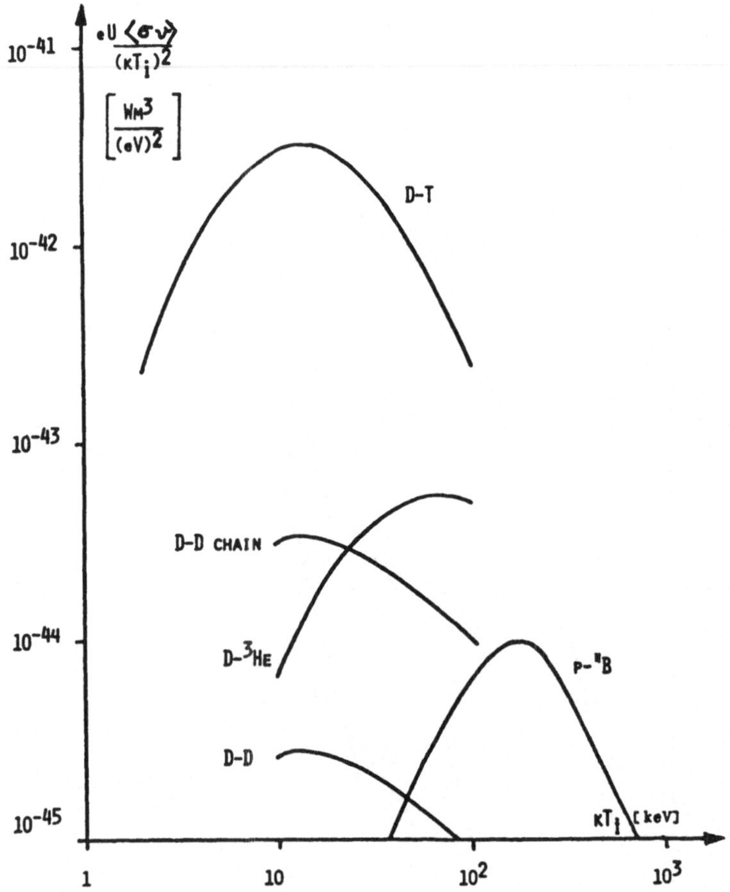

Fig. 2. Fusion power density, normalized as in fig. 1, for constant plasma pressure (increasing T is compensated by decreasing n).

be much smaller and the ionization energy would increase to 2.7 keV. For a D-T plasma this would lead to an ignition temperature of 3.5 keV due to incomplete ionization.

This beneficial effect has to be compared, however, with the production energy for μ-mesons. These are usually produced by bombarding pions with protons of 0.5-1 GeV energy which process requires 10 GeV per produced μ^- at the minimum. Present μ-meson factories work at 10^9 GeV per μ^- which is fully unacceptable if compared to the 17.6 MeV released per fusion reaction. These numbers show that even for the minimum production energy of 10 GeV one would need to catalyze much more than 10^3 fusion reactions during the life time of μ^- which is 2.2 μs. If at all, such figures are only conceivable for inertial confinement where the extremely high density leads to rather high reaction rates.

One should, however, keep such possibilities in mind. It is not fully excluded that high mass, negatively charged, sufficiently stable particles may be discovered in future. If this would happen and if such particles could be produced with low enough energy, it would considerably ease the fusion reactor.

Properties arising from fuel selection

It is interesting to compare the properties of fusion power generation for the four examples of fuel quoted above. From this it will become apparent that many byproducts are not necessary in principle but have to be accepted if the high cross-section of the D-T reaction is to be exploited.

As seen from fig. 2 the D-T reaction at a given pressure has the highest fusion power density at the lowest temperature. 10 MW/m^3 will be achieved for a pressure of 11 bar.

But tritium, one of the fuel elements, is an unstable isotope and decays via the reaction

$$T \rightarrow {}^3He + e\ (< 5.7\ keV >) \tag{4}$$

with a decay time of 12.3 a. It is therefore not available in nature but has to be bred. Two reactions, both using the neutron resulting from the fusion reaction, are usually favoured for this purpose

$$\begin{aligned} {}^6Li + n &\rightarrow {}^4He + T \quad &+ 4.8\ MeV \\ {}^7Li + n &\rightarrow {}^4He + T + n \quad &- 2.5\ MeV \end{aligned} \tag{5}$$

The ${}^6Li$-reaction is an exothermic one and adds to the fusion power, but it is the ${}^7Li$ reaction which by the secondary neutron allows to reach or even to exceed the breeding ratio of one already without further neutron multiplier. 7.4% of natural lithium is ${}^6Li$. If a higher fraction of ${}^6Li$ were needed, enrichment is rather cheap.

The neutron economy for achieving a breeding ratio larger than one requires the breeding blanket to be located immediately behind the first wall and a rather complicated technology is required to extract the bred tritium continuously.

80 % of the fusion power arising from D-T reactions is carried by high energy (14 MeV) neutrons. Their interaction with the surrounding materials leads to helium generation within the crystal structure via (n, α)-reactions, it leads to radiation damage and finally to swelling and activation of the material.

Most of the above effects are specially connected with the use of D-T as fusion fuel and are not an inherent property of fusion, i.e. they can be avoided in principle. This will become clear when looking at the other examples.

The D-D reaction has two branches which occur with about equal probability. The main advantage of using this kind of fuel rests in that it has only one constituent which, in addition, is stable, everywhere available and provides no transport problem (like hydrogen). There is no need for fuel breeding and therefore the blanket is much simpler, does not contain lithium and is only needed for power conversion.

The D-D reaction is producing reaction products which increase its power output by order of magnitude. Tritium is burnt via the D-T and ^{3}He via the D-^{3}He reaction. The tritium occurring as an intermediate reaction product provides only minor problems. It is burnt on a fast time-scale due to its high reaction cross-section. It is unfortunate that there are still neutrons produced from both the D-D and the D-T reactions so that there is only some quantitative improvement as far as neutron induced problems are concerned.

The D-D reaction needs a plasma pressure of about one order of magnitude higher than for D-T to achieve the same power density. The working point is also in the neighbourhood of the bremsstrahlung limit. Whether this limit can be shifted upwards by lowering the electron temeprature but keeping the ion temperature high is not yet clear.

The D-^{3}He reaction is a very interesting candidate because both reaction products, ^{4}He and p, are stable and electrically charged. The reaction power would then mainly go into the electrons and occur as bremsstrahlung which could easily be absorbed in the surrounding walls. This would lead to a blanket of much smaller dimensions than needed for the other cases discussed so far. The blanket dimensions would be determined by thermohydraulics rather than by neutron absorption. With such properties, such reactors could be conceived to be located also in regions of rather dense population. But also the D-^{3}He reaction requires a plasma pressure of about one order of magnitude higher than for D-T if the same power density were to be achieved. But there are two caveats:

First, the resources of ^{3}He are extremely small, practically zero, although ^{3}He is a stable isotope. Only 10^{-4} of the already rare helium is ^{3}He. It has to be produced therefore.

The preferred method of producing large quantities of ^{3}He would be by beta decay of tritium. The necessary tritium production could be a subtask of remotely located D-T plants installed for covering the base load. Shipping of ^{3}He is no problem at all. Such scenarios would allow to place D-^{3}He reactors also in densely populated regions.

The second caveat is more serious. Under usual conditions, not only D-^{3}He but also a considerable number of D-D reactions would occur and thus tritium and neutrons as reaction products. This would spoil all the above-mentioned advantages. These effects are unwanted, they do not even increase the power output significantly.

At least in principle, there might be a loophole to save the advantages by nuclear spin orientation between the constituents(3). This method would increase the fusion reaction rate by factors around two or three,which holds also for the D-T reaction. Much more important, however, seems to be the possibility of reducing the number of D-D reactions by orders of magnitude. Spin orientation seems to be extremely stable within magnetic confinement systems, even against wall collisions. It is not clear at present whether efficient methods for creating nuclear spin oriented nuclear beams can be found. The required throughput would be of the order of kg/day per reactor.

The alternative to this method would be the ^{11}B-p reaction. Its obvious advantage is that it involves only stable particles, that no breeding of fuel is necessary, that the fuel resources are plentiful, and that there are no neutrons involved in the process. All this looks extremely positive, except that the cross-section is very small. A pressure which is higher by two orders of magnitude is required to obtain the same fusion power density as from D-T. Already the bremsstrahlung would be too large against the power production. It is very difficult to imagine how such a plasma could be kept burning or even could be ignited except (perhaps) for very sophisticated systems operating far from thermal equilibrium.

All these considerations clearly lead to the conclusion that the only near term solution is the D-T process because of its highest power producing capability. The drawbacks connected with this choice have to be accepted at present. But it should be kept in mind as a challenge that, in principle, many of them could be avoided by other fuel systems.

In the following, we will restrict ourselves to the D-T reaction and only to consider plasmas being close to thermal equilibrium.

A fundamental criterion which has to be observed is the Lawson criterion which requests that the fusion power output has to exceed the bremsstrahlung losses at least. For optimum temperature, this condition reads

$$n\tau > 10^{20}\ \mathrm{m}^{-3}\mathrm{s}\ . \qquad (6)$$

Exact thermal equilibrium is not possible, however, in a limited volume with boundary conditions set by an acceptable environment. This leads necessarily to power and particle sources and sinks and gradients in parameters. Essential macroscopic effects are resulting from these deviations of thermal equilibrium, like electric currents,mass flows, etc.. For special questions, however, the assumption of thermal equilibrium is a good approximation but its validity has to be carefully checked from case to case.

In general, there are two possibilities from eq. (6) to generate limited fusion power in a limited volume. Either to increase the particle density so far that inertia is sufficient for a small enough volume. This is the concept of inertial confinement which produces fusion power in a series of microexplosions. The other method consists of a drastic reduction of the free expansion velocity and the particle density selected such that the plasma pressure becomes acceptable. This is the concept of magnetic confinement which has the potential of steady state operation.

Inertial confinement

The concept of inertial confinement, discussed in this volume by R. Haas, is based on essentially two components, a carefully designed pellet and a highly efficient driver.

Assuming for the pellet an instantaneous heating to fusion temperatures, one has to request the expansion time of the high pressure plasma formed to be longer than the fusion reaction time

$$\rho \cdot r_{pellet} \gtrsim \frac{\sqrt{8mkT}}{<\sigma v>} = f(T) \qquad (7)$$

with the temperature at optimum and ρ being the mass density. The minimum energy required of the driver to reach ignition is obtained for the smallest possible number of pellet particles compatible with the requirements of eq. (7). This number is proportional to r^3_{pellet} on the one hand, and, from the time scale of power generation, to $1/\rho^2$ on the other hand. These considerations led to the concept of high pellet compression, and reductions of the required driver energy by orders of magnitude became possible this way.

For effective pellet compression, it is essential to deposit the driver energy on the pellet surface and to achieve the compression by shock-waves partly created by a rocket effect produced by ablated particles. Central heating then leads to ignition. For this process to work it is important to prevent preheating of the pellet in front of the shock-wave by anomalous electron heat conduction because this would reduce the compression factor.

A reactor based on inertial confinement is conceived as a sphere of a radius of 5 m or so, in the centre of which the microexplosions take place. The inner wall of the sphere is covered by a blanket. The driver typically has to deliver 10^6 J within leass than one ns to the pellet surface. The power has to increase during the pulse to reach about 10^{14} W at the end. This power has to be introduced into the burn chamber and to be well focussed upon the surface of the tiny pellet.

Drivers of various types are under consideration: (i) Lasers of various wave lengths, requiring neutron radiation resistant optical systems at the entrance into the reaction chamber, (ii) electron beams for which the space charge effects provide some difficulties for efficient focussing, (iii) light and (iv) heavy ion beams which have the advantage of a short stopping length and are susceptible to some space charge compensation by electrons.

Magnetic confinement

The concept of magnetic confinement aims at steady state operation under reactor conditions, although present experiments are run in a pulsed mode. System studies have clearly shown that pulsed reactors would experience rather severe difficulties from fatigue behaviour of materials, from the time varying electromagnetic circuits and last but not least from achieving a steady turbine operation.

The concept of magnetic confinement aims at using the magnetic field to reduce heat conduction and convection, and particle transport. In principle, there is no need to achieve also high particle confinement. If the plasma energy was well confined and if refuelling worked like a counterstream heat exchanger, it would be sufficient. But the physics of confinement of charged particles by magnetic fields does not allow full separation of the two properties.

For the two degrees of freedom perpendicular to $\mathbf{B}$ the magnetic field is very efficient to confine particles and their energy. Problems arise with the third degree of freedom, parallel to $\mathbf{B}$. Increasing field strength provides a repelling force $\mu \cdot \nabla B$. This force acts on both positive and negative charges in the same direction. from the electric field, $e\,\mathbf{E}$, depend on the sign of charge. From this follows that end systems to be effective have to use a rather
from the electric field, $e\,\mathbf{E}$, depend on the sign of charge. From this follows that end systems to be effective have to use a rather
complicated combination of gradients of both magnetic field and potential.

The improvement factor required is rather large. Eq. (6) requires confinement times of the order of a second at the pressures which can conceivably be confined by magnetic fields. If one allowed free streaming between the ends and took $L \approx 100$ m as a resonable length of the confinement device, the improvement factor would have to be

$$\frac{\tau_{conf}}{L} \cdot v_i \approx 10^4 \tag{8}$$

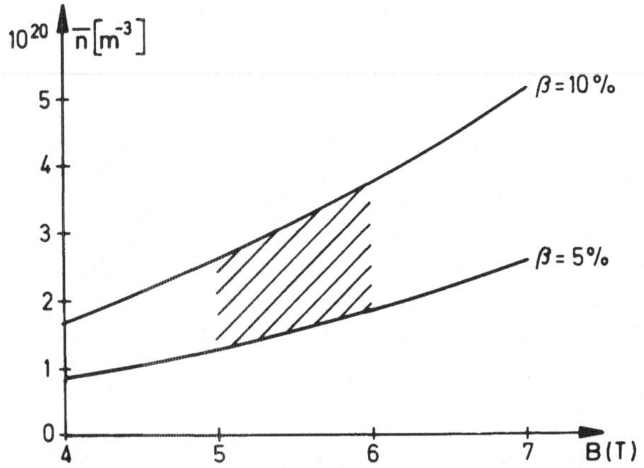

Fig. 3. Plasma density vs. magnetic field for beta values of 5 and 10 %. The dashed area indicates the probable range of operation.

The use of magnetic confinement also leads to a further limitation of the useful parameter space, mainly with respect to the practically achievable fusion power density. This can best be explained with the help of fig. 3. Technical reasons (forces) limit practical fields to values of 5 to 6 T. The maximum fraction $\beta = 2\mu_0 nkT \cdot B^{-2}$ usable for stably confining the plasma is determined by physics, and for toroidal confinement beta values around 5- 10 % are conceivable according to present estimates. Fig. 3 then shows that the operating density will be around $2 \cdot 10^{20}$ m^{-3}. When evaluating the fusion power density from these data, it has to be considered that beta does not only contain the pressure of the D-T fuel but also the pressure of the electrons, the helium ash and the impurities.

For the above-mentioned cases, the conditions of toroidal confinement were used as example. For mirror machines the beta in the control cell is higher but B, being determined by the conditions in the end cells, is lower. Thus, about the same operating density results for the central cell.

Conclusions for magnetic confinement

Already from these very general considerations the plasma parameters are rather narrowly defined for a fusion reactor working after the concept of magnetic confinement and using D-T as fuel:

Optimum temperature	: $T \approx 12 - 14$ keV	
From limitations on B (technical) and ß (physical)	: $n \approx 2 \cdot 10^{20}$ m^{-3}	(9)
From Lawson criterion	: $\tau \approx 1$ s	

These parameters also fix the achievable fusion power density to

$$P \approx 1 \frac{MW}{m^3} \qquad (10)$$

According to eq. (3), the fusion power density is proportional to $n^2_{D,T}$ and thus, for fixed plasma temperature,

$$P \sim \beta^2 B^4 \qquad (11)$$

Therefore, significant further increases of the fusion power density would become possible if physics achievements would allow much higher beta values or technical development the use of higher magnetic fields. In addition, if one was able to maintain plasmas with large deviations from a Maxwell-Boltzmann distribution function, one could arrive at higher output powers for less perpendicular pressure. Achievements in these directions would even open up the future use of different fuel systems as discussed earlier.

In the following, the two main lines of magnetic confinement, mirrors and toroidal systems, will be discussed in some detail.

Mirror concept

Fig. 4 illustrates the main features of the basic mirror concepts in the sequence of their historic development.

(a) is a simple mirror. Plasmas embedded in such mirror fields, however, are MHD unstable because the curvature of the magnetic lines has the wrong sign. Stable configurations need a magnetic well, or $|B|$ increasing in all directions. Ioffe bars, carrying current of opposite direction in neighbouring bars, were the means used first to achieve this property. Such a system is shown in (b). Later improvements lead to baseball and to Yin - Yang coils. They are shown in (c) and (d).

These systems are still sensitive to micro-instabilities arising from particle interactions with plasma waves and leading to the low energy particles lost faster than the high energy ones. The loss cone thus formed results in an inversion of the population and in

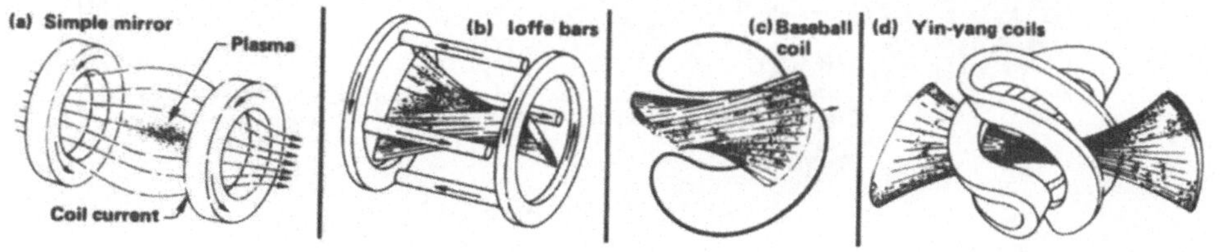

Fig. 4.[4] Mirror concepts.

the generation of the drift cyclotron loss cone instability. As a countermeasure streams of cold plasma are introduced from both ends into the system in order to replenish the distribution function. Stability achieved this way was demonstrated in the 2 x II B device but the loss rates were still uncomfortably high. Recovery of the energy of the particles leaving the system through the ends would be essential for arriving at a power balance positive enough.

An essential step forward was the idea of tandem mirrors. This concept connects two mirrors by a homogeneous magnetic field forming a central cell. The mirrors are used as end-plugs and provide the stability of the central cell plasma. Fig. 5 gives a sketch of a tandem mirror arrangement.

Energetic neutral beams are introduced into the mirror cells and the plasma is allowed to diffuse into the central cell. This way the electron temperature is much higher in the mirror regions than in the central cell and a positive potential forms leading to electrostatic confinement of the ions. Under these conditions the power is still lost by electron heat conduction towards the end. This channel can be blocked by insulating the mirror electrons from those of the central cell by a thermal barrier. This completes the most favoured concept of today's mirror research.

Fig. 6 shows the axial variation of the magnetic field strength, the plasma potential and the plasma density for a tandem mirror device.

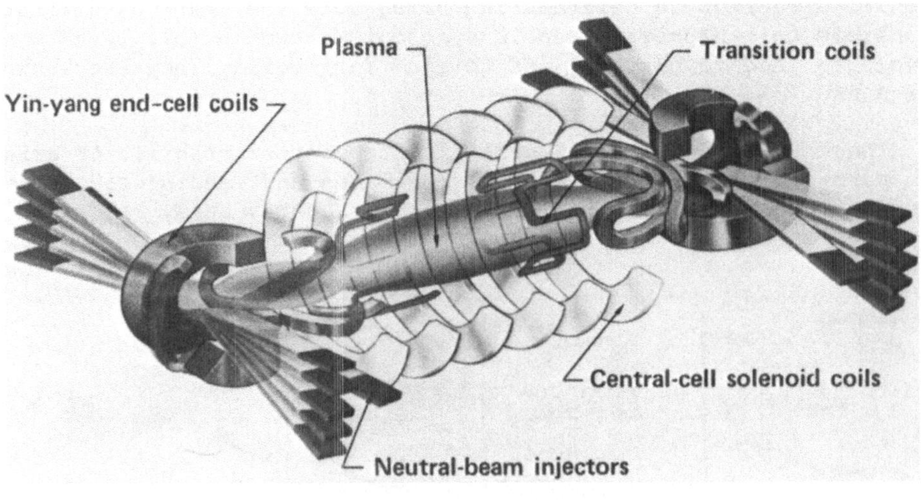

Fig. 5[4]. Tandem mirror.

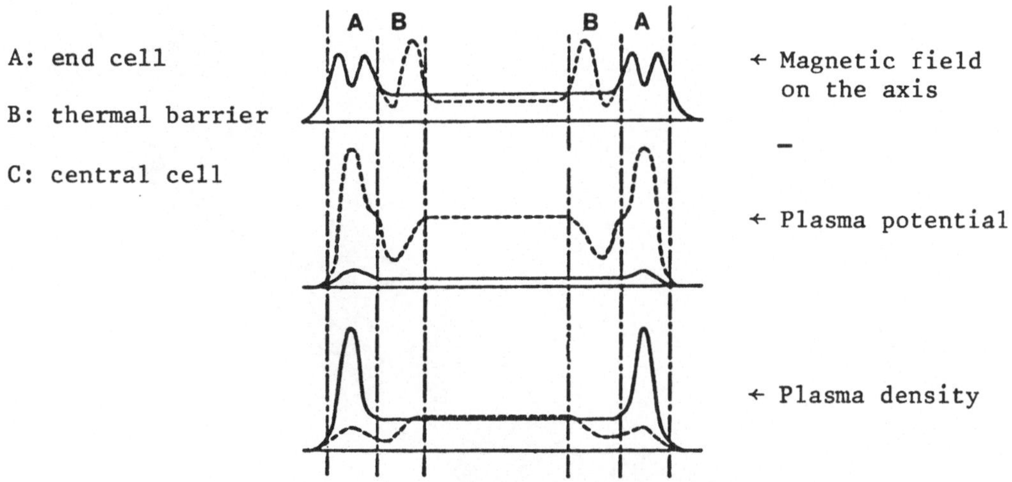

Fig. 6[4]. Axial variation of B, plasma potential and density for a tandem mirror arrangement with (dotted line) and without (full line) thermal barrier.

Comparing the two cases illustrated in fig. 6 shows that adding a further mirror and decreasing the density in that mirror yields the required rise in the plasma potential, whereas a density dip in the second mirror results in a potential dip as required for confinement. This density dip has to be maintained in steady state by charge exchange of injected particles.

The mirror line is aiming at a step by step demonstration of the various properties. The present discussion is centered around the stability of the center cell where the connection to the ends via particles is needed for stability by good curvature of the field lines. But the connection is disliked at the same time for the benefit of the thermal barrier.

Fig. 7 gives a sketch of a tandem mirror machine.

Toroidal magnetic confinement

Purely toroidal magnetic fields obey the relation

$$\oint B dl = \mu_o I \tag{12}$$

from which follows for axisymmetry

$$B \sim \frac{I}{R} \tag{13}$$

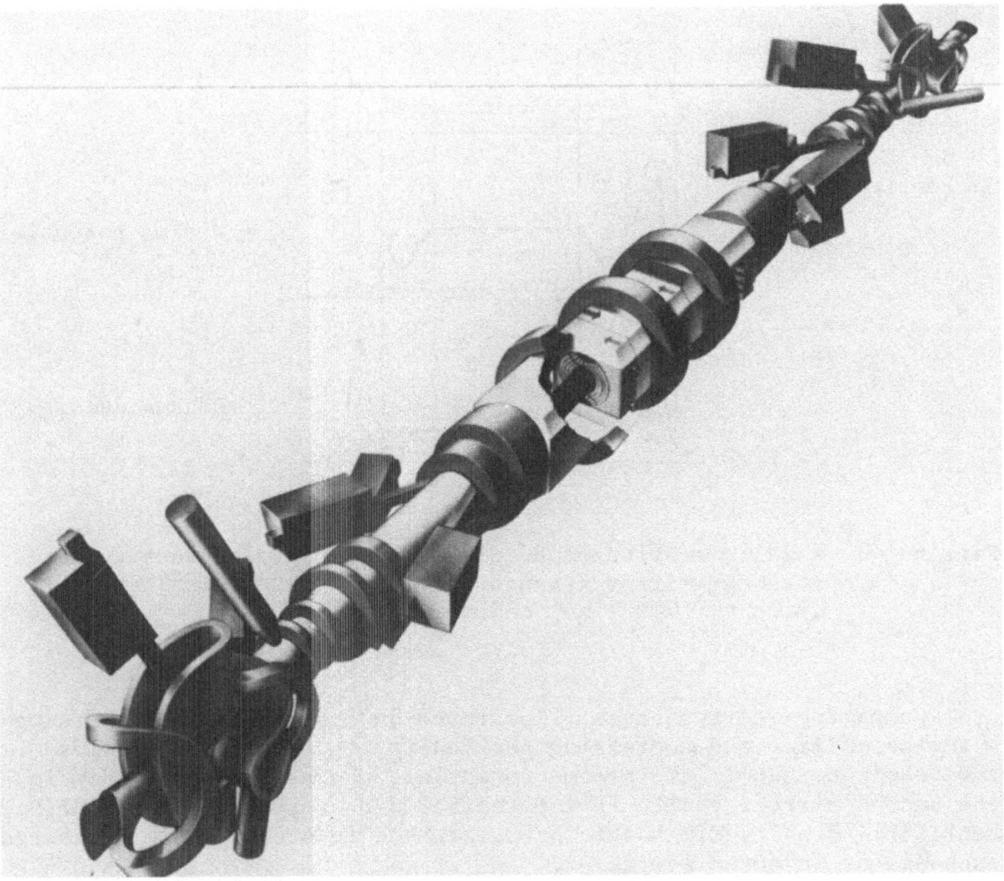

Fig. 7[4]. TASKA, tandem mirror with thermal barrier, KfK Karlsruhe. Parts with arrows indicate removable test facilities for technology development. Boxes represent heating equipment.

Particles moving in such a magnetic field experience a centrifugal and a ∇B-force leading to a drift motion parallel to the axis of symmetry and with the velocity

$$v_D \approx \frac{eU}{RB} \qquad (14)$$

where eU is the particle mean energy and R the radius of curvature of the field lines. Due to this drift the particle would be lost on a fast time-scale.

In principle, there are two potential countermeasures. The formation of twisted magnetic lines and nested magnetic surfaces would allow for a compensation of the drift motion by motion parallel to $\underset{\sim}{B}$. Or from $\underset{\sim}{j} \times \underset{\sim}{B} = \underset{\sim}{\nabla} p$, div $\underset{\sim}{j} = 0$ could be achieved by allowing currents parallel to $\underset{\sim}{B}$, such that div $(\underset{\sim}{j}_{\perp} + \underset{\sim}{j}_{\parallel}) = 0$. This measure is still compatible with axisymmetry but then requires a net current in the plasma to generate the twist of the magnetic lines (like a rope) necessary for forming magnetic surfaces.

The alternative measure would use non-axisymmetry and aim at making the average drift zero by alternating the sign of the curvature when going around the device. Again div $(\underset{\sim}{j}_{\perp} + \underset{\sim}{j}_{\parallel})$ would have to be zero but here $\underset{\sim}{j}_{\parallel}$ would be an oscillatory function of the azimuthal angle. This method is particularly powerful if also combined with twisted magnetic lines and nested magnetic surfaces. Non-axisymmetry allows generating these configurations without net plasma currents.

It seems to be the existence of nested magnetic surfaces which allowed the rapid progress in toroidal magnetic confinement. Fig. 8 is a sketch showing how the magnetic surfaces are formed by superposition of toroidal and poloidal fields. Each of the surfaces of flux tubes has to be conceived to be formed by one single field line moving around in a twisted fashion and covering the whole surface. Only for resonant cases would the field line close upon itself after a limited number of revolutions.

Magnetic surfaces combine two advantageous properties: Along $\underset{\sim}{B}$ they allow fast communication within a surface so that differences in pressure and potential would tend to be reduced. Perpendicular to $\underset{\sim}{B}$ they provide high terminal insulation from surface to surface.

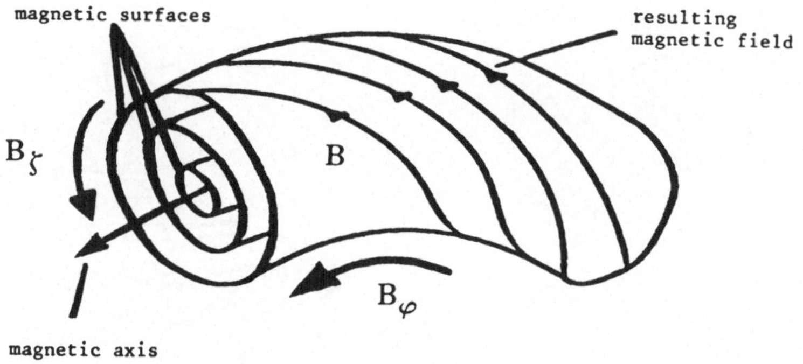

Figure 8. Nested magnetic surfaces formed by superposition of toroidal and poloidal magnetic field components.

There are just two classes of toroidal configurations with nested surfaces: For "internal" confinement net currents in the plasma are indispensable for creating the configuration. Tokamaks are the main representatives of this class. The configuration has the advantage of being axisymmetric but due to the plasma current it is coupled to the plasma position, and also the current distribution is difficult to influence from the outside.This makes this concept susceptible to current disruptions. It also needs two independent coil systems at least.

For "external" confinement the configuration is established by external coil currents only and is thus necessarily non-axisymmetric. Stellarators are the main representatives of this class. The influence of the plasma on the configuration is limited to beta effects. This provides a high positional stability. Only one coil system is needed but with coils of non-planar shape.

Fig. 9 gives the principles of Tokamak and Stellarator configurations, and Fig. 10 is a schematic drawing of JET, the largest European Tokamak at present.

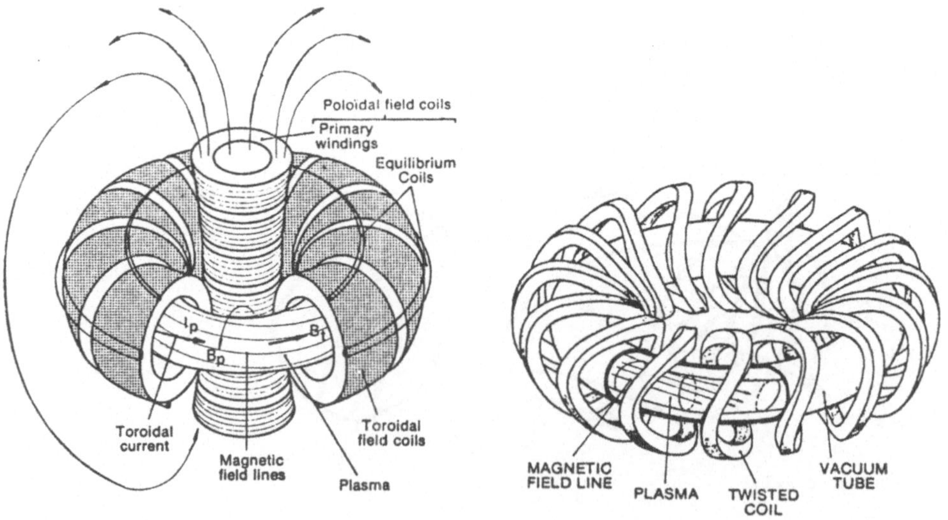

Fig. 9. Principles of Tokamak (left) and Stellarator (right).

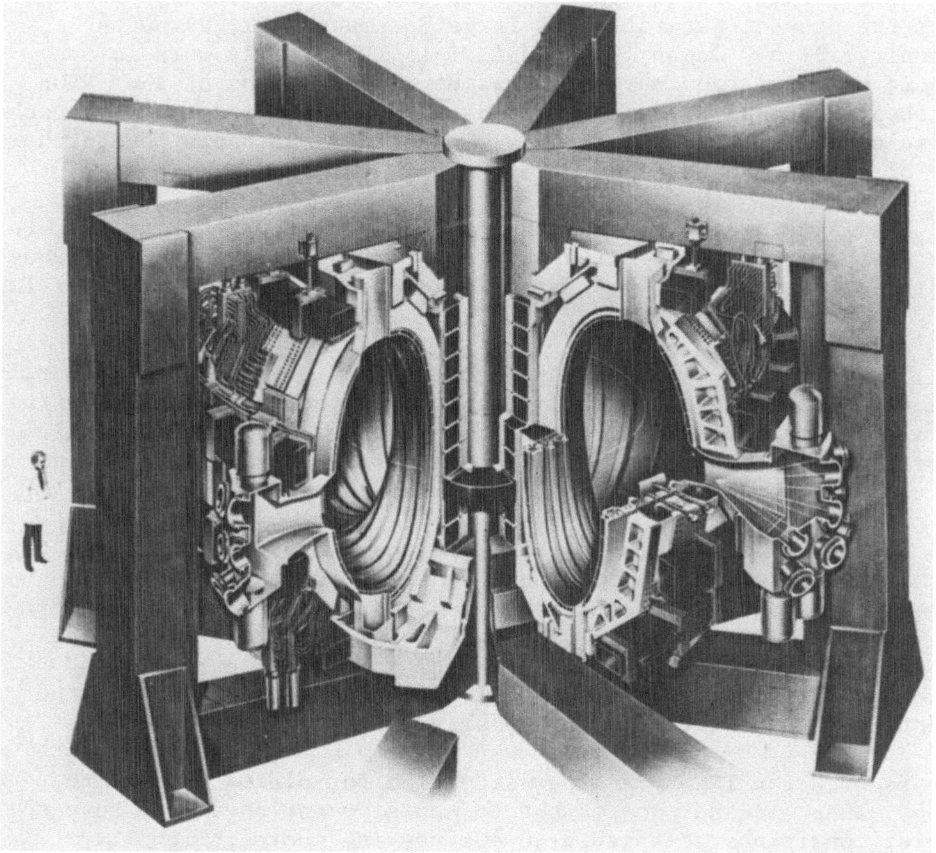

Fig. 10. The JET device.
Major diameter of torus: 5.9 m; horizontal plasma diameter: 2.5 m; vertical plasma elongation: 4.2 m; overall diameter: 14.8 m; overall height: 11.5 m.

2. INTOR

INTOR (International Tokamak Reactor) is a project being prepared in international cooperation under the auspices of the IAEA. The participants are the European Community, Japan, the USA, and the USSR. The task of this undertaking is the definition, the design, the construction and the operation of the next step device coming after the present generation of large Tokamaks (JET, European Community; JT-60, Japan; TFTR, USA; T-15, USSR). The work is done by home bases of the four partners. Regular sessions of the INTOR Workshop in Vienna allow comparisons between the different approaches and selection of the optimum ones. These form the basis for further work of the home base teams.

The INTOR project is running in phases, with the first two, Data Base Assessment(5), and Definition and Conceptual Design(6), already concluded. At present, the project is close to the end of phase IIa, Study of Critical Issues.

The advantage of having a conceptual design which includes all aspects is to be able to evaluate all interdependencies between the various reactor components. Consistency was widely achieved, all major problem areas were identified, and short-term R and D needs defined.

The INTOR objectives in short are the demonstration of reactor plasma physics and the provision of a test bed for the development of reactor technology. Therefore, in connection with the aims of this paper, INTOR can ideally be used to assess the state of the art of plasma physics and the prospects for reliable reactor operation. This paper will not deal with technology except to the extent it sets requirements to physics.

To work for INTOR is a new situation for plasma physicists, indeed. They have not to ask but to answer questions. They have to provide conditions acceptable to engineering where reliability determines the choice of solutions. The inclusion of too high safety factors, margins, and redundancy very soon becomes too expensive.

Physics has to provide to INTOR a reacting plasma steadily burning for at least 100 s per pulse. The neutron wall loading should be sufficient for technology investigations (≈ 1.3 MW m^{-2}) and fluence be collected on a not too slow time-scale (duty cycle > 70%, availability up to 0.4). The device should run reliably for up to 10^6 discharges and its requirements with regard to engineering should not be too complicated to allow practical solutions.

There were two major decisions of large influence and far ranging consequences already early in the INTOR work:

It was concluded that only with a divertor could the impurity content and the wall bombardment reliably be kept low enough and the ash exhaust facilitated so that the required operating conditions and availability could be reached. A single null divertor was considered the optimum solution with regard to the overall aspects.

Maintenance considerations lead to the requirement that all poloidal field coils must not be interlinked with the toroidal field coils.

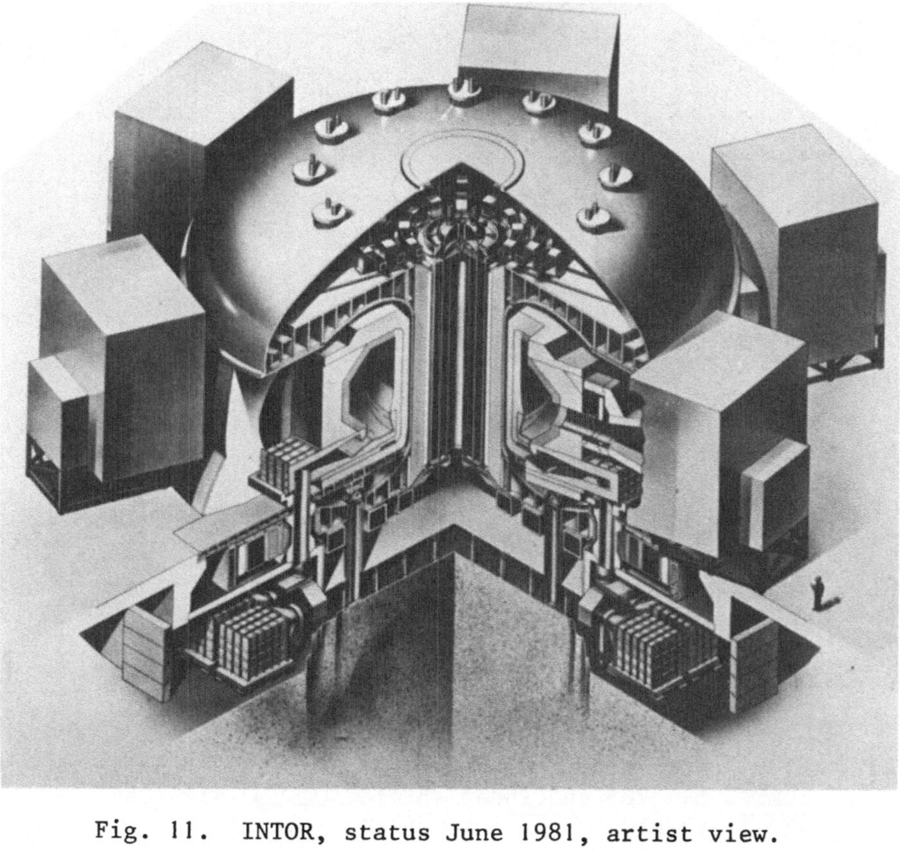

Fig. 11. INTOR, status June 1981, artist view.

The major phase-I INTOR parameters are collected in the following table. Please note that phase IIa will lead to some modifications Fig. 11 gives an artist view of the INTOR concept of June 1981. Fig. 12 is a closer view of the cross-section of INTOR.

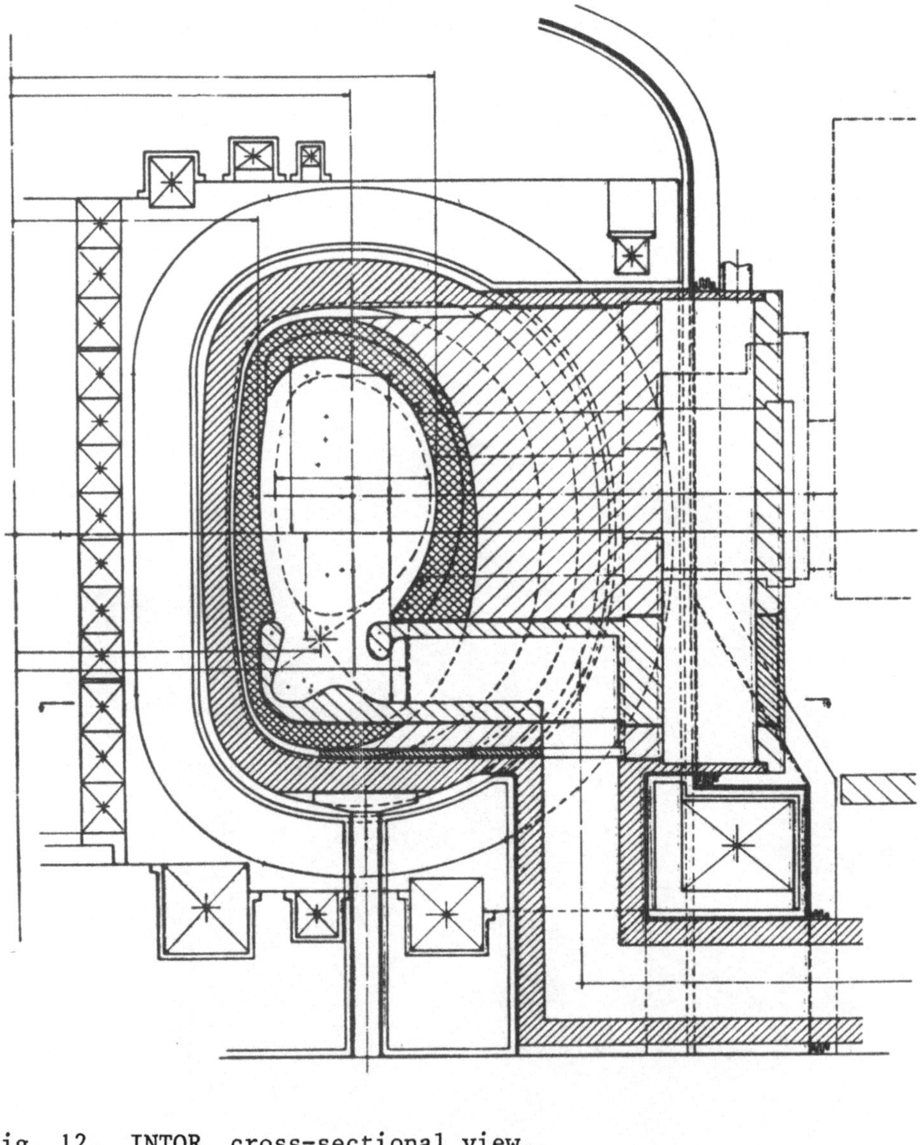

Fig. 12. INTOR, cross-sectional view.
[cross-hatched]: blanket and inner shield; [hatched]: divertor cassette
[hatched]: shield; [ ]: toroidal field coil
[⊠]: poloidal field and transformer coils

Table 1. INTOR Parameters

Major radius	5.3 m
Plasma radius	1.2 m
Elongation	1.6
Triangularity	≈ 0.25
Toroidal magnetic field	5.5 T
Plasma current	6.4 MA
Safety factor q_I	2.1
DT power	620 MW
Average ß (burn)	5.6 %
Average DT density	$1.4 \cdot 10^{20}$ m^{-3}
Average DT temperature	10 keV
Neutron wall load	1.3 MW m^{-2}
Burn time	100 → 200 s
Duty cycle	up to 0.8
Neutral beam energy	175 keV
Neutral beam power	75 MW

Equilibrium configuration and poloidal field coils

As already pointed out, a divertor was considered necessary for INTOR. With a plasma elongated in vertical direction to increase the stable beta, it is the natural solution to locate divertors at the top or bottom by using the existing stagnation points. Half of the components can be saved, however, when going to a single null divertor, selecting the lower stagnation point at the bottom as the active one. The inactive null point at the top should then be moved away far enough to avoid unnecessary broadening of the scrape-off layer there.

The poloidal and toroidal coil systems have then to satisfy the following requirements: (i) Continuous divertor operation requests effective position control of the separatrix also during start-up. (ii) The closer one gets to the burn beta, the closer the ellipticity and the triangularity of the magnetic surfaces have to be to the values required for high beta operation. (iii) In order to allow access to the blanket and the divertor, the respective regions have to be kept free of coils. (iv) For reason of maintenance, all poloidal field coils have to be outside the toroidal field coils. (v). The toroidal field coils have to be small for reasons of force and cost but not too small, to keep the ripple in acceptable limits.

Fig. 13 is a sketch of the poloidal and toroidal coils relative to the plasma shape ($ß_{poloidal}$ = 2.6 in this case).

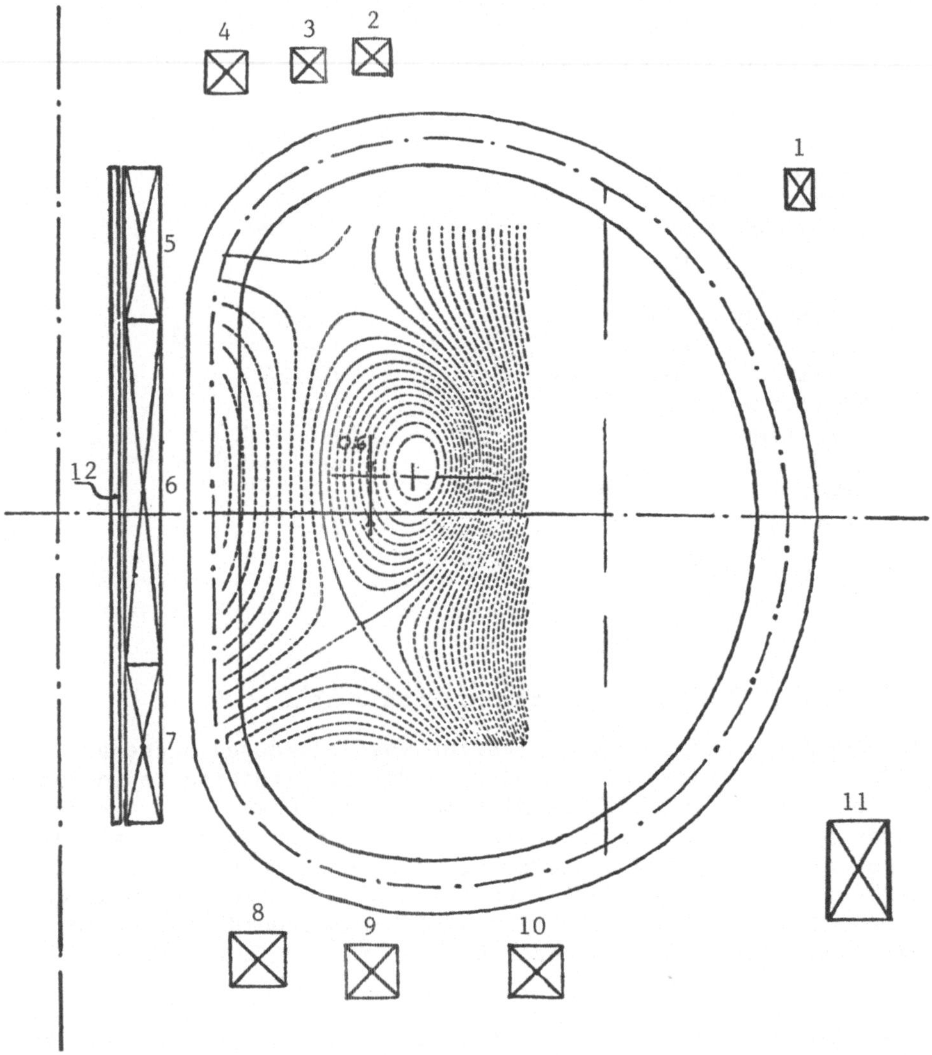

Fig. 13. INTOR coil arrangement for single null divertor operation.

The plasma is slightly shifted in vertical direction for effective use of space. The cross-section of the poloidal field coil is proportional to the current to be carried by the various coils. The sum of these currents is about 100 MA-turns. The comparison with the plasma current of 6.4 MA demonstrates the high price to be paid for requesting that all poloidal fields should be outside the toroidal ones. All coils contribute to the ohmic heating flux swing.

Stability limits

The strength of the magnetic field in INTOR has been chosen to the limit of present or soon-to-develop technology. Within this limit burn requires a $\beta_{D,T} \gtrsim 4$ %. If, in addition, one allows a concentration of thermal α-particles of 10 % and the presence of some impurities, one arrives at an operating $\beta \approx 5.6$ %, or at a poloidal beta, $\beta_I \approx 2.6 = 0.6$ R/a for $q_I = 2.1$. This seems to be a reasonable extrapolation from present experiments and will be checked by JET once it operates. The present judgement relies on the confidence that the elongation of 1.6 will enter linearly in the achievable beta, that the triangularity of 0.25 be sufficient, that the aspect ratio, being different from today's machines, will have no unfavourable consequence, that the safety factor of 2.1 can be reached, and that the necessary pressure and current profiles can be established.

Vertical position control

Large vertical elongation of the plasma cross-sections requires effective position control. A conductive shell would be fully effective only if it were closer to the plasma than about 30 cm. This is not possible for reasons of maintenance and assembly, frequent access to the interior, and tritium breeding efficiency. The feedback control, on the other hand, has to use the outer coils and the field has to penetrate the massive blanket structure. It is therefore essential to build a passive loop system of 12 sections (Fig. 14) into the first wall so that the response time of the active feedback control can be reduced to about 50 ms. This is an area where much more detailed studies of the effectiveness of such systems are necessary.

Energy confinement

During the INTOR Data Base Assessment phase ALCATOR scaling was the preferred basis for estimating energy confinement times

$$\tau_E \approx 0.5\, n_e\, a^2 \qquad (15)$$

In the meantime, a much more refined data base is available for clean plasmas with medium q, for ohmically heated plasmas, and for additional heating. There is a large scatter of the data points yielding coefficients for eq. (15) perhaps a factor of two smaller for OH plasmas, but this is not so clear for plasmas mainly heated by other means. There is also a variety of different scaling laws proposed but for the INTOR operating regime none of them leads to values very different from eq. (15) so that the ALCATOR scaling still is a reasonable assumption. Again the JET generation of experiments will answer the question of size scaling.

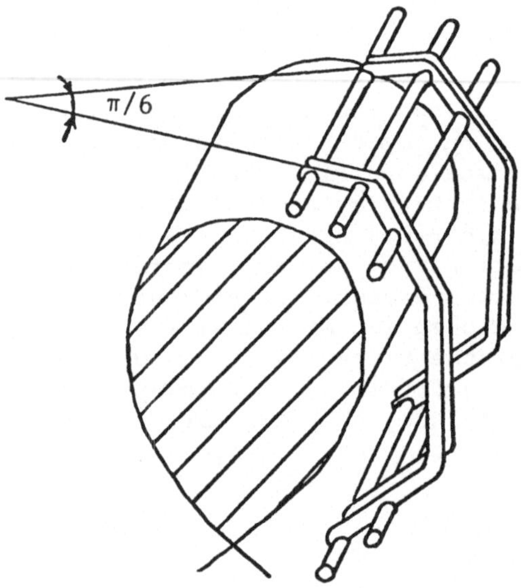

Fig. 14. Passive loop system for vertical position control.

It would be very interesting if the transport losses would show a steep increase when approaching the beta limit. Such an effect could be used as an automatic burn control freely provided by nature. Any active scheme considered up to now looks a bit clumsy.

The ion heat conduction is assumed to be three times the neoclassical value in the INTOR design basis and there is no experimental indication towards larger values. Altogether there seems to be a safety margin of two with respect to the INTOR design assumptions but the present uncertainties are larger than this factor.

Magnetic field ripple

A ripple in the magnetic field arises from the finite number of toroidal field coils and spoils the axisymmetry prevailing otherwise. Ripple trapping leads to losses particularly of the high energy particles including fractions of the fusion generated α-particles. Particularly susceptive are also the ions from near perpendicular injection or from ion cyclotron heating. The theory is still evolving and seems to settle down at manageable levels.

There are important consequences for the design, though, because the losses tend to occur rather locally and also provide high power densities particularly during start-up. Earlier ideas to use a variable ripple for burn control seem to require a too high variation of coil currents to be technically interesting.

Density limit

Progress since the Data Base Assessment phase has increased the density limit by about a factor 2, particularly for experiments with additional heating. There is a good chance that the other factor of 2 improvement needed for the foreseen operation of INTOR will be achieved in time.

Prospects are rather low, on the other hand, for the complete avoidance of disruptions already from the beginning of INTOR operation. This being the case, INTOR has to be designed disruption-proof at least for a limited number of disruptions, e.g. for about 1 % of the discharges.

Proper design of INTOR against disruptions requires a characterization of major disruptions which is a difficult task for the physicist but indispensably needed by the engineers. The best estimate is as follows: 30% of the energy will be uniformly distributed, and 70% will go to 30% of the chamber surface - mainly inboard-with a peaking factor of 2. The total energy dissipated is 220 MJ and the decay time 20 ms. The disruption frequency could be limited to $5 \cdot 10^{-3}$ during the early operating phase of INTOR and to 10^{-3} afterwards. This is a critical field where much more experimental information is needed including the dependence on parameters.

Minor disruptions are not so dangerous for the INTOR structure but they are critical as well. Sudden losses of up to 30 % of the energy could occur during minor disruptions and this will be sufficient to quench the burn. In addition, the action of the control systems to avoid them might cause early fatigue of the structural materials. Via the release of impurities, they might even trigger major disruptions. It is an urgent task for the experimental groups, therefore, to explore the reactor regime for stable islands.

INTOR operating scenarios

Fig. 15 gives a gross impression of the INTOR operating scenario. It consists of four phases, start-up, burn, shutdown where the power has to be driven to zero without disruption, and dwell where the next pulse is prepared.

Start-up

For technical reasons, it is essential to limit the breakdown voltage. Experiments have shown that 35 V loop voltage are sufficient, indeed, if the gas pressure is in the range of $1 - 3 \cdot 10^{-4}$ torr and stray magnetic fields kept smaller than

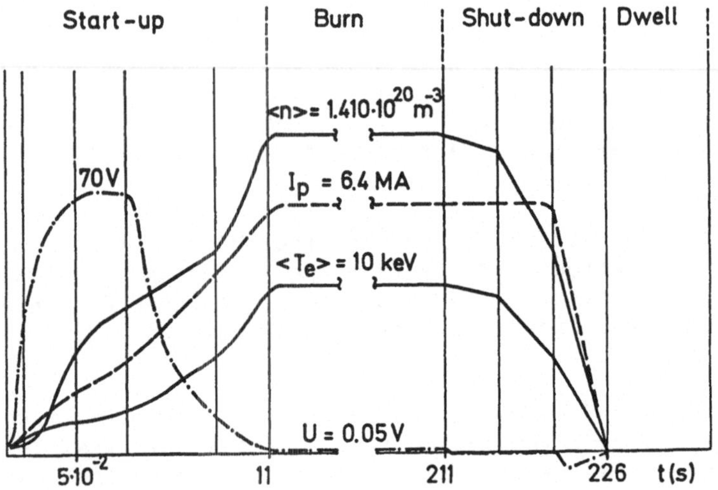

Fig. 15. INTOR operating scenario for density, plasma current, plasma temperature, and loop voltage.

$2\cdot 10^{-3}\cdot B_{toroidal}$. During this part, the losses are mainly determined by radiation (70 %) and charge exchange (20 %) and the time scale involved allows the corona model to be applied.

To assist penetration of the radiation barrier, electron cyclotron heating with a power of 5 - 10 MW is foreseen. This should allow a localized power deposition, bulk heating of electrons (to replace the losses directly) without extra generation of impurities.

Subsequently, the current has to be risen to 5.4 MA within 5 s. Energy losses by minor disruptions have to be particularly avoided during this phase and $\dot{I} < I/\tau_{skin}$ has to be guaranteed for proper profile formation. Thus $\dot{I} = 6$ MA/s has been selected for $I \leq 0.6$ MA and $\dot{I} = 1$ MA/s for $0.6 < I < 5.4$ MA.

The density evolution will show a linear increase of $\langle n \rangle$ up to $5\cdot 10^{19}$ m^{-3} during this phase. The expected fraction of impurities of $n_{oxygen} \approx 2\cdot 10^{17}$ m^{-3} and $n_{Fe} \approx 5\cdot 10^{16}$ m^{-3} will be sufficient for edge cooling.

It seems not to be possible to have the divertor properly working during this early phase. Therefore, 1 m wide strips are arranged at the upper and lower outboard side to bridge the gap by acting as limiter. 10 MW with 0.3 MW/m^2 and a peaking factor of 1.5 have to be accepted by these limiters.

After the plasma current has reached 5.4 MA, 75 MW of 175 keV neutral injection will be switched on to heat the plasma to ignition. Neutral injection is still considered as the method of highest predictability but radio frequency heating, especially ion cyclotron heating, is also under consideration,and very probably the project will go in this direction.

During the heating phase the beta is increased from 0 to 4 % and via flux conservation the plasma current increases towards the final 6.4 MA, too.

The average density rises from $5 \cdot 10^{19}$ to $1 \cdot 10^{20}$ m^3 mainly by trapping of the injected particles. In order to reach a 50/50 D-T mixture at the ignition point, one has to start from a pure T-plasma if one uses D-injection only which is a bit simpler than T-injection. Shine through problems when starting the injection lead to a localized wall load of about 1 MW/m^2 which disappears with increasing density. Towards the end of the heating phase beam ripple losses start for those injected particles which are trapped too close to the plasma edge.

With the start of neutral injection heating the divertor action has to start and the position of the separatrix to be controlled within 5 cm also.

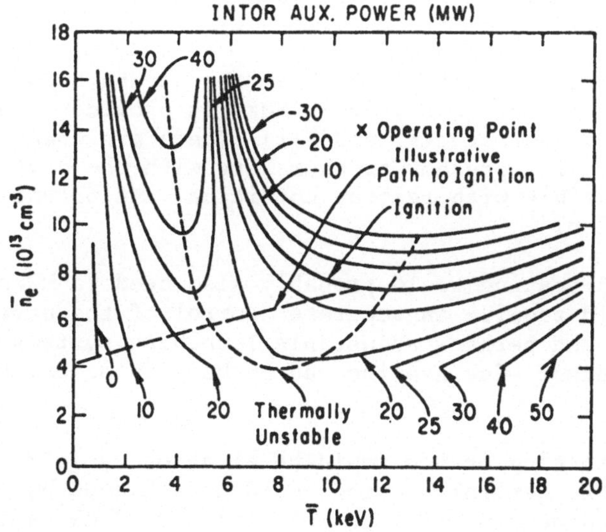

Fig. 16. INTOR auxiliary power for reaching ignition (see text).

With the help of fig. 16 the path to ignition can be optimized and an illustrative path is indicated. This path would require only 25 MW but no account is taken of radiation or any other of such losses. Other evaluations are arriving at powers of up to 60 MW so that the choice of 75 MW gives only a small margin.

Burn phase

A duration of 100 - 200 s is intended for the burn phase which makes burn temperature control mandatory. A controllable toroidal ripple would work but probably require a too high control amplitude. Beta limit control would be ideal but here experimental information is still lacking. Nevertheless, it is still the preferred method for INTOR. Feedback stabilization by compression and decompression and combined with additional heating would be a well understood method but would require additional and expensive space and provide difficulties with separatrix control for proper divertor action.

Alternative heating methods

As already mentioned, neutral injection was selected for first priority for INTOR heating to ignition because its properties are the most predictable ones. There are, however, also a number of drawbacks connected with its use. The deposition profile depends on the plasma density profile which leads to shine through at the beginning of heating and to edge heating towards the end. It has a low efficiency (without energy recovery) and adds a large volume to be shielded against radiation.

For ion cyclotron heating, 50 MW at the second harmonic of ω_D would suffice. The frequency is 85 MHz. It provides central heating largely independent of n(r) and T(r). There are chances for a better efficiency. The present problems are with the development of antennas compatible with reactor conditions and potential extra impurity generation.

Lower hybrid heating would probably also need 75 MW, and work at 2 GHz. It would require an accurate control of the parallel refraction index and perhaps adjustable launching systems. It is not clear yet whether edge heating can be kept in acceptable limits.

A very interesting method would be electron cyclotron heating working at 150 GHz. Central heating could be achieved by this method largely independent of n(r) and T(r), and 50 MW heating power would therefore be sufficient. Here the question is whether power sources can be manufactured at low enough prices.

Plasma edge and divertor

The physics of the plasma boundary is one of the most determining parts for the operating conditions of the reactor core. Not only does it determine the transport of heat and particles and thus the extraction of power and ashes, it also determines the reaction of the wall surrounding the plasma. Contact with the wall leads to reflection and recombination and, depending on edge density and temperature to sputtering causing erosion and release of impurities. High local power increases the wall temperature and together with the cooling of the wall from behind stresses occur from the temperature gradient.

High power densities at moderate particle densities lead to high erosion rates. Frequent exchange of these wall parts is necessary. These considerations stimulated the preference for the inclusion of a divertor system into the concept.

Local divertors, like bundle divertors, do not seem to be compatible with other requirements. A poloidal divertor extending all around the machine looks much more feasible. Fig. 17 compares a double null divertor (left) with a single null divertor (right). In a double null divertor the inner and the outer scrape-off layers are decoupled and will probably have the same width. In a single null divertor flux conservation determines the ratio between the

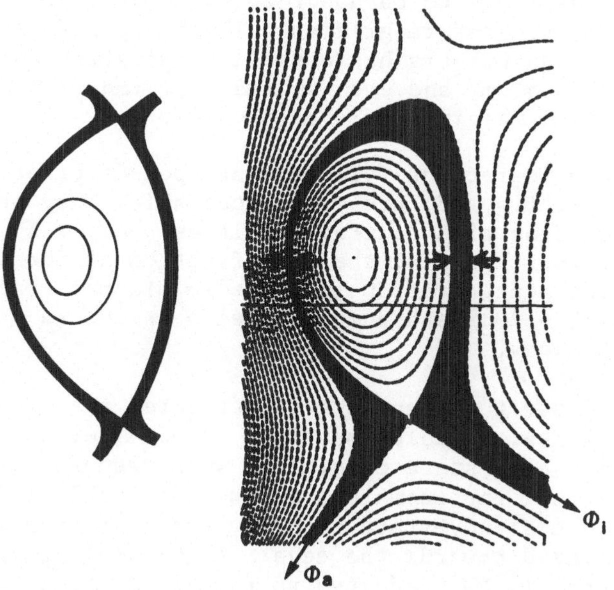

Fig. 17. Comparison between double and single null divertors (see text).

widths of the outer and inner scrape-off layers. Although the space thus needed for the single null divertor is larger, it was selected because it requires only half as many divertor modules.

The special requirements of INTOR, namely not to tolerate poloidal field coils inside the toroidal field coils allow only moderate curvatures of the magnetic lines and thus relatively open divertor configurations. Practically, the divertor plates are part of the discharge chamber. But it turned out that such systems have very interesting properties.

There are three input conditions to be observed by the plasma edge and the divertor physics: (i) The power fraction generated by the α-particles, P_α = 124 MW, has to be removed across the separatrix with an average heat flux of $\langle Q_{th} \rangle = 0.37$ MW/m^2. (ii) The corresponding helium exhaust flux is $\Gamma_{\alpha,ex} = 2.2 \cdot 10^{20}$ s^{-1}. (iii) The average density in the plasma column is $\langle n \rangle = 1.4 \cdot 10^{20}$ m^{-3}.

Several experiments are now producing results but for power fluxes and plasma densities smaller than necessary for INTOR. Therefore, rather strong theoretical support is necessary for plasma edge modelling.

A case with high edge density, $n \approx 5 \cdot 10^{19}$ m^{-3}, and low temperature,$T \approx 100$ eV, is of particular interest. If the heat is to be carried to the divertor by heat conduction, the temperature in front of the divertor plates has to be about 25 eV in order to allow a sufficiently large temperature gradient. The scrape-off layer characterized by these numbers has a small penetration depth for neutrals, $\lambda_{D,T} \approx 3 - 5$ cm, and even shorter for impurities so that sufficient impurity screening would be provided.

From these considerations follows that open divertor configurations, as shown in fig. 18, are very acceptable. They provide high recycling and thus high neutral densities. Only 1 % of the arriving helium flux is pumped, the same fraction as for D,T, leading to a 5 % helium concentration in the edge plasma. Fractional burn-up of up to 10 - 20 % and pumping speeds as low as $2 \cdot 10^5$ l/s can be reached this way.

Carbon, iron, and tungsten were investigated for their properties as particle sinks. Calculations were carried out with the BALDUR code until steady-state conditions were reached. These calculations show that for anomalous and neoclassical diffusion in the edge region, the steady-state impurity concentration is acceptable and concentrated towards the edge. 110 MW would be radiated to the wall and only 12 MW conducted to the divertor plates. But these results strongly depend on the quality of the data and of the model and here are still large uncertainties so that conclusions should not be drawn too early.

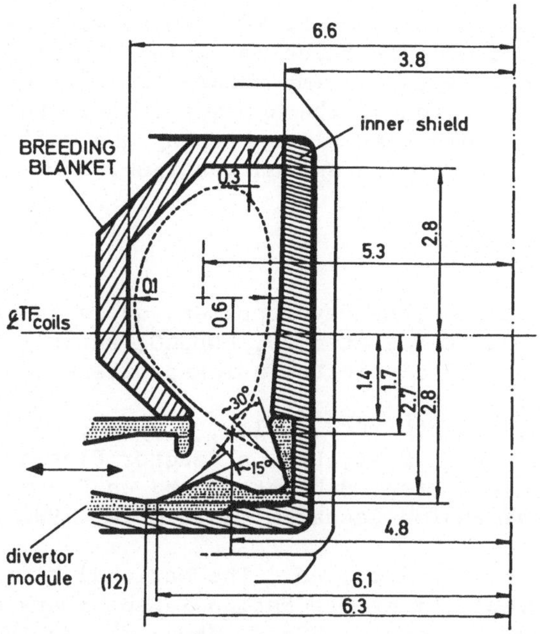

Fig. 18. INTOR cross-section with retractable divertor module.

If the conditions could be adjusted as described above, also a pumped limiter could be possible again. 12 MW could probably be handled without large difficulty. There is still the difference in properties, the divertor being decoupled from the plasma core and less sensitive to positional fluctuations of the plasma and the limiter being located within the confinement region and thus more sensitive to such fluctuations. In essence, however, both the divertor and the pumped limiter offer the same possibilities but the divertor has the advantage of better control. It is essential to substantiate these results by more basic data and more relevant experiments.

Conclusions

The most important conclusions for this Institute are that the fusion programme will very probably be successul in establishing the properties necessary for INTOR for plasma confinement and stability, efficient plasma heating to ignition, etc.. The largest uncertainty still to be faced is about the generation, behaviour, and influence of impurities or, better, of $Z > 0$-elements. Not only that we have to keep their presence in bounds, we rather have to use them in well defined quantities to taylor the plasma edge condition for a high density, low temperature to achieve low sputtering rates and the main fraction of the α-power transported to the wall via radiation. This would then allow keeping the wall thickness low which increases its resistance against fatigue.

REFERENCES

1. G. Grieger, D. Palumbo, Le dêveloppement de la fusion nucléaire, Impact: Science et Sociêté, 29, UNESCO, Paris (1979): 87
2. J. Meyer-ter-Vehn, Katalysierte Fusionsprozesse, Physikal.Blätter, 35, 5 (1979): 211
3. R.M. Kulsrud, H.P. Furth, E.J. Valeo, R.J. Budny, D.L. Jassby, B.J. Michlich, D.E. Post, Fusion Reactor Plasmas with Polarized Nuclei, Proc. 9th Int. Conf. on Plasma Physics and Contr. Nuclear Fusion Research, Baltimore, 1982: IAEA-CN-41..., to be published
4. R.R.Borchers, C.M. Van Atta, ed., The National Mirror Fusion Program, Lawrence Livermore National Laboratory and Office of Fusion Energy, Livermore, USA (1980): 33, 35 (UCAR-10042-80)
5. INTOR Group, International Tokamak Reactor - Phase Zero, IAEA, Vienna (1980), STI/PUB/556
6. INTOR Group, International Tokamak Reactor - Phase One, IAEA, Vienna (1982), STI/PUB/619

GENERAL PRINCIPLES OF MAGNETIC CONFINEMENT

John T. Hogan

Oak Ridge National Laboratory
P. O. Box Y
Oak Ridge, TN 37830

1. INTRODUCTION

These lectures are intended to prepare the way for succeeding detailed discussion of the role of atomic and molecular physics in confinement research. Hence, we will begin with a brief discussion of basic magnetic confinement issues, chiefly single particle confinement, finite pressure equilibrium and stability, and transport. Then we proceed to a description of the major approaches to magnetic confinement; tandem (ambipolar) mirrors with their associated auxiliary barriers, tokamaks and stellarators. The leading alternatives, EBT and Reversed Field Pinch, are also treated. The evolution equations for particle, energy and, where relevant, field diffusion, are presented and discussed. This is the context for atomic and molecular processes relevant to confinement.

There are many texts and review articles dwelling on basic considerations of plasma physics and of confinement experiments. A selection is given in the Appendix. Familiarity will be assumed with the elemental ideas of magnetic fusion, such as containment of charged particles by nonuniform magnetic and electrostatic fields, and the nature of simple mirror and toroidal confinement geometries.

1.1 Background

The lack of success with simple mirrors and toroidal (stellarator) configurations in the 1960's led to the elaboration of both open and closed approaches. The conceptual basis for analyzing the merit of new configurations relies on distinguishing the problems of equilibrium, stability, and transport.

By equilibrium, we mean the confinement of a finite pressure assembly of particles for many gyro-periods in a state of macroscopic balance, possibly with a non-Maxwellian distribution of velocity. This is a necessary condition for a confinement geometry, but it leads to a further requirement: perturbations to this equilibrium state should either damp away or saturate, ending in a new equilibrium configuration. (While most theoretical analyses of stability assume infinitesimal disturbances, stability to finite perturbations is the physically relevant test.) Conditions for equilibrium and stability have been most significant for establishing the criteria for confinement configurations. If these criteria can be met, then the success of the approach depends on the rate of loss of confined plasma to the external world on a slower, transport, timescale. The equilibrium, stability, and transport timescales are quite disparate, as shown by the entries in Table I.

1.2 Major Approaches and Rationale

1.2.1 Tandem (Ambipolar) Mirror. To combat the MHD interchange instability of the simple mirror configuration, additional windings, or an asymmetric arrangement of noncircular coils, are used to provide a so-called minimum-B configuration. Figure 1b shows the arrangement of the central cell of the TMX experiment at LLNL which has this feature. The configuration as shown, however, suffers from the classic loss-cone of the open system (Fig. 1a). Alternatively, the min-B configuration can be produced by complex "yin-yang" coils, as in the 2XIIB experiment at LLNL. This latter configuration has inherent micro-stability problems: the drift cyclotron loss cone (DCLC) instability[1] has been identified, arising from the open loss cone and the concomitant velocity space gradient. 2XIIB experiments showed, however, that a thermal plasma stream, injected into the plasma to fill the loss cone, produced microstability and led to diffusion timescale losses.[2] An idea for the marriage of the straight and "yin-yang" mirror systems was proposed independently.[3,4] Referred to as the tandem or ambipolar mirror, the emerging flux from the loss cone of the central cell is used to provide the needed thermal stream into the loss-cone of the "yin-yang" systems, which now serve as end plugs (Fig. 2). The central cell end loss serves to maintain an electrostatic potential in the end cells, (Fig. 1a) and this potential, in turn, enhances the confinement of particles in the central cell. Intense neutral beam injection in the end plugs is required to maintain the confining potential, and beam or radio frequency heating can be used on the central cell to raise its temperature to fusion conditions. The attraction of the configuration is that the length of the central cell can be increased without the need to scale up the size of the end-cells, and benefits of scale accrue.

TABLE-I
Timescales for Equilibrium, Stability, Transport

Process	Timescale	Magnitude	Description
Loss of Equilibrium	a/v_{the}	10^{-2} μsec	Direct particle flight to walls
Micro-Stability (velocity space)	ω_{ci}^{-1}	10^{-1} μsec	Drift cyclotron loss-cone mode (e.g., DCLC in mirror)
Macro-Stability (ideal kink)	a/v_A	1-10 μsec	Ideal free-boundary kink mode (tokamak)
Macro-Stability (resistive)	a/v_R	10-50 msec	Resistive tearing mode (tokamak)
Cross-field transport (particles, energy)	a^2/D	~1-10 msec	Cross-field transport caused by collisions or low-frequency fluctuations (tokamak, and cells of tandem mirror)

a: Typical spatial dimension

v_{the}: Electron thermal speed

v_A: Alfven speed = $B/\sqrt{2\mu_o\rho}$

ρ: Mass density $\tau_A = a/v_A$

v_R: Tearing time $\tau_R = \tau_S^{3/5}\,\tau_A^{2/5}$ $\tau_S = \frac{a^2}{\eta}$

η: Resistivity

D: Diffusivity

ℓ: Larmor radius

ω_{ci}: Ion-cyclotron frequency

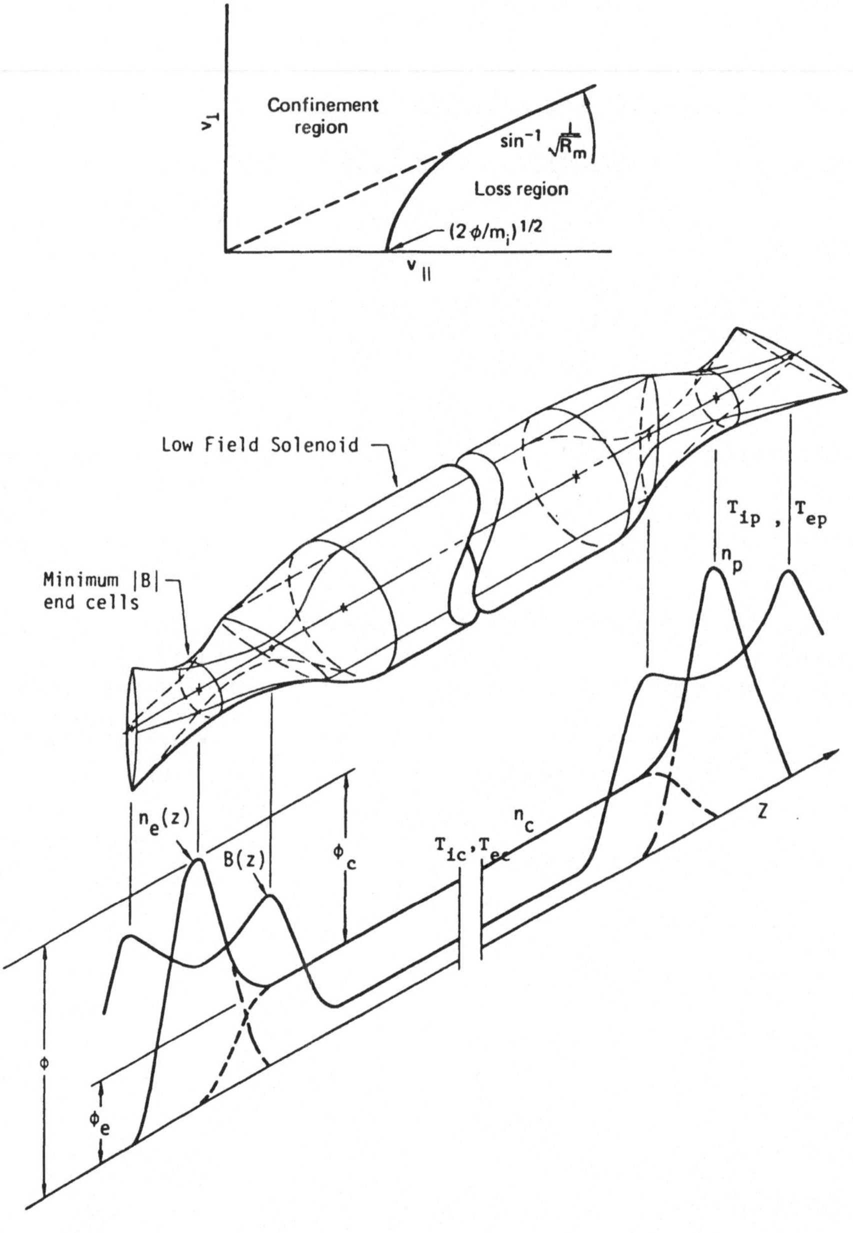

Fig. 1. (Top) velocity space for solenoid ions without potential, (dotted lines), the loss cone (revolved about the $v_\parallel$ axis) determined by strength of the mirror ratio R_m). Electrostatic confinement is provided for low-energy particles with ϕ. (Bottom) minimum-B coil arrangement, field, potential, density and temperature variation along field lines.

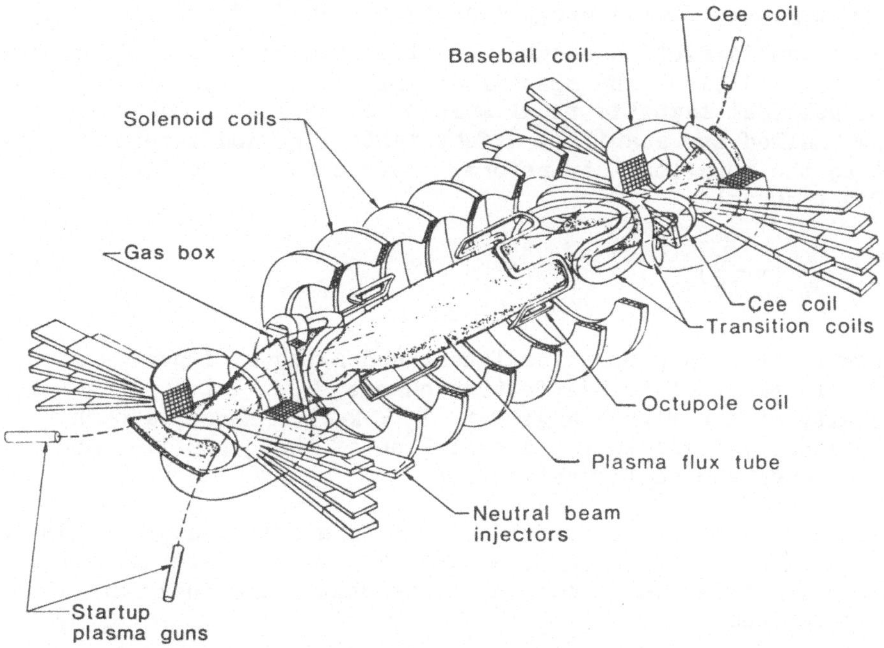

Fig. 2. Tandem mirror configuration with neutral injection to sustain the end cell density, solenoid and yin-yang coils.

1.2.2 Tokamaks. The provision of a strong toroidal field was shown to prevent the kink (Kruskal-Shafranov) instability of the plasma column. A series of detailed studies of magnetic fluctuations[5] and soft x-ray emissions[6] showed that residual effects from saturated instabilities are present, and can influence energy confinement.[7] Nonetheless, the basic tokamak configuration has proven successful, and the temperature appears to be limited only by the availability of heating power.[8-11] Attention has turned to providing features which will make the tokamak an attractive reactor: high pressure (β) and steady state.

Requirements for high pressure (expressed as the ratio of plasma to magnetic field energy density $\beta \equiv \int p dV / \int \frac{dVB^2}{2\mu_o}$) (p: pressure, B: magnetic-field strength) involve shaping of the minor cross section. In addition, the spatial distribution of $\partial\chi/\partial\psi$ (χ: toroidal, ψ: poloidal magnetic flux) must be controlled. However, this quantity (called the stability safety factor, q, and inversely related to the rotational transform) evolves on the resistive diffusion timescale.

$$\frac{\partial q}{\partial t} \simeq \frac{\partial}{\partial V}\left(\eta \frac{\partial q}{\partial V}\right) \tag{1}$$

The resistivity η is proportional to the density of impurity ions, expressed in terms of the "effective charge" $Z_{eff} \equiv \Sigma\, n_k Z_k^2 / \Sigma\, n_k Z_k$ (n_k: density of ion with charge k-1, Z_k = k-1). Hence, a high impurity density leads to rapid evolution of the q-profile, and lack of control over the configurations.

Maintenance of the toroidal current on a steady-state basis is another challenge confronting the tokamak scheme. Several techniques are being tried to produce a steady state, and their attributes will be described.

1.2.3 Stellarator. Success in understanding details of tokamak processes, and the evident reactor advantages of a steady state scheme, have combined to rekindle interest in the stellarator approach. New results from W7A[12] and Heliotron-E[13] have been cited as encouraging. True (zero current) stellarator operation remains to be thoroughly explored, however.

The chief distinction between stellarators and tokamaks, other than the absence of current, is the sensitivity of stellarator magnetic surfaces to breakup into ergodic regions.[14] Most calculations assume the existence of good surfaces, and, if this is the case, stellarator transport calculations resemble their tokamak counterparts.[15] Figure 3a,b shows typical tokamak and stellerator configurations while Fig. 4a,b compares their magnetic surfaces.

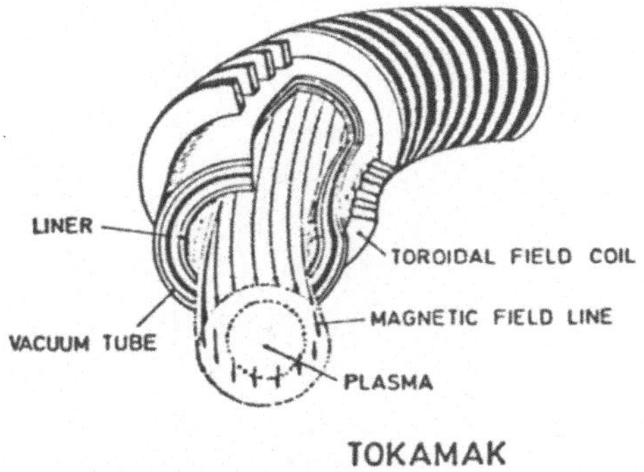

Fig. 3a. Tokamak plasma and field configuration. The vacuum (toroidal) magnetic field is supplemented by a plasma-produced (poloidal) field.

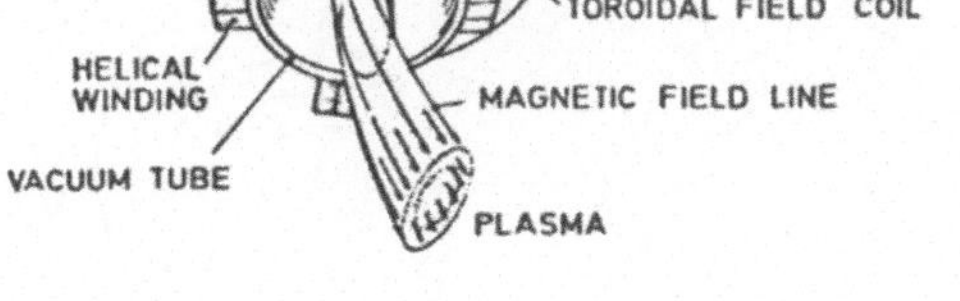

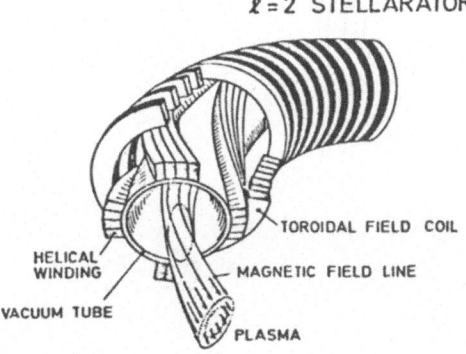

Fig. 3b. External helical coils produce a non-axisymmetric plasma configuration.

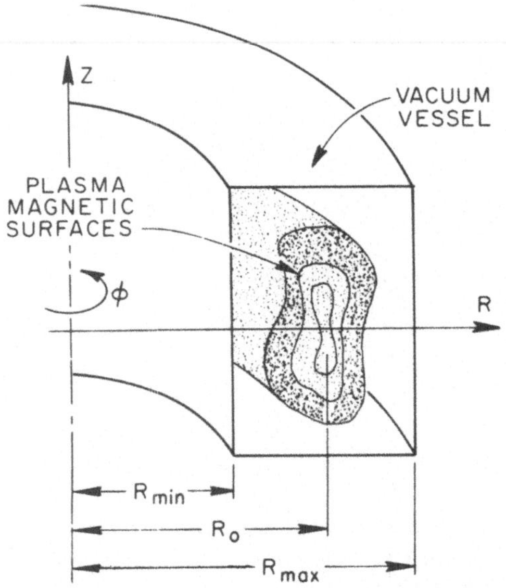

Fig. 4a. The magnetic surfaces in an axisymmetric tokamak configuration.

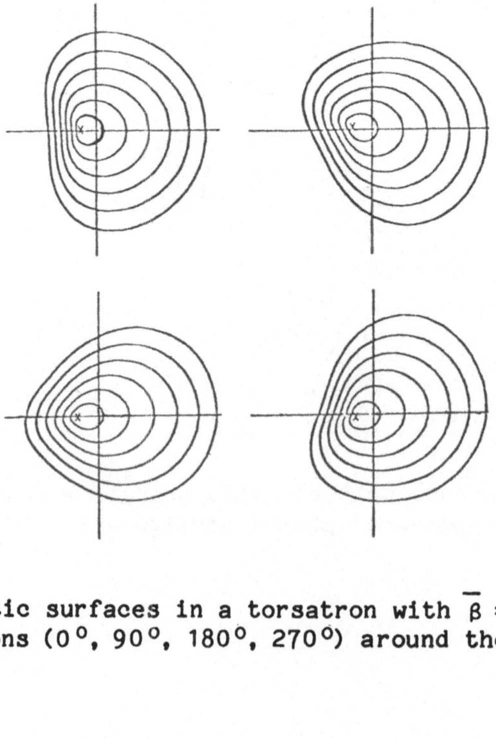

Fig. 4b. Magnetic surfaces in a torsatron with $\bar{\beta}$ = 1% for various sections (0°, 90°, 180°, 270°) around the torus.

1.2.4 Alternatives. Somewhat further removed from reactor consideration at the moment are the EBT and Reversed Field Pinch approaches. EBT consists of a toroidal set of linked simple mirrors, each embedded in a microwave cavity. Microwave heating of relativistic electrons cause the formation of annular, high β electron rings in each cavity, and these rings provide stability for the simple mirror configuration.[16]

In the reversed field pinch, the major (toroidal) field is programmed to reverse direction on a short timescale, thus producing a $B_\phi(r)$ profile which promises a higher β than the tokamak.[17]

2. BASIS FOR CONFINEMENT ANALYSIS

Equilibrium and stability considerations provide the bases for transport timescale analysis.

2.1 Equilibrium and Stability

Particle motion is computed assuming that the magnetic field is known. For low-β plasmas, the magnetic field is very nearly given by the vacuum fields of external conductors, with little effect from the plasma. (The "poloidal" β is the relevant quantity for this comparison for tokamaks, with the magnetic pressure contributed by the current-produced (or poloidal) field being used in the β definition.) For high-β conditions the plasma- induced fields must be calculated self-consistently. The equilibrium conditions are found from the pre-Maxwell equations (displacement current is neglected) and the steady state momentum balance:

$$\nabla \cdot \underline{\underline{\pi}} = \underline{j} \times \underline{B} \qquad \nabla \cdot \underline{B} = 0 \qquad \mu_0 \underline{j} = \nabla \times \underline{B} \tag{2}$$

$\underline{\underline{\pi}}$ is the stress tensor which, for non-isotropic velocity distributions has the Chew-Goldberger-Low form:

$$\underline{\underline{\pi}} = p_\parallel \hat{b}\hat{b} + (\underline{\underline{I}} - \hat{b}\hat{b})p_\perp \tag{3}$$

$p_\perp$, $p_\parallel$ are the plasma pressures, respectively, along and normal to the magnetic field, $\hat{b} \equiv \frac{\underset{\sim}{B}}{|\underset{\sim}{B}|}$; $\underline{\underline{I}}$: identity tensor.

For non-symmetric mirrors, stellarators and EBT, (2) as posed is a three-dimensional nonlinear elliptic boundary value problem, which must be solved numerically for realistic configurations.

For cases with symmetry (axisymmetric mirrors, tokamaks, reversed field pinch), the problem becomes two-dimensional. With $\nabla \cdot \underline{B} = 0$, magnetic fluxes may be introduced

$$\underline{B} = f\nabla\phi + \nabla\phi \times \nabla\psi \tag{4}$$

(ϕ: toroidal angle; $f \equiv rB_\phi$; ψ: poloidal flux) and the spatial form of the magnetic surfaces (ψ = constant) may be determined from

$$\Delta^{*}\psi \equiv r^2\nabla \cdot (r^{-2}\nabla\psi) = \frac{\partial^2\psi}{\partial r^2} - \frac{1}{r}\frac{\partial\psi}{\partial r} + \frac{\partial^2\psi}{\partial z^2}$$

$$= -r^2\sigma^{-1}\frac{\partial P_\parallel}{\partial\psi} - \sigma^{-2}\frac{\partial}{\partial\psi}\left[\frac{1}{2} r^2\sigma^2 B_\phi^2\right] - \sigma^{-1}\nabla\psi \cdot \nabla\sigma \tag{5}$$

where $\sigma \equiv 1 + \frac{2\mu_0}{B^2}(p_\perp - p_\parallel)$.

A survey of the magnetic equilibrium problem has appeared.[19] Special discussion of non-axisymmetric problems is available[20] for stellarator configurations.

The computation of equilibria is strongly coupled to the question of stability of the resulting solution, since numerical convergence schemes often mimic natural dynamic modes of the time-dependent equations. For confinement studies, we will assume that the equilibria which are used satisfy criteria which prohibit exponential growth of micro- or macroscopic modes. If this condition is not met, the atomic processes are typically too weak to affect the system's development.

Residual effects from the stability timescale will enter confinement analysis, however. Nonlinear saturated modes may produce local changes of symmetry. "Pinch buckling" transforms an interior region of the tokamak into a helically deformed shape. Radiofrequency fluctuations due to the DCLC mode in the tandem mirror end-plugs may substantially heat plasma in the central cell.

2.2 Transport Considerations for Tandem Mirrors

2.2.1 General Features of Tandem Mirror Transport. Particles bouncing between high field regions in a simple mirror experience a steady azimuthal drift caused by the radial field gradient. The drift is confined to within a Larmor radius (r_L) of a magnetic flux surface, and the magnitude of deviation from a surface is limited by the existence of the longitudinal invariant

$$J(\varepsilon,\mu,r) = \int_{S_1}^{S_2} ds \ \sqrt{\varepsilon - \mu B - e\psi} = \text{constant} \tag{6}$$

ε, μ: particle energy, magnetic moment; ϕ: electrostatic potential; S, S_1, S_2: distance along field, turning points.

The azimuthal step/bounce is

$$\Delta\psi \simeq \frac{L}{R} r_L$$

L: length of central cell; R: plasma radius.

For reactor conditions (high temperature), $\Delta\psi \ll 1$ for electrons, and $\Delta\psi \gtrsim 1$ for fuel ions and α- particles.

Motion in non-axisymmetric geometry is more complex, and is best expressed in terms of the perturbation caused by asymmetry to the angular average longitudinal "invariant"

$$\delta J \equiv J(\varepsilon,\mu,r,\psi) - \frac{1}{2\pi} \int d\psi \ J(\varepsilon,\mu,r,\psi) \tag{7}$$

In order to have small radial (cross-field) excursions, we must have

$$\alpha \left(\equiv \frac{\partial J}{J_0}\right) \ll 1 \tag{8}$$

so that drift trajectories remain close to their starting points (a property called 'omnigeneity').[21]

The contribution to α from the strong gradient region is proportional to $(R/L)^2$. Contributions from regions whose coils are rotated by 90° with respect to each other nearly cancel. In this case, $\alpha \sim (R/L)^4$. In such a case ($\Delta\psi \ll 1$) the radial excursions of untrapped (axially lost) particles are

$$\frac{\delta r}{r_{max}} \sim \alpha \qquad (\delta r \equiv r_{max} - r_{min}) \tag{9}$$

while, for trapped (mirror confined) particles

$$\frac{\delta r}{r_{max}} \sim \sqrt{\alpha} \tag{10}$$

Thus, the trapped particles contribute the largest part of the radial diffusion. (Borrowing from tokamak terminology, this is called 'neo-classical' diffusion).

For $\Delta\psi \gtrsim 1$, the successive azimuthal excursions will, in general, randomize with the result that

$$\delta r \sim r_L \frac{R}{L} \qquad (11)$$

There is, however, a class of 'resonant' particles, for which

$$\Delta\psi = (2k + 1) \frac{\pi}{2} + \varepsilon; \quad \varepsilon << 1 \quad . \qquad (12)$$

For these particles the displacements steadily accumulate and

$$\delta r \sim \sqrt{\frac{a}{\left|\frac{\partial \Delta\psi}{\partial r}\right|}} \qquad a \simeq r_l \frac{R}{L} \qquad (13)$$

leading to $\delta r >> a$. In addition, resonances may overlap, so that 'stochastic diffusion' [22] produces a flux of particles to the wall in the absence of collisions.

Analytical[23] and numerical calculations based on these elementary concepts have led to a model spanning the various $\Delta\psi$ regimes, giving diffusion rates in the neoclassical, resonant, and stochastic regions.[24]

The confinement picture for tandem mirrors is summarized in the radial balance equations for particle and energy confinement, and for the electrostatic potential.[25] The plug and central cell (solenoid) ions and electrons are treated separately.

2.2.2 Plug ions. Particle balance is given by:

$$\frac{\partial}{\partial t} n_p(r,t) = J_p - \frac{n_p}{\tau_{\parallel_p}} - \frac{n_p}{\tau_{\perp_p}} \qquad (14)$$

n_p: midplane ion density

J_p: neutral beam ionization source

$\tau_{\perp,\parallel_p}$: Radial, axial plug confinement times

(e.g., $\tau_\parallel$ for classical ions with charge exchange loss)

$$\frac{\partial}{\partial t}(n_p E_p) = J_p \left[E_{p_s} + \frac{\sigma_{cx}}{\sigma_{ion}}(E_{p_s} - E_p)\right] - \frac{n_p E_{pL}}{\tau_{\parallel_p}} - P_{p\to e} \qquad (15)$$

E_p: plug ion energy/particle

$\sigma_{cx,ion}$: charge exchange and ionization potentials for beam neutrals

E_{p_s}: neutral beam energy

E_{p_L}: average energy of axial ion loss

$P_{p\to e}$: classical coulomb energy exchange from protons to electrons[26]

2.2.3 Central cell ions. Particle balances

$$\frac{\partial}{\partial t} n_i(r,t) = J_i \frac{n_i}{\tau_{\parallel_i}} - \frac{1}{r}\frac{\partial}{\partial r}(r\Gamma_i) \qquad (16)$$

n_i: central cell ion density (for each of the species)

J_i: ionization source current density

$\tau_{\parallel_i}$: axial loss time

Γ_i: radial particle flux.

The energy balance:

$$\frac{3}{2}\frac{\partial}{\partial t} n_i T_i = J_i\left[E_{i_s} + \left.\frac{\sigma_{cx}}{\sigma_{ion}}\right|_i \left(E_{i_s} - \frac{3}{2}T_i\right)\right] - \frac{n_i E_{iL}}{\tau_{\parallel_i}}$$

$$- \frac{1}{r}\frac{\partial}{\partial r}(rQ_i) + \left[P_{e\to i} + \sum_j P_{j\to i}\right]$$

$$+ P_i^{ext} - Z_i \Gamma_i \frac{\partial \phi_c}{\partial r} \qquad (17)$$

where

T_i: temperature of the (near Maxwellian) solenoid ions

E_{i_s}: source ion energy

σ^i_{cx}, σ^i_{ion}: charge exchange and ionization cross-sections

Q_i: radial energy flux; $= \frac{5}{2} \Gamma_i T_i + q_i$, where q_i is the radial heat flux.

$P_{e,j \to i}$: classical Coulomb energy equipartition[26]

P_i^{ext}: radiofrequency heating of central cell by DCLC (plug) fluctuations.

The models for Γ_i, Q_i are drawn from the relevant (collisional, resonant, stochastic) regime, while $\tau_{\parallel_i}$ models have been presented in Refs. 23 and 24.

2.2.4 Electrons. The electron density is constrained by charge neutrality. In the plug

$$n_{e_p} = n_p Z_p \tag{18}$$

while in the central cell

$$n_{e_c} = \sum_i n_i Z_i \; . \tag{19}$$

A radial electron energy balance equation is found by averaging over field lines; the average density is thus

$$N_e = \frac{2L_p}{B_p} n_{e_p} + \frac{L_c}{B_c} n_{e_c} \; . \tag{20}$$

L,B are lengths, midplane field strengths in the plug and central cell.

The electron energy balance is

$$\begin{aligned} \frac{3}{2}\frac{\partial}{\partial t}\left(N_e T_e\right) = & \left[\left(\frac{2L_p}{B_p}\right) J_{ep} E_{e_{ps}} + \left(\frac{L_c}{B_c}\right) J_{ec} E_{e_{cs}}\right] \\ & - \frac{N_e E_{eL}}{\tau_{\parallel e}} - \left[\left(\frac{2L_p}{B_p}\right)\frac{1}{R}\frac{\partial}{\partial R}\left(R Q_{ep}\right) + \left(\frac{L_c}{B_c}\right)\frac{1}{r}\frac{\partial}{\partial r}\left(r Q_{ec}\right)\right] \\ & + \left[\left(\frac{2L_p}{B_p}\right) P_{p \to e} - \left(\frac{L_c}{B_c}\right) \sum_i P_{e \to i}\right] \\ & + \left[\left(\frac{2L_p}{B_p}\right) \Gamma_{ep} \frac{\partial \phi_p}{\partial R} + \left(\frac{L_c}{B_c}\right) \Gamma_{ec} \frac{\partial \phi_c}{\partial r}\right] . \end{aligned} \tag{21}$$

The electron and ion source currents are related by:

$$J_{e_p} = J_p Z_p \tag{22}$$

$$J_{e_c} = \sum_i J_i Z_i \tag{23}$$

and

E_{eps}, E_{ecs}: Plug, solenoid electron source energies

$\tau_{\parallel e}$: electron axial confinement time

E_{eL}: axial electron energy loss

Q_e: radial electron energy flow. $Q_e = \frac{5}{2}\,\Gamma_e\, T_e + q_e$, where q_e is the radial electron heat flux.

R,r: plug, central cell flux tube radii: $(B_p R^2 \simeq B_c r^2)$

$P_{i \to j}$: classical coulomb energy exchange from species i to j.

Models described in 2.2.1 are used for $\tau_{\parallel}$, q and Γ.

The energy and particle balance is completed by finding the self-consistent electrostatic potential.

2.2.5 Electrostatic Potential. For a Maxwellian distribution, with uniform temperature, the potential at the plug and solenoid midplanes are related by

$$\phi_p = \phi_c + T_e \,\ell n \frac{n_{e_p}}{n_{e_c}} \tag{24}$$

$\phi_c(r,t)$ will depend on the detailed model. If radial and axial loss rates are assumed balanced on each flux tube:

$$\left(\frac{2L_p}{B_p}\right)\frac{n_p Z_p}{\tau_{\parallel p}} + \frac{L_c}{B_c}\sum_i z_i\left[\frac{n_i}{\tau_{\parallel_i}} + \frac{1}{r}\frac{\partial}{\partial r}\left(r\Gamma_i\right)\right]$$

$$= \frac{N_e}{\tau_{\parallel e}} + \frac{2L_p}{B_p}\frac{1}{R}\frac{\partial}{\partial R}\left(R\Gamma_{ep}\right) + \frac{L_c}{B_c}\frac{1}{r}\frac{\partial}{\partial r}\left(r\Gamma_{ec}\right) \tag{25}$$

where Γ_i can depend on ϕ_c and must be found self-consistently. Equations (14)-(25) describe the manifold inter-related processes which govern tandem mirror behavior on the diffusive timescale.

2.3 Tokamak Confinement

2.3.1 General considerations. Many of the collisional diffusion concepts referred to in Section 2.2 were based on analogy with earlier tokamak calculations. The major features of collisional diffusion are briefly reviewed.

Passing particles (those which can traverse the 1/R variation in field strength) suffer relatively small excursions from a magnetic surface (ψ = constant). For them

$$\frac{\delta\rho}{a} \simeq \frac{\hat{r}_L}{R} \tag{26}$$

a,r: Minor, major radii

$\hat{r}_L$: Larmor radius in the poloidal magnetic field.

Particles trapped in the field minima, however, have

$$\frac{\delta\rho}{a} \simeq \sqrt{\frac{R}{a}}\,\frac{\hat{r}_L}{R} \, . \tag{27}$$

Hence, collisionally induced pitch angle scattering will change passing to trapped particles, and give a large contribution to spatial diffusion. A comprehensive survey of collisional diffusion theory has been given[27] and recent developments reviewed.[28]

The collisional (neoclassical) model has had some success in describing ion behavior in tokamaks, but does not model electron behavior. Anomalous transport models, based on electrostatic fluctuations or magnetic field line flutter, have been proposed. A number of empirical models, describing gross trends in energy confinement, have been attempted.

3.3.2 Relaxation of time scales. One feature of the neoclassical calculation should be singled out, because of subsequent discussions of stellarator confinement. The ion heat flux in a near-Maxwellian plasma is[26]

$$\underset{\sim}{q_i} = -k_\parallel \nabla_\parallel T_i - k_\wedge \hat{b} \times \nabla T_i - k_\perp \nabla_\perp T_i \tag{28}$$

where

$$\nabla_\parallel \equiv \hat{b}\hat{b} \cdot \nabla \qquad \nabla_\perp \equiv \nabla - \hat{b}\hat{b} \cdot \nabla$$

and $k_\parallel : k_\wedge : k_\perp = 1 : (\omega_c \tau_i)^{-1} : (\omega_c \tau_i)^{-2}$ (ω_c: ion gyrofrequency, τ_i:

ion-ion collision frequency; typically $\omega_c \tau_i \sim 10^4$).

These widely varying timescales for $\parallel$, drift and perpendicular heat flux yield certain features which are likely to be preserved in any confinement model, and thus are of general interest. On the fastest of the timescales, the energy balance is

$$\nabla \cdot \underset{\sim}{q_i} = 0 \tag{29}$$

which requires that $T_i^{(0)}(r) = T_i^{(0)}[\psi,(r)]$. That is, rapid equilibration levels the temperature on a magnetic flux surface.

To next order

$$\underset{\sim}{B} \cdot \nabla T_i^{(1)} = \frac{Bk_{\wedge}}{k_{\parallel}} rB_{\phi} \frac{\partial T_i^{(0)}}{\partial \psi} \left(1 - B^2 \langle B^2 \rangle^{-1}\right) \tag{30}$$

In axisymmetry $\underset{\sim}{B} \cdot \nabla a = B_{pol} \frac{\partial a}{\partial \ell p}$ in the minor cross section. Small flux surface variations are thus permitted in $T^{(1)}$ ($\lesssim a/R$). In the next order, on the slowest timescale, the diffusive cross-field transport equation for $T_i^{(0)}(\psi,t)$ is found. A simular evolution to the slow timescale proceeds for the particle and field diffusion equations.

2.3.3 Transport Equations. The solution of the coupled particle and energy balance equation must be combined with the evolution of the magnetic equilibrium. Hence, there is an interaction between the equilibrium (and associated stability properties) and the transport processes.

These equations are[29]

$$\frac{\partial}{\partial t} \left[V'(\psi)\, n_e\right] = \frac{\partial}{\partial \psi} \left(V' D \langle |\nabla \rho|^2 \rangle \frac{\partial n_e}{\partial \psi}\right) + V'(\psi) \langle \Sigma_e \rangle \, . \tag{31}$$

Σ_e = ionization source from neutral hydrogen, impurities, and beam particles.

$$\frac{3}{2} \frac{\partial}{\partial t} \left[(V')^{5/3} P_e\right] + (V')^{2/3} \frac{\partial}{\partial \psi} \left[V' \left(q_e + \frac{5}{2} T_e \Gamma_e\right)\right] = (V')^{5/3}$$

$$\times \left[P_{i \to e} + P_{oh}\right] + (V')^{5/3} \times \left(\langle Q_e \rangle^{inj,\alpha} - \langle P \rangle_z\right) \tag{32}$$

$$P_{oh} = \eta j(j - j_s); \quad j = \frac{f}{A} \frac{1}{\rho} \frac{\partial}{\partial \rho} \left(\rho \left\langle \frac{|\nabla \rho|^2}{R^2} \right\rangle \tilde{B}/f\right) ,$$

j_s = external current density source

where $Q_e^{inj,\alpha}$ are the energy sources from neutral injection and alpha particle thermalization, and P_z represents the loss due to ionization of impurities and to in elastic electron collisions with impurities leading to radiated energy loss. $P_{i\to e}$ is the Coulomb equilibration term.

$$\frac{3}{2}\frac{\partial}{\partial t}\left[P_i(V')^{5/3}\right] + (V')^{2/3}\frac{\partial}{\partial\psi}V'\left(q_i + \frac{5}{2}T_i\Gamma_i\right) = (V')^{5/3}$$

$$\times\left[P_{E\to i}\right] + \left(\langle Q_i^{inj,\alpha}\rangle - \langle Q_{cx}\rangle\right)^{\bullet}(V')^{5/3} . \tag{33}$$

$$\frac{\partial\tilde{B}}{\partial t} = \frac{\partial}{\partial\rho}\left\{\frac{n}{\mu_o}\left[\frac{f}{A}\frac{1}{\rho}\frac{\partial}{\partial\rho}\left(\rho\left\langle\frac{|\nabla\rho|^2}{R^2}\right\rangle\frac{\tilde{B}}{f}\right) - j_s\right]\right\} \tag{34}$$

where $Q_i^{inj,\alpha}$ are the ion energy sources from injection and alpha particle thermalization, and Q_{cx} is the charge-exchange transfer term. (For small, relatively low density plasmas, this is a loss term throughout the plasma volume.) j_s is the external current density source, $\tilde{B}$ is related to the stability safety factor q by $q = \rho B_\phi/RB$. Then one-dimensional (flux surface average) equations must be combined with the equation describing macroscopic force balance

$$\Delta^*\psi = -r^2\left[P_T' + r^2\left(\frac{\omega^2}{2T}\right)'\right]e^{\frac{r^2\omega^2}{2T}} - ff' \tag{35}$$

where

$\omega = V_\phi/r$: toroidal rotation frequency
P_T: thermal component of pressure = $\sum_i n_i T_i$
$f : rB_\phi$, $' \equiv \frac{\partial}{\partial\psi}$.

2.2.3 Stability Considerations. The stability properties of the magnetic flux surfaces are embodied in the geometric characteristics of the cross-section shape. For example, the magnetic well depth, which provides stability against magnetic interchange, is related to the vertical elongation and triangularity:[30]

$$W'' = -2\frac{1}{RB_T^2\sqrt{1-e^2}}\left[1 - \frac{3}{2}e\frac{1+e}{2+e} + 6eQ\frac{1-e}{2e}\right] \tag{36}$$

where coordinates of the magnetic surface are

$$\psi = \rho_o^2 + Q\varepsilon\rho^2 \cos 3\theta$$

(ρ, radius, θ, angle. Q: triangularity; e: elongation, $\varepsilon = r/R$.)

Further, the location and effectiveness of poloidal magnetic divertor separatrices are determined by the flux surface shape as it evolves under the action of the processes described in Eqs. (31)-(34).

The conditions for occurrence of saturated nonlinear resistive instabilities also depend on the q-profile. The Δ' criterion[31] can be applied to find the width of a so-called magnetic "island": a region of essentially helical symmetry with parameter varying as $e^{i(m\theta + n\phi)}$ (θ, ϕ: poloidal, toroidal angle; m, n low order integers; m = 2, n = 1, m = 3, n = 2 are typical pairs.

2.3 Stellarators

The relaxation of timescales for stellarators proceeds in essentially the same fashion as for axisymmetric geometry. However, the solvability constraint[30] is a key difference in nonsymmetric geometry.

To solve this equation in a nonsymmetric geometric geometry requires repeated integration around the torus along a field line and introduces the problem of resonant denominators for perturbations matching the harmonic variation of the magnetic field. While the KAM theorem guarantees the existence of solutions for a class of perturbations ($T_1^{(1)}$ in this case) the solutions so obtained are pathological.[14] The physical problem must by formulated with the *a priori* assumption that possible ergodic regions are widely separated. If this is done, the stellarator transport problem can be reduced to an equivalent one-dimensional system with essentially the same form as the tokamak transport equations. However, the electrostatic potential enters, and must be computed, and the work done against it added to the energy equation. In addition, flux surface averaged quantities must now be obtained from a three-dimensional equilibrium. The calculation is much more complex and studies in this area are just beginning.

Kinetic calculations give the neoclassical fluxes for an $\ell = n$ stellarator geometry as,[32] for example,

$$\Gamma_j = - \nu_j n_j \left(\frac{r}{R}\right)^2 \sqrt{2\varepsilon_n}\, r_{Lj}^2\, V_j^2 Q_j^2 \left[\frac{1}{n_j}\frac{\partial n_j}{\partial r}\right.$$

$$+\left(\frac{Q_s^j}{Q_2^j}-\frac{3}{2}\right)\frac{1}{T_j}\frac{\partial T_j}{\partial r}-\frac{e_j}{T_j}\frac{\partial\phi}{\partial r}\Big] \tag{37}$$

E_n: helical field ratio; Q_j^i; numerical constants. Hence, the electrostatic potential must be determined self-consistently from charge neutrality and ambipolarity.

2.4 EBT

The EBT geometry is stabilized by relativistic electron annuli so that the core plasma is then governed by drift and diffusion processes, along with the atomic physics related impurity radiation and charge exchange. The radial transport balance is[33]

$$\frac{\partial n_j}{\partial t}=-\frac{1}{r}\frac{\partial}{\partial r}\left(r\Gamma_j\right)+nn_0\langle\sigma v\rangle_{ion} \tag{38}$$

$$\frac{3}{2}\frac{\partial}{\partial t}\left(nT_i\right)=P_{e\to i}-\frac{1}{r}\frac{\partial}{\partial r}\left(rQ_i\right)-nn_0\langle\sigma v\rangle_{cx}\frac{3}{2}\left(T_i-T_0\right)$$

$$+e\Gamma_iE_r+\frac{3}{2}nn_0\langle\sigma v\rangle_{ion}T_0 \tag{39}$$

$$\frac{3}{2}\frac{\partial}{\partial t}\left(nT_e\right)=2nm_e\langle D_M\rangle-\frac{1}{r}\frac{\partial}{\partial r}\left(rQ_e\right)+P_{i\to e}$$

$$+nn_0\langle\sigma v\rangle_{ion}E_{ion}-eE_r\Gamma_e \tag{40}$$

$$\frac{E_\perp}{e}\frac{\partial E_r}{\partial t}=\Gamma_e-\Gamma_i=\left(D_{n_i}-D_{n_e}\right)\frac{\partial n}{\partial r}+D_{T_i}\frac{\partial T_i}{\partial r}$$

$$-D_{T_e}\frac{\partial T_e}{\partial r}-n\left(\mu n_i-\mu n_e\right)E_r \tag{41}$$

where

E_r: radial electrostatic field

$\Gamma_{e,i}$: radial electron, ion fluxes

$Q_{e,i}$: radial electron, ion energy fluxes

$P_{i \to j}$: Coulomb equilibration

$\langle D_M \rangle$: ECH resonant power absorption

μ_{nj}: $(e_j/kT_j)D_{nj}$; $\mu_{Ej} = (e_j/kT_j)K_{nj}$

and the neoclassical fluxes are:

$$\Gamma_j = - D_{nj} \frac{\partial n_j}{\partial r} - D_{T_j} \frac{\partial T_j}{\partial r} + n \mu_{nj} E_r \tag{42}$$

$$Q_j = - K_{nj} \frac{\partial n_j}{\partial r} - K_{Tj} \frac{\partial T_j}{\partial r} + n_j \mu_{Ej} E_r \tag{43}$$

The solution for the self-consistent potential in EBT shows that several theoretically possible solutions exist, according to neoclassical theory, although one is thermally unstable. A similar situation is expected to occur in the stellarator transport problem as well.

2.5 Reverse Field Pinch

The time evolution of the magnetic field is a central concern for the RFP, as processes which govern field diffusion occur on the same timescale as those for particle and energy loss. The particle energy and field balance for radial RFP transport are[34]:

$$\frac{\partial n(r,t)}{\partial t} = - \frac{1}{r} \frac{\partial}{\partial r} (rnV_r) \tag{44}$$

$$\frac{\partial B_\theta}{\partial t} = - \frac{\partial}{\partial r} \left(V_r B_\theta - \frac{c}{en} R_z\right) \tag{45}$$

$$\frac{\partial B_z}{\partial t} = - \frac{1}{r} \frac{\partial}{\partial r} \left[r\left(V_r B_z + \frac{c}{en} R_\theta\right)\right] \tag{46}$$

$$\frac{3}{2} \frac{\partial}{\partial t} (nT_e) = - \frac{1}{r} \frac{\partial}{\partial r} \left[r\left(q_e + \frac{5}{2} \Gamma_e T_e\right)\right] + \frac{\underline{R} \cdot \underline{j}}{en} - Q_e \tag{47}$$

$$\frac{3}{2}\frac{\partial}{\partial t}(nT_i) = -\frac{1}{r}\frac{\partial}{\partial r}\left[r\left(q_i + \frac{5}{2}T_i\Gamma_i\right)\right] + Q_i \tag{48}$$

$$\frac{\partial P}{\partial r} = -\frac{1}{\mu_o}\left[B_z\frac{\partial B_z}{\partial r} + \frac{B_\theta}{r}\frac{\partial}{\partial r}(rB_\theta)\right] \tag{49}$$

where $P = n(T_e + T_i)$ and q, Γ, and R are assumed to have the classical collisional value.[26] Note the evolution of pinch velocity and major(B_z) field on the same timescale as the other transport variables.

Impurity radiation has played a determining role in many fast pinch experiments, although optimum choice of vacuum liner material has done much to reduce losses and prolong the timescale for decay of plasma parameters.

3. CURRENT STATUS OF THE APPROACHES

Each approach faces difficulties with regard to eventual reactor application. In most cases, solutions have been proposed which require further understanding of underlying physical processes. These are described.

3.1 Tandem (Ambipolar) Mirrors

Experimental results from the Tandem Mirror Experiment have confirmed the working elements of this new concept: the tandem mirror configuration has been produced, electrostatic plugging has been demonstrated, improved electron confinement (e.g., with respect to 2XIIB) has occurred, and the rf activity in the plug obeys 2XIIB steam-stabilization scaling.[35]

Reactor studies based on these concepts show some possible difficulties, though. Neutral beams with energies 0.5-1.2 MeV and superconducting coils with a 15-20 T field strength could be needed to produce Q (fusion power out/power input) > 5. Hence, new designs embodying features consistent with previous mirror theory, have been developed.

The addition of several auxiliary cells is being studied, in order to improve plug confinement and reduce field and beam requirements. In one version, "thermal barriers" are to be interposed between the central cell and the end plugs to shield hot electrons in the plug from the colder central cell electrons.[36]

New theoretical doubts have been voiced, however, about the possible enhanced transport due to magnetically trapped particles.[37] Instabilities driven by such trapped particles were previously predicted for tokamaks in low collisionality regimes, but these modes proved to saturate at low levels. Design considerations for auxiliary cells are being modified to lessen the impact of these modes.

3.2 Tokamaks

Neutral beam heating has proven to be effective, and temperatures have advanced well into the low collision frequency regime needed for reactors. Chief concerns now are related to practical features needed for reactor systems: high power density (β), long pulse current (steady state) and stable, steady-state heating (ICRF or NB).

Recent experiments on the ISX-B tokamak have shown a saturation in β with rising beam power, with $\beta \approx 2$-2.5% obtained with $P_{beam} \simeq 2$ MW.[38] The PDX tokamak, with 7-8 MW (in a large volume device) has obtained $\beta \simeq 2.5$-3.0% without saturation.[39] While the mechanisms for saturation on ISX-B has not been finally established, a promising model has been proposed involving the stimulation of small scale fluctuations caused by resistive pressure-driven modes.[40] Such a mechanism, which diminishes (with the resistivity $\sim T^{-3/2}$) at high T, should not limit reactor β values to presently achieved limits.

The maintenance of steady state introduces entirely new considerations. While a neutral beam-driven current has been observed[41] and could be used as a basis for reactor, the conditions for a large current from this source are best at low $\overline{n}_e$ and high Z_{eff}; and these are the opposite of those desired in a reactor. There has been recent success in the PLT experiment in which up to 200 kA current has been driven with lower hybrid heating.[42] The density ($\overline{n}_e \sim$ mid 10^{12} cm^{-3}) is very low in these trials, and must be raised for reactor application. Preliminary trials have been made with injection of MeV relativistic electron beams into a tokamak.[43] Further experiments on larger devices are needed to evaluate this technique. Table 2 summarizes the current drive schemes.

3.3 Stellarators

A parallel effort in obtaining a steady state toroidal reactor has received impetus from new stellarator experimental results.

The W7A device (with $\ell = 2$ and 5 field periods) has recorded high temperatures and densities[12] but impurity densities and radiation, felt to be introduced with the injected beam, play a significant role. Analysis and diagnostic work relies heavily on atomic data in this area. The Heliotron-E (an $\ell = 3$ stellarator) has also

TABLE II

Steady-State Current Schemes

Type	Description	References
Neutral beam	Distortion of background plasma by fast ions produces current when $Z_{eff} \neq 1$, $j_b = n_b \Gamma \left(1 - \frac{1}{Z_{eff}}\right)$	T. Ohkawa[53] DITE group[41]
Lower hybrid	Wave at the lower hybrid frequency $\omega^2 = \omega_{ce}\omega_{ci}$ maintains a distortion at high energy on the electron distribution	N. Fisch[54]
Relativistic electron beam	MeV beam of electrons is periodically pulsed, stored beam slowly decays	J. Benford et al.[55]

shown encouraging early results[13]. Zero current operation, the true test of stellarator confinement, has been tested in two modes: in the first, a toroidal current is formed and then allowed to decay. In the second, ECH is applied, with no loop voltage. The resulting confinement properties are similar and encouraging in both cases, although possible residual effects of high energy electrons (from the startup voltage and resonant ECH, respectively) could cause problems with interpretation.

4. CASE STUDY: H^o CHARGE EXCHANGE WITH MULTIPLY CHARGED IONS

To emphasize the role of atomic physics in magnetic confinement research, it is useful to select an example of the impact made in the past, and the continuing change which new atomic data makes in confinement models.

This case study is drawn from tokamak confinement research, and began with design considerations for the new generation of large devices (TFTR, JET, JT-60, T-15). It was feared that injection of 120 keV D^o into a plasma with impurity densities typical at that time (1974-76, $Z_{eff} \approx 2\text{-}4$) would lead to a beam-trapping instability with the beam being deposited preferentially on the exterior[44] thus limiting the effectiveness of these machines. Detailed theory and measurement (reviewed in Ref. 45) showed that charge transfer would be the dominant trapping process at this energy, and that the cross-section increases only linearly with impurity charge, rather than as Z^2, as assumed in modelling calculations.

As discussed at the previous NATO ASI[45] this finding led to improved understanding of current experiments, as well as to increased confidence in large machine design. The trapping process

$$H^o + A^{n+} \rightarrow H^+ + A^{(n-1)+} \qquad A = \text{IMPURITY ION}$$

is a recombination process for the target impurity ion. Hence, with intense neutral beam injection a strong recombination to lower charge states occurs. This, in turn, raises the radiated power, especially for low-Z materials commonly found in tokamaks (C,O) which would be completely stripped and nonradiative without beam-induced recombination. As noted earlier, this newly available atomic data thus provided a fresh insight into the limitations on neutral beam heating. This development has been intensively exploited by the DITE[46] and other tokamak groups.[47,48]

Development on the atomic physics side has continued, and new results have been found, both theoretical and experimental, on the $H^o + C^{n+} \rightarrow H^+ + C^{(n-1)+}$ reactions at lower energies, which are characteristic of the thermal plasma (as distinguished from injected beam ions).[49-51] The Z-scaling which was valid for neutral beam scale energies has been shown to fail for lower energies. Important detailed differences have been revealed. For example, the cross sections for $C^{4+} \rightarrow C^{3+}$ electron capture are ~ 5 times higher than those for $C^{3+} \rightarrow C^{2+}$. (The Ryufuku-Watanabe Z-scaling[52] would predict a 30% change.) Thus, a large recombination flux produces the Li-like CIV ions, which are strongly radiative, and a diminished flux issues from this state to lower charge states.

A further consideration in the new application of these data is the actual spatial distribution of H^o in tokamaks. The charged particle parameters are thought to be axisymmetric, and to vary in space only across magnetic surfaces, because of the rapid thermalization described in Section 2. The path of neutral hydrogen atoms, however, is unaffected by the magnetic field, and is localized near external sources: limiters, gas inlet valves, neutral beam ducts, etc. Figure 5 shows calculated spatial distributions of $n_{H^o}(r,\phi,Z)$ in an axisymmetric tokamak.

The steady state charge balance for carbon ions is

$$\nu_{D} \cdot \nabla n_c^k = n_e \left[n_c^{k-1} I_{k-1} + n_c^{k+1} R_{k+1} - n_c^k (I_k + R_k) \right]$$

$$- n_o n_c^k \langle \sigma v \rangle_c^{cx} + n_o n_c^{k+1} \langle \sigma v \rangle_{k+1}^{cx} \qquad (50)$$

where ν_D is the drift (radial, and along field lines) of carbon ions of charge k, I_k is the electron impact ionization rate, R_k is the sum of radiative and dielectronic recombination rates and $n_o n_c^k$ $\langle \sigma v \rangle_{cs}^k$ is the charge exchange recombination rate. Because neutral densities are highly localized, and heavy impurities move more slowly along the field than protons, the spatially localized H^o density produces a spatially localized impurity charge density.

This fact, yields a relatively simple solution to a paradox which has plagued tokamak confinement theory for some years. As described in the earlier ASI, impurity radiative emission has often been seen to be asymmetrically distributed. There was speculation that basic assumptions about rapid thermalization on flux surfaces would have to be reformulated, since neoclassically consistent asymmetries were limited to ~ 30% variation, while factors of 2 were commonly observed. With the new charge transfer data, however, a more natural explanation is possible. Figure 6a,b shows the CV density predicted both with and without impurity charge exchange effects. It is seen that the asymmetric H^o density produces an asymmetric CV density (and corresponding radiative emission). The ion densities are, however, all constant on a magnetic flux surface, in accord with basic theory.

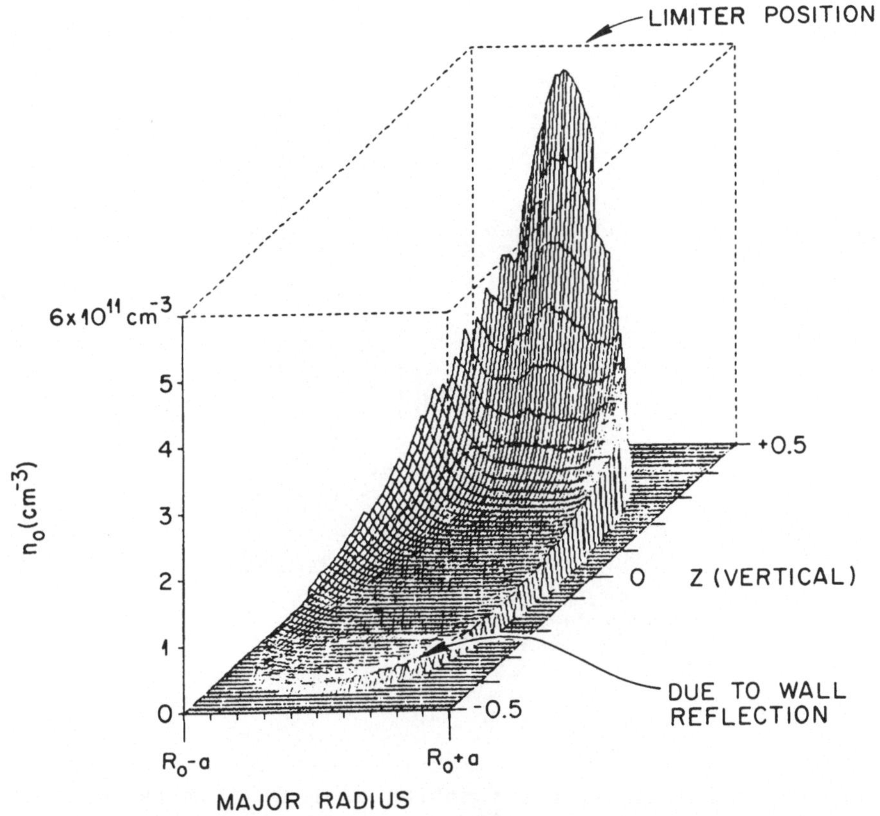

Fig. 5a. Neutral density variation in a tokamak with a limiter source. Here, the density averaged on ϕ is shown. The limiter position is at the top, center.

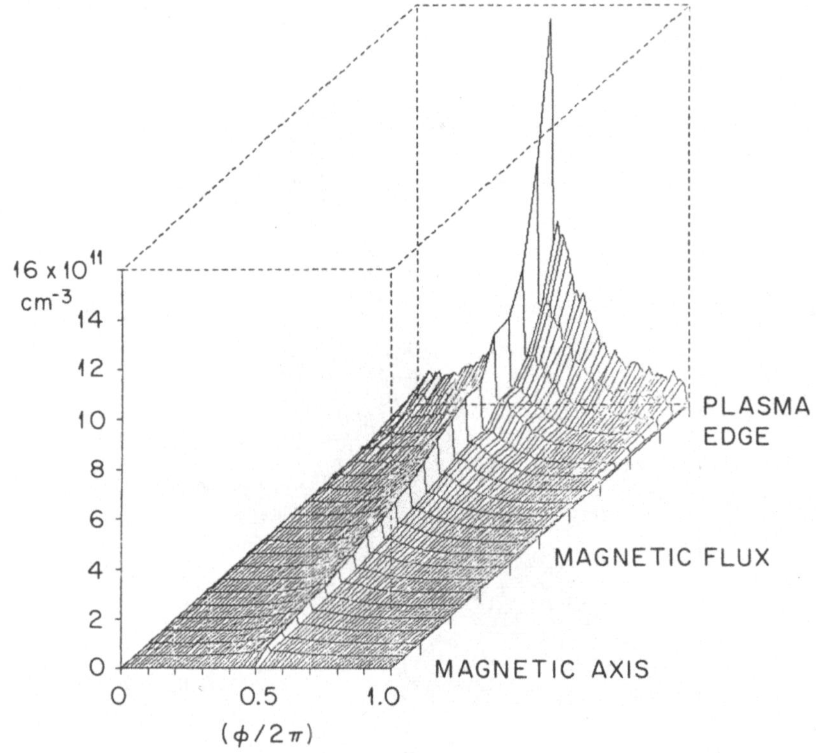

Fig. 5b. Neutral density as a function of ψ, ϕ. Limiter position is at the edge, $\phi = 180^\circ$.

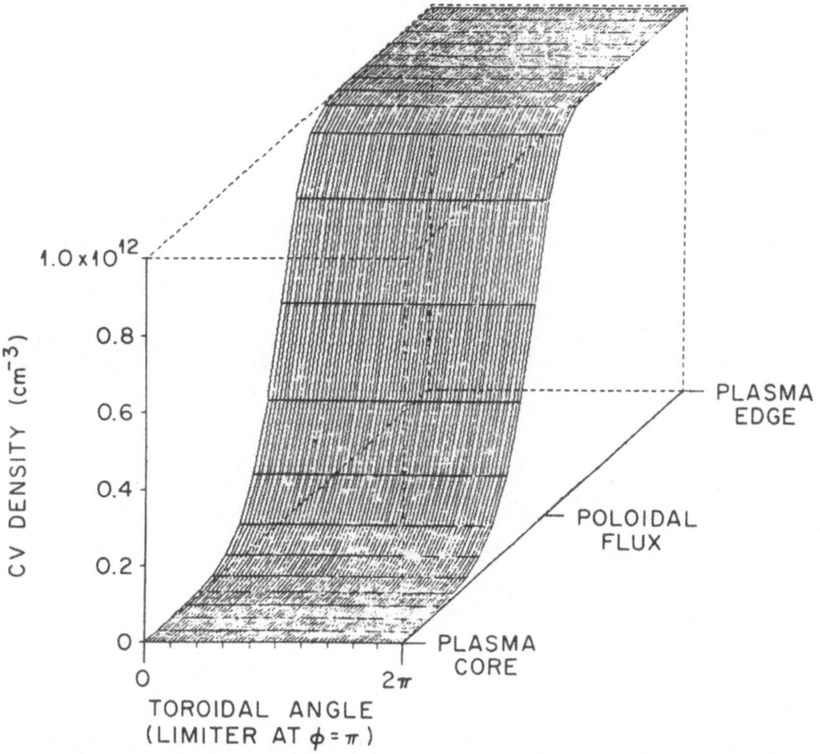

Fig. 6a. Calculated spatial density of CV ions, assuming a uniform overall carbon density and $T_e = T_e(\psi)$. In steady state this density depends only on T_e, and hence does not vary with toroidal angle.

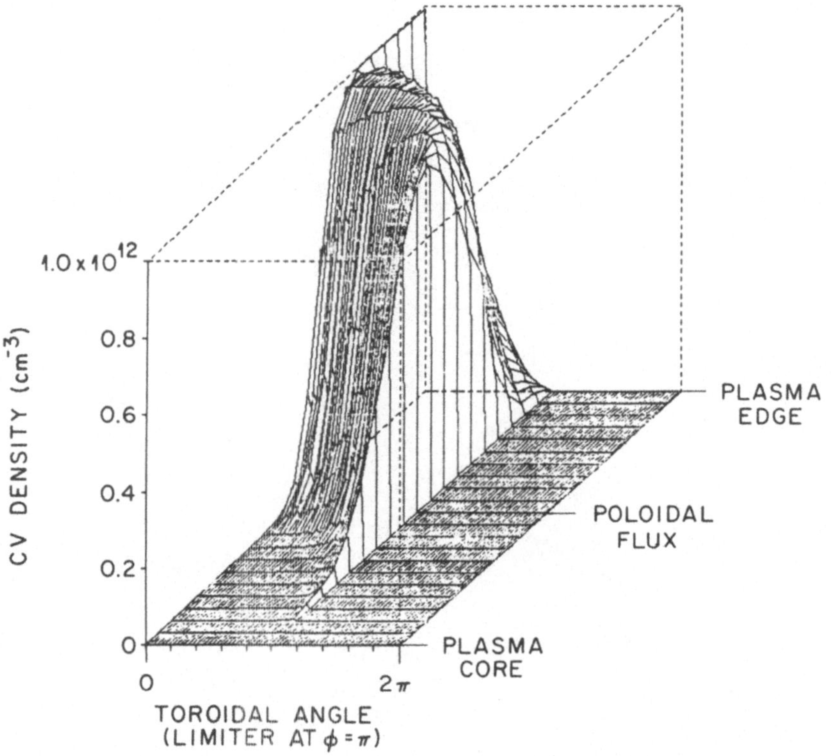

Fig. 6b. As with 6a, except charge exchange effects due to the n_{H^o} density (seen in Figs. 5a-b) are included. (The right half of the profile is removed for clarity). A strong variation in the CV density is thus expected because of charge exchange effects.

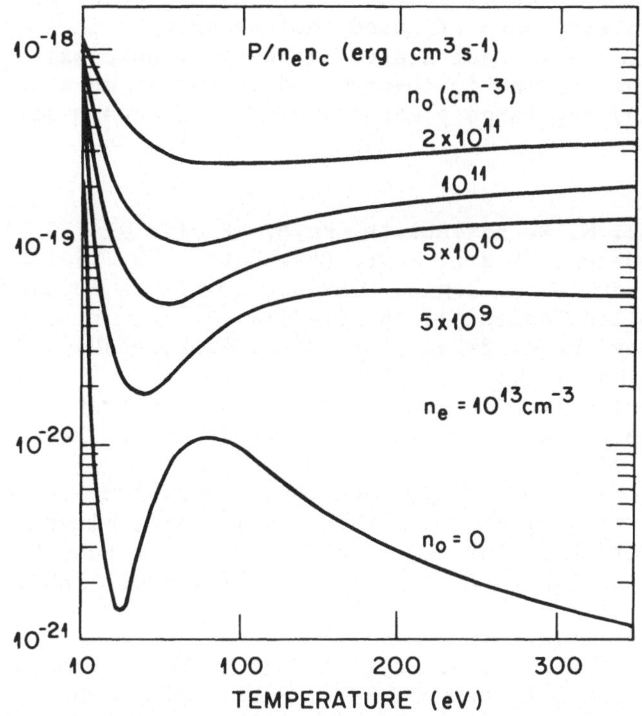

Fig. 7. Calculated total radiative power density for carbon as a function of T_e and n_H0. For $n_H0 = 0$ the results exclude CX effects. For typical edge densities, especially with variations such as those in Figs. 5a-b, 30-fold ennancement of the radiative output is found. Cases are all computed with a typical edge electron density $n_e = 10^{13}$ cm^{-3}.

In present experiments, very good results have been obtained with carbon-based limiters. The impurity charge exchange data may perhaps explain why. As shown in Fig. 7 there is a great increase in the calculated radiative emission from carbon ions when neutral densities are assumed at levels typical of the plasma edge. Hence, power efflux consists predominantly of relatively benign edge radiation, and sputtering and evaporation of the limiter is minimized.

Thus, data originally sought to evaluate a potential beam trapping instability have effected radical changes in models for the core energy balance in beam heated discharges, have explained a puzzling paradox in kinetic theory, and may serve as a basis for new schemes for handling large power fluxes from future reactors.

REFERENCES

1. R. F. Post, M. N. Rosenbluth, Phys. Fluids 9, 730 (1966).
2. F. H. Coensgen, et al., Phys. Rev. Lett. 35, 1501 (1975).
3. T. K. Fowler, B. G. Logan, Comments on Plasma Physics and Controlled Fusion 11, 167 (1977).
4. G. I. Dimov, V. V. Zakaidakov, M. E. Kishinevskii, Sov.J. Plasma Phys. 2, 326 (1976).
5. S. V. Mirnov, I. B. Semenov, Sov. Atomic Energy 30, 22 (1971).
6. S. Von Goeler, W. Stodiek, N. Sauthoff, Phys. Rev. Lett. 33, 1201 (1974).
7. S. V. Mirnov, Plasma Phys. and Cont. Nucl. Fus., Proc. of 7th Intl. Conf., Innsbruck, 1978, Publ. IAEA, Vienna, 1979.
8. L. Berry et al., Plas. Phys. and Cont. Nucl. Fus. Res., Proc. 6th Intl. Conf., Berchtesgaden, 1976, Publ. IAEA, Vienna, 1977.
9. Equipe TFR, ibid.
10. H. Eubank et al., Plas. Phys. and Cont. Nucl. Fus. Res., Proc. 7th Intl. Conf., Innsbruck, 1978, Publ. IAEA Vienna, 1979.
11. W. Stodiek et al., Plas. Phys. and Cont. Nucl. Fus. Res., Proc. 8th Intl. Conf., Brussels, 1980, Publ. IAEA Vienna, 1981.
12. W7A Team: Plasma Phys. and Cont. Nucl. Fus., Proc. of 8th Intl. Conf. Brussels, 1980, Publ. IAEA Vienna, 1981.
13. A. Iiyoshi et al., Phys. Rev. Lett. 48, 745 (1982).
14. H. Grad, Phys. Fluids 10, 137 (1967).
15. H. Grad, Ann. NY Acad. Sci. 357, 223 (1980).
16. R. Dandl et al., Plasma Phys. and Cont. Nucl. Fus., Proc. of 5th Intl. Conf., Tokoyo, 1974. Publ. IAEA Vienna, 1975.
17. D. Baker et al., Plasma Phys. and Cont. Ncul. Fus., Proc. 8th Intl. Conf., Brussels, 1980, Publ. IAEA Vienna, 1981.
18. G. F. Chew, M. C. Goldberger, F. E. Low, Proc. Roy. Soc. (London) A236, 112 (1956).
19. B. McNamara, in Methods of Computat. Phys., V. 16, J. Killen, Ed., Academic Press. NY, 1976.

20. F. Bauer, O. Betancourt, P. Garabedian, "A Numerical Method in Plasma Physics," Springer, NY, 1978.
21. L. Hall, B. McNamara, Phys. Fluids 18, 552 (1975).
22. D. Ryutov, G. Stupakov, JETP Lett. 26, 174 (1977).
23. V. P. Pastukhov, Nucl. Fusion 14, 3 (1974).
24. R. Cohen, M. Rensink, T. Cutler, A. Mirin, Nucl. Fusion 18, 1229 (1978).
25. M. Rensink, R. Cohen, A. Mirin, in "Physics Basis for MFTF-B," PTII, Lawrence Livermore Laboratory Report UCID-18496, 1980.
26. S. Braginskii, in "Reviews of Plasma Physics," Vol. I (Ed. M. A. Leontovich) Consultants Bureau, NY, 1965.
27. F. Hinton, R. Hazeltine, Rev. Mod. Phys. 48, 239 (1976).
28. S. Hirshman, D. Sigmar, Nucl. Fusion 21, 1079 (1981).
29. J. Hogan, Nucl. Fusion 19, 753 (1979).
30. O. Pogutse, N. Chudin, E. Yurchonko, Sov. J. Plas. Phys. 6, 341 (1981).
31. R. White, D. Monticello, M. Rosenbluth, B. Waddell, Phys. Fluids 20, 800 (1977).
32. L. Kovryzhnick, Sov. Phys. JETP 29, 475 (1969).
33. E. Jaeger, D. Spong, L. Hedrick, Phys. Rev. Lett. 40, 866 (1978).
34. E. Caramana, F. Perkins, Nucl. Fusion 21, 23 (1981).
35. TMX Group, "Summary of results from TMX Experiment," Lawrence Livermore National Laboratory Report UCRL-53120, 1981.
36. D. E. Baldwin, B. G. Logan, Phys. Rev. Lett. 43, 1318 (1979).
37. M. N. Rosenbluth, H. L. Berk, D. E. Baldwin, H. V. Wong, T. M. Antonsen, Sherwood Theoretical Meeting, Santa Fe (1982).
38. D. W. Swain et al., Nucl. Fusion 81, 1409 (1981).
39. R. Fonck and PDX group, Proc. 5th International Conference on Plasma Surface Interactions, Gatlinburg, 1982 (to be published in J. Nucl. Mater.).
40. B. Carreras, P. Diamond, Sherwood Theory Conference, Santa Fe, NM, (1982).
41. W. Clark et al., Phys. Rev. Lett. 45, 1101 (1980).
42. W. Hooke et al., Bull. Am. Phys. Soc. 26, 975 (1981).
43. G. Proulx, B. Kusse, Phys. Rev. Lett. 48, 749 (1982).
44. J. Girard, D. Marty, P. Moriette, Plasma Phys. and Cont. Nucl. Fusion, Proc. of 5th Intl. Conf. Tokyo, 1974, Publ. IAEA Vienna, 1975.
45. Proc. NATO Advanced Study Institute, on At. and Molecular Proc. in Cont. Fus. Res., M. McDowell, A. Ferendici; Eds., Plenum Press, NY, 1980.
46. K. B. Axon et al., Plasma Phys. and Cont. Nucl. Fusion, Proc. 8th Intl. Conf., Brussels, 1980, Publ. IAEA Vienna, 1981.
47. R. Hulse, D. Post, D. Mikkelsen, J. Phys. B. 13, 3895 (1980).
48. R. Isler et al., Phys. Rev. A 24, 2701 (1981).
49. L. Gardner et al., Phys. Rev. A 21, 1397 (1980); D. Crandall, R. Phaneuf, F. Meyer, Phys. Rev. A 19, 504 (1979);

R. Phaneuf, F. Meyer, K. McKnight, Phys. Rev. A 17, 534 (1978).
50. R. Phaneuf, Phys. Rev. A 24, 1138 (1981).
51. T. Heil, S. Butler, A. Dalgarno, Phys. Rev. A 23, 1100 (1981); S. Bienstock, T. Heil, C. Bottcher, A. Dalgarno, Phys. Rev. A 25, 2850 (1982).
52. H. Ryufuku, T. Watanabe, Phys. Rev. A 19, 1538 (1979).
53. T. Ohkawa, Nucl. Fusion 10, 185 (1970).
54. N. Fisch, Phys. Rev. Lett. 41, 873 (1978).
55. J. Benford, B. Ecker, V. Bailey, Phys. Rev. Lett. 33, 574 (1974).

APPENDIX

Useful texts and reviews:

TEXTS

L. Spitzer "Physics of Fully Ionized Gases"
Wiley-Interscience, NY, 1962

N. Krall and
A. Trivelpiece "Principles of Plasma Physics"
McGraw Hill, 1973

K. Miyamoto "Plasma Physics for Nuclear Fusion"
MIT Press, Cambridge, MA 1980

REVIEWS

Mirrors

D. E. Baldwin, B. G. Logan, T. C. Simonen, Eds.
"Physics Basis for MFTF-B"
Lawrence Livermore Laboratory Report
UCID-18496 (I and II) January 1980

Lectures by R. Post and D. Ryuto
in the "Physics of Plasmas Close to
Thermonuclear Conditions"
CEC Document EUR FU BRU/XII/476/80

TMX Group "Summary of Results from TMX"
Lawrence Livermore National Laboratory Report
UCRL53120, February 1981

Stellarators	Joint US-Euratom Report "Stellarators: Status and Future Directions" Published by Max Planck Institut fur Plasma Physik Report IPP-2/254, July 1981
Tokamaks	L. A. Artsimovich, Nucl. Fusion 12, 2651 (1972)
	H. Furth, Nucl. Fusion 15, 487 (1975)
	INTOR - International Tokamak Reactor Phase I Study published by IAEA Vienna, 1982
EBT	R. A. Dandl et al., "EBT Experimental Results," in Plasma Physics and Controlled Nuclear Fusion Research, (Proc. 7th Conf., Innsbruck, 1978) IAEA Vienna, 1979
RFP	D. A. Baker et al., "Initial Reversed Field Pinch Experiments on ZT-40 and Recent Advances in RFP Theory," (Proc. XIII Conf. on Plasma Physics and Controlled Nuclear Fusion Resarch, Brussels, 1980) IAEA Vienna, 1981
A/M PHYSICS AND CTR	Proc. of IAEA Tech. Comm. Mtg. on A/M Data for Fusion. Published in Physica Scripta 23, No. 2, 1981, (H. Drawin, K. Katsonis, Eds.)
	Proc. of NATO Advanced Study Institute on A/M Processes in Controlled Fusion Research (M. McDowell, A. Ferendici, Eds.) Plenum Press, NY 1980.

GENERAL PRINCIPLES OF INERTIAL CONFINEMENT

Roger A. Haas

University of California
Lawrence Livermore National Laboratory
Livermore, California

1. INTRODUCTION

Over the past thirty years, mankind has witnessed a worldwide effort by scientists and engineers to conceptualize and develop an economical technology for conversion of nuclear fusion energy to electricity. Nuclear fusion energy is released when the nuclei of light elements are fused together at high temperature to produce more tightly bound, heavier nuclei. Technical requirements for fusion fuel ignition and thermonuclear burn favor the deuterium-tritium (DT) fusion reaction which releases an enormous amount of energy : 337 MJ/mg of fuel reacted or burned. The resources of this fuel are virtually inexhaustible. For comparison, the combustion of fuel oil releases 46 J/mg and its resources are being depleted rapidly.

The central scientific problem in producing electricity by nuclear fusion is the generation and confinement of a sufficiently dense, ultrahigh temperature DT plasma. Two generic technical approaches, known as magnetic confinement fusion[1,2] and inertial confinement fusion[3-6], have emerged as potential solutions to this problem. Theoretical studies indicate that for either approach to be successful, the DT fuel must be heated to a temperature of order 10keV, and confined so that the product of the nuclei or ion density, n, and their confinement time, τ, within the reaction volume is in the range $n\tau \simeq 10^{14}$ to $10^{15} cm^{-3}$ sec. In magnetic confinement fusion this is accomplished by using a magnetic field to confine a large, hot, dilute (10^{14} to $10^{15} cm^{-3}$ DT plasma so that the residence time of the deuterium and tritium ions is long enough (1 to 10 sec) for fusion to occur. In inertial confinement, a small pellet of DT fuel is compressed to ultrahigh density (10^{25} to $10^{26}cm^{-3}$) and a central temperature of 5 to 10 keV by irradiating a target, containing the DT, with a pulsed laser or particle beam. Under these conditions, the DT rapidly reacts on a time scale of approximately 50psec, creating a microexplosion. In this approach, at ultrahigh density, the DT fuel burns before it can hydrodynamically disassemble. The DT fuel is confined by its own inertia. The physical conditions required for magnetic and inertial confinement fusion are remarkably different. This situation presents mankind with a maximal opportunity to succeed at developing a nuclear fusion power plant. To date, great progress has been made in each approach and it is now believed that within the next decade both will demonstrate scientific feasibility.

Atomic and molecular processes relevant to the concept of inertial confinement fusion (ICF) fall into several broad categories. Processes associated with: (1) target dynamics and the fusion microexplosion, (2) production and transport of driver beam power to the target, (3) conversion of microexplosion energy into electricity in the reactor, (4) fabrication of targets and (5) diagnostics of inertial fusion microexplosions. The purpose of this article is to discuss the general principles and technical progress made toward the development of ICF. Only the target, driver and reactor aspects of the ICF concept are considered. The role of atomic and molecular processes is discussed in a general way. A more detailed discussion of atomic processes in fusion microexplosions is given by More[10,11] in this book. Due to limitations of space the target fabrication[7] and diagnostics[8,9] aspects of ICF are not considered.

Many of the atomic and molecular processes important to the target, driver, and reactor components of ICF are summarized in Table 1. The ranges of density and temperature over which these processes are important are also indicated. The ultimate success of the ICF concept depends fundamentally on understanding and controlling a multitude of atomic and molecular processes under conditions never before explored in the laboratory. This article is organized in the following manner. In Sec. 2 the energetics of an inertial fusion power plant are dicussed. Requirements on target energy gain (thermonuclear energy released/driver input energy) and driver efficiency are developed. In Sec. 3 the general physical principles of ICF are discussed. The relationship between target gain and driver energy is developed. The various target and driver options are outlined. In Secs. 4 and 5 the performance characteristics of laser and ion beam driven targets are discussed. In Sec. 6 the nominal performance requirements for inertial confinement fusion drivers are summarized. In Sec. 7 inertial fusion reactor concepts are discussed. Finally, in Secs. 8 and 9 laser and ion beam driver technologies are discussed.

Table 1. Atomic and Molecular Processes Important to the Development of Inertial Confinement Fusion Technology

Range of physical parameters ($10^{15} \lesssim n(cm^{-3}) \lesssim 10^{26}$, $0.025 \lesssim T(eV) \lesssim 100k$, $\lambda \lesssim 2\mu m$, particle beam energy $\lesssim$ 20GeV)

Data	Element	Media
Atomic/molecular structure Energy levels, Spectroscopic constants, Radiative lifetimes	D,T,R	S,L,G,P
Photoabsorption cross-sections	D,T,R	S,L,G,P
Electron collision cross-sections Dissociation, Attachment, Ionization, Excitation, Recombination	D,T,R	G,P
Excited state kinetics	D,T	S,G,P
Ionic recombination	D	G
Ion charge changing and stripping Gas, Ion	D,T,R	G,P
Optical damage mechanisms (surface-states) Avalanche, Multiphoton	D,T	S

Abbreviations: D-driver, T-target, R-reactor, S-solid, L-Liquid, G-Gas, P-plasma

2. POWER REACTOR CONSIDERATIONS

Thermonuclear Reaction Cycles

The first thermonuclear fusion reactors will make use of one or more of the following reaction cycles[12,13]:

Deuterium-Tritium (DT) Cycle

$${}_1D^2 + {}_1T^3 \rightarrow {}_2He^4(3.5MeV) + {}_0n^1(14.1MeV) \qquad \varepsilon_b = 337MJ/mg$$

Deuterium-Deuterium (DD) Cycle

$${}_1D^2 + {}_1D^2 \rightarrow {}_1T^3(1.01MeV) + {}_1H^1(3.03MeV) \qquad \varepsilon_b = 95.8MJ/mg$$

$${}_1D^2 + {}_1D^2 \rightarrow {}_2He^3(0.82MeV) + {}_0n^1(2.45MeV) \qquad \varepsilon_b = 78.3MJ/mg$$

Deuterium - Helium Three (DHe^3) Cycle

$${}_1D^2 + {}_2He^3 \rightarrow {}_2He^4(3.67MeV) + {}_1H^1(14.67MeV) \qquad \varepsilon_b = 351MJ/mg$$

The quantity ε_b is the total energy released by the nuclear reaction per unit mass of fuel burned. Both the DD and DT cycles release an energetic neutron. Fusion reactors based on these cycles will therefore require a blanket capable of converting the neutron's kinetic energy to thermal energy. If $_3L^6$ is used in the DT cycle, tritium can be bred with a net energy yield or enhancement of 4.6 MeV, making 22.2 MeV the total yield for the DT cycle. A great potential advantage of the DHe^3 cycle is that the reaction products are charged particles which offer the potential of direct conversion. The DHe^3 cycle produces neutrons only through concomittant DD fusion reactions. This neutron production level will be much lower than in the DT cycle, alleviating many neutron related technology problems, such as materials radiation damage and induced radioactivity.

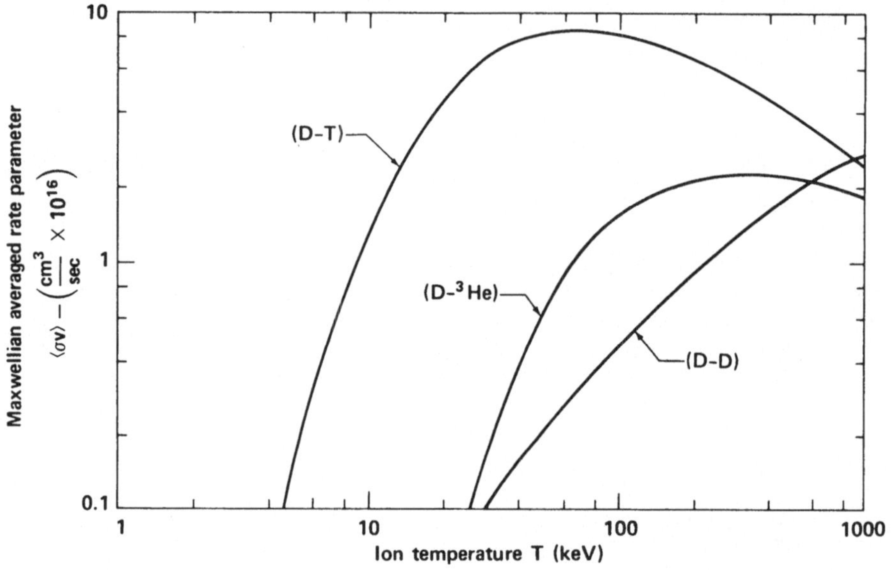

Fig. 1. Nuclear fusion reaction rates.

The choice between these potential fuel cycles for the DT reaction in the near term becomes clear by examining the reaction rate <σv> for the various fusion reactions shown in Fig. 1. The reaction rate is derived by averaging the appropriate reaction cross section σ over a Maxwellian distribution of speeds for both incident particles. The DT reaction rate is largest and peaks at the lowest temperature of 70keV. Therefore, it appears likely that the first fusion reactor will employ the DT cycle. It is also apparent from Fig. 1 that the reaction rate for the fusion reaction becomes large only at high energies, and thus it is necessary to heat the reacting species to temperatures where the fusion rate becomes reasonably large. For the DT cycle, this means temperatures of order 10keV. In the remainder of this article the discussion specializes to the DT cycle, although many of the ideas and considerations also apply to other cycles.

Inertial Confinement Fusion Concept

The generation of thermonuclear energy by ICF is illustrated in Fig. 2. Upon direct irradiation of a small sphere of solid DT with laser or particle beams, the sequence of events illustrated in Fig. 2 occurs. First a low density atmosphere, extending to several target radii, is created by ablating the target surface with a driver prepulse. This atmosphere is then irradiated more or less uniformly from all sides by a second, much more intense

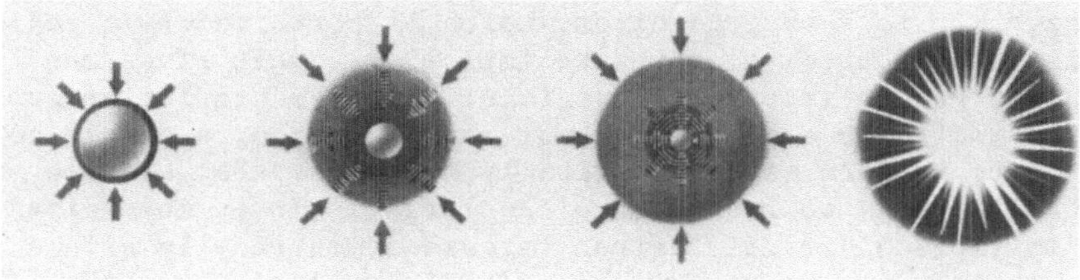

(1) (2) (3) (4)

Fig. 2. In the ICF concept the driver energy (dark arrows) is converted to thermal energy (crosslined arrows) which drives the four stages of the implosion: (1) atmosphere formation, (2) compression, (3) ignition, and (4) burn.

driver beam pulse. Absorption of the driver energy in the outer atmosphere heats the electrons. The atmosphere and the target surface are heated by electron diffusion and transport. As the electrons move inward through the atmosphere, scattering, and solid angle effects increase the spherical symmetry. Violent ablation and blow-off of the target surface generate the pressure which implodes the target. The driver pulse is shaped in time so the ablation pressure generates a sequence of weak spherical shock waves that nearly isentropically compress the central DT fuel region to an ultrahigh density, 10^3 to 10^4 times solid density, Fermi degenerate state. Toward the end of the implosion process, these shock waves coalesce at the center of the compressed fuel region creating a small, central hot spot or spark there. The spark region temperature is near 10keV and initiates the thermonuclear reactions. A thermonuclear burn front then propagates radially outward from the central hot spot, heating and igniting the surrounding dense DT fuel which is confined by its own inertia. After the fuel burns, it expands and disassembles by hydrodynamic motion. Energy released by this fusion microexplosion is carried by 14MeV neutrons, charged particles, and electromagnetic radiation and ultimately appears as thermal energy. This energy is many times the driver input energy. To generate electricity the fusion thermal energy must be converted to electricity. Approximately 3GW of thermal energy must be generated to produce 1GW of electricity. If fusion microexplosions with an energy gain of 60 are produced at a repetition rate of 10Hz, then a 5MJ driver is required. Each microexplosion burns approximately 1mg of DT fuel and produces 300MJ of energy.

In general the configuration of the DT fuel and its attendant structure is more complicated than the small sphere of DT referred to above. There are two basic target designs[14] for inertial confinement fusion: direct drive targets and hohlraum targets. Direct drive targets absorb the driver beam in an outer layer as illustrated in Fig. 2. Hohlraum targets absorb the driver energy and convert it to x-ray radiation, which is contained by a radiation case and is used to drive the implosion. This gives the hohlraum target an important advantage: improved implosion symmetry. Symmetry is achieved without requiring the beams of the driver to be uniform and symmetrically arranged. For example, in some designs just two beams drive the target. It is more difficult to implode directly driven targets symmetrically with a few beams. Another important feature of hohlraum targets is that they may couple efficiently to several driver candidates, possibly including CO_2 lasers.

Power Flow in an ICF Reactor

In its simplest form, an ICF reactor[15] or power plant consists of a combustion chamber to contain the fusion microexplosion, a target factory to make and project targets containing solid DT into the center of the chamber at a rate of several per second, a driver system to irradiate the targets, and a steam turbine generator system to produce electricity. The elements of this system are shown in Fig. 3. The economical viability of such a power plant depends fundamentally on the recirculated power fraction which determines the overall efficiency of the plant. In order to reduce the cost of electricity, the capital cost of the plant must be reduced by reducing the size of the driver and auxiliary equipment within the plant (i.e. by reducing the recirculating power fraction).

Using Fig. 3 it can be shown[15] that the recirculating power fraction P_{in}/P_g and overall plant efficiency $\eta_s=P_e/P_t$ are determined by

$$P_{in}/P_g = (P_a+P_d)/(P_e+P_a+P_d) = (f_a/\eta_t)+(1/\eta_t\eta_d GM) \qquad (2\text{-}1)$$

and

$$\eta_s/\eta_t = 1-(P_{in}/P_g) = 1-(f_a/\eta_t)-(1/\eta_t\eta_d GM) \qquad (2\text{-}2)$$

The quantity f_a is the fraction of the thermal power generated that is required to power the auxiliary equipment in the plant: liquid metal pumping, vacuum pumping, boiler feed, etc. In an ICF reactor it is estimated[15] that $f_a \simeq 3\%$, or from 7 to 8% of the electrical power generated is used to run auxiliary equipment. The blanket energy multiplication M is defined as the ratio of blanket thermal energy to the thermonuclear energy released by the target. Neutron reactions in lithium and steel produce effective blanket gains[15] $M \simeq 1.15$. Higher values of M are possible in fusion-fission hybrid systems[16] in which fissionable material is placed in a blanket around the fusion microexplosion chamber. The high energy, fusion neutrons boost the fission process, even in depleted uranium, producing large multiplications of fusion energy[16] and fissile material[17].

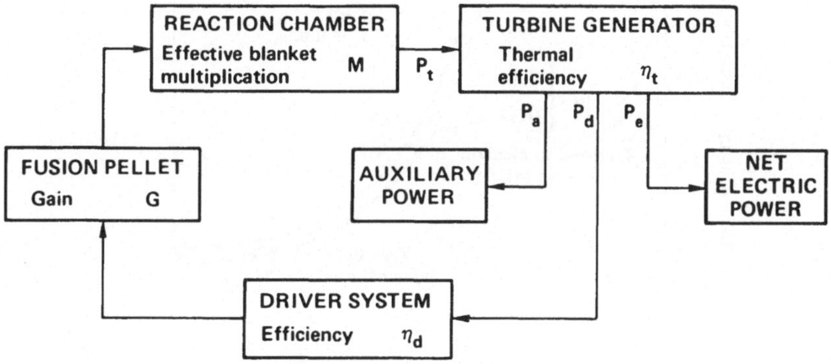

Fig. 3. Power flow diagram for an ICF power plant.

Eqs. (2-1) and (2-2) indicate that at low recirculating power fractions, the overall efficiency of the plant approaches the efficiency η_t of conversion of thermal energy to electricity. The key factor in making this happen is the fusion energy gain. The fusion energy gain Q is equal to the product of the driver efficiency η_d and the target or microexplosion energy gain G: $Q=\eta_d G$. Fig. 4 shows how the overall system efficiency depends on fusion energy gain. The inverse of Eq. (2-2) is plotted because the cost of electricity is inversely proportional to the net electricity generated and, therefore, to the overall system efficiency.

Two thermal efficiencies are shown in Fig. 4: $\eta_t=39\%$ corresponds to a superheated steam system typical of a high temperature liquid metal cooled system and $\eta_t=34\%$ to a saturated steam system typical of a water cooled reactor. The figure illustrates how the relative cost of inertial fusion drops as the fusion energy gain is increased, thereby reducing the recirculated power fraction. A fusion energy gain of ten is necessary to bring the cost of electricity to within roughly 40% of the cost at zero recirculated power. This corresponds to a recirculated power fraction of approximately 25%. Present power plants have about 5% recirculating power fractions. Fig. 4 shows that for a $Q<10$, the cost of electricity increases very rapidly, and consequently the inertial fusion community has generally taken the minimum useful value of Q as 10. This condition places serious constraints on the allowable driver efficiency for a given pellet gain, namely $\eta_d > 10/G$ (i.e. if the target gain is 100 the driver efficiency must exceed 10% in order to make an economically

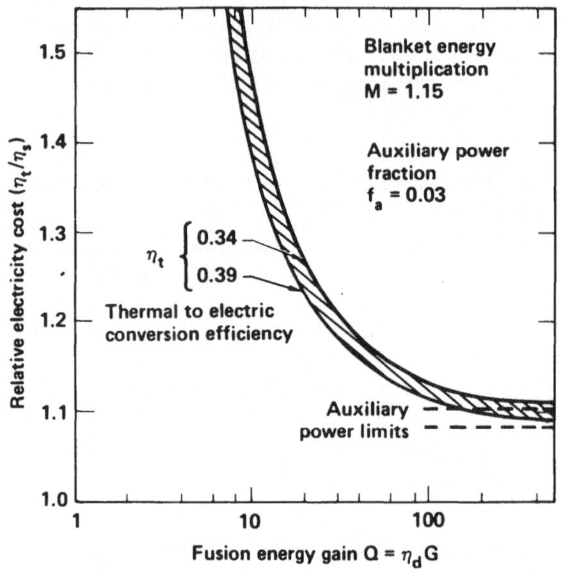

Fig. 4. As shown by Monsler, et. al.[15], increasing the fusion energy gain decreases the cost of electricity by reducing recirculating power.

viable power reactor). Laser and particle beam drivers appear capable of initiating high gain microexplosions and operating at an efficiency comparable to or greater than 10%.

3. TARGET GAIN CHARACTERISTICS

DT Gain, Lawson, and Ignition Conditions

There are several general figures of merit[12,13] that have been used to characterize the condition of a reacting fusion plasma: DT gain, Lawson condition, and ignition condition. The thermal energy of an equilibrated reacting DT plasma is $3n\theta$, where θ is the product of temperature T times Boltzmann's constant. Due to charge neutrality, the electron density n is twice the equal deuteron and triton densities. In this plasma, fusion energy is released at a rate $(n/2)^2\langle\sigma v\rangle q$. For DT plasma $q=17.6$MeV. If the thermal energy is confined for a characteristic time τ, then the DT or fuel gain is

$$G_f = (n/2)^2 \langle\sigma v\rangle q / 3n\theta/\tau = n\tau(\langle\sigma v\rangle/12\theta)q \tag{3-1}$$

or

$$n\tau = G_f(12\theta/\langle\sigma v\rangle q) \tag{3-2}$$

The DT gain defined by Eq. (3-1) is the ratio of the thermonuclear energy released by the plasma to the thermal energy invested in the DT fuel. Curves of $n\tau$ versus T are plotted in Fig. 5 for different values of DT gain. For a given DT fuel gain, the minimum product of plasma density and the plasma energy confinement time occurs at approximately 25keV, and it increases sharply for T<10keV because the fusion reaction rate decreases rapidly.

The requirement for a magnetic fusion plasma to approximately achieve energy breakeven is often given as the Lawson criterion. To derive this condition, plasma loss mechanisms must be accounted for. Lawson[18] assumed that bremsstrahlung radiation was the main loss mechanism. This energy loss rate is given for hydrogenic plasma by $P_b(W/cm^3) = C_b n^2(cm^{-3})\theta^{1/2}(keV)$ where $C_b = 4.8x10^{-31}\ Wcm^3/(keV)^{1/2}$. Assuming that the fusion reaction power plus the bremsstrahlung and plasma thermal power is available for conversion to electricity at an overall efficiency η_L, the Lawson condition arises from the following power balance on a unit volume of the plasma

$$(3n\theta/\tau)+P_b = \eta_L[(n/2)^2\langle\sigma v\rangle q+P_b+(3n\theta/\tau)] \tag{3-3}$$

or, solving for $n\tau$

$$n\tau = 3\theta(1-\eta_L)/[(\langle\sigma v\rangle/4)q\eta_L - C_b\theta^{1/2}(1-\eta_L)] \tag{3-4}$$

which is a function of plasma temperature only. Lawson assumed $\eta_L=1/3$. For this value of η_L, the minimum $n\tau$ product required to meet the Lawson criterion is approximately $5x10^{13}cm^{-3}sec$ at 25keV. However, it is usually quoted at a temperature of 10keV where it attains the value of $10^{14}cm^{-3}sec$. In general, the Lawson criterion is too optimistic since the choice of η_L is quite large. As will be shown later, this is especially true in the case of inertial confinement fusion due to the relatively low efficiency of implosions. In addition, a plasma that meets the Lawson condition is not necessarily ignited and self-sustaining because the analysis does not include the possibility of direct deposition in the plasma of some or all of the energy of the fusion reaction products. The condition which characterizes an ignited and self-sustaining plasma is thus another important criterion.

The ignition condition for a magnetic fusion plasma is obtained by balancing the power lost from the plasma due to bremsstrahlung and thermal losses in a time τ against the power deposited in the plasma from the slowing down of fusion reaction products. Letting q_d denote the fusion energy deposited in the plasma the balance is

$$(n/2)^2\langle\sigma v\rangle q_d = (3n\theta/\tau)+P_b \qquad (3\text{-}5)$$

and the ignition condition on $n\tau$ becomes

$$n\tau = 12\theta/(\langle\sigma v\rangle q_d - C_b\theta^{1/2}) \qquad (3\text{-}6)$$

For the DT cycle, $q_d \simeq 3.5$MeV, the energy of the alpha particle which would be contained in the plasma. This criterion has a minimum $n\tau$ at approximately $2x10^{14}cm^{-3}sec$. Thus, it is more difficult to meet the ignition criterion. The temperature at which the energy deposited by nuclear fusion within the reacting system just exceeds the energy loss by bremsstahlung is often referred to as the ideal ignition temperature. For the DT and DD cycles these temperatures are 4 and 36keV, respectively, and represent the lowest possible operating temperatures for a self-sustaining thermonuclear plasma[12,13].

For ICF plasmas, the ignition condition[19,20] (to be discussed in the next subsection) is more complex because of the transient temporal and finite spatial nature of the implosion and explosion process. In ICF the DT fuel is ignited by creating a central hot spot. The density, size, and temperature of this hot spot must be large enough to produce and capture alpha particles created by the DT fusion reactions. Since the DT fuel is usually surrounded by a thick (hot) pusher-tamper, bremsstahlung losses are reduced by about a factor of two. Under these conditions, both electron conduction and bremsstrahlung radiation losses are important. However, electron conduction losses become dominant as the minimum ignition condition is approached. The minimum ignition condition for an inertially confined DT plasma[19,20] requires that the central hot spot have a density-radius product, ρR_h, of $0.3g/cm^2$ and a temperature of 5keV. For lower values of ρR_h and temperature energy losses exceed gains and the plasma will not self-heat to high temperature where rapid and efficent DT fuel burn-up occurs. The approximate ignition curve[14] for inertial fusion DT plasmas is plotted in Fig. 5.

To compare the above criteria with what has been achieved experimentally and is planned in magnetic and inertial confinement, Fig. 5 also shows the regime of operation of several well known thermonuclear facilities. The remarkable flexibility of laser irradiation systems is illustrated by the diverse conditions achieved with Shiva. For example, using the Shiva laser facility target plasma temperature was traded for imploded fuel density and densities as high as[9,21] approximately 100 times solid DT density (ρ_s=0.2g/cm^3) were achieved at ion temperatures of about 0.4keV. It is expected[14] that the Nova facility will compress DT to 1000 times solid density.

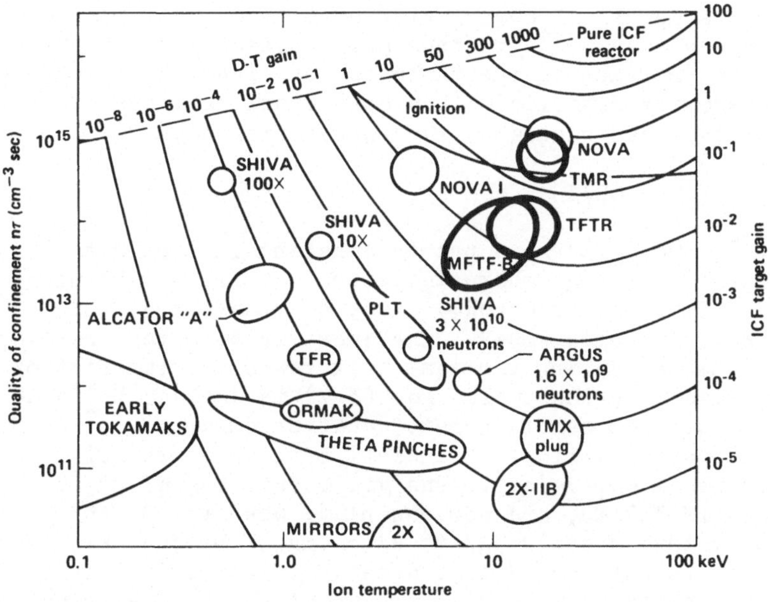

Fig. 5. Thermonuclear conditions have been achieved in several fusion experiments[9,21].

Target Gain

A description of the relationship between the performance of a complex fusion target and driver irradiation characteristics (power, energy, wavelength, etc.) requires large scale computer calculations. For example, the LASNEX code developed by Zimmerman[22] has been used extensively to simulate experiments and design targets. The conclusions drawn from such calculations are presented at the end of this Sec. and in Secs. 4 and 5. However, to provide some feel for the complexity of the problem and some direct insight into the important physical processes in fusion targets, consider a simple target model studied by Nuckolls, et. al.[3,4], Brueckner[23], Kidder[20,24], Bodner[25], and Meyer-Ter-Vehn[26].

In the previous section several thermonuclear plasma conditions were related to constraints on $n\tau$. In inertial confinement fusion, however, the confinement time is $\tau \simeq R_f/4c_s$, and the number density n is replaced by the mass density ρ. The quantity τ is the time required for a rarefaction wave travelling at the isothermal sound speed c_s to significantly disassemble a compressed spherical fuel mass of radius R_f. The factor of 4 arises because[3,4], in a sphere, approximately half the mass is beyond 80% of the radius. Thus, for inertially confined thermonuclear microexplosions, the confinement parameter is $\rho R_f = (4m_i c_s)n\tau$ where m_i is the average ion mass of the fuel and $c_s{}^2 = 2\theta/m_i$.

During the spherical compression of a fixed fuel mass M_f, the density increases as R_f^{-3} and thus ρR_f increases as $[3M_f\rho^2/4\pi]^{1/3}$. The resulting increase in the thermonuclear reaction rate compared to the disassembly rate leads to a corresponding increase in the fraction of the fuel burned or the burn efficiency[4]

$$\phi = \rho R_f/(\rho R_f^* + \rho R_f) \tag{3-7}$$

where $(\rho R_f)^* = 8m_i c_s/<\sigma v>$ is nearly constant in the 20 to 70 keV temperature range characteristic of efficient DT microexplosions. In this range $(\rho R_f)^*$ is approximately 6, and increases sharply below 20keV. The quantity ϕ is the fraction of the fuel burned during the time τ it takes the rarefaction wave to dissassemble the fuel. Eq. (3-7) is derived by integrating the fuel ion reaction rate equation $dn/dt = -(n^2/2)<\sigma v>$ over the dissassembly time. Efficient burn up of the DT fuel, $\phi>1/3$, requires a $\rho R_f > 3g/cm^2$. The yield from 1mg of DT fully burned ($\phi=1$) is 337 MJ, and consequently, when $\phi=1/3$, the yield is 112MJ. To achieve a $\rho R_f \simeq 3g/cm^2$ with 1mg of DT fuel requires that the fuel be compressed to a density of approximately 1600 times solid density. Efficient burn-up of the DT fuel requires that it undergo a large volumetric compression to a high density. Therefore, burn efficiencies much in excess of 1/3 are difficult to achieve. Generally, the DT fuel may be configured as a solid sphere or a spherical shell. For a given fuel mass, the spherical shell is preferred[4] because for a given amount of compression work it undergoes a larger volume change and requires a lower driving pressure (i.e. driver intensity). It stores kinetic energy during the implosion and then converts it to compressional energy at stagnation with a power multiplication. Of course, in the final stages of the compression the shell evolves into a solid spherical fuel mass.

The thermonuclear energy output from a burned-up DT fuel mass is

$$E_{tn} = \phi M_f \epsilon_b \tag{3-8}$$

If the fuel is uniformly heated to a temperature θ by the driver then

$$\eta E_d = M_f \epsilon_h \tag{3-9}$$

where $\epsilon_h(J/g) = 3\theta/m_i \simeq 1.1 \times 10^8 \theta(keV)$ is the specific thermal energy of the fuel and η is the efficiency of transfer of driver energy into fuel energy. In general, η is the product of the absorption efficiency η_a, the hydrodynamic efficiency η_h and the transformation efficiency η_s. The fraction of the driver energy absorbed by the target is η_a. The fractional conversion of absorbed driver energy into kinetic energy of the imploding shell is the hydrodynamic efficiency. The transformation efficiency is the efficiency of conversion of implosion kinetic energy into compression and heating of the fuel. The product $\eta_h \eta_s$ is often referred to as the implosion efficiency. Typically η is in the range 0.10 to 0.15. If the fuel is uniformly heated, then the target gain is

$$G = E_{tn}/E_d = \eta\phi(\varepsilon_b/\varepsilon_h) \quad (3\text{-}10)$$

To obtain ignition and significant fuel burnup requires $\theta > 5$keV, and thus for $\eta < 0.15$ and $\phi \simeq 1/3$ Eq. (3-10) indicates that target gain G will be <30. For power plant service such a gain requires a driver efficiency $\eta_d > 10/G \simeq 0.30$. Currently only particle beam drivers appear capable of accessing this efficiency range.

Higher target gains may be achieved by noting that the specific energy required to isentropically compress the fuel to a nearly Fermi degenerate state is small compared to the specific energy required to heat it to the ignition temperature. Specifically, the specific energy required to compress a Fermi degenerate electron gas is $\varepsilon_c(J/g) = (3\varepsilon_F/5m_i) \simeq 1.1\times10^5 \eta_c^{2/3}$ where ε_F is the Fermi energy $(h^2/8m)(3n/\pi)^{2/3}$ and $\eta_c = \rho/\rho_s$ is the compression. Thus at $\eta_c = 10^3$ the compressional energy $\varepsilon_c \simeq 1.1\times10^7$J/g which is approximately 2% of the specific thermal energy at 5keV. Compression of the fuel to a Fermi degenerate state requires that its temperature be maintained much less than the Fermi temperature $T_F(ev) = 2\varepsilon_F/5 \simeq 2\eta_c^{2/3}$. For a compression $\eta_c = 10^3$, the temperature must be much less than 200ev. The DT will not ignite at such low temperatures and a low isentrope compression and an additional ignition technique is needed.

A solution to this problem may be found by noting that if initially ρR_f is sufficiently large, the reaction products may be trapped by the burning fuel, thereby heating it and reducing ignition requirements. The range of 14MeV neutrons[27] in DT is sufficiently large, $\rho R_n \simeq 4.6$g/cm^2, that for practical conditions they escape from the compressed fuel region. However, the range of 3.5MeV alpha particles[27] depends on plasma properties, but is in the range $\rho R_\alpha \simeq 0.3$ to 0.5 g/cm^2 for DT at a temperature of 5 to 10keV and a density of 10^3 to 10^4g/cm^3. Consequently, it is possible to achieve ignition by heating only a small central portion of the compressed fuel with $\rho R > 0.5$g/cm^2 to a temperature of 5 to 10keV while keeping the surrounding compressed fuel at a much lower temperature. In this case, trapping the alpha particles rapidly heats the central reaction zone to temperatures into the 20 to 70keV range. The energy output from the small central hot spot (consisting of α-particles, thermal conduction, shocks, etc.) subsequently heats the adjacent cold compressed material so that it can burn. This sequential burning of larger and larger annular regions of the cold fuel is called propagating thermonuclear burn[3,4,22-26]. In this approach, the driver input energy required to compress the fuel is

$$\eta E_d = \alpha M_f \varepsilon_c + M_h \varepsilon_h \tag{3-11}$$

where $\alpha>1$ denotes the deviation from isentropic compression on the Fermi degenerate adiabat. The quantity M_h is the mass of the spark region. The target energy gain is then

$$G = \eta\phi\varepsilon_b/[\alpha\varepsilon_c + (M_h/M_f)\varepsilon_h]. \tag{3-12}$$

To compare this process to the uniformly heated fuel approach, a parameter β may be introduced to account for alpha particle self-heating and thermonuclear propagation so that the average specific thermal energy required for ignition is approximately ε_h/β, then

$$\varepsilon_h/\beta = \alpha\varepsilon_c + (M_h/M_f)\varepsilon_h \tag{3-13}$$

where

$$\beta^{-1} = \alpha(\varepsilon_c/\varepsilon_h) + (M_h/M_f). \tag{3-14}$$

The target gain is then

$$G = \eta\beta\phi(\varepsilon_b/\varepsilon_h) \tag{3-15}$$

or β times the gain of a target with uniformly heated fuel. In magnetic confinement fusion, β is approximately one. However, in inertial fusion, by exploiting central fuel ignition and thermonuclear propagation values much greater than one may be achieved. If M_h/M_f is very small and the compression is isentropic to $\eta_c \simeq 10^3$ on the Fermi degenerate adiabat ($\alpha=1$), then β would achieve its maximum value $\beta_{max} \simeq \varepsilon_h/\varepsilon_c \simeq 50$. In general, several practical considerations limit β to smaller values. First, due to preheat by high energy electrons and x-rays and shock waves produced by driver target interaction, the compression is not degenerate and $\alpha>1$. Secondly, in order to achieve ignition the central region must be large enough to confine the α-particles produced there. This requires $R_h > R_\alpha$ or $(M_h/M_f) > (\rho R_\alpha/\rho R_f)^3$. Practical considerations limit $\rho R_f < 4 g/cm^2$. Thus, if $\rho R_\alpha \simeq 0.4 g/cm^2$ and the compression is $\eta_c \simeq 10^3$, then $(M_h/M_f)_\alpha \simeq 10^{-3}$. Thus the constraint of α-particle ignition does not prevent achievement of maximal values of β. The most serious limitation on β comes from considerations of symmetry of the implosion. The symmetry of the implosion is determined largely by driver illumination uniformity and Rayleigh Taylor[3,4,28] instability . If R_0 is the initial (precompressed) fuel radius, then if R_h/R_0 is too small, the required symmetry of the implosion can be too severe to be practical. There is a lower bound[25] to R_h/R_0 given the symmetry of the implosion velocity which in turn is given by the symmetry in the applied pressure on the fuel shell. If the shell velocity is $\underline{v} = -v_r \underline{e}_r + \delta \underline{v}$, then roughly, $(R_h/R_o) > |\delta \underline{v}|/v_r \equiv \varepsilon$. Given this lower bound on R_h/R_0 one can solve for the lower bound on M_h/M_f for a fuel shell[25]

$$\frac{M_h}{M_f} = \frac{1}{3}\frac{\rho}{\rho_0}\left(\frac{R_h}{R_0}\right)^3\left(\frac{R_0}{\Delta R_0}\right) \gtrsim \frac{1}{3}\eta_c \varepsilon^3 \left(\frac{R_0}{\Delta R_0}\right) \tag{3-16}$$

The initial solid fuel density and thickness of the fuel shell are ρ_0 and ΔR_0, respectively. The implosion symmetry limit on β for high aspect ratio, $R_0/\Delta R_0$, spherical shells may be written

$$\beta^{-1} \gtrsim \alpha(\varepsilon_c/\varepsilon_h) + \eta_c \varepsilon^3 (R_0/\Delta R_0)/3. \tag{3-17}$$

The non-isentropic compression and symmetry constraints are comparable when

$$\varepsilon \simeq \left[(3\alpha/\eta_c)(\Delta R_0/R_0)(\varepsilon_c/\varepsilon_h)\right]^{1/3} \tag{3-18}$$

For example, if a fuel shell with an aspect ratio of 10 is compressed to 10^3 solid density on the α=2 adiabat and ignited at 5keV, the compression and symmetry terms will be comparable if $\varepsilon \simeq 2\%$. These combined compression and symmetry constraints limit the value of β to approximately 13 for this example. For the same coupling (η<0.15) and burn up efficiency ($\phi \simeq 1/3$) constraints imposed on the uniformly heated fuel example, this target would produce $G \lesssim 390$.

Detailed computer simulations[3,4] of target implosions indicate that central ignition can be achieved by compression shock convergence controlled by driver pulse shaping during implosion. These calculations reveal that practical limits on implosion symmetry set a maximum on β of approximately 30 which occurs for $\rho R_f < 3g/cm^2$. Other models[29] of central spark ignition suggest even larger values of β may be achieved. However, these predictions have not been substantiated by detailed numerical simulations[30]. In addition, the ignition temperature is usually approximately 10keV in order to avoid the relatively long time to self-heat from the ideal ignition temperature. If the compressed fuel is such that $\rho R_f \gg 0.3g/cm^2$, then only about $0.3g/cm^2$ in the central region need be heated to approximately 10keV in order to initiate a radially propagating burn front which ignites the entire fuel. In this case the energy released from the spark region, about one fifth of this energy is in alpha particles, is sufficient to heat three times more DT to 10keV. In general, the ideal minimum compression energy cannot be achieved due to electron preheat or the approximately 1Mbar initial shock produced by the driver prepulse.

If the approximation is made that ρR_f is somewhat less than $(\rho R_f)^*$ then

$$\rho R_f \simeq (6\theta/q)(G/\beta\eta)(\rho R_f)^* \tag{3-19}$$

and the corresponding confinement quality factor is

$$n\tau \simeq (12\theta/\langle\sigma v\rangle q)(G/\beta\eta) \tag{3-20}$$

which is analogous to Eq. (3-2). From Eq. (3-20) at the ignition temperature of 10keV the Lawson condition, $n\tau = 10^{14}cm^{-3}sec$, requires $G/\beta\eta \simeq 2$. Thus from Eqs. (3-19) and (3-7), this corresponds to a fractional burn-up of approximately 0.3%. For "energy breakeven" G=1, if $\beta\eta \simeq 0.05$, the confinement parameters must be $n\tau \simeq 10^{15}cm^{-3}sec$ and $\rho R_f \simeq 0.16g/cm^2$ leading to a fractional burn up of approximately 3%. Alternatively, if a self-heating factor $\beta>10$ is achieved, then energy breakeven occurs at or below the Lawson condition.

Using Eqs. (3-19) and (3-8), the driver energy requirement may be written[23]

$$E_d \simeq (4\pi\theta/m_i\rho_o^2)(6\theta(\rho R_f)^*/q)^3 G^3/(\beta\eta)^4\eta_c^2 \tag{3-21}$$

For the DT fuel cycle if a Li^6 blanket is used, and Eq. (3-21) is evaluated at 10keV, then[23-26]

$$E_d(MJ) \simeq (4G)^3/(\beta\eta)^4\eta_c^2 \tag{3-22}$$

The corresponding pellet mass and confinement time requirements[23-26] are $M_f(mg) \simeq 4G^3/(\beta\eta)^4\eta_c^2$ and $\tau(ns) \simeq G/\beta\eta\eta_c$.

Although this is a simple model of inertial fusion pellet performance, it reveals the strong sensitivity of driver requirements to driver absorption fraction, ablation efficiency, fuel compression, self-heating and propagation, and to pellet gain requirements. According to Eqs. (3-21) and (3-22), the driver energy requirements decrease inversely as the square of the compression and the fourth power of the product of the self-heating factor and the driver to fuel energy conversion efficiency. Correspondingly, fuel mass requirements decrease as the target coupling and compression increase.

Over a wide range of compressed pellet parameters ($10^2<\rho(g/cm^3) < 10^3$; $1/2 < \rho R_f<2$), a more accurate analysis[4] has been performed for laser driven targets leading to an expression for the laser driver energy

$$E_d \propto \eta^{-a}\eta_c^{-b} \tag{3-23}$$

at constant gain, where $a \simeq 4/3$ to 2 and $b \simeq 2$. These results indicate that compared to the simpler model above, the dependence on compression is at least as strong and generally stronger than the dependence on the absorption and implosion efficiencies. Using these results, Fig. 6 shows the variation of gain with compression and laser light energy. The curves are normalized to computer simulations of implosion and burn. The solid curves include propagation whereas the dashed curves assume uniform ignition and no propagation. The importance of self-heating and propagation is readily apparent. Gains approaching 100 are predicted for laser energies of 1MJ. At compressions less than a thousand, the gain increases strongly with increasing compression because of increasing burn efficiency and self-heating of the fuel. The gain decreases with compressions greater than 10^4 because of the depletion of the DT, increased ablative energy losses, and because the energy of compression (against degeneracy pressure) becomes dominant. Achievement of this performance requires nearly isentropic compression in which the laser wavelength and pulse shape may be varied[3,4].

Ablative Compression

The simple model[23-25] outlined above overestimates the target gain. The spark and cold compressed fuel have nearly the same pressure[26], not density, at peak compression. The ignition point with maximum ρR_f and central temperature is generally reached only after the shock, emerging from the center after shell

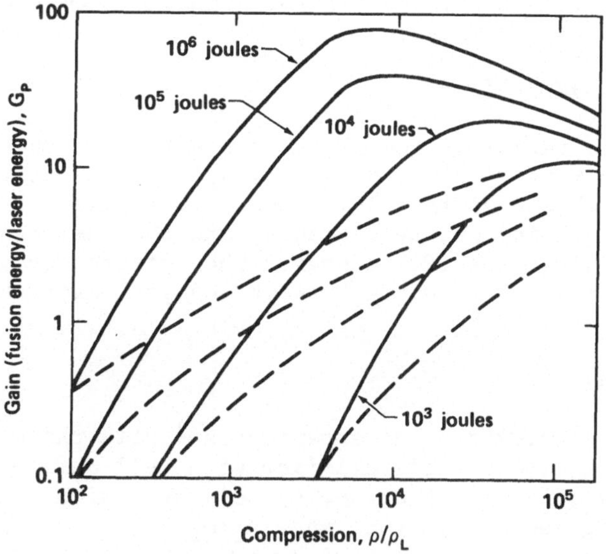

Fig. 6 Computer calculations by Nuckolls, et. al.[3,4] of pellet gain as a function of density compression η_c for various laser driver energies. Solid curves include effects of propagation; dashed curves assume uniform ignition (no propagation). The curves are computed for optimum pellet mass and laser pulse shape.

collapse, has passed through the fuel. At this time almost all the inward going kinetic energy has been converted to internal energy and the pressure is nearly uniform over the fuel. This situation may be used to estimate the required implosion velocity v, namely,

$$v^2/2 \simeq \alpha\varepsilon_c + (M_h/M_f)\varepsilon_h = \varepsilon_h/\beta \qquad (3\text{-}24)$$

For the spark ignition example, $\beta \simeq 13$ cited above, the average fuel specific energy is approximately 4.4×10^7 J/g and requires an implosion velocity of 3×10^7 cm/sec.

The driver deposition required to achieve the implosion velocity v by target ablation may be estimated by using a simple one dimensional rocket analogy[31]. Generally, the fuel shell is surrounded by an ablator shell composed of material which optimizes the driver energy deposition. If the aspect ratio of the combined fuel-ablator single shell target is sufficiently large, the one-dimensional rocket analogy is reasonable, otherwise spherical convergence effects must be accounted for[32]. In this model, the target (rocket) of mass M and velocity v is accelerated by the steady-state blow off (exhaust) of the ablator (propellant) at a large constant velocity u defined in the accelerated target (rocket) frame of reference. The velocity u is approximately equal to the ion acoustic speed in the nearly isothermal blow off plasma. At any given time, the rate of change of momentum of the target is then

$$d(Mv)/dt = (-dM/dt)(u-v) \qquad (3\text{-}25)$$

The mass M of the target is the DT fuel mass M_f plus the ablator mass M_a which is time varying; $M(t) = M_f + M_a(t)$. The initial ablator mass is chosen so that after the driver pulse of duration τ_d is over, the ablator has been used up and the fuel mass has attained the required specific energy. The integration of Eq. (3-25) produces the relationship between the implosion velocity v and the target parameters

$$v = u\ln(M_0/M) \qquad (3\text{-}26)$$

Hydrodynamical calculations and experiments[31] with planar targets have shown that a steady state ablation with well defined velocity u is set up very quickly. From Eq. (3-26) the maximum kinetic energy is transferred to the fuel when $M_0/M_f \simeq 7.5$.

The rate of delivery of driver energy to the ablation surface $d(\eta_t\eta_a E_d)/dt$ must be balanced by the energy dissipated in the ablation and acceleration of the target

$$\frac{d}{dt}(\eta_t\eta_a E_d) = \frac{d}{dt}(Mv^2) + \frac{1}{2}\left(\frac{-dM}{dt}\right)(u-v)^2 = \frac{-u^2}{2}\frac{dM}{dt} \qquad (3\text{-}27)$$

The quantity η_t is the efficiency of transport of absorbed driver energy from the absorption region to the ablation region. It corrects for heating of the blow-off through which energy is transported. Typically, $\eta_t \simeq 0.5$ for short wavelength lasers and ion beam drivers[30]. The hydrodynamic efficiency η_h is obtained by using Eqs. (3-26) and (3-27)

$$\eta_h = Mv^2/(2\eta_a E_d) = \eta_t (v/u)^2/(e^{v/u}-1) \qquad (3\text{-}32)$$

The hydrodynamic efficiency maximizes at approximately 0.3 when the target velocity is comparable to the ablation velocity, $\eta_t \simeq 0.5$, and 80% of the initial mass is ablated away.

The absorbed specific driver energy required to drive the implosion is

$$\varepsilon_d \simeq \eta_a E_d/(M_o - M) = u^2/2\eta_t \qquad (3\text{-}29)$$

For a maximum efficiency implosion the implosion velocity $v \simeq 1.6u$. From the previous example if $v \simeq 3 \times 10^7$ cm/sec, then $\varepsilon_d \simeq 3.5 \times 10^7$ J/g. Computer simulations of implosions[19] indicate that efficient implosions may be driven with specific driver energy depositions in the range of 20 to 50MJ/g. The ablation pressure required to drive the implosion is

$$P_a = (M/A)(dv/dt) = -ud(M/A)/dt \qquad (3\text{-}30)$$

where A is the area of the target. The ablation pressure is approximately

$$P_a \simeq -u(dM_a/dt)/4\pi r^2 \simeq uM_a/4\pi r^2 \tau_d \simeq uv(\Delta r/r)\rho_a/2 \qquad (3\text{-}31)$$

where ρ_a is the density of the ablator material and is usually comparable to the fuel density. The quantity r is the target radius. The implosion of a thin shell $r/\Delta r \simeq 10$ at maximum efficiency to an implosion velocity of 3×10^7cm/sec requires an ablation pressure of approximately 5.6×10^{12}erg/ cm^3 or 5.6Mbar. Generally, ablation pressures in the range[19] of 5-40Mbar are required to drive high gain targets. The ablation pressure is related to the driver intensity by

$$P_a \simeq 2\eta_a \eta_t I_d/u. \qquad (3\text{-}32)$$

The driver intensity required to drive the implosion may be estimated by combining Eqs. (3-26), (3-31), and (3-32)

$$I_d \simeq (\rho_a v/\eta_a \eta_t)(u/2)^2(\Delta r/r) \qquad (3\text{-}33)$$

For the previous example, if the driver energy is absorbed with an efficiency of 80% and the implosion hydrodynamic efficiency is near maximum, then $I_d \simeq 1.3 \times 10^{13}$W/cm^2. Practical implosions[19] are less efficient and require driver intensities $\simeq 2\times10^{14}$W/cm^2. The blow off velocity u is approximately the sound speed $c_a \alpha \theta_e^{1/2}$ in the hot atmosphere where the driver energy is deposited. Drivers such as ion beams and short wavelength lasers deposit their energy at higher density near the ablation surface and create higher ablation pressures and more efficient implosions than, for example, long wavelength lasers[32]. The driver pulse length must equal the implosion time

$$\tau_d \simeq 2r/v \simeq [(2M_0/\pi)(r/\Delta r)/\rho_a v^3]^{1/3} \qquad (3\text{-}34)$$

which is typically in the range of 10 to 20ns for target masses of several mg.

This discussion has outlined the performance characteristics for idealized spherical DT targets. Over the past decade it has become apparent, through experiment and computation, that to these simple approximations must be added several additional considerations. In general, particularly for lasers, driver-plasma coupling involves complex plasma physics[32]. Driver pulse shaping is needed to set and maintain the proper isentrope for the fuel and ignition mass; timing is a delicate matter. Preheat[33] from long range particles such as electrons will raise fuel entropy so it resists compression. A Rayleigh-Taylor type fluid instability[3,4,28] may develop at the interfaces, such as the ablation surface, where a low density medium is pushing on a higher density medium. Overall spherical symmetry must be maintained to permit radial convergence of $\simeq$300, requiring better than 0.5% driving pressure uniformity. When these effects are properly included, the target performance predicted by simple models is degraded substantially. Consequently, to overcome these limitations much more sophisticated targets and computational tools[22] have been developed. For example, the code LASNEX[22] provides the capability to study the complex influence of absorption, transport, implosion, thermonuclear burn and instability processes on target designs and experiments.

It was recognized[4] that implosion of hollow targets containing one or more concentric shells reduces substantially the driver power requirements relative to those for a solid spherical pellet. As shown in Fig. 7, two generic types of direct drive targets have been evolved: single and double shell targets[30]. These targets incorporate layers of different materials to control the deleterious physical processes mentioned above. During the implosion of these direct illumination targets, the driver beam energy is absorbed in a thin ablator or beam deposition layer. Inside this layer is a preheat shield consisting of a low Z material doped with a high Z material such as TaCOH. For the single shell target, a layer of solid DT fuel is placed inside the preheat shield. When this target is imploded, the initial shock wave reflects off of the inner DT surface creating a puff of hot

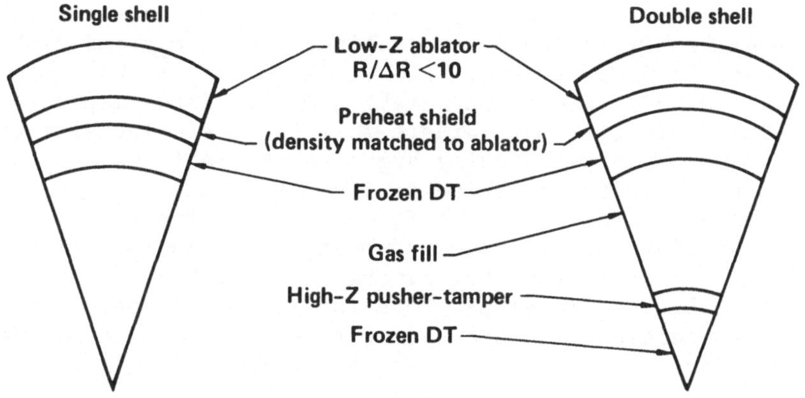

Fig. 7 Schematic cross sections of two generic types of direct drive target, single and double shell[30], used in inertial confinement fusion.

DT gas. This gas is compressed on a higher adiabat than the main fuel layer, and forms the initial hot spot which ignites the compressed, nearly degenerate fuel. Although this single shell target is simple in design, computer calculations indicate that it is sensitive to pulse shape, and requires a high peak power. The single shell target stresses driver technology.

Inside the preheat shield, the double shell target also has a layer of solid DT fuel. However, suspended at the center of this outer multilayer shell is a small spherical shell. This inner shell consists of a central shell of solid DT coated with a high Z (Z is the ion charge) material which serves as a tamper during the implosion process. When the double shell target is irradiated by driver beams, the massive outer shell is accelerated slowly inward. Eventually it collides with the inner shell to achieve the velocity multiplication required to compress and ignite the inner DT shell. This inner DT shell then serves as the spark which ignites the main DT fuel inside the massive outer shell. The double shell target is complex and difficult to fabricate. However, it is relatively insensitive to driver pulse shape, and requires a lower peak power (1/2X) than the single shell target. It thus relaxes driver requirements at the consequence of target fabrication difficulty.

As discussed by More[10,11], atomic processes play an important role in the dynamics of these targets. The absorption and transport of energy within these targets is strongly dependent on the collisional and radiative properties of the matter involved. The ionization kinetics of the high Z layers brings up many problems of atomic physics. Non-LTE effects are often very significant[10,11].

Several potential inertial confinement fusion driver technologies have been identified which can deliver the energy and peak power required: lasers, heavy ions (10-20GeV, 10kA, $^{238}U^{+}$), light ions (10MeV, 20MA $^{2}H^{+}$), electrons (2MeV, 100MA). Experiments and theoretical calculations indicate that there are two serious difficulties associated with using high energy electrons. The fuel preheat problem is severe due to the long range of high energy electrons, and space charge forces prevent the required focusing of the high current electron beams. At the present time, the leading driver candidates are lasers, and heavy and light ions. In the following three sections the technical issues and requirements associated with the use of these driver technologies will be discussed.

Table 2. Major Laser Fusion Facilities

		1.06μm Wavelength		
LLNL, USA	Novette	30 kJ 25 TW	Nova	150 kJ 125 TW
NRL, USA	Pharos	1 kJ 2 TW		
Univ. of Rochester, USA	GDL	0.12 kJ 0.5 TW	Omega	4 kJ 12 TW
KMSF, USA	Chroma I	1 kJ 2 TW		
Limeil, France	Octal	4 kJ 4 TW	Phebus	30 kJ 25 TW
Osaka Univ., Japan	GEKKO IV	2 kJ 4 TW	GEKKO XII	20 kJ 40 TW
AWRE, England	Helen	1.2 kJ 3 TW		
Rutherford Lab, England	Vulcan	1.2 kJ 2 TW		
Lebedev, USSR	Del'fin (108-Beam)	5 kJ 2-7 TW		
Inst. of Gen. Phys., USSR	UMI-35	8-10 kJ 4-6 TW		
		10.6μm Wavelength		
LASL, USA	Helios	5-10 kJ 10-20 TW	Antares	40 kJ 40-80 TW
Osaka Univ., Japan	LEKKO II	1 kJ 1 TW	LEKKO III	10 kJ 10 TW

Energy quoted reflects performance at long pulses
Power quoted reflects performance at short pulses

4. LASER DRIVEN ICF

For the past decade lasers have been primarily the driver of choice for conducting inertial confinement fusion experiments because of their ability to generate and focus the energy and power levels required. Table 2 lists several laser fusion target facilities that are operating or under construction. The experimental work carried out with these facilities has been directed toward understanding the laser-plasma interaction process and imploding matter to high temperature and, alternatively, high density, see Fig. 2.

Laser-Plasma Interaction

The implosion models presented in Sec. 3 revealed that driver coupling efficiency is as important as compression in inertial fusion implosions. Preheat is also important since it can preclude achievement of the very large compressions needed for ignition and burn. For lasers coupling efficiency and preheat generation are strongly dependent on the nature of the laserplasma interaction process[32-37], which is a function of laser intensity and wavelength. Great progress has been made in understanding the laser-plasma interaction process. Primarily because of the coherent nature of the laser beam, it has been found that many different plasma processes influence the coupling of laser light. Table 3 lists these processes and indicates the region of the absorption layer in which they occur. The critical density $n_c = \omega^2 m/4\pi e^2 \sim 10^{21} cm^{-3}/\lambda^2(\mu m)$ is the density above which electromagnetic radiation will not propagate. The quantity ω is the radian frequency of laser light and e is the electron charge. Therefore, laser radiation incident on the target is absorbed or reflected before or at the critical density surface. Energy transfer to the ablation layer occurs by electron and radiation transport.

The optimum absorption mechanism is inverse bremsstrahlung or collisional absorption because it preferentially heats the slow electrons in the distribution. The electron-ion collision frequency varies as v^{-3} where v is the electron velocity. The heating rate is the oscillatory energy times the electron-ion collision frequency ν_{ei} and equals the damping rate of the light wave[37]

$$\nu E_L^2/8\pi = \nu_{ei}\, nmv_{os}^2/2 \tag{4-1}$$

where $v_{os} = eE_L/m\omega$, E_L is the electric field of the light, and ν is an effective damping rate. Then

$$\nu = (n/n_c)\,\nu_{ei} \simeq 3\times10^{-6} \ln\Lambda\, [nZ/\theta_e(ev)^{2/3}](n/n_c) \tag{4-2}$$

where θ_e is the electron temperature, and Λ is the usual expression determined by the minimum and maximum collisional cut-off parameters. Note that ν increases with density and charge state, and decreases as the plasma heats.

The intensity-wavelength window for efficient inverse bremsstrahlung absorption has been estimated by Kruer[37]. The absorption will be >70% when the size, L, of the underdense plasma is greater that an absorption length: $\nu L/c > 1$. The plasma temperature in the absorption region is approximately related to the laser light intensity by the free streaming condition[32-37]; $\eta_a I_d = fnv\theta_e$. For a Knudsen gas model $f \simeq 0.6$, however, ICF experiments have indicated that electron heat transport is strongly reduced or inhibited (i.e. $f \simeq 0.02$ to 0.1). This reduction in heat flow is not presently understood and limits target performance[10,11,38]. For moderately inhibited electron transport and efficient absorption $f \simeq 0.1$ and $\eta_a \simeq 1$, respectively. Combining these estimates gives a bound on the intensity

$$I_d(W/cm^2) < 5x10^{14} Z\, L(cm)/\lambda^4(\mu m). \qquad (4\text{-}3)$$

At higher intensities, the plasma becomes too hot and the collisional absorption mechanism is ineffective. For reactor targets $L > 0.5$cm, and consequently for intensities $I_d(W/cm^2) < 2.5x10^{14} Z/\lambda^4(\mu m)$ inverse bremsstrahlung may be an efficient absorption mechanism. Since targets require a drive intensity of order 10^{14} W/cm^2, this means that lasers with $\lambda < 1\mu m$ are desirable. The wavelength enters strongly since $\nu_{ei} \simeq n/\theta_e^{3/2}$. As the wavelength decreases, the light deposits its energy at higher density, and the plasma is cooler since more electrons absorb the incident laser energy.

In addition to collisional absorption, Table 3 indicates that a fusion plasma may absorb or scatter laser light due to collective effects. Due to the long range of the coulomb force, the plasma supports several waves. Of these various collective modes of oscillation, two waves are of primary importance. The first is a high frequency electron plasma wave which occurs at the natural frequency with which electrons oscillate: $\omega_{pe} = 4\pi ne^2/m$. The second is a low frequency, $\omega_{ia} = 2\pi(\theta_e/m_i)^{1/2}/\lambda_a$, ion acoustic wave. The laser light can resonantly excite these waves which, in turn, heat the plasma and/or scatter light. The frequency matching conditions, Table 3, determine the region of plasma density where these processes occur. Resonant absorption is the conversion of laser radiation to electron plasma waves at the critical density surface. Only incident laser radiation with an electric field component along the plasma density gradient drives this absorption mechanism. Resonant absorption is important in small scale length plasmas where other absorption and scattering mechanisms are less important. However, for short wavelength laser radiation interacting with long scale length reactor plasmas, inverse bremsstrahlung will reduce the amount of light reaching the critical surface, and therefore the importance of resonant absorption. Ion-acoustic decay instability leads to the decay of a light wave into an ion acoustic wave and an electron plasma wave very near the critical density surface. This process will also be relatively unimportant in short wavelength laser driven reactor targets.

Table 3. Laser-Plasma Coupling Processes[37]

Collisional heating $\nu_{ei} \simeq Zn_c/\theta_e^{3/2}$

Absorption and scatter by plasma waves	(nonthermal distributions)
o Resonance absorption	$\omega \rightarrow \omega_{pe}$ at $n = n_c$
o Ion acoustic decay instability	$\omega \rightarrow \omega_{pe} + \omega_i$ at $n \simeq n_c$
o Raman instability	$\omega \rightarrow \omega_{sc} + \omega_{pe}$ at $n \lesssim 1/4n_c$
o Brillouin instability	$\omega \rightarrow \omega_{sc} + \omega_i$ at $n \lesssim n_c$
o $2\omega_{pe}$ instability	$\omega \rightarrow \omega_{pe} + \omega_{pe}$ at $n \simeq 1/4n_c$
Light Filamentation	at $n \lesssim n_c$

Three resonant wave processes and the filamentation instability are particularly important in long scale length plasmas characteristic of reactor targets. They may prevent the incident laser light from reaching the high density critical surface region where it may be efficiently absorbed by inverse bremsstrahlung. In addition, processes which excite electron plasma waves generate hot electrons which produce fuel preheat. The resonant wave processes are: stimulated Raman and Brillouin scattering and $2\omega_{pe}$ instability. The $2\omega_{pe}$ instability is excited at near $n_c/4$ density when a light wave decays into two electron plasma waves. The Raman instability occurs when a light wave excites an electron plasma wave and scatters. This process occurs at densities lower than $n_c/4$. Brillouin instability is a similar scattering process which excites ion acoustic waves for densities $\lesssim n_c$. Brillouin scattering is of special concern because it can lead to reduced absorption by efficiently back scattering the incident laser light. Filamentation occurs when the local laser light intensity pushes the plasma aside by the ponderomotive force. In the region of reduced plasma density, the plasma refractive index is larger and the laser light refracts into this region, further increasing the light intensity and expelling more plasma. The net result of this chain of events is the break-up of the incident beam. Filamentation is also believed to be responsible for hot electron generation, and may also produce highly non-uniform target drive.

Computer simulations[32,37] of $2\omega_{pe}$ and Raman instabilities in large regions of plasma with density $\lesssim n_c/4$ show more than 10% absorption into electrons with a characteristic temperature of order 100keV. Both instabilities have been observed experimentally. The density gradient determines the minimum threshold intensity I_T for these instabilities, because it limits the region over which resonant coupling can occur[37]:

$$I_T(\mathrm{W/cm^2}) \simeq 10^{13}\left(\frac{\theta_e(\mathrm{ev})}{L(\mu m)\ \lambda(\mu m)}\right) \qquad 2\omega_{pe}\ \text{instability} \tag{4-4}$$

and

$$I_T(\mathrm{W/cm^2}) \simeq \frac{5\times10^{17}}{[L(\mu m)]^{4/3}[\lambda(\mu m)]^{2/3}} \qquad \text{Raman instability} \tag{4-5}$$

where L is the local density gradient length. Provided $\theta(ev) < 5\times10^4[\lambda(\mu m)/L(\mu m)]^{1/3}$ the $2\omega_{pe}$ instability appears first. The threshold for $2\omega_{pe}$ instability under this condition, when the maximum gradient length near $n_c/4$ is taken as the inverse bremsstrahlung absorption length is, from Eq. (4-4)

$$I_T \simeq 3\times10^{13}Z/[(\theta(ev))^{1/2}\ (\lambda(\mu m))^3] \tag{4-6}$$

For 1/2μm light in a high Z plasma, this threshold appears in a regime of interest $\simeq3\times10^{14}$ W/cm^2. The above estimates are conservative for high Z plasmas irradiated by short wavelength laser light because collisions also damp unstable waves.

Experiments by Ripin, et. al.[39], with 1.0μm light at high intensity, $\simeq10^{16}$ W/cm^2, have found that about 40% of the light is back reflected when a prepulse is used to prepare a long scale-length plasma. Kruer[37] has shown that for low Z plasmas irradiated by short wavelength light, the Brillouin and filamentation intensity thresholds are comparable and approximately

$$I_T\ (\mathrm{W/cm^2}) \simeq 10^{14}\left(\frac{Z}{26}\right)^{3/4}\frac{1}{[\lambda(\mu m)]^{11/4}} \qquad \text{Brillouin Scatter, filamentation} \tag{4-7}$$

For 1/3μm light and Z=3 this threshold is $\simeq4\times10^{14}$ W/cm^2 and again, in a regime of interest for reactor targets. The instability threshold estimates of Kruer, given above, suggest that at short wavelengths collisional absorption is efficient over a wide range of intensities. Recent experiments[35,36], Figs. 8 to 10, with short wavelength light and small underdense plasmas show a significant regime of high absorption and reduced Brillouin scattering and hot electron generation (as exhibited by reduced high energy x-ray emission). Minimizing collective effects for reactor size plasmas imposes more substantial restraints on intensity. However, there still appears to be a regime of interest for reactor target design in which short wavelength (<1/2μm) laser light at intensities of order 10^{14}W/cm^2 will be absorbed efficiently by inverse bremsstrahlung and collective effects minimized. In addition, recent theoretical studies[40-42] suggest that parametric instabilities such as stimulated Raman and Brillouin scattering can be controlled by using finite bandwidth ($\delta\lambda/\lambda$>0.01 to 0.10) laser radiation. Also it has recently been suggested[43] that very uniform illumination of laser irradiated targets may be achieved by inducing a controlled amount of transverse spatial incoherence on the output beam of a broadband laser. This incoherence may also suppress certain parametric instabilities.

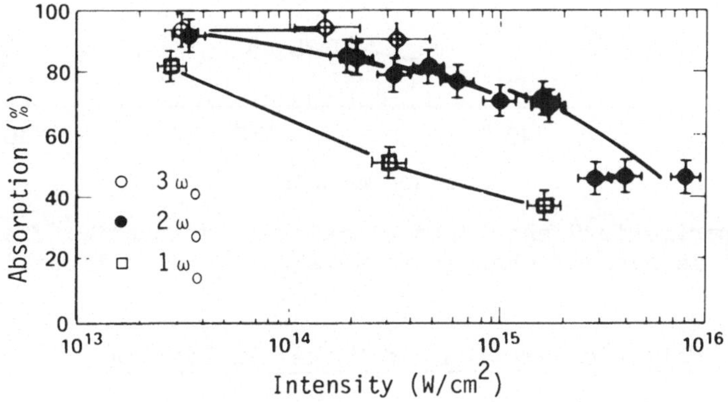

Fig. 8 Laser Light Absorption by low Z disk targets has been observed[36] to increase dramatically for short wavelength laser light.

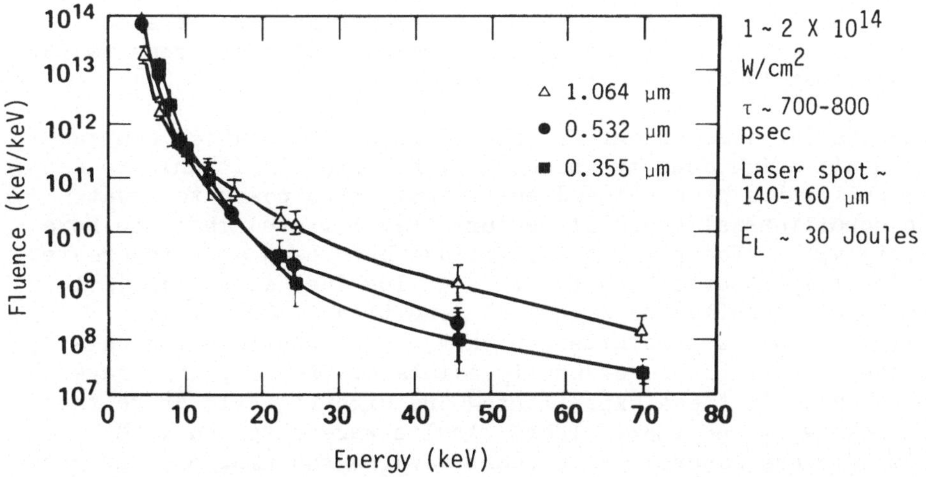

Fig. 9 Experiments[36] show that the generation of hot electrons and, consequently, hot x-rays is greatly reduced when short wavelength laser light is used.

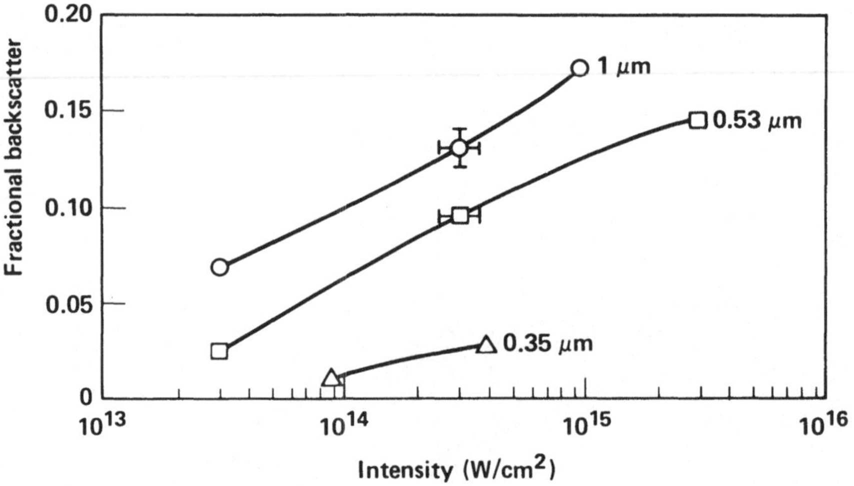

Fig. 10 Experiments[36] show that stimulated backscatter from low Z disk targets is reduced when short wavelength laser light is used.

Laser Driven Foil Acceleration and Target Implosions

In addition to enhanced absorption and reduced scattering and hot electron generation, recent experiments[31,32] have shown that ablation pressure increases at shorter wavelength. For example, it has been found that at an intensity of 3×10^{14}W/cm^2 that the ablation pressure is approximately 10Mbar at 1.0μm, 30Mbar at 0.5μm, and 60Mbar at 0.35μm for laser irradiated low Z disk targets. At shorter wavelengths the laser light is absorbed at higher density and thus nearer to the ablation surface. The coupling efficiency between the critical surface and ablation surface is much better leading to increased ablation pressure for a given input laser drive intensity.

Successful development of laser driven ICF requires compression to ultrahigh density and central hot spot ignition. To date, laser facilities have not had sufficient drive power to create these conditions although it is currently believed that the Nova facility will. The present facilities[9] have been able to create high ion temperatures (several keV) at low fuel densities (1 to 2 times liquid density) or high fuel densities ($\lesssim$100 x liquid density) at low temperatures ($\simeq$0.4keV). Calculations indicate that the DT fuel pressure-density adiabats, Fig. 11, that have been achieved in these experiments[9] are significantly above the Fermi degenerate adiabat. These results were obtained with 1.06μm Nd:glass lasers. With short wavelengths (2ω, 3ω) and careful pulse shaping, it is believed that next generation Nova system will achieve higher densities and much more nearly degenerate fuel compressions. However, as shown in Fig. 12, very impressive compressions have been achieved at several major laboratories.

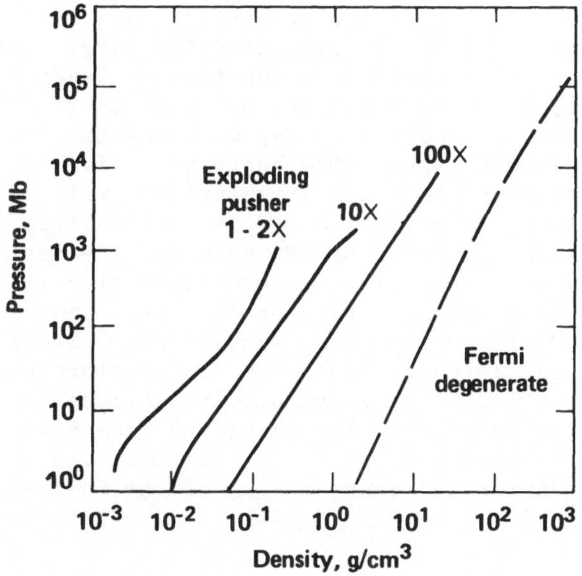

Fig. 11 DT fuel adiabats computed[19] for high density target implosion experiments[9].

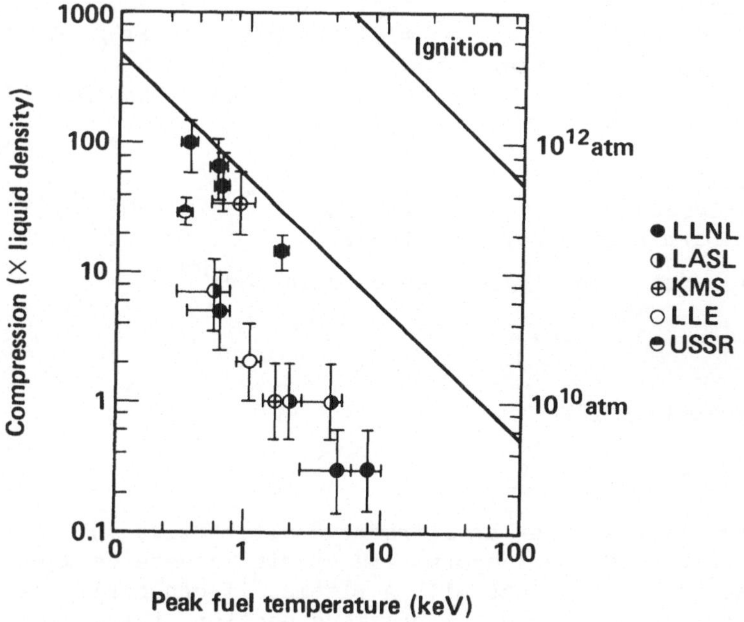

Fig. 12 High DT fuel densities have been achieved in laser driven implosion experiments[9,14,21].

On the basis of laser fusion experiments and extensive computer simulations, Lindl[19] has projected the target gain and laser peak power requirements as a function of laser input energy. Figs. 13 to 15 present results of these projections for both single and double shell targets, assuming the driver is a 1/4μm short wavelength laser. The heavy ion driver curves developed by Bangerter, Mark and Thiessen[44,45] will be discussed in the next section. The ideal curve is derived from target design calculations using the LASNEX code[22]. The best estimate bands denoted by the dashed lines are derived from the dotted ideal curves. The difference between the ideal and best estimate band is due to uncertainty in the calculations. These uncertainties are listed in Table 4. The correction factors endeavor to take account of potential degradations in compression and burn efficiency due, respectively, to deviations from Fermi degenerate compression and asymmetries and fluid instabilities which arise during the compression and cause mixing of the fuel and the pusher and tamper material.

Table 4. Assumptions Used in Target Gain Collections[19]

Parameter	Ideal	Best Estimates
Absorption	100%	80%
Transport Inhibition	f≃0.03	f≃0.03
Fast Ion Losses	Negligible	Negligible
Ablation Efficiency	≃15% (LASNEX)	≃15% (LASNEX)
Entropy of Compressed Matter	$\alpha \lesssim 2$	$\alpha \lesssim 2$
Preheat	Not Significant	Not Significant
Pulse Shaping	Optimum	Optimum
Asymmetries	No effect	70% (LASNEX)
Fluid Instabilities	*	*
Ignition Efficiency	LASNEX	50% (LASNEX)
Propagation/Burn Efficiency	LASNEX	70% (LASNEX)

*Growth not worse than single mode calculations[28].

5. ION BEAM DRIVEN ICF

Ion Beam - Plasma Interaction

Ion beam drivers offer an attractive alternative to lasers. Primarily because of its temporal and spatial coherence when an intense laser beam interacts with a plasma, it coherently excites plasma waves. As seen from the previous section, these plasma waves produce deleterious effects by scattering the laser beam and/or accelerating electrons to high energies. Alternatively, when an ion beam passes through a fusion plasma, calculations[11,46]

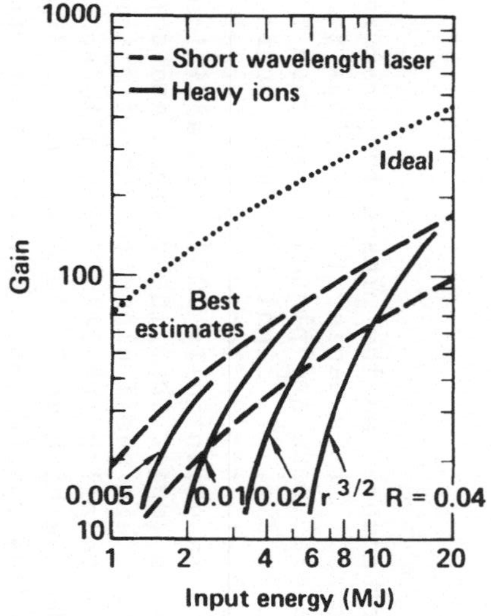

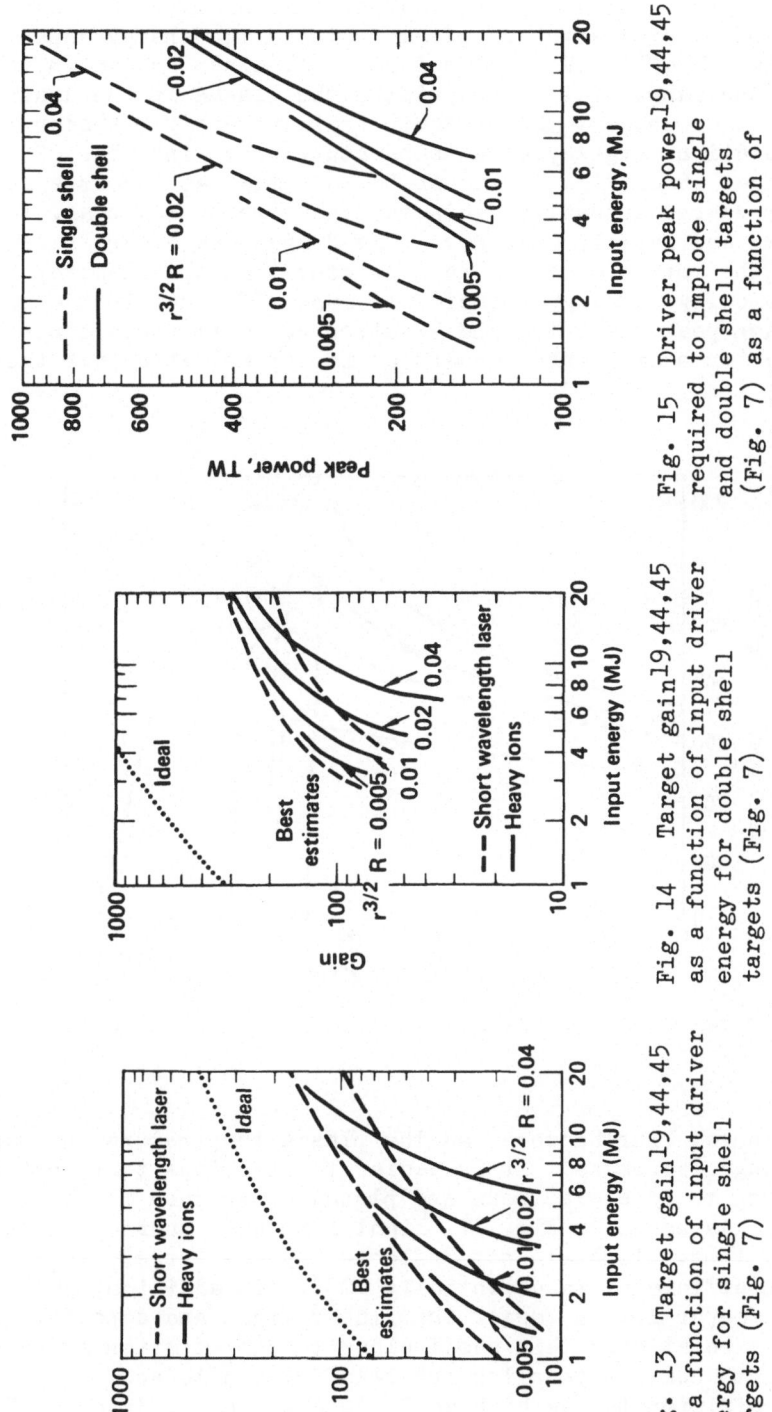

Fig. 13 Target gain[19,44,45] as a function of input driver energy for single shell targets (Fig. 7)

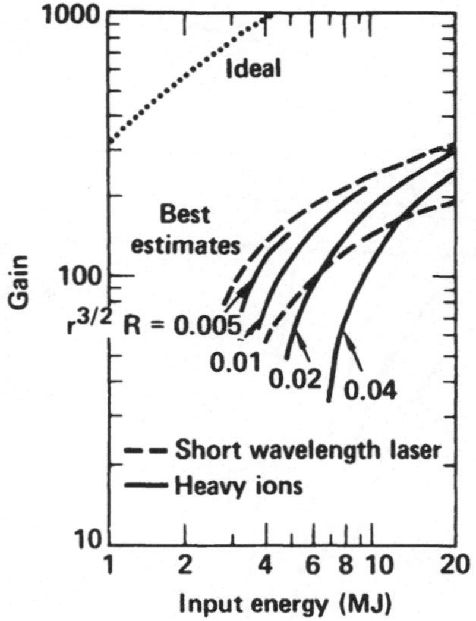

Fig. 14 Target gain[19,44,45] as a function of input driver energy for double shell targets (Fig. 7)

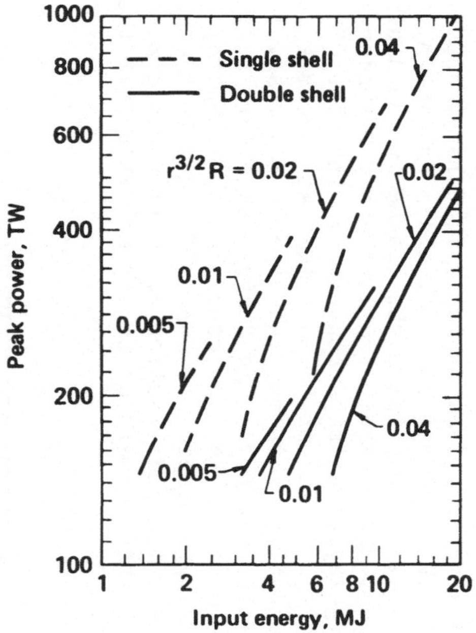

Fig. 15 Driver peak power[19,44,45] required to implode single and double shell targets (Fig. 7) as a function of input driver energy

suggest that it loses energy by coulomb collisions with electrons and ions and by excitation of plasma waves. The energy deposition characteristics of 10GeV uranium ions in aluminum are shown in Fig. 16. The range of the ions within the plasma is the distance over which they deposit 90% of their incident energy. Under ion beam irradiation target plasma temperature rises into the 200 to 300 ev range. From Fig. 16 it can be seen that as the ions slow down their energy deposition rate increases reaches a peak, the Bragg peak, and rapidly declines. The Bragg peak occurs because initially the ion velocity is much greater than the electron thermal velocity. Energy transfer is more efficient in the region of the Bragg peak where ion and electron velocities are comparable. The ion beam energy deposition is concentrated near the end

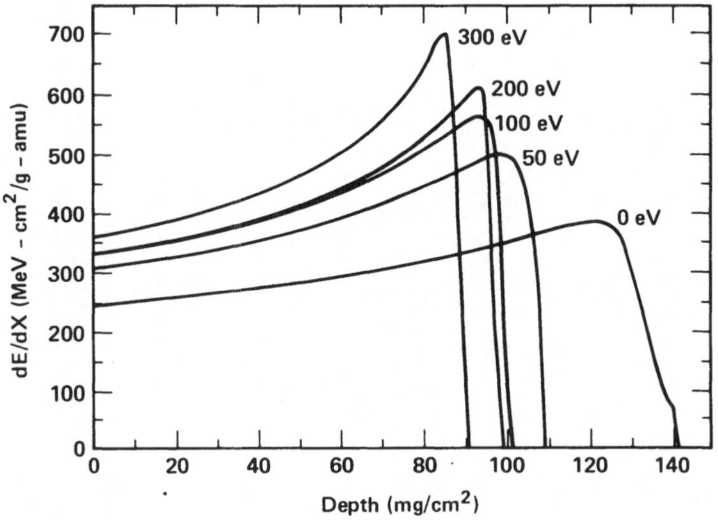

Fig. 16 Energy loss characteristics of a 10GeV Uranium ion in aluminum[46]

of their range. Furthermore, as the plasma temperature increases, the ion range decreases. It is easier for the ions to transfer their energy to free electrons and plasma waves than to bound electrons. As shown in Fig. 17, light ions have shorter ranges than heavy ions. Ions are not reflected by the target, so the absorption efficiency is essentially 100%. In addition, ions penetrate deeper into a target than laser light and deposit their energy in a relatively thin shell close to the ablation surface. These effects combine to allow ion beam drivers to achieve overall implosion efficiencies as high as 15 to 20%, compared to the 5 to 10% efficiencies characterizing laser driven implosions.

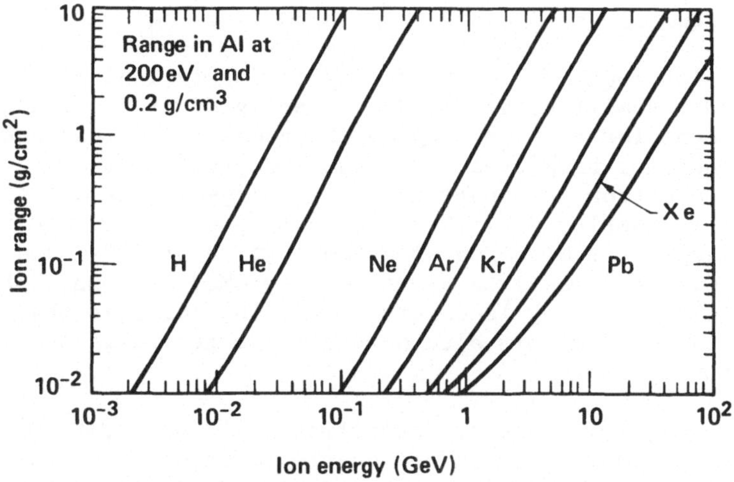

Fig. 17 Ion range in hot aluminum as a function of incident ion energy for several ion species

In contrast to the coherence of the laser-plasma interaction, ion beams interact with the plasma on an essentially independent particle basis. Since the beam particles have no fixed phase relative to one another, strong collective excitation is unlikely. For example, a $100TW/cm^2$, 2MeV proton beam incident on a target with an electron density of 10^{22} to $10^{24}cm^{-3}$, has a beam interparticle separation of 10^{-6}cm and a Debye shielding distance in the target of 10^{-8}cm. Hence the 100 shielding lengths between particles means they interact as independent particles. At higher ion energies, such as 10GeV uranium, the beam ion density is proportionally lower and the interparticle spacing 17 times larger than for 2MeV protons. These considerations have led to the notion that inertial fusion ion beams are not really "intense" and that their interaction with the target is "classical" as opposed to the "anomalous" absorption which often characterizes the interaction for all but short wavelength laser light. Although collective effects are possible, a recent study[47] suggests that they are benign on the time scales of interest.

Ion Beam Driven Implosions

The specific driver energy deposition, Sec. 3, required to cause an inertial fusion implosion in a reactor is roughly 20 to 50 MJ/g. Since the driver energy is deposited within an ion range $R = \rho\Delta r$ for a spherical target of radius r, this specific energy deposition is

$$\varepsilon_d \simeq E_d/4\pi r^2\rho\Delta r = E_d/4\pi r^2 R \simeq 20 \text{ to } 50 \text{ MJ/g}. \qquad (5\text{-}1)$$

This expression requires $R \simeq .02$ to $.2g/cm^2$ for typical target radii of order a millimeter and driver energies in the megajoule regime. Ranges longer than $0.2g/cm^2$ increase the driver energy required, whereas, ranges shorter than $0.02g/cm^2$ increase reliance on electron transport to carry energy to the ablation surface. For example, Fig. 17 shows that to achieve a nominal range of $0.1g/cm^2$ in aluminum at a temperature of 200eV and a density of $0.2g/cm^3$ requires approximately 8MeV protons, 35MeV alpha particles, or 9GeV lead ions. Fig. 16 indicates that roughly 10GeV uranium ions also have a range near $0.1g/cm^2$ for the above conditions.

Since the specific energy deposition ε_d is an important target drive parameter, it is reasonable to expect that for a fixed driver energy, target gain would depend on the product r^2R instead of r and R independently. However, as shown in Figs. 13 to 15, target gain[44,45] as a function of driver energy, for ion beam fusion, depends parametrically on $r^{\varepsilon}R$, where $\varepsilon \simeq 3/2$ instead of two. The curves are valid for beam radii such that: $0.1 < r/E^{1/3} < 0.2$. The solid curves in these figures are best estimate target gains for different values of $r^{3/2}R$, where r may also be interpreted as the ion beam focal radius (in cm) and R is the ion range (mg/cm^2). For example, a 10GeV lead ion-beam focused to a radius of 0.2cm gives $r^{3/2}R \simeq 0.01$. Thus the 0.01 gain curve is roughly appropriate for this case.

6. NOMINAL PERFORMANCE REQUIREMENTS FOR ICF DRIVERS

By choosing a fusion energy gain $Q=G\eta_d$ required for economical power reactor operation (Fig. 4), and simplifying the target gain curves (Figs. 13 and 14), it is possible to construct a reactor performance map[15]. Fig. 18 shows such a map for a 1GW electric power plant. This plant has a thermal to electrical conversion efficiency of 39%, an overall efficiency of 32%, a single micro-explosion chamber, and a blanket multiplication factor of M=1.15. The yield is defined as the product of pellet gain and driver energy. The microexplosion or driver repetition rate corresponding to each pellet yield is given by the right ordinate. Slightly higher repetition rates would be needed for a thermal to electric conversion efficiency lower than 39%. A 300MW electric plant would need only 30% of the indicated repetition rate.

The optimistic and conservative gain bands synthesize the ideal and best estimate predictions from Figs. 13 and 14. The lower boundaries correspond to several different driver efficiencies. These curves denote the lower limit of usefulness for a driver with a given efficiency. For example, a driver with an

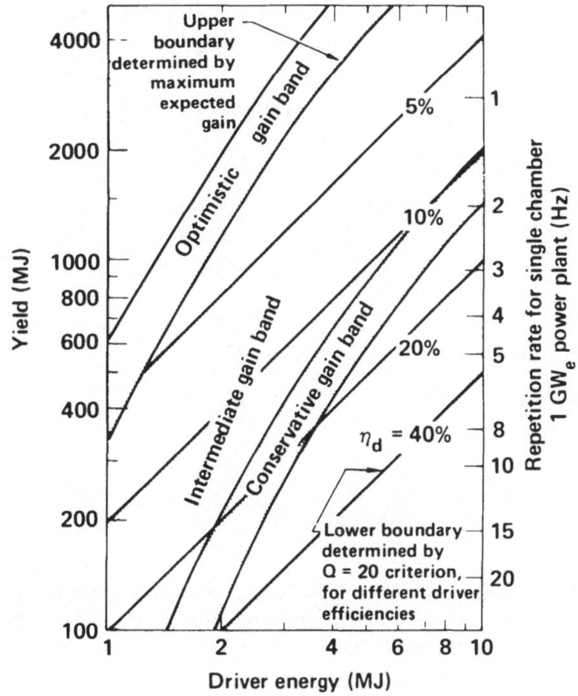

Fig. 18 Performance map for inertial fusion power plants derived by Monsler, et. al.[15]

efficiency of 10% can only be used in this reactor to drive targets with yields above the 10% boundary line. The curves presented in Fig. 18 show that increasing pellet gains and driver efficiencies both enlarge the accessible area of parameter space. Targets with performance in the conservative band require 10 to 20% efficient drivers, repetition rates <10Hz, and driver energies greater than 2MJ. If drivers with only 5% efficiency can be developed, quite large pellet gains, near the optimistic boundary, will be required. These large gains will produce high yields (>500MJ), and require low repetition rates (<5Hz). The driver requirements derived from performance maps such as Fig. 18 are sensitive to the value of Q chosen. If higher electricity costs are acceptable, then driver requirements are relaxed. On the basis of these considerations, Table 5 lists the nominal performance requirements[48] for short wavelength laser, light and heavy ion beam ICF drivers for operation within the conservative and optimistic gain band regimes. These results also assume that double shell targets are utilized. For single shell targets, the peak power requirement is approximately two times larger. The extension of this table to other ion species may be based on the specific energy deposition considerations discussed earlier.

Table 5. Nominal Performance Requirements for Inertial Confinement Fusion Drivers[48]

Parameter	Conservative	Optimistic
Driver efficiency (%)	≳6	≳1
Wavelength or ion voltage	≲0.3μm 10GeV U 5MeV p	≲2μm 20GeV U 20MeV p
Energy (MJ)	3-5	>1
Peak Power (TW)	200	>100
Pulse rate (Hz)	>4	>1

7. ICF REACTOR CONCEPTS

The reaction chamber in an inertial fusion power plant, based on the DT cycle, must breed tritium and survive the products of the microexplosion. Although the momentum of the pellet debris is small, large surface stresses may be created in the structure due to energy deposition by x-rays and particle debris from the pellet. The kinetic energy of the 14MeV neutrons is deposited through large material volumes, and may seriously weaken and activate the chamber structure. These phenomena can seriously threaten the economic and environmental viability of the plant. The challenge of protecting the reaction chamber wall from these various forms of energy deposition has led to several diverse concepts[15]: dry walls (bare walls and sacrificial liners), magnetically protected walls, gas-filled chambers, wetted walls, and thick liquid-metal walls. A rough performance map for these various concepts is illustrated in Fig. 19. Although the operating boundaries for each concept are not presently well defined. no one chamber concept is appropriate for the entire range of target yields, repetition rates, and drivers. It is clear, however, that a wide range of options is available. The chamber background density determines propagation windows[49] for the three driver types: lasers, light ions and heavy ions.

Optical breakdown and beam refraction limit laser operation to chamber environments and pressures of less than a few torr. Studies[49] suggest that light-ion beams can propagate through an ionized channel created by a laser beam in a suitable gas having pressures in the range of 5 to 50 torr. It is most likely that heavy ions could be used in two regimes. At pressures below 10^{-3} to 10^{-4} torr, ballistic propagation appears feasible. For pressures in the 10^{-1} to 1 torr range, a pinched propagation mode may be possible.

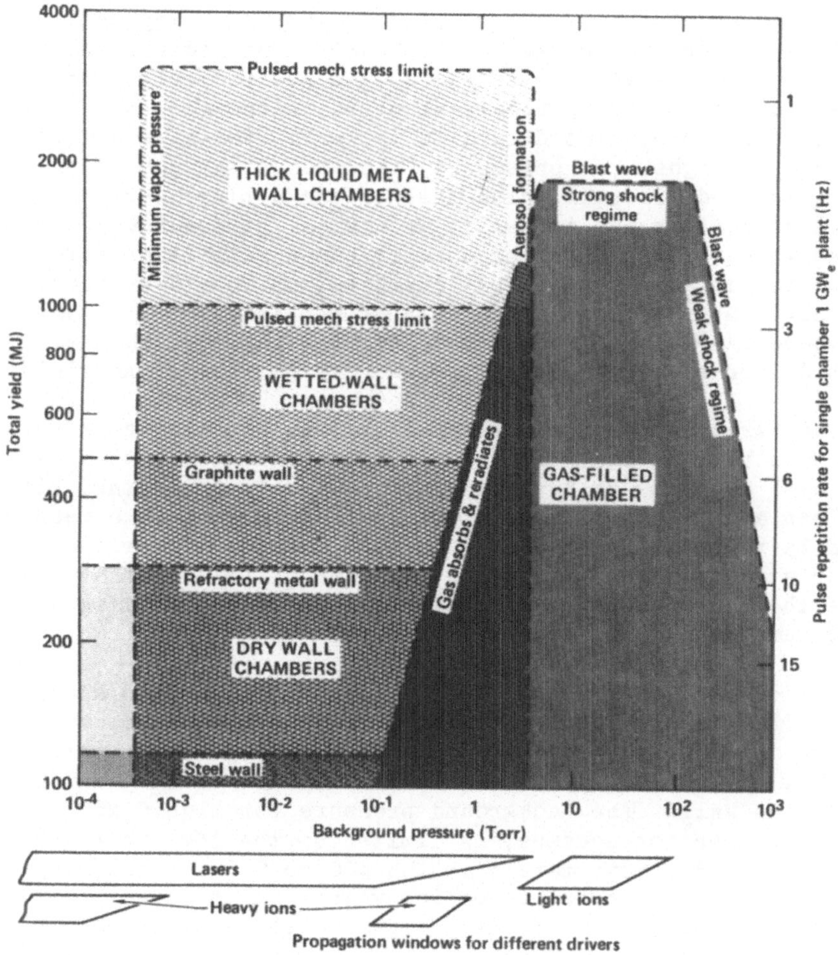

Fig. 19 Performance map for different chamber and driver concepts derived by Monsler, et. al.[15]

The performance map of Fig. 19 reflects the physical constraints imposed by wall protection considerations, and the necessity to reestablish ambient conditions for pellet and driver beam injection. A 10m chamber radius limit was assumed. Dry-wall chambers are yield limited, due to the effects of energy deposition by x-rays and pellet debris, by surface evaporation and/or sputtering. Although target designs may be used to minimize these effects they are serious. Studies of sacrificial liners suggest that graphite has many attractive features: reduced sputtering, high thermal conductivity, heat capacity, and thermal resistance, relatively low cost and ease of fabrication. The approximate yield limits for steel, refractory metal, and carbon bare-wall chambers are believed to be on the high side.

Some chamber designs[15] suggest that performance may be enhanced by either adding gas to the chamber, or redirecting the debris ions by using a magnetic field. In gas protection schemes, the x-rays and ions are partially absorbed in the gas, and their energy is mainly re-radiated back to the wall over a relatively long time. At higher gas pressures, the absorption creates a centrally located fireball which then drives a blast wave. The chamber must sustain the transient mechanical stress induced by the blast wave. The gas filled chamber has a repetition rate limit due to gas exchange and pellet injection considerations. Gas-filled chambers seem particularly appropriate for ion beam fusion where the gas is necessary to form propagation channels, and high driver efficiencies allow lower target gains and yields.

If the first wall surface is a liquid, it can ablate and re-establish without damage. Studies[15] indicate that a wetted wall chamber with a porous transpiration cooled surface can take higher yields than a dry wall. However, the pulsed mechanical stresses produced by rapid ablation of the surface liquid are serious limits on structure integrity. In addition, in all of the above concepts the neutrons pass through the first wall, creating distributed damage which limits its lifetime.

The thick liquid-metal wall concepts are designed to absorb x-rays, pellet debris, and neutrons. Fig. 20 shows a schematic diagram of the HYLIFE reactor concept[50]. These concepts permit the highest pellet yield, and are limited by both liquid and gas impact on the wall. The background pressure for liquid metal, first wall protection schemes, is limited on the low end by the vapor pressure of liquid metals in the 400 to 500°C range, and on the upper end by aerosol formation which limits driver beam propagation. Conditions currently required to reestablish ambient

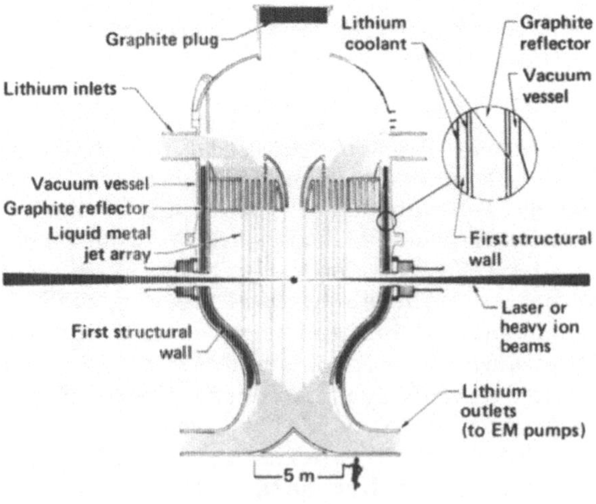

Fig. 20 Schematic diagram of the HYLIFE Reaction chamber concept[50]

chamber conditions limit liquid metal concepts to the lowest repetition rates. However, it may be feasible to design thick liquid-metal walls, such as the HYLIFE reactor, that protect the first solid wall of the reactor from significant radiation damage for the useful life of the power plant.

The INPORT reactor concept[46] offers the advantages of a liquid metal wall and higher repetition rates. The first wall consists of woven SiC tubes through which a lead-lithium alloy flows. The tubes are porous to the liquid metal and a layer about 1mm thick protects the tubes from target debris damage. This approach avoids the problem of liquid metal wall breakup and permits higher repetition rates and lower liquid metal flow rates.

The performance limits outlined in Fig. 19 are approximate, and will undoubtedly shift with the advent of deeper understanding, new innovations and concepts. It is clear, however, that the characteristic freedom of chamber design in inertial fusion gives the diversity needed for covering a wide operational range of yields and repetition rates with various drivers.

8. SHORT WAVELENGTH LASER DRIVER TECHNOLOGY

Short wavelength laser systems concepts capable of the performance levels called for in Table 5 have been addressed in terms of two generic types: (1) lasers based on electronic transitions of atoms and molecules and (2) so called "free" electron lasers. All of these laser concepts can be scaled to the energy and power levels required for power plant service. The principle issues are cost and efficiency. The study of driver costs is not yet sufficiently developed to be reviewed here. However, the technical issues that influence scalability and efficiency have been seriously engaged and will be briefly discussed within this section.

Electronic Transition Lasers

The interaction of laser radiation with an atomic or molecular electronic transition is characterized, on a microscopic level, by the stimulated emission cross-section σ_s. For a homogeneously broadened transition with a Lorentzian lineshape $\sigma_s \simeq (\lambda/2\pi n)^2(1/\tau_s\Delta\nu)$ where τ_s is the spontaneous lifetime of the transition, n is the index of refraction, and $\Delta\nu$ is the linewidth (FWHM). Laser action is produced by preferentially exciting the upper level of the transition and creating a population inversion. Efficient laser operation requires that: the upper level be excited efficiently, negligible decay of the upper level occur during excitation, and the lower level be maintained at a very low density relative to the upper level during optical energy extraction. In addition, it is desirable to have the lower level close to the ground state so that the quantum efficiency (photon energy/excitation energy per excited state) of the laser transition be as large as possible. For example, this occurs naturally for excimer lasers where the lower laser level is the dissociative ground state. When these circumstances occur, amplification of the laser light is governed approximately by the coupled radiation transport and upper laser level population N_u rate equations.

$$(n/c)(\partial I/\partial t + \partial I/\partial z) = (\alpha - \gamma_a)I \tag{8-1}$$

and

$$\partial N_u/\partial t = - \alpha I/h\nu + R_p - N/\tau_u \tag{8-2}$$

The quantity $\alpha = \sigma_s N_u$ and $\gamma_a = \Sigma_a \sigma_a N_a$ are the gain and non-saturable photoabsorption coefficients; respectively. The quantities σ_a and N_a are the absorption cross section and density of species a, respectively. The gain coefficient α is saturable because as the laser intensity increases N_u and consequently α decrease due to stimulated emission. The absorption coefficient γ_a is called nonsaturable because it is independent of intensity. Nonsaturable photoabsorption losses are serious in short wavelength excimer lasers because they reduce the efficiency of the laser. The quantities R_p and τ_u are the pump rate and lifetime of the upper laser level, respectively.

Table 6. Typical Fusion Laser Media Parameters

Laser	$\lambda(\mu m)$	$\sigma_s(cm^2)$	$N_u(cm^{-3})$	$\Delta\nu(cm^{-1})$	$\tau_s(s)$	Type
Nd:Glass	1.06	2.7×10^{-20}	3×10^{18}	340	360×10^{-6}	Storage
CO_2*	10.6	6×10^{-19}	10^{18}	0.33	4.7	Storage
I	1.3	2×10^{-19}	10^{17}	0.56	0.13	Storage
KrF	0.249	2.5×10^{-16}	4×10^{14}	400	6.7×10^{-9}	Nonstore

* P(20) line at 1800 torr, 3:1/4:1, He;N_2:CO_2 mix

The properties of several fusion laser media are summarized in Table 6. In general fusion electronic transition laser media may be categorized as two basic types[51]: (1) energy storage media in which τ_u is much larger (typically >1μs) than the laser pulse length and (2) nonstorage media in which τ_u is a few nanoseconds and consequently comparable to the laser pulse length.

As a class storage laser media are particularly attractive for the ICF application because they may be pumped on a time scale (>1μs) that is long compared to the laser pulse length. The peak pump power is then much less than the laser peak power. The laser medium pumping and optical energy extraction processes are decoupled. This allows the use of relatively inexpensive slow power conditioning technologies: pulse forming networks, rotating machinery, etc. In addition, simple relatively compact optical systems Fig. 21, such as regenerative and multipass laser architectures may be utilized. Pulse shaping is easily accomplished. Eq. (8-2) that governs the laser population during optical energy extraction simplifies to

$$\partial N_u/\partial t = -\alpha I/h\nu \qquad (8\text{-}3)$$

for storage laser media. Using Eq. (8-3), Eq. (8-1) may be integrated over the pulse width to yield

$$d\Gamma/dz = \alpha_o\Gamma_s(1-e^{-\Gamma/\Gamma_s}) - \gamma_a\,\Gamma \qquad (8\text{-}4)$$

where $\Gamma = \int I dt$ is called the pulse energy fluence, $\alpha_o = \alpha(I\rightarrow 0)$ is the small signal gain coefficient and $\Gamma_s = h\nu/\sigma_s$ is the saturation fluence. The energy storage density in the laser medium

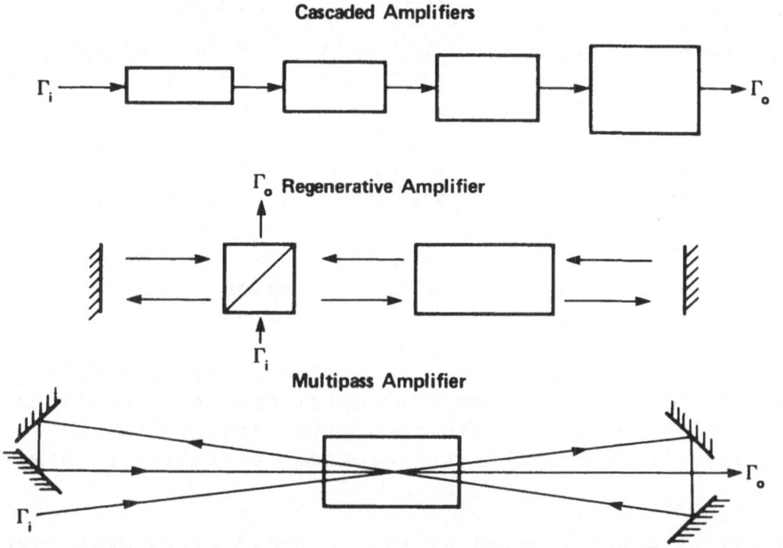

Fig. 21 In the multipass and regenerative laser architectures the laser beam passes through the medium several times, permitting a simpler more compact optical train. This gives these approaches technical and economic advantages over the conventional cascaded amplifier or MOPA (master oscillator power amplifier) architecture.

prior to extraction is $\varepsilon_{st} = \alpha_o\Gamma_s = h\nu N_{uo}$. The extraction efficiency is $\eta_{ext} = (\Gamma_{out} - \Gamma_{in})/\alpha_o L\Gamma_s$ where L is the length of the amplifier. In the small signal regime ($\Gamma \ll \Gamma_s$) the laser beam fluence increases exponentially as it passes through the medium. As Γ

approaches Γ_s the beam enters the efficient extraction regime, the signal increases linearily with position then plateaus out when $\Gamma/\Gamma_s \simeq \alpha_0/\gamma_a$. The plateau occurs when gain due to stimulated emission equals loss due to nonsaturable photoabsorption. The extraction efficiency increases in the small signal regime, maximizes, then decreases to zero in the plateau regime. Several requirements for storage laser media are summarized in Table 7. Efficient extraction requires that the output fluence be $\gtrsim 2\Gamma_s$ and less than the loss limit. Efficient amplifier performance requires that $\alpha_0/\gamma_a \gtrsim 20$. The output fluence must also be less than the optical damage fluence. The damage fluence depends on the material, optical coating type, and radiation wavelength. For example, the damage fluence[52,53] for AR (antireflection) coatings is $\Gamma_d(J/cm^2) \simeq 12\tau_L(ns)^{1/2}$ for wavelengths near 1μm. It is presently somewhat less for HR (high reflection) coatings.

Table 7. Energy Storage Laser Medium Requirements

Property	Symbol	Value	Constraint
Gain Coefficient	$\alpha_0=\sigma_s\Delta N$	$\gtrsim 10^{-2}cm^{-1}$	Beam quality
Loss Coefficient	γ_a	$\lesssim 10^{-3}cm^{-1}$	Efficient Extraction
Saturation Fluence	Γ_s	$\lesssim 20J/cm^2$	Optics Damage
Laser Cross-Section	σ_s	$\gtrsim 2x10^{-20}cm^2$	Efficient Extraction
Population Inversion	ΔN	$\gtrsim 5x10^{17}cm^{-3}$	Device Size
Inversion Lifetime	τ	$\gtrsim$1msec	Minimize Pulse Power

At shorter wavelengths optical coating technology is less developed and all optical damage limits[54] are much less. The causes of optical damage are not presently understood. However, it is likely that atomic processes at surfaces and thin film interfaces are important. Other considerations, not quantified in Table 7, related to excited state absorption and parasitics suppression also influence the technological viability of a given laser medium.

In general, the achievement of high excited state densities and long lifetimes requires unusually stable collisional properties for gas phase media. For example, the constraints in Table 7 require an excited state-excited state quenching rate constant $<2x10^{-15}cm^3/sec$, which is far below gas kinetic. To date, the photolytically pumped atomic iodine laser[55] is the only gas phase electronic transition laser to exhibit storage capability. The properties of the atomic iodine laser are ideal with the exception that known pumping techniques are so inefficient that the overall efficiency of the iodine laser is <1%.

Alternatively, rare earth and transition metal ions in solids exhibit long lifetimes suitable for the storage mode of operation[56]. This situation is due to the characteristic metastability of these ions and their weak interaction with each other and the host medium. The emission wavelengths of these media are characteristically in the near infrared; 0.7 to 1.5μm. Harmonic conversion is required to bring their wavelengths into the visible and near ultraviolet. Fortunately, techniques have been developed using KDP and its crystal isomorphs, which permit efficient (≃70%) conversion of near infrared frequencies to their second and third harmonics[56,57]. This approach has the technical advantage that it subjects only the harmonic conversion and target irradiation optics to intense ultraviolet radiation.

The Nd:glass laser is a prototypical example of a solid state storage laser[56]. The laser medium is excited by radiation emitted by electrically excited Xe flashlamps. Xe flashlamps radiate 75% of their input electrical energy. This radiation is absorbed over a broad spectral range by the Nd^{3+} doped glass. The energy absorbed by the Nd^{3+} ions rapidly (<1ns) relaxes to the $^4F_{3/2}$ upper laser level which has a radiative lifetime of 360μsec in ED-2 glass. Typically the flashlamps pump for a duration of 600μsec. Large aperture amplifiers are constructed by placing several (2 to 4) solid Nd:glass disks within a cavity excited by the flashlamps. The disks are oriented at Brewsters angle relative to the laser beam propagation direction in order to minimize reflection losses. This configuration also is beneficial for flashlamp pumping. Typically the lamps are arrayed on the two opposite sides of the disks. The other walls of the pump cavity are silvered to create high reflectivity for the pump radiation. The Nd:glass laser has exhibitied exceptional flexibility and has been highly valued as a means to experimentally explore inertial fusion target performance. For reactor service the Nd:glass laser has two significant limitations: low efficiency (<5%) and pulse repetition rate capability. The dominant mechanisms limiting efficiency are incomplete absorption of flashlamp radiation and spontaneous decay of population inversion during pumping. The $^4F_{3/2}$ level lifetime is too short for efficient flashlamp pumping. Average power operation is limited by the low thermal conductivity of glasses.

The pump efficiency limitation can be overcome[56] by choosing laser ions (activators) which have broad absorption bands and long radiative lifetimes (>1ms). The unusual thermal and mechanical properties required for average power operation may be satisfied by using crystalline hosts. Repetition rates up to 10Hz appear feasible[56]. A generic class of media that satisfy these requirements are transition metal ions doped into crystalline hosts. A specific example is the divalent vanadium ion (V^{2+}) in crystalline magnesium fluoride (MgF_2). The phonon terminated energy level scheme[58,59] for this system is illustrated in Fig. 22. Flashlamp pumping of the broad absorption bands of V^{2+}:MgF_2 lead to predictions[60] the overall system efficiency could exceed

10%. Recent experiments[61] with $V^{2+}:MgF_2$ indicate that the upper laser level remains radiative with a lifetime of 2.5ms for V^{2+} densities up to $7x10^{20}cm^{-3}$ and temperatures of 200°K. Unfortunately, during laser experiments an unidentified excited state absorption process was found which seriously reduces the extraction efficiency and increases the thermal loading of the medium. Although this renders $V^{2+}:MgF_2$ unacceptable theoretical arguments[56] suggest that $V^{2+}:KMgF_2$ may have an energy level structure which will avoid excited state absorption. Other Vanadium ion/host combinations are also currently being investigated[56].

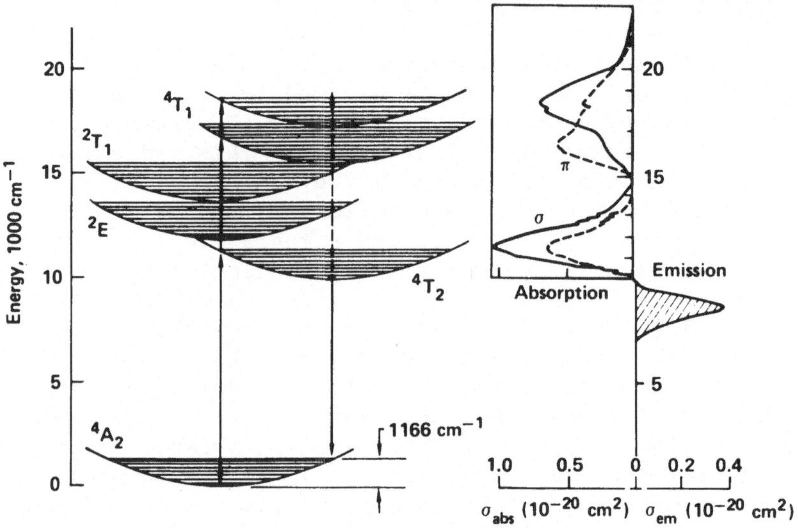

Fig. 22 Energy level scheme and absorption and emission band structure of the $V^{2+}:MgF_2$ phonon terminated laser[58,59].

Sensitizer/activator schemes[56] may also be used to increase flashlamp light absorption, but avoid excited state absorption. A sensitizer ion provides for efficient broadband absorption of pump radiation and rapidly transfers excitation energy to the activator ion. The latter is chosen to be free of energy levels that could lead to excited state absorption at either the pump or laser wavelengths. An example of this approach is the Yb^{3+}/Cr^{3+} activator/sensitizer pair[62]. In addition to flashlamps short pulse surface discharges[63] may be used to pump systems with visible and near UV wavelength transitions, where the activator ion upper laser level lifetime is only a few tens of microseconds. Another approach that has been proposed[56] uses efficient (>25%) narrowband semiconductor laser pump sources to directly excite activators. For example, the Nd^{3+} ion may be pumped by the GaAs semiconductor diode laser. Recent advances[56,64] in diode laser array technology make this a potentially interesting approach. The essence of this discussion is that there is good reason to believe that efficient (>10%), high average power (<10Hz) solid state lasers based on rare earth and transition metal ions doped into crystalline media may be feasible.

Nonstorage lasers are characterized by media which have high gain and short excited state lifetimes (<10ns). The pulse length for efficient operation is, therefore, considerably longer than required for the ICF application. Under these conditions, the amplification Eqs. (8-1) and (8-2) may be combined to yield steady state amplification

$$dI/dz = (\alpha - \gamma_a)I \qquad (8\text{-}5)$$

and

$$\alpha = \alpha_0/(1+I/I_s) \qquad (8\text{-}6)$$

where the small signal gain coefficient is $\alpha_0 = \sigma_s R_p \tau_u$ and $I_s = h\nu/\sigma_s\tau_u$ is the saturation intensity. The extraction efficiency is $\eta_{ext} = (I_{out} - I_{in})/\alpha_0\, LI_s$. In the small signal regime ($I<<I_s$) the laser beam intensity increases exponentially as it passes through the medium. As I approaches I_s the beam enters the efficient extraction regime. When $I>I_s$ stimulated emission exceeds upper laser level collisional and radiative relaxation losses. The signal increases nearly linearly with position then plateaus out at $I/I_s = (\alpha_0/\gamma_a)-1$. In the plateau regime stimulated emission equals photoabsorption. The extraction efficiency increases in the small signal regime maximizes and then decreases to zero in the plateau regime.

The rare gas halide (RGH) excimer lasers[65-67] are nonstorage lasers. The most promising RGH excimer for fusion is the KrF laser. The potential curves[68] for KrF are shown in Fig. 23. The KrF molecule can be regarded as a Kr^+F^- ion pair complex that emits to the repulsive covalent KrF ground state by transferring an electron from the F^- to the Kr^+. Gain at 248nm is provided by the strong $B_{1/2}{\rightarrow}X_{1/2}$ transition sometimes referred to as the $III(1/2) \rightarrow I(1/2)$ or $B^2\Sigma(1/2) \rightarrow X^2\Sigma(1/2)$ transition. The radiative lifetime of this transition is 9ns and the stimulated emission cross section is $2.5\times10^{-16}cm^2$. Recent experiments[69] indicate that vibrational relaxation processes may play an important role in energy extraction and the transition may not be homogeneously broadened for narrow band pulses.

The KrF molecule is synthesized by pumping an Ar rich mixture of $Ar/Kr/F_2$ (0.97/0.06/0.003) with a high energy electron beam. A schematic diagram of the kinetic chain[65-67] leading to KrF is shown in Fig. 24. The KrF molecule is formed primarily via the ion recombination reaction $Kr_2^+ + F^- + Ar \rightarrow KrF + Kr + Ar$. In addition to the kinetic processes responsible for the creation and destruction of KrF, several nonsaturable photoabsorbers[65-67] are created during the production of KrF. The most important nonsaturable photoabsorbers are F_2, F^-, Ar^{**}, Kr^{**}, Ar_2^+, Kr_2^+ and Ar_2F. The trimmer Kr_2F also absorbs but is saturated when the

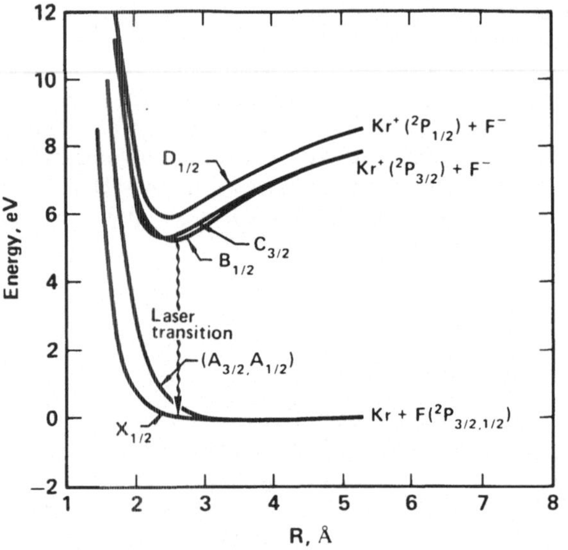

Fig. 23 Calculated potential curves for the low lying states of KrF[68].

density of KrF is reduced by stimulated emission. Measurement of the photoabsorption processes in excimer lasers is a continuing and critical area of research[70]. These loss processes along with the kinetic processes depicted in Fig. 24 determine the optimum operating conditions for the KrF laser in terms of $Ar/Kr/F_2$ fractions, other possible mixtures, total pressure and electron beam power deposition. Detailed studies[71] of KrF laser performance indicate that 20 to 25% of the electron beam energy deposited in the medium is channeled into KrF molecule formation. Values of α_o/γ_a in the range of 15 to 20 are produced for atmospheric pressure operation. These values of α_o/γ_a limit the output and extraction efficiency of practical devices to approximately $10MW/cm^2$ and 50% respectively. The combined production and extraction efficiencies produce an intrinsic efficiency (laser energy/e-beam energy deposited in the laser medium) of approximately 10%. The e-beam can be produced with an efficiency of about 70% so the electrical efficiency[65-67,71] of a KrF laser is in the range of 5 to 7%. If nonsaturable photoabsorption processes could be controlled higher KrF laser efficiencies could be achieved. Similar limitations exist for other rare gas halide excimer lasers.

Unfortunately, if short excitation pulses could be readily generated, the KrF amplifier output fluence would be only a fraction of a Joule per square centimeter. The application of a non-storage laser such as KrF to the generation of 5 to 20ns pulses requires an efficient method for conversion of KrF pulses of duration in excess of 100ns to the shorter pulses desired. In

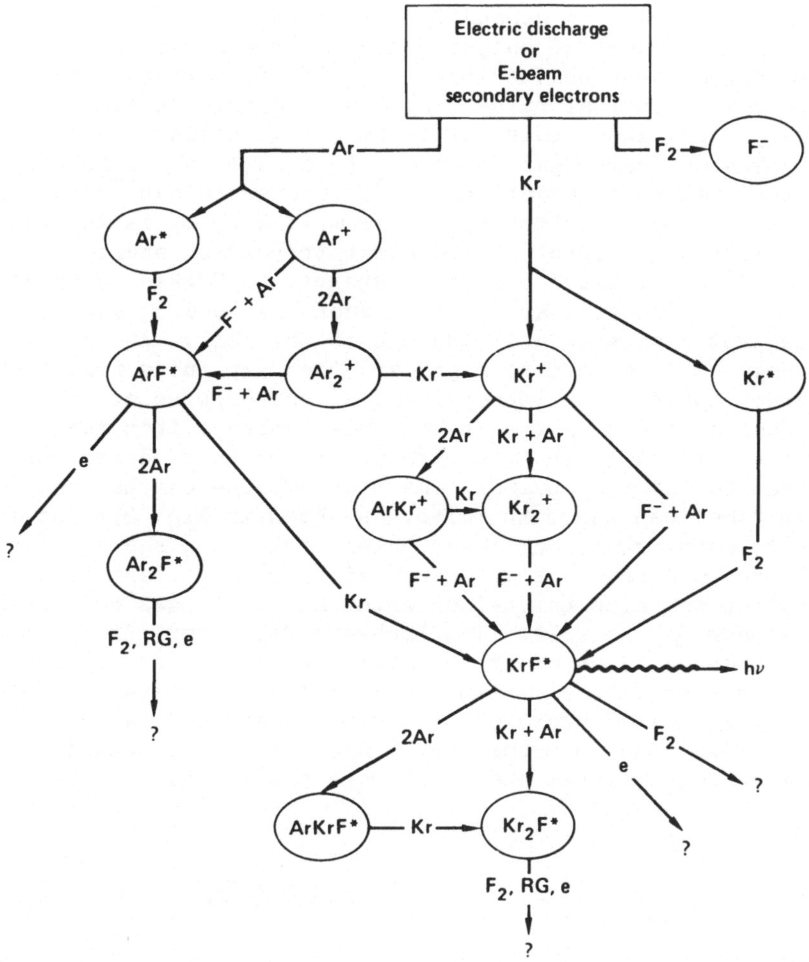

Fig. 24 Kinetic channeling in the electron beam pumped KrF excimer laser[65-67].

addition, some optical means of intensification of the KrF laser radiation is required (especially at shorter pump pulse lengths) in order to operate at laser output fluences close to damage limits. Two generic optical pulse compression techniques, Fig. 25, have been proposed: pulse stacking[72] and backward stimulated Raman scattering[73].

One pulse compression approach, Fig. 25a, is to extract energy from an RGH amplifier using a sequence of short (≈10nsec) pulses propagating through the amplifier at slightly different angles and different times with respect to each other. These angle-coded time sequenced pulses could then be applied to the target simultaneously (pulse stacked) by adjusting the propagation time to the target for each pulse. A second technique for effectively compressing a long (>100nsec) output pulse of an excimer laser is that of backward Raman scattering. Fig. 25b illustrates the principle involved in optical pulse compression using a backward Raman amplifier. A long laser pulse of duration τ_p, called the pump pulse, propagates from right to left into a cell containing a Raman active medium (interaction zone), such as methane[73] for KrF, producing gain at a frequency $\omega_p - \omega_r - \omega_s$. Here ω_p is the pump frequency, ω_r the frequency of the Raman transition excited, and ω_s is called the first Stokes frequency. A Stokes pulse at ω_s is injected into the cell at the time when the leading edge of the pump pulse just starts to exit the end of the cell. As the Stokes pulse propagates from left to right and the pump pulse from right to left, the Stokes pulse is amplified at the expense of the pump pulse. After a time $\tau_p/2$ the Stokes pulse emerges from the cell as an amplified pulse traveling in the backward direction with respect to the pump pulse. The Stokes pulse can be much shorter in time than the pump pulse, as shown in Fig. 25b, so the result is to compress a long laser pulse at ω_p to a short pulse at ω_s. There is a net loss of energy since $\omega_s < \omega_p$. Part of the pump radiation may also not be converted and will pass out through the exit window of the cell. The backward Raman amplifier can also be considered as an energy storage laser, but one in which the energy is stored in an intense collimated pump beam of photons rather than in excited atoms or molecules. Gain and saturation can then be discussed in terms very similar to those used to describe more conventional energy storage amplifiers.

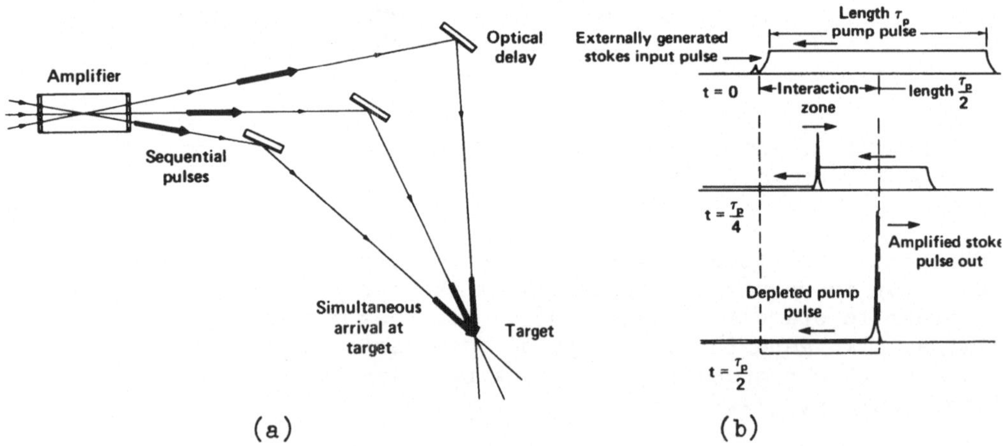

Fig. 25 Methods of optical pulse compression: (a) pulse stacking[72] and (b) backward stimulated Raman scattering[73].

The overall efficiency of the KrF laser is estimated to be in the range of 5 to 7%. This efficiency range is on the low side of the desirable range for a shortwavelength ICF driver. Currently the biggest leverage area for improvement in KrF laser efficiency is control of nonsaturable photoabsorption loss. The search for other more efficient nonstorage lasers should continue.

Free Electron Lasers

The free electron laser (FEL) is a promising new technology in which the energy of a relativistic electron beam is converted directly to coherent short wavelength radiation[74]. Such an approach circumvents the photoabsorption and optical transport difficulties that limit electronic transition lasers. As shown in Fig. 26, several electromagnetic generators have been developed that efficiently convert electron beam energy into radiation in the radio and microwave bands. Long wavelength radiation is now produced with an efficiency near 70% by radio tubes, gyrocons and klystrons. Gyrotrons and traveling wave tubes have operated at efficiencies of 20 to 40% in the microwave region. The FEL is a short wavelength analogue of these devices that calculations indicate may have a 20 to 40% conversion efficiency[75-78].

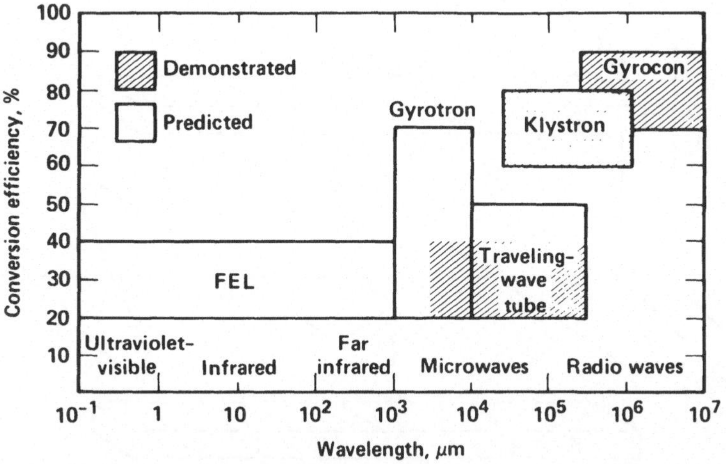

Fig. 26 Several electromagnetic radiation generators convert electron beam energy to radiation with high efficiency. Overall system efficiencies are less than shown here. For an FEL, the overall system efficiency is estimated to be in the 10 to 20% range[75-78].

Several FEL concepts have been identified. Currently the most promising approach consists of an electron beam, an input laser beam, and a transverse periodic magnetic field structure, usually called a wiggler, as illustrated in Fig. 27. The laser beam propagates in the same direction as the electron beam. The periodic magnetic field B_w gives the electrons a velocity component parallel to the laser field so that the laser radiation field and the electrons may exchange energy. If a precise relationship[75-78]

$$\lambda = (\lambda_w/2\gamma^2)[1 + (eB_w\lambda_w/2\pi mc)^2/2] \tag{8-7}$$

is maintained between the energy of the electron $\gamma = E/mc^2 = [1-v^2/c^2]^{-1/2}$, the period λ_w and strength B_w of the magnetic field an the wavelength of the laser field, then the laser beam's radiation field acts to retard the electron's oscillatory motion as it passes through the wiggler. The kinetic energy lost by the electron as it decelerates is transferred to the laser beam thereby amplifying it. At non-relativistic energies ($\gamma \simeq 1$), the synchronism condition, Eq. (8-7), specifies that the period of the wiggler field must be approximately the same as the wavelength of the laser light. This causes very serious technological difficulties for laser wavelengths less than several millimeters. These difficulties are overcome by using a relativistic electron beam (γ>>1).

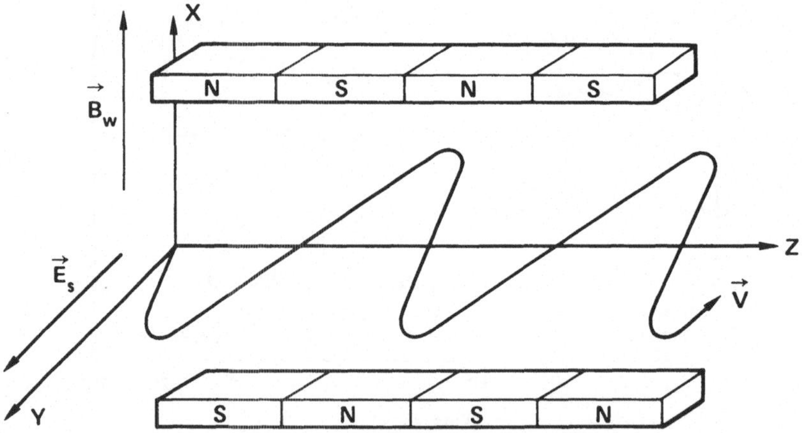

Fig. 27 Energy transfer in the FEL is accomplished by giving the electrons a velocity component parallel to the laser electric field.

The synchronism condition, Eq. (8-7), raises an important design issue. If the wiggler period and field are constant then as the electron beam loses energy and slows down, it no longer satisfies the synchronism condition. Only a small percentage

(<1%) of the electron beam's energy may be transferred to the laser beam. To increase the extraction or energy conversion efficiency to the level required for inertial confinement fusion, the wiggler parameters must be adjusted to ensure that the synchronism condition is maintained as the electron beam loses energy. This approach is known as a variable parameter or tappered wiggler FEL[76].

When the synchronism condition is maintained with a tappered wiggler moving potential wells, called buckets are formed. These buckets trap and bunch the electron beam. Energy is extracted from the entire bunch. Inside the tappered wiggler, the buckets decelerate as the electron beam loses energy. Calculations[77,78] suggest that greater than 80% of the electrons can be trapped in bunches and 50% deceleration of these bunches is possible. This process is identical to that responsible for trapping electrons in accelerating buckets in a linac. The size of the potential well and the resultant input laser intensity required to create the buckets is directly related to the quality of the input electron beam's energy spread and emittance[77]. Higher quality beams require smaller buckets.

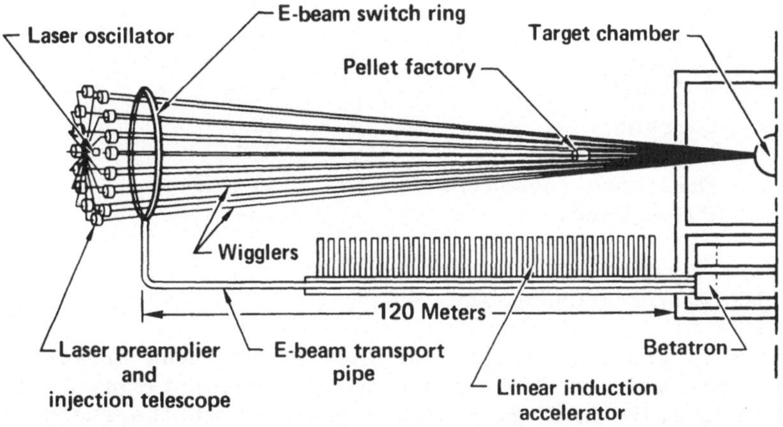

Fig. 28 Free electron laser-reactor configuration (one-half of a 3MJ driver; Table 8)[78].

A sketch of one-half of a 3MJ, 250nm FEL fusion driver[78] is shown in Fig. 28. The laser amplifier consists of 15, 120m long wigglers, each producing an output of 100kJ. The FEL amplifiers are arranged in a cone shaped distribution around a pellet factory and injector. The electron source consists of a 1.1GeV betatron and a 50 to 100MeV linear induction accelerator (linac) which serves as the injector for the betatron. An alternative source is a 1.1GeV linac. The long (270ns) accelerator pulse is transported to the electron beam switch ring where fast magnetic switches divide the pulse into 15, 15ns, flat top e-beam pulses. These are then magnetically guided into the wigglers. The FEL amplifier laser output proceeds directly to the target. The only optical components are in the laser preamplifiers and injector telescopes which are over 130 meters from the reactor vessel. Fast acting values prevent reactor contaminants from traveling back through the wiggler to the preamplifier window. The principle e-beam and laser characteristics of each beam line, as determined by 1D numerical calculations, are shown in Table 8. This system is estimated to have an efficiency of approximately 15%. After extensive analysis, several experiments are now underway to test the scientific feasibility of the tappered wiggler FEL concept.

Table 8. 250nm FEL Fusion Driver Characteristics[78]

Amplifier	
Length	120m
Wiggler Period	15cm
Wiggler Field (variable)	2.5-5kG
I_{in}	10 GW/cm^2
I_{out}	34 TW/cm^2
E_{out}	100 kJ (15ns, .5cm diameter)
Accelerator	
Voltage	1.1GeV
Current	20kA
Energy Spread	0.5%
Emittance (normalized)	.1 rad-cm
Pulse Length	.5-.6μs

9. ION BEAM DRIVER TECHNOLOGY

In order to provide reactor service, an ion beam driver must deliver several hundred TW of power to the fusion target. In addition, to achieve high gain target performance the energy of the particles must be chosen so that their range and specific energy deposition in the target are approximately 0.1mg/cm^2 and 20 MJ/g, respectively. This requires high currents of low energy light ions or low currents of high energy heavy ions. For example, to achieve the above deposition conditions and deliver 200 TW to the target, 40MA of 5MeV protons, 1MA of 200MeV Ne ions or 20kA of 10GeV U ions are required. The technologies required to generate, accelerate transport and focus these light and heavy ion beams are vastly different[5,6,46,49,79-96]. For energies below about 10MeV, pulsed diode accelerator technology[80-83] is well developed and capable of producing beam currents of about a MA. For higher energies micro-wave linear accelerators and sychrotrons are capable of producing very high kinetic energy, but are generally limited to relatively small beam currents (<10mA). These capabilities do not directly meet the needs of ion beam fusion, although the possibility of accumulating particles in a high current storage ring appears feasible[5,6,46,89-96]. An alternative approach is direct acceleration in a linear induction accelerator[5,6,89-96]. Recently a new accelerator concept, known as PULSELAC[82], for intermediate mass ions has been proposed. PULSELAC combines the ion diode and induction accelerator technologies to accelerate high currents of intermediate mass ions.

Light Ion Beam Generation and Transport

Light ion beam diode generator technology[5,6,49,80-84] contains four basic elements: A capacitive energy storage system, pulse forming lines, a high power transmission line and a diode. With the exception of the diode, the technology is very similar to diode electron beam generator technology. The capacitive energy storage system is typically a Marx generator that charges capacitors in parallel, and then discharges them in series to achieve the high voltages required for particle acceleration. The rise time of the Marx generator pulse is slow (≃500ns) and therefore the voltage pulse must be compressed in time. To accomplish this, the Marx generator, switched by a UV laser triggered gas switch, pulse charges a transmission or pulse line. The line usually contains a water dielectric, but options such a solid dielectrics and other liquids are available. The pulse energy is compressed in stages to the desired width by a sequence of pulse forming lines (PFL's). These stages are coupled by closing switches. Water switches are used in contemporary high-power generators. However, they are not capable of repetitive operation. In the future very high power, repetitive, efficient, long life gas or magnetic switches will be required. High power gas switching is currently under development at power levels of a few TW, and repetitive gas switching at peak power levels, about a GW, have been achieved. Gas switches have demonstrated lifetimes of 10^6 to 10^7 shots.

The shaped, high voltage pulse is transmitted to a final output line that carries the power to a magnetically insulated vacuum transmission line (MITL). The output line is separated from the MITL by an insulator. The power per unit area that can be transmitted through the vacuum insulator is a key limitation to the system since this electromagnetic intensity is much less than that required at the diode. The MITL tapers down as it transports the high voltage pulse to the diode. The MITL works in the following way. If the voltage pulse on a vacuum transmission line creates an electric field less than about 25MV/m, only the displacement current flows. The line behaves as a classical transmission line. However, if the voltage pulse creates an electric field greater than 25MV/m, the surface of the transmission line emits electrons, and a conduction current flows across the vacuum gap. If the voltage source has sufficiently low impedance, the current increases rapidly to a critical value at which the electrons are deflected in their self-magnetic field. This self-magnetic field terminates the initial current flow. Once the loss current is established, the power flow in the MITL is nearly 100% efficient. For example, in PBFA-I[80] the same power that would require 4000cm^2 of insulator surface in a traditional transmission line can be carried by 50cm^2 of MITL at an electrical stress of 200MV/m and an intensity of 16GW/cm^2. When field emission occurs in an MITL, a surface plasma forms that provides the necessary current. This plasma eventually expands accross the vacuum electrode gap shorting the MITL. This phenomenon limits MITL performance and is a subject of current research. In general, the transmission of electromagnetic energy from the capacitive energy store to the diode is expected to be very high (approximately 70 to 80%).

The high voltage electromagnetic energy from the MITL is converted into ion beam energy in a diode. The technology of light ion diode accelerators is evolving very rapidly[5,6,49,80-84]. In a time span of 5 years, ion diode power levels have increased from 100MW to greater than 1TW. The efficient generation of an intense ion beam depends critically on the ability to accelerate ions with an electrostatic field while constraining the electrons. For example, if the ion diode is operated in a space charge limited mode with bipolar flow, electron and ion currents, the ratio of the ion current density to the electron current density is $(m/m_i)^{1/2}(Z/A)^{1/2}$ where Z is the ion charge and A its atomic number. For protons this means that the backward proton current carries about 2.3% of the total power flow. Thus for efficient ion diode operation, the net electron flow must be almost completely suppressed. To date, three generic ion diode types, shown in Fig. 29, have been developed.

In the reflex diode[49,81-83], the net electron current is

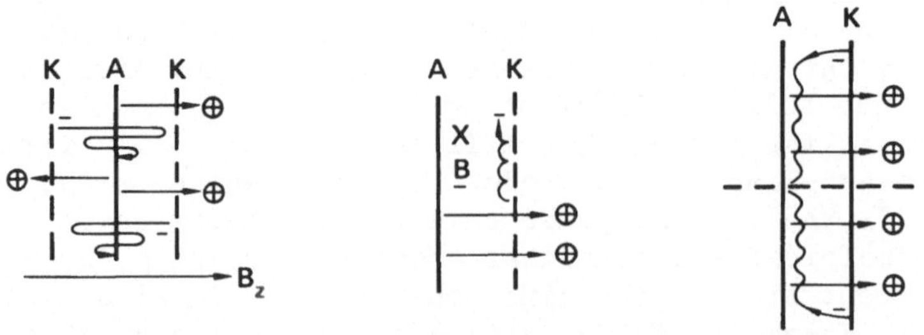

(a) Reflexing-electron (b) Magnetically-insulated (c) Pinched electron beam

Fig. 29 Ion Diode Concepts[49,81,82]

suppressed by having the electrons reflex in a potential well centered on the anode. Electrons emitted by one of the cathodes pass through the anode and lose a small amount of energy so they cannot reach the opposing cathode. The electrons continue to oscillate with decreasing amplitude until captured by the anode. In this manner the electrons provide charge neutralization for the ions, but their net current is severely reduced. The electrons hitting the anode create a plasma that emits the ions. The scheme shown in Fig. 29a accelerates ions in both directions. The energy loss due to the second beam can be overcome by creating a virtual cathode as employed in the reflex tetrode ion injector.

In the magnetically insulated diode[49,81,82] a transverse magnetic field prevents the electrons from flowing across the acceleration gap. The field strength is choosen so that the ions are essentially unaffected. The simplified version shown in Fig. 29b would only work for a short time until negative charge accumulation at the top produced electrical breakdown. The ExB drift of electrons can be used to create a cylindrical drift motion about the acceleration axis. This avoids charge accumulation. In the magnetically insulated diode, the ions are created by a surface discharge. The cathode is a thin foil or mesh transparent to ions.

In the pinched electron beam diode[49,81,82], the electron current exceeds the value, $I(A) = 17{,}000\beta\gamma(r/2d)$. The quantity d is the anode to cathode spacing, and r is the cathode radius. The selfmagnetic field of the electron beam becomes so large that the electron flow becomes largely radial as shown in Fig. 29c. The net electron flow can therefore be much less than the bipolar electron value. However, the electron density still contributes to the needed space charge field.

After the ion beam is generated, it must be transported to the target. The various ion beam propagation modes have been characterized by Olson[49] and are illustrated schematically in Fig. 30. Studies suggest[49,83] that for light ion fusion beam propagation in a preformed channel is preferred. A high conductivity current carrying channel is created prior to injection of the ion beam. The channel may be produced by an electric discharge that is initiated along a preionized path created by laser photoabsorption. Several methods[85-87] of laser initiated plasma channel formation have been identified: (1) initiation and guiding of an electrical discharge by CO_2 laser heating of a molecular gas[85,87] (NH_3, C_2H_4, etc.), (2) tunable dye laser ionization of alkali atoms (Li, Na, etc.) based on resonance saturation[86] and (3) direct (ArF laser at 193mm) or multiphoton (UV/visible) photoionization of lithium[84]. Ideally the ion beam is charged and current neutralized as it propagates through the discharge plasma channel. The net current is just the original discharge channel current. Then the magnetic field associated with this current confines the beam.

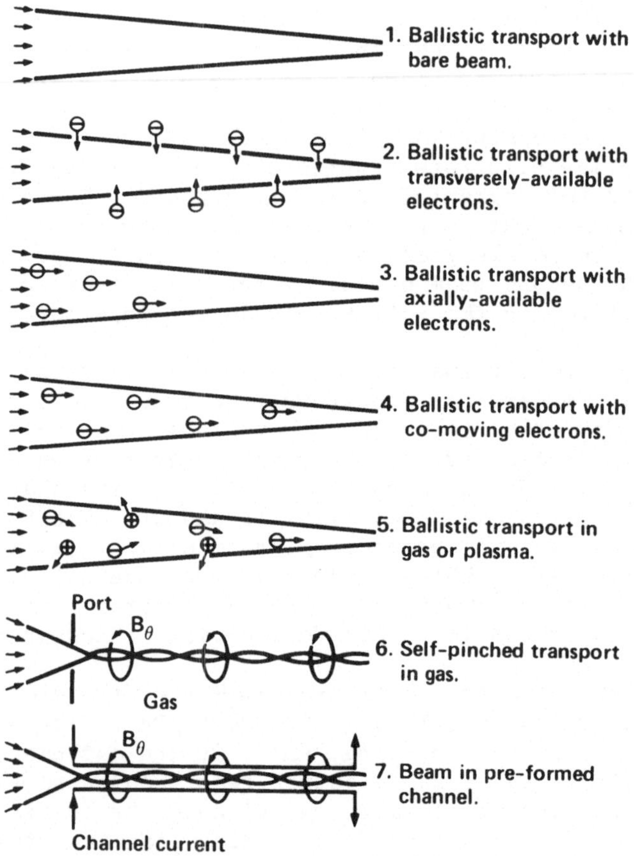

Fig. 30 Ion Beam Propagation modes[49]. The small circles with the minus (plus) signs represent electrons (ions).

Fig. 31 shows a conceptual schematic of a light ion pinch reflex diode[83], its focusing-drift region and transport channel. The pinch reflex diode is an advanced version of the pinched electron beam concept shown in Fig. 29c. Ions are accelerated axially across the diode vacuum gap by the applied electric field, and are accelerated radially inward by their azimuthal selfmagnetic field. The transmission cathode separates the acceleration region from a low pressure, gas filled drift section. Beam ionization of the gas reduces the self-magnetic field by about two orders of magnitude so that the ion orbits in the drift section are nearly straight. The anode and cathode structures are shaped so that ions leaving the diode converge to a focus at the entrance to the transport channel. Beam ions are magnetically confined in the centimeter diameter discharge channel and transported to within a few centimeters of the target. The maximum injection angle of the ions into the channel must be <0.2 rad so that the discharge current can be kept below instability limits

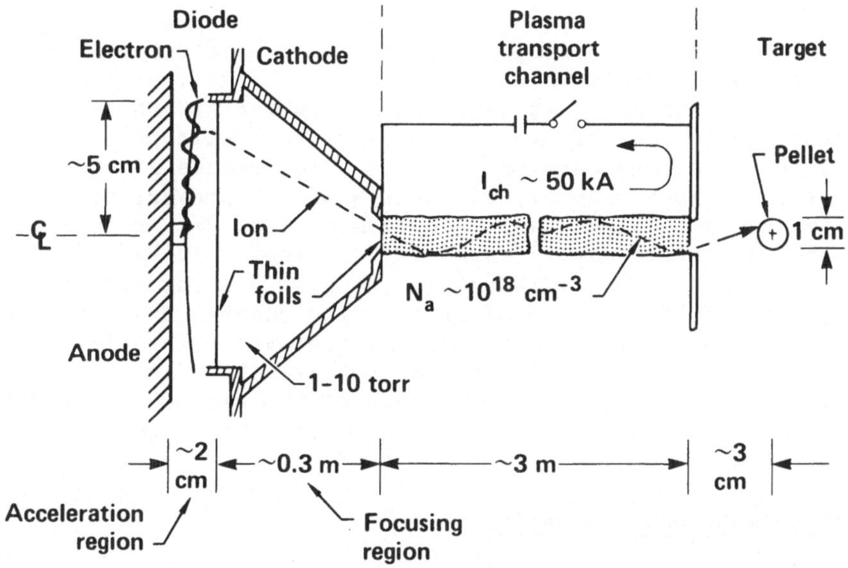

Fig. 31 Conceptual schematic of a light ion pinch-reflex diode with focusing drift region and preformed transport channel[83].

(<50kA), yet suffice to confine the beam[5,6,83,87]. Keeping the injection angle small also prevents excessive ion beam expansion between the channel exit and the pellet. For such injection and current conditions, light ions can propagate in a range of channel-plasma density conditions limited from below by neutralization requirements and above by excessive beam energy loss. Since the ions are not relativistic, increasing the diode acceleration voltage during the ≈50ns injected ion pulse permits power multiplication by a factor of 5, during transport by axial bunching, to the ≈10ns required to drive the target. Substantial power multiplication by bunching requires small injection angles and axial velocity dispersion during transport[83,88]. Further power multiplication can be achieved by overlapping the ion beams on target[88]. To date pinch reflex diode experiments[49,82,83] have produced proton currents up to 700kA at 1.3MV with pulse lengths of ≈50ns. Ballistic focusing through a gas filled drift region has produced 300kA/cm^2, or approximately 0.4TW/cm^2. The protons were produced with an efficiency of 66%. In present experiments the anode foil is destroyed on every shot and therefore for repetitive operation a method must be found to repetitively create the anode ion plasma. Other diode concepts[6,49,81,82] such as the AMPFION (auto-magnetic, plasma filled, ion) diode are being considered.

The transport of proton beams in CO_2 laser initiated discharge channels, in 4 to 20 torr ammonia gas, has recently been studied[85,87]. Total current transport efficiencies up to 50%

were achieved in 13 to 45kA, one meter long discharges. At currents above 30kA the transport efficiencies became less reproducible, apparently due to a kink mode of discharge instability that was observed to occur at these higher currents. In these experiments, 50 to 70kA of 0.7 to 0.8MeV protons with pulse durations of 60 to 80ns were transported. To produce a total power on target, in a reactor, of 200TW with 20 nonoverlapping 5MeV proton beams requires that each beam deliver 2MA or 10TW to the target. Alternatively, to achieve $100TW/cm^2 = 10^{14}W/cm^2$ of beam incident on a target, $2TW/cm^2$ ion sources are required if temporal bunching by a factor of five during channel transport and a beam overlap gain factor of ten can be achieved. The pinch reflex diode experiments cited earlier achieved $0.4TW/cm^2$. If preformed channel transport efficiencies of 50% can be achieved at the power levels required for reactor service, then light ion driver efficiencies near 20% will be attained[84].

Intermediate Ion Beam Generation

Recently a new ion accelerator concept called the PULSELAC[82] or pulsed linear induction accelerator, was proposed. In this approach, the ion source produces a plasma of intermediate mass ions such as Neon. The ions are extracted from the plasma and accelerated down an annular accelerator consisting of a series of induction driver accelerator gaps. Inbetween the acceleration gaps the beam is neutralized to permit high current transport. The acceleration gaps are magnetically insulated to resist electron current buildup and permit efficient ion beam acceleration. This approach represents a novel synthesis of high current magnetic insulated diode technology and high energy induction linear accelerator technology. The ion source and accelerator can be operated at the repetition rates required for ICF. Once produced, the ion beam must be focused, bunched, and transported to the pellet. A recent reactor study[84] indicates that a PULSELAC accelerator to produce a 100TW Neon ion beam with a current of 300kA would be about 50m long. As with other electromagnetic particle accelerators, the efficiency of PULSELAC drivers is predicted to be greater than 20%. Experimental development of the PULSELAC concept is presently in progress[82].

Heavy Ion Generation and Transport

A typical heavy ion fusion driver consists of several distinct elements[5,6,89-96]: ion source and preaccelerator, low velocity (low β) accelerator, high velocity (high β) accelerator, beam manipulators (injectors, stackers, bunchers, accumulators and compressors) and a final focusing system. The ion source usually consists of a discharge produced plasma from which the desired ions are extracted by an applied electric field of several tens of kilovolts. Generally, the velocity of these ions is too low and their space charge repulsion too high to permit direct injection into the main accelerator stage. Thus the ion source is followed by a preaccelerator, usually a Cockroft-Walton, which accelerates the ions to around 1MeV. The low β accelerator further boosts the ion energies. The high β or main accelerator is the subsystem which gives the ions the largest fraction of their final kinetic energy. Accelerator systems are often classified according to their main accelerator type. Three accelerator schemes have been considered for heavy ion fusion driver service: rf linac, synchrotron and induction linac. After the ions have been accelerated and bunched to their final energy and current, they are transported and focused onto the target through the reactor chamber. As noted, above several different types of electromagnetic devices are used to manipulate the ion beam, to couple the various acceleration stages and increase the current density of the ion beam.

Although the ion currents (>10kA) required for heavy ion fusion are much less than those required (≃1MA) for light ion fusion, they are large currents by conventional accelerator powers. At high currents of interest for heavy ion fusion, the beam plasma frequency is large enough so that the Laslett tune shift phenomenon[6], the longitudinal resistive instability[6] and other non-neutral collective effects[6] must be seriously considered. The requirement to transport and control the high current and power ion beams with a magnetic quadrupole system, due to the practical constraint on field strength, places a limit[94] (Masche limit) on the transportable beam power. This maximum transportable beam power scales as $(\gamma-1)(\beta\gamma)^{5/3}$ and is a steep function of ion kinetic energy. This limit appears when the electrostatic selfrepulsion of the beam ions becomes comparable to the magnetic restoring force of the magnetic transport lenses. For heavy ions this constraint usually limits individual beam powers to the range of 10 to 20TW. Fortunately, accelerator scientists[95] have found that this limit can be satisfied and the cost of the accelerator reduced by accelerating clusters of beamlets which thread the same accelerator structure and are separately focused onto the target.

Of the general accelerator types only the rf linac and linear induction accelerator are currently considered as serious candidates. The heavy ion synchrotron has recently been ruled out[93] because it needs a high (≃2GeV) injection energy. Consequently, it has a small energy gain which does not outweigh the extra beam manipulations required to couple it into a driver system.

The rf linear accelerator, Fig. 32a, consists of a series of cylindrical resonant cavities excited at frequencies up to several hundred MHz. The ions are accelerated by the high frequency electric field which is uniform and axial in the region of the beam. Each cavity is a series of small cylinders or drift tubes through which the beam passes. These cylinders shield the beam from the high frequency field during its decelerating phase. The low β

section usually consists of several Wideröe linacs operated at frequencies in the 10 to 50MHz range. The high β main accelerator consists of several linearly coupled Alvarez linacs with a current of 100mA. rf linacs are inherently low current devices. Following this point, a compression system must be used to amplify the beam current the required factor of 10^5 to 10kA. In one system[46,96], Fig. 32a, the beam is transferred to ten storage rings via five intermediate condenser rings. Inside these rings, the beam is stacked in both vertical and horizontal phase planes. The current is further increased and pulse length decreased by rf bunchers in the rings. An important consideration during this rather lengthly period (≃1ms) is that the particles remain bunched for many revolutions. During this period beam ions may be lost due to ion-gas collisions and ion-ion collisions (within a bunch) of the charge exchange type[89-92]: $2A^+ \rightarrow A + A^{2+}$. Currently these reaction processes are being considered for many ion types in order to establish the optimum ion species.

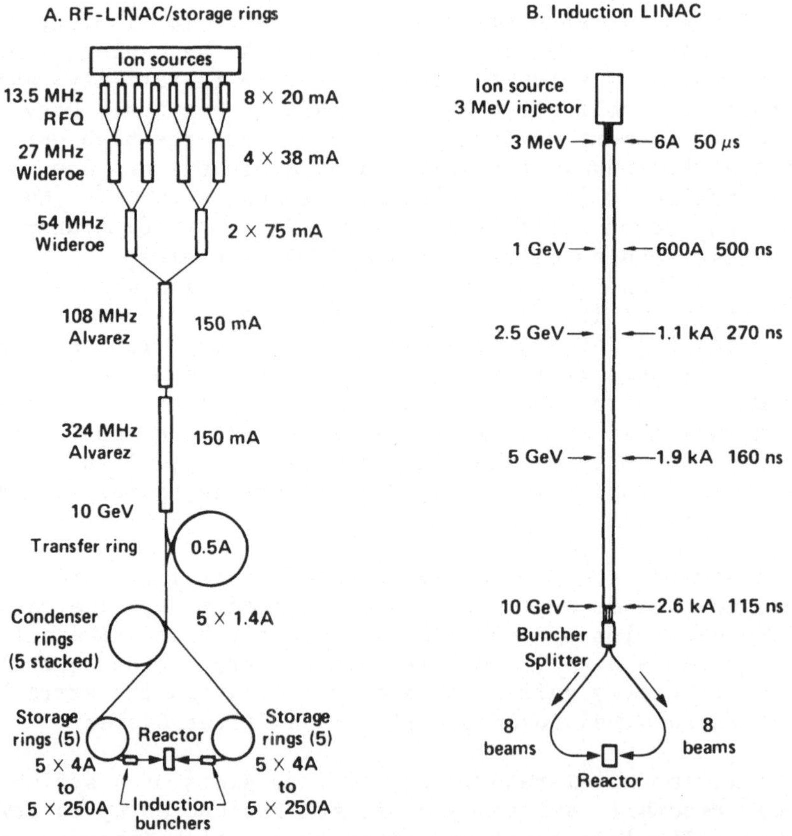

Fig. 32 Schemes of two proposed[46,95] heavy ion fusion drivers: (a) a 7MJ, rf-linac/storage ring type system[46,89-94,96] first proposed by Masche and, (b) a single pass 3MJ induction linac system[89-95].

After accumulation the ten beams are simultaneously extracted from the storage rings and directed to a pulse compressor and then to the target. The pulse compressor consists of a ramped voltage induction linac.

The second type of linear accelerator capable of heavy ion fusion service is the induction linac[89-95], Fig. 32b. The induction linac consists of a sequence of accelerator modules. The accelerating electric field in each module is produced by a changing magnetic field inside a ferromagnetic ring which surrounds the beam. The accelerating action is analogous to the operation of a transformer consisting of a one turn primary (ferromagnetic core) and a one turn secondary winding (ion beam). Induction linacs are capable of high current single pass acceleration. Amplification of the ion beam current pulse can be produced[95] by ramping the accelerating voltage pulse as the beam passes. This accelerates the tail of the beam so that it catches up with the front of the beam. Current amplification of a factor of 500 is believed possible by this method.

Both the rf and induction linacs may be operated at high repetition rate (>10Hz) and high efficiency (20-30%). The key technology issue for this approach appears to be achievement of high peak power and low cost. After acceleration, the heavy ions must be transported to the target. Of the transport mechanisms outlined in Fig. 30, it appears[49] that ballistic focusing with or without charge neutralization is feasible. If the background gas pressure is the 10^{-1} to 1 torr range, stable propagation may also be possible. Calculations[49] also suggest that in the 1 to 10 torr range, a self-pinched mode of propagation may occur.

Although no target experiments have been performed with heavy ions, it has recently been proposed by the heavy ion fusion community that an ion induction accelerator with an energy of 2 to 5 kJ be built. Such a device would possibly use 100-200MeV Na or K ions and produce plasma temperatures of ≈50 to 100eV in a small spot. Such a device would test some accelerator issues and validate the important issues of ion energy deposition in hot matter. In view of the promise of heavy ion fusion, the importance of these experiments cannot be underestimated.

10. SUMMARY AND CONCLUSIONS

Over the past decade the evolution of inertial confinement fusion has been very rapid. The original concept, which utilized a laser to drive the implosion, has been greatly expanded to include ion beam drivers as well. In inertial confinement fusion, the reactor and driver are physically separated leading to many novel reactor concepts. Because of their ability to flexibly provide the driving conditions, lasers have been used extensively to experimentally explore the feasibility of inertial confinement fusion. Laser driven implosions have achieved compressions of 100 times the density of solid DT. The next generation of laser, Nova[14], and light ion beam, PBFA-II[80], drivers are expected to closely approach and possibly achieve scientific breakeven. The results of these experiments will permit target designers to predict with confidence a target and a driver that will be required to achieve the high gain required for reactor service. Presently it is believed that achievement of high gain will require a megajoule class driver with high enough efficiently (>10%) to make electricity economically. Current driver efficiency and knowledge of energy deposition favors heavy ion beams. However, experimental validation of these prejections remains to be accomplished. Short wavelength lasers also produce nearly classical drive conditions and several novel concepts, such as the FEL, appear capable of high efficiency operation. The successful development of inertial fusion clearly depends on the development of several diverse technologies; targets, drivers, reactor chambers, etc. Development of these technologies depends on understanding many different atomic and molecular collisional and transport processes occuring under a very large range of physical conditions of density and temperature, Table 1.

11. ACKNOWLEDGEMENT

The author's understanding of ICF is due in large measure to many discussions with his colleagues at the Lawrence Livermore National Laboratory and in the ICF community. The author is grateful to J.T. Hunt for his support and to R.M. More for his encouragement and helpful advice during the preparation of this manuscript. The author also thanks D.L. Cook, J.D. Lindl, M.J. Monsler, and J. W.-K. Mark for helpful discussions during the preparation of this manuscript. The author especially thanks Nancy Lusby for her patience and care in preparation of the text.

This work was performed under the auspices of the U.S. Department of Energy by the Lawrence Livermore National Laboratory under contract NO. W-7405-ENG-48.

12. REFERENCES

1. J.T. Hogan, "General Principles of Magnetic Confinement", this volume.
2. E. Teller, ed., Magnetic Confinement Fusion, Vol. 1, Parts A and B, Academic Press, 1981.
3. J. Nuckolls, L. Wood, A. Thiessen, and G. Zimmerman, Nature, 239, 139 (1972).

4. J. Nuckolls, in Laser Interaction and Related Phenomena (H. Schwartz and H. Hora, eds.) Vol. 3B, 399, Plenum Press, 1974.
5. J.J. Duderstadt and G.A. Moses, Inertial Confinement Fusion, John Wiley & Sons, 1982.
6. D. Keefe, Ann. Rev. Nucl. Part. Sci., 32, 391 (1982).
7. Laser Program Annual Reports-1977 to 1981, Law. Liv. Natl. Lab Rpts. UCRL-50021-77 to 81.
8. D.T. Attwood, IEEE J. Quant. Electron., QE-14, 909 (1978).
9. H.G. Ahlstrom, Physics of Laser Fusion Vol. II, "Diagnostics of Experiments on Laser Fusion Targets at LLNL", UCRL-53106, 1982.
10. R.M. More, "Atomic Processes in Dense Plasmas", this volume.
11. R.M. More, "Atomic Processes in Inertial Confinement Fusion", in Applied Atomic Collision Physics (H.S. Massey, ed.) Vol II, Academic Press, 1982 and Law. Liv. Natl. Lab. Rpt. UCRL-84991.
12. S. Glasstone and R.H. Lovberg, Controlled Thermonuclear Reactions, Van Nostrand, 1960.
13. R.W. Conn, in Magnetic Confinement Fusion (E. Teller, ed.), Vol. 1, Part B, Ch. 14, p. 194, Academic Press, 1981.
14. J.H. Nuckolls, Physics Today, 35, 24 (1982).
15. M.J. Monsler, J. Hovingh, D.L. Cook, T.G. Frank, and G.A. Moses, Nucl. Tech./Fusion, 1, 302 (1981).
16. J. Maniscalco, Nucl. Tech., 28, 98 (1976).
17. D.H. Berwald and J.A. Maniscalco, Nucl. Tech./Fusion, 1, 137 (1981).
18. J.D. Lawson, Proc. Phys. Soc. (London) B70, 6 (1957).
19. J.D. Lindl, private communication.
20. R.E. Kidder, Nucl. Fusion, 19, 223 (1979).
21. J.M. Auerbach, W.C. Mead, E.M. Campbell, D.L. Matthews, D.S. Bailey, C.W. Hatcher, L.N. Koppel, S.M. Lane, P.H.Y. Lee, K.R. Manes, G. McClellan, D.W. Phillion, R.H. Price, V.C. Rupert, V.W. Slivinsky, and C.D. Swift, Phys. Rev. Lett., 44, 1672 (1980).
22. G.B. Zimmerman and W.L. Kruer, Comm. Plasma Phys. Cont. Fusion, 2, 85 (1975).
23. K.A. Brueckner and S. Jorna, Rev. Mod. Phys., 46, 325 (1974).
24. R.E. Kidder, Nucl. Fusion, 16, 405 (1976).
25. S.E. Bodner, J. Fus. Energy, 1, 221 (1981).
26. J. Meyer-Ter-Vehn, Nucl. Fusion, 22, 561 (1982).
27. G.S. Fraley, E.J. Linnebur, R.J. Mason, and R.L. Morse, Phys, Fluids, 17, 474 (1974).
28. J. Lindl and W. Mead, Phys. Rev. Lett., 34, 1273 (1975).
29. Yu. Y. Afanasev, N.G. Basov, P.P. Volosevich, E.G. Gamalii, O.N. Krokhin, S.P. Kurdyumov, E.I. Levanov, V.P. Rosanov, A.A. Samarskii, and A.N. Tikhonov, JETP Lett., 21, 68 (1975).
30. J.D. Lindl, "Low Aspect Ratio Double Shell Targets For High Density and High Gain and a Comparison with Ultra Thin Shells", Law. Liv. Natl. Lab. Rpt. UCRL-79735, Sept. 30, 1977.
31. B.H. Ripin, R. Decoste, S.P. Obenshain, S.E. Bodner, E.A. McLean, F.C. Young, R.R. Whitlock, C.M. Armstrong, J. Grun, J.A. Stamper, S.H. Gold, D.J. Nagel, R.H. Lehmberg, and J.M. McMahon, Phys. Fluids, 23, 1012 (1980).

32. C.E. Max, The Physics of Laser Fusion Vol. 1, "Theory of the Coronal Plasma in Laser Fusion Targets", Law. Liv. Natl. Lab. Rpt. UCRL-53107, Dec. 1981.
33. J.D. Lindl, Nucl. Fusion, 14, 511 (1974).
34. C.E. Max and K.G. Estabrook, Comm. Plasma Phys. Cont. Fusion, 5, 239 (1980).
35. C. Garbon-Labaune, E. Fabre, R. Fabbro, F. Amiranoff, and M. Weinfeld, in Rapport d'Activite 1979, GRECO Interaction Laser-Matiere (Ecole Polytechnique, Palaiseau, France), p. 64.
36. E.M. Cambell, W.C. Mead, R.E. Turner, D.W. Phillion, C.E. Max, F. Ze, K. Estabrook, G. Tirsell, B. Pruett, V.C. Rupert, P.H. Lee, W.L. Kruer, C.W. Hatcher, G. Hermes, 3rd U.S.-Japan Seminar on Theory and Application of Multiply Ionized Plasmas Produced by Laser and Particle Beams, Nara, Japan, May 3-7, 1982, Law. Liv. Natl. Lab. Rpt. UCRL-87219.
37. W.L. Kruer, Comm. Plasma Phys. Cont. Fusion., 6, 167 (1981).
38. W.L. Kruer, Comm. Plasma Phys. Cont. Fusion., 5, 69 (1979).
39. B.H. Ripin, F.C. Young, J.A. Stamper, C.M. Armstrong, R. Decoste, E.A. McClean, and S.E. Bodner, Phys. Rev. Lett., 39, 611 (1977).
40. W.L. Kruer, K.G. Estabrook, and K.H. Sinz, Nucl. Fusion, 13, 959 (1973).
41. J.J. Thomson, W.L. Kruer, S.E. Bodner, and J.S. DeGroot, Phys. Fluids, 17, 849 (1974).
42. J.J. Thomson, Nucl. Fusion, 15, 237 (1975).
43. R.H. Lehmberg and S.P. Obenschain, "Use of Induced Spatial Incoherence for Uniform Illumination of Laser Fusion Targets", Naval Research Laboratory (to be published).
44. R.O. Bangerter, J.W.-K. Mark, and A.R. Thiessen, Laser Program Annual Report-1980. Law. Liv. Natl. Lab. Rpt. UCRL-50021-80, p. 3-17.
45. J.W.-K. Mark, Proceedings of the Symposium on Accelerator Aspects of Heavy Ion Fusion, Gesellshaft für Schwerionenforschung, Darmstadt, West Germany, Mar. 29-Ap. 2, 1982.
46. B. Badger, F. Arendt, K. Becker, K. Beckert, R. Bock, D. Bohne, I Bozsik, J. Brezina, M. Dalle Donne, L. El-Guebaly, R. Engelstad, W. Eyrich, R. Frohlick, N. Ghoniem, B. Goel, A. Hassanein, D. Henderson, W. Hobel, I. Hoffman, E. Hoyer, R. Keller, G. Kessler, A. Klein, R. Krentz, G. Kulcinski, E. Larsen, K. Lee, K. Long, E. Lovell, N. Metzler, J. Meyer-Ter-Vehn, V. Von Mollendorff, N. Moritz, G. Moses, R. Muller, K. O'Brien, R. Peterson, K. Plute, L. Pong, R. Sanders, J. Sapp, M. Sawan, K. Schretzmann, T. Spindler, I. Sviatoslavsky, K. Symon, D. Sze, N. Tahir, W. Vogelsang, A. White, S. Whitkowski, and H. Wollnik, HIBALL-A Conceptual Heavy Ion Beam Driven Fusion Reactor Study, Univ. of Wis. Rpt. UWFDM-450, 1981.
47. J.A. Swegle, Comm. Plasma Phys. Cont. Fusion, 7, 141 (1982).
48. J.H. Nuckolls, LLNL Laser Program Annual Report 1979, UCRL-50021-79, 2, p 3-1, Mar. 1980.
49. C.L. Olson, J. Fus. Energy, 1, 309 (1982).

50. M.J. Monsler, J. Maniscalco, J.A. Blink, J. Hovingh, W.R. Meier, and P.E. Walker, "Electric Power From Fusion - The HYLIFE Concept", Proc. IECEC, San Diego, Calif., Aug. 1978, p. 264; also "Laser Program Annual Report-1977 to 1981", Law. Liv. Natl. Lab. Repts., UCRL-50021-77 to 81.
51. W.F. Krupke, E.V. George, and R.A. Haas, in Laser Handbook (M.L. Stitch, ed.) Vol. 3, p. 627, North Holland Publ. Co., 1979.
52. W.H. Lowdermilk and D. Milam, IEEE J. Quant Electron., QE-17, 1888 (1981).
53. J.E. Swain, W.H. Lowdermilk, and D. Milam, Appl. Phys. Lett., 41, 782 (1982).
54. F. Rainer and T.F. Deaton, Appl. Opt., 21, 1722 (1982).
55. K. Hohla and K. Kompa, in Handbook of Chemical Lasers (R.W.F. Gross and J.F. Batt, eds.) p. 667, John Wiley & Sons (1976).
56. J.L. Emmett, W.F. Krupke, and J.B. Trenholme, Physics of Laser Fusion Vol. 4, "The Future Development of High-Power, Solid State Lasers", Law. Liv. Natl. Lab. Rpt. UCRL-53344, Nov., 1982.
57. R.S. Craxton, IEEE J. Quant. Electron., QE-17, 1771 (1981).
58. L.F. Johnson and J.H. Guggenheim, J. Appl. Phys., 38, 4837 (1967).
59. D.E. McCumber, Phys. Rev., 134, A299 (1964).
60. W.F. Krupke, Proc. Intl. Conf. on Lasers '80, Dec. 1980, p.511.
61. P. Moulton, Appl. Phys., B28, 233 (1982).
62. G.A. Bogomolova, D.N. Vylegzhanin, and A.A. Kaminskii, Zh. Eksp. Teor. Fiz, 69, 860 (1975); Sov. Phys.-JETP, 42, 440 (1975).
63. R.E. Beverly III, R.H. Barnes, C.E. Moeller, and M.C. Wong, Appl. Opt., 16, 1572 (1977).
64. D.R. Scifres, R.D. Burnham, and W. Streifer, Appl. Phys. Lett., 41, 118 (1982).
65. J.J. Ewing, in Laser Handbook (M.L. Stitch, ed.) Vol. 3, p. 135, North Holland Publ. Co., Amsterdam (1979).
66. C.A. Brau, Excimer Lasers, Top. Appl. Phys., 30, 87 (1979).
67. M. Rokni, J.A. Mangano, J.H. Jacob, and J.C. Hsia, IEEE J. Quant. Electron., QE-14, 464 (1978).
68. T.H. Dunning and P.J. Hay, Appl. Phys. Lett., 28, 549 (1976).
69. J.H. Jacob, D.W. Trainer, M. Rokni, and J.C. Hsia, Appl. Phys. Lett., 37, 522 (1980).
70. H.T. Powell and K.S. Jancaitis, "Electron Beam Excited Absorption in Rare Gas Mixtures Using a SRS Probe", Laser Program Annual Report-1981, Law. Liv. Natl. Lab. Rpt. UCRL-50021-81.
71. R.A. Haas, W.L. Morgan, D. Eimerl, A. Siegman, A. Szoke and C. Turner, Laser Program Annual Report-1979, UCRL-50021-79, Vol. 3, p. 7-2, Mar. 1980.
72. J.J. Ewing, R.A. Haas, J.C. Swingle, E.V. George, and W.F. Krupke, IEEE J. Quant. Electron., QE-15, 368 (1979).
73. J.R. Murray, J. Goldhar, D. Eimerl, and A. Szoke, IEEE J. Quant. Electron., QE-15, 342 (1979).

74. D.A.G. Deacon, L.R. Elias, J.M.J. Madey, G.J. Ramain, H.A. Schwettman, and T.I. Smith, Phys. Rev. Lett., 38, 892 (1977).
75. D. Prosnitz, in Handbook of Laser Science and Technology (M.J. Weber, ed.), Vol. 1, p. 425, CRC Press, 1982.
76. N.M. Kroll, P.L. Morton, and M.N. Rosenbluth, IEEE J. Quant Electron., QE-17, 1436 (1981).
77. D. Prosnitz, A. Szoke, and V.K. Neil, Phys, Rev. A24, 1436 (1981).
78. D. Prosnitz, R.A. Haas, L.G. Schlitt, L.G. Seppala, and J.C. Swingle, Laser Program Annual Report-1980, UCRL-50021-80, p. 8-55, June 1981.
79. J.E. Leiss, IEEE Trans. on Nucl. Sci., NS-26, 3870 (1979).
80. G. Yonas, Proc. 10th Eur. Conf. Controlled Fusion Plasma Phys. (Moscow, Sept.), in press.
81. G.W. Kuswa, J.P. Quintez, J.R. Freeman, and J. Chang, Applied Charged Particle Optics, Part III, Advances in Electronics and Electron Physics, Suppl. 13C (A. Septier, ed.) Ch. 5, Academic Press (1983).
82. S. Humphries, Nucl. Fusion, 20, 1549 (1980).
83. D. Mosher, D.G. Colombant, and S.A. Goldstein, Comm. Plasma Phys. Conf. Fusion, 6, 101, 1981.
84. W. Allen, "The EAGLE Light Ion Beam Reactor", Trans. Am. Nucl. Soc., 43, 183 (1982).
85. J.N. Olsen, J. Appl. Phys., 52, 3279 (1981).
86. R.M. Measures, N. Drewell, and P. Cardinal, J. Appl. Phys., 50, 2662 (1979).
87. J.N. Olsen and R.J. Leeper, J. Appl. Phys., 53, 3397 (1982).
88. P.A. Miller, D.J. Johnson, T.P. Wright, and G.W. Kuswa, Comm. Plasma Phys. Cont. Fusion, 5, 95 (1979).
89. R.O. Bangerter, W.B. Herrmannsfeldt, D.L. Judd and L. Smith, eds., ERDA Summer Study of Heavy Ions for Inertial Fusion, Law. Berk. Lab. Rpt. LBL-5543, 1976.
90. L.W. Smith, ed., Proc. Heavy-Ion Fusion Workshop, Brookhaven Natl. Lab., Brookhaven Natl. Lab. Rpt. BNL-50769, 1977.
91. R.C. Arnold, ed., Proc. Heavy-Ion Fusion Workshop, Argonne Natl. Lab., 1978, Argonne Natl. Lab. Rpt. ANL-79-31, 1979.
92. W.B. Herrmannsfeldt, ed., Proc. Heavy-Ion Fusion Workshop, Berkeley, 1979, Law. Berk. Lab. Rpt. LBL-10301, 1980.
93. L.C. Teng, D.L. Judd, F.E. Mills, and D.F. Sutter, Proc. Heavy-Ion Fusion Workshop, Argonne Natl. Lab., 1978, Argonne Natl. Lab. Rpt. ANL-79-31, p. 159, 1979.
94. A.W. Maschke, Brookhaven Natl. Lab Rpt. BNL-20297, 1975.
95. A. Faltens and D. Keefe, Proc. 1981 Linear Accel, Conf., Santa Fe, NM, Oct. 1981, Los Alamos Natl. Lab. Rpt. LA-9234-C, p. 205, 1982.
96. L.C. Teng, ERDA Summer Study of Heavy Ions for Inertial Fusion, Lawrence Berkeley Lab. Rpt. LBL-5543, p. 13, 1976.

PART II

THE CALCULATION AND MEASUREMENT OF ATOMIC AND MOLECULAR PROCESSES RELEVANT TO FUSION

RECENT PROGRESS IN THEORETICAL METHODS FOR ATOMIC COLLISIONS

C.J. Joachain

Physique Théorique, Faculté des Sciences
Université Libre de Bruxelles, Belgium

1. INTRODUCTION

In this lecture I shall review recent developments which have occurred in theoretical methods for calculating atomic collision cross sections. The lecture is an up-dated (but self-contained) version of the review I gave at the previous NATO Advanced Study Institute on "Atomic and Molecular Processes in Controlled Thermonuclear Fusion"[1]. It is meant to provide a general introduction to subsequent lectures by other speakers, who will deal in more detail with the theoretical analysis of specific atomic and molecular collision processes of interest in fusion research.

I shall start by considering electron scattering by atoms and ions, and then proceed to discuss atom (ion)-atom collisions. All quantities will be expressed in atomic units.

2. ELECTRON SCATTERING BY ATOMS AND IONS

Let us begin by discussing the main theoretical methods which have been proposed to study electron-atom (ion) collisions. It will be convenient to divide the energy range for the incident electron into low, intermediate and high energy regions. In the low energy region the velocity of the incident electron is of the same order or less than the velocity of the target electrons which are taking an active part in the collision. In this region only a few target states can be excited, or in other words only a few channels are open. In the intermediate energy region the projectile electron has a velocity extending to several times the velocity of the active target electrons. This is the most difficult region to treat theoretically, since there is an infinity of open channels (including continuum ones due to the possibility of ionizing the target) and yet the velocity of the projectile is not large enough for perturbation methods to be rapidly convergent. Finally, the high-energy region is characterized by projectile electron velocities significantly larger than the velocity of the active target electrons, so that perturbation theory is rapidly convergent. A general survey of theoretical methods for electron-atom collisions is contained in the review article of Burke and Williams[2], while the reviews of Robb[3] and Henry[4] deal with the electron impact excitation of positive ions. Detailed discussions of electron-atom collisions at intermediate and high energies may be found in the review articles of Bransden and McDowell[5] and Byron and Joachain[6].

2.1. Low-energy scattering of electrons by atoms and ions.

As indicated above, in the low-energy region only a few channels are open. In this case all the open channels can be included explicitly in the total electron-atom (ion) scattering wave function, which can be accurately represented in terms of a sum of configurations in a way similar to the configuration interaction expansions used for bound state calculations.

The close-coupling approximation

The simplest scattering wave function of that kind is given by the well known close-coupling[7,8] expansion

$$\Psi = \mathcal{A} \sum_{n=1}^{M} F_n(q_{N+1})\ \psi_n(q_1,q_2,\dots q_N) \qquad (2.1)$$

where $q_j \equiv (\mathbf{r}_j,\sigma_j)$ denotes the combined space and spin coordinates of the jth electron and $\mathcal{A}$ is the antisymmetrization operator whose presence is required by the Pauli exclusion principle. The functions F_n describe the motion of the scattered electron in channel n and the ψ_n are a finite number (M) of wave functions corresponding to low-lying target states. The atomic eigenstates ψ_n retained in the truncated expansion (2.1) usually include all the open channels and possibly also some closely coupled closed channels. By projecting the Schrödinger equation $(H-E)\Psi = 0$ onto the functions ψ_n, or upon substitution of the expansion (2.1) into the Kohn variational principle[9], coupled integro-differential equations - known as the close-coupling equations - are obtained for the functions F_n.

As an example, let us consider the case of electron scattering by a one-electron atom or ion of nuclear charge Z. The close-coupling expansion for the spatial part of the wave function then reads

$$\Psi^{\pm}(\mathbf{r}_1,\mathbf{r}_2) = \sum_{n=1}^{M} [F_n^{\pm}(\mathbf{r}_1)\psi_n(\mathbf{r}_2) \pm F_n^{\pm}(\mathbf{r}_2)\psi_n(\mathbf{r}_1)] \quad (2.2)$$

where the superscripts + and - refer to the singlet (S = 0) and triplet (S = 1) cases, respectively, and $\psi_n(\mathbf{r})$ is an eigenfunction of the one-electron target corresponding to an eigenenergy w_n, namely,

$$(-\frac{1}{2}\nabla^2 - \frac{Z}{r} - w_n)\ \psi_n(\mathbf{r}) = 0 \quad (2.3)$$

The system of close-coupling equations for the functions F_n is given by

$$(\nabla^2+k_n^2)\ F_n^{\pm}(\mathbf{r}) = \sum_{m=1}^{M} U_{nm}^{\pm}(\mathbf{r})\ F_m^{\pm}(\mathbf{r}) ,\qquad n = 1,2,\ldots M \quad (2.4)$$

where $k_n^2 = 2(E-w_n)$, E being the total energy. The potential energy operators $U_{nm}^{\pm}$ which appear in (2.4) are such that

$$U_{nm}^{\pm}(\mathbf{r})\ F_m^{\pm}(\mathbf{r}) = 2[V_{nm}(\mathbf{r})\ F_m^{\pm}(\mathbf{r}) \pm \int W_{nm}(\mathbf{r},\mathbf{r}')F_m^{\pm}(\mathbf{r}')d\mathbf{r}'] \quad (2.5)$$

where V_{nm} is a local direct potential given by

$$V_{nm}(\mathbf{r}) = -\frac{Z}{r}\delta_{nm} + \int \psi_n^{*}(\mathbf{r}')\ \frac{1}{|\mathbf{r}-\mathbf{r}'|}\ \psi_m(\mathbf{r}')\ d\mathbf{r}' \quad (2.6)$$

and $W_{nm}(\mathbf{r},\mathbf{r}')$ is the non-local exchange kernel

$$W_{nm}(\mathbf{r},\mathbf{r}') = \psi_n^{*}(\mathbf{r}')\ \psi_m(\mathbf{r})\ [\frac{1}{|\mathbf{r}-\mathbf{r}'|} + w_n + w_m - E] \quad (2.7)$$

After a partial wave decomposition[10], the close-coupling equations may be solved by various methods, to which we shall return shortly. It is found that the close-coupling approximation gives a good account of strong transitions between low-lying states well separated in energy from all other states. This is the case, for example, for the ns-np transition in the alkalis. In other instances the convergence of the close-coupling expansion (2.1) is often slow and the method must be improved.

Modified close-coupling expansions. Pseudostates. Correlation functions

One way of improving the close-coupling approximation is to add on the right of (2.1) a second expansion in terms of pseudostates $\bar{\psi}_n$, which represent the distortion of the atomic eigenstates in the field of the unbound electron. We recall in this connection that polarization effects, which are characteristic of the interaction between a charged particle and a neutral polarizable system, give rise in the elastic channel to an attractive long range potential varying like r^{-4} at large distances. The approach of adding polarized pseudostates, first proposed for e - H scattering by Damburg and Karule [11], is a development of the polarized orbital method of Temkin[12] and an orthogonal pseudostate expansion discussed by Burke and Schey[13]. The Damburg-Karule approach has been extended to many-electron atoms by Vo Ky Lan, Le Dourneuf and Burke[14], who have developed a general method allowing these polarized pseudostates to be calculated by solving variationally the first order perturbation equation for the target within a superposition of configurations framework.

A second way of obtaining improvements over the close-coupling approximation is to add to the truncated expansion (2.1) a certain number of (N+1)-electron bound state, square integrable (L^2) functions χ_n. These functions, which vanish unless all the electrons are close together, allow for electron-electron correlations. This approach is particularly useful when the remaining strongly coupled channels are all closed, so that they can be exactly represented in terms of square integrable functions. Excellent results can be obtained in this way if enough L^2 functions are retained. For example, in the case of e^- - H scattering, where Hylleraas-type L^2 functions can be used, Schwartz[15], Armstead[16] and Rudge[17] have calculated very accurate elastic scattering phase shifts by using this method and solving the resultant equations variationally. The continuum Bethe-Goldstone hierarchy of equations considered by Mittleman[18] and by Nesbet[19,20] also belongs to this approach. In this method, the total scattering wave function is expanded in terms of a set of orthogonal basis orbitals, which include those occupied in the target states as well as additional unoccupied orbitals. The expansion consists of a sum of Slater determinants which includes terms where up to two electrons are excited into the unoccupied orbitals.

The foregoing discussion shows that a general electron-atom collision wave function suitable for the low-energy range, and from which the methods discussed above can be deduced as particular cases, may be written in the form[2]

$$\begin{aligned}\Psi = \mathcal{A} \sum_{n=1}^{M} F_n(q_{N+1})\, \psi_n(q_1,q_2,\dots q_N) \\ + \mathcal{A} \sum_{n=M+1}^{M+P} F_n(q_{N+1})\, \overline{\psi}_n(q_1,q_2,\dots q_N) \\ + \sum_{n=1}^{R} c_n\, \chi_n(q_1,q_2,\dots q_{N+1}) \qquad (2.8)\end{aligned}$$

Upon substitution of the expansion (2.8) into the Kohn variational principle, coupled integro-differential equations are obtained for the radial part of the functions F_n, which are coupled to linear equations for the coefficients c_n. These coupled equations must be solved accurately and speedily in order to obtain the collision cross sections. In the case of electron scattering by light atoms and ions, where relativistic effects are unimportant, considerable progress has been made recently, and general computer programs have been developed for this purpose, using a linear algebraic method[21], variational techniques[22], a non-iterative integral equation method[23] and the R-matrix method[24], which will be described below. As an example, we mention the recent calculations of electron impact induced transitions between the n = 1,2,3 levels of atomic hydrogen, carried out by Hata et al.[25] by using an algebraic variational close-coupling code.

L^2 methods

In recent years there has also been considerable interest in methods which include only L^2 functions, i.e. only the third expansion on the right of eq. (2.8), to describe the total scattering wave function. The basic idea is to avoid the specification of channels and asymptotic forms of wave functions by using an expansion in terms of functions which vanish asymptotically. Of course, the oscillatory behaviour of the wave function at large distances cannot be reproduced but at short distances the wave function can be accurately represented[26,27]. This approach is particularly useful in determining the positions and widths of resonances[28-30], since the wave function corresponding to a resonant state has a large amplitude at short distances. A discretisation of the Hamiltonian in the chosen L^2 basis then yields a discretisation of the continuous spectrum, from which scattering information can then be extracted. A significant development in L^2 methods is also the J-matrix mathod of Heller and Yamani[31], which has been extended to treat multichannel scattering and photoionization of complex atoms by Broad and Reinhardt[32]. It is worth noting that since the L^2 approach essentially uses bound state methods, it presents the important advantage of allowing the calculation of cross sections by adapting standard bound state computer programs.

The R-matrix method

This approach, originally introduced in nuclear physics by Wigner and Eisenbud[33], was first applied to atomic collisions by Burke and co-workers[34,35] and reviewed some time ago by Burke and Robb[36]. The basic idea is to realize that the dynamics of the projectile-target system depends on the relative distance r between the two colliding partners. As a result, configuration space describing the electron-atom (ion) complex may be divided into an

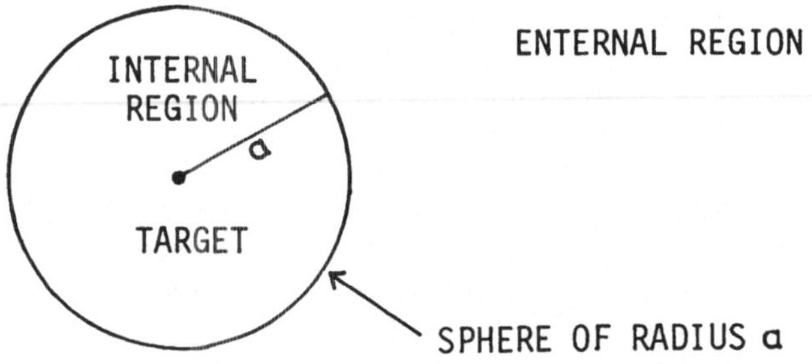

Fig. 1. Division of configuration space in R-matrix theory.

internal region ($r < a$) and an external region ($r > a$) as illustrated in Figure 1. For electron-atom (ion) collisions r is the radial distance between the unbound electron and the target nucleus, and a is chosen so that the charge distribution of the target is contained within the sphere $r = a$. In the internal region the electron interaction with the target is strong, electron exchange and correlation effects are important and the intermediate complex behaves very much like a bound state, so that bound state methods can be used in this region in a way similar to the L^2 methods discussed above. On the other hand, in the external region, the electron interaction with the target is simple and analytic solutions are often available. This occurs for example in electron-ion scattering, where it is often possible to just include in the external region the Coulomb interaction between the scattered electron and the ion core.

The link between the two regions is provided by the R-matrix which is defined as follows. We start by imposing logarithmic boundary conditions on the surface of the internal region

$$\frac{a}{F_i(a)} \left. \frac{d\, F_i(r)}{dr} \right|_{r=a} = b_i \tag{2.9}$$

where $F_i(r)$ are the radial functions describing the motion of the scattered electron in channel i, and b_i are arbitrary constants which in the work on electron-atom (ion) scattering are usually taken to be zero[36]. With these boundary conditions the energy spectrum of the electron-target system in the internal region consists of discrete energy levels E_k, and the corresponding eigenstates Ψ_k form in this region a complete set. For $r < a$ one may therefore expand the wave function for any energy as

$$\Psi_E = \sum_k A_{Ek} \Psi_k \tag{2.10}$$

By substituting this expansion into the Schrödinger equation $(H-E)\Psi_E = 0$ and using the boundary conditions (2.9) satisfied by the radial parts of the functions Ψ_k, one finds that

$$F_i(a) = \sum_j R_{ij}(E)\left(a \frac{dF_j}{dr} - b_j F_j\right)_{r=a} \tag{2.11}$$

where

$$R_{ij}(E) = \frac{1}{2a} \sum_k \frac{w_{ik}(a)\, w_{jk}(a)}{E_k - E} \tag{2.12}$$

is the R-matrix. The surface amplitudes $w_{ik}(a)$ which appear in the above equations are the values of the radial parts of the functions Ψ_k on the boundary. The central problem in the R-matrix approach is thus to calculate the surface amplitudes $w_{ik}(a)$ and the eigenenergies E_k. From the knowledge of the R-matrix (2.12), the K-matrix (or S-matrix) and cross sections can be obtained by using the solution in the external region. It should be noted that the main part of the work required in calculating the R-matrix, which is the diagonalisation of the Hamiltonian in the internal region, must only be performed once in order to determine the R-matrix at all energies. The evaluation of the matrix elements of the Hamiltonian in the internal region is similar to bound state calculations, so that standard bound state procedures can be used for that purpose with little modification, as indicated above. In practice, the expansion (2.12) is slowly convergent, and a correction for the omitted far-away poles suggested by Buttle[37] must be included; a variational correction proposed by Zvijac et al.[38] can also be applied. We also remark that if the quantities b_i in eq. (2.9) are non-zero, then the Hamiltonian is not Hermitian owing to a non-zero surface term. However, as shown by Bloch[39], this surface term can be eliminated by the introduction of an additional operator.

Quantum defect theory

The fact that the R-matrix is a real meromorphic function of the energy, with simple poles on the real axis, provides the basis of the quantum defect theory[40-42], in which the analytically known properties of electrons moving in a pure Coulomb field are used to analyze electron-ion scattering in terms of a few parameters. In particular, this theory describes the behaviour of cross sections in the presence of an attractive Coulomb interaction and enables for example the resonance structure below threshold to be predicted from the knowledge of the scattering amplitude above threshold.

Many-body methods

The many-body Green's function methods developed by Taylor et al.[43] are also capable of giving accurate results for electron-atom scattering. For elastic scattering the generalised random phase approximation has been used to calculate the response function of the target. For inelastic collisions the approach developed by Taylor et al. is similar to the distorded wave approximation considered below.

Resonances and the Feshbach projection operator formalism

Before leaving the subject of low-energy electron-atom (ion) collisions, it is important to stress the fundamental role played by resonances. The general theory of resonances has been reviewed by Burke[44] and a recent review of electron-atom (molecule) resonances has been given by Golden[45]. Resonances in electron-ion collisions are also discussed in the review articles of Robb[3] and Henry[4].

It is useful to distinguish two categories of resonances, namely open-channel (or shape) resonances and closed-channel (or Feshbach) resonances. Open channel resonances arise when the interaction between the incident electron and the target is such that a quasi-bound intermediate state of the electron-target system can be formed without changing the configuration of the target. Usually, the effective potential between the projectile electron and the target must have a characteristic shape, with an inner well and an outer barrier, so that open-channel resonances are also referred to as shape resonances. We remark that shape resonances occur above the threshold of the channels to which they are most strongly coupled, and that they may occur even if only a single channel is involved (as in potential scattering).

On the other hand, closed-channel or Feshbach resonances occur for incident energies just below the threshold for the excitation of some state. This may happen when a compound, autoionizing state of the (N+1)-electron system (incident electron plus target) coincides with an electron in the continuum of the N-electron system. This autoionizing or doubly excited state will either decay radiatively to a bound state of the (N+1) electron system, or it will autoionize into the continua associated with the target N-electron states. In the latter instance we have a closed-channel resonant behaviour. Since open-channel resonances are above the threshold energies of the target states to which they are most strongly coupled (so that there is an additional decay mode nearby which is open), they are usually much broader than closed-channel resonances, which lie below the thresholds corresponding to the target states with which they couple strongly.

The projection-operator formalism of Feshbach[46,47] is particularly convenient to study various aspects of low-energy electron-atom collisions, especially the closed-channel resonances. Of course the other low-energy methods we have described above are also capable of giving a detailed account of resonances in low-energy electron collisions with atoms and ions.

2.2. Intermediate and high-energy electron-atom (ion) collisions

We shall now examine various theoretical methods which have been proposed to deal with electron-atom (ion) collisions at intermediate and high energies. We shall begin by discussing extensions of the low-energy methods considered in the previous section and then analyze various approaches in which perturbation methods are used.

Modified close-coupling expansions

Let us first consider the energy region extending from the ionization threshold to a few times the ionization energy of the target. In this energy region there is an infinite number of open channels, and the approximations based on the low-energy wave function (2.8) are no longer obviously applicable. However, several methods have been proposed to modify the low-energy approach in order to deal with this problem. In eq. (2.8), pseudostates ψ_n were added in order to take into account important long range interactions due to polarization effects. In the present case, an additional important physical effect is the loss of flux from the channels retained in the first sum on the right of eq. (2.8). In view of this, Burke and Webb[48] and Callaway and Wooten[49] extended the pseudostate approach to intermediate energies, where the pseudostate channels are open and hence can carry away flux. Extensive calculations using the pseudostate method have been performed by Callaway et al.[50] for e^- - H scattering and by Willis and McDowell[51] for e^- - He collisions. The method, however, presents undesirable features due to the unphysical pseudoresonances associated with the pseudostates and with the short range correlation terms. For example, in the R-matrix calculation of electron impact excitation of helium performed by Fon et al.[52], the energy range from 30 to 80 eV had to be omitted due to the presence of pseudoresonances. Another example of pseudoresonances arising from correlation terms is shown in Fig. 2. The full curve represents an R-matrix calculation[53] of the collision strength (which is proportional to the cross section) for electron impact excitation of the $1s^2\ 2s^2\ {}^1S^e$ - $1s^2\ 2s2p\ {}^3P^0$ transition in C^{2+}. The resonances below 2 Ryd are real, but those between 2 and 11 Ryd are pseudoresonances. At first sight one would be tempted to disregard this energy region, but Burke et al.[53] have shown recently that useful information can be obtained by a suitable averaging of the scattering amplitude over the pseudoresonances. The

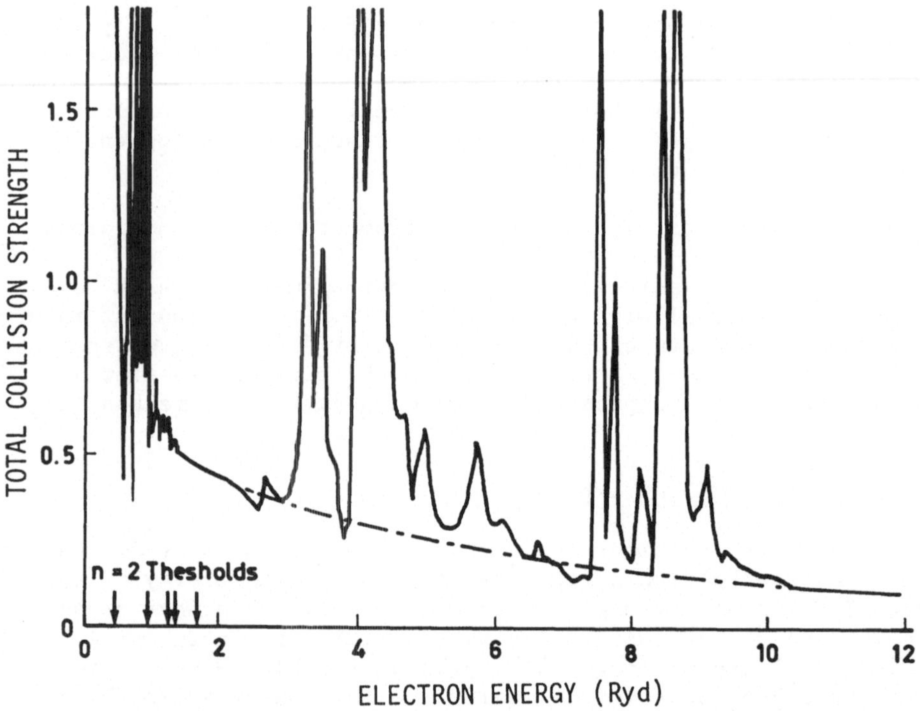

Fig. 2. Total collision strength for the $1s^2\ 2s^2\ {}^1S^e - 1s^2 2s2p\ {}^3P^0$ transition in $e^- - C^{2+}$ scattering. [From Burke et al., ref. 53, by permission of the Institute of Physics].

collision strength obtained in this way is represented by the dash-dot curve in Fig. 2. Although the work of Burke et al.[53] represents an important step in solving the pseudoresonance problem, further work in this direction is required.

Another modification of the close-coupling method, called the second order potential (SOP) method, has been applied by Bransden et al.[54] to a variety of electron-atom scattering processes. The modified close-coupling equations [compare with (2.4)] read

$$(\nabla^2+k_n^2)\ F_n^{\pm}(\underset{\sim}{r}) = \sum_{m=1}^{M} [U_{nm}^{\pm}(\underset{\sim}{r}) + K_{nm}^{\pm}(\underset{\sim}{r})] F_m^{\pm}(\underset{\sim}{r}) \qquad (2.13)$$

where $U_{nm}^{\pm}$ is defined by (2.5) and $K_{nm}^{\pm}$ is a potential matrix which accounts approximately for the coupling with the states $n' \geq M+1$ omitted from the close-coupling expansion. A detailed discussion of the SOP method may be found in the review article of Bransden and McDowell[5], and we also mention the recent elastic e^- - H calculations performed by Scott and Bransden[55] in the energy range 1 - 100 eV by using this approach.

L^2 methods

Some progress in the treatment of electron-atom collisions at intermediate energies has also been accomplished by using L^2 integrable functions[56]. The extraction of the scattering information has been made by extrapolation or continuation of the T-matrix from complex energies[57-60], by Fredholm analytic continuation[61] and by Stieltjes imaging and moment T-matrix methods[62-64]. However, although these methods are in principle quite general, they have until now been limited in practice to the scattering of electrons and positrons by atomic hydrogen.

The Born series

At higher energies it is reasonable to try an approach based on perturbation methods. We shall first discuss direct elastic and inelastic scattering. We denote respectively by $\mathbf{k}_i$ and $\mathbf{k}_f$ the initial and final momentum of the projectile and by $\boldsymbol{\Delta}$ the momentum transfer. The free motion of the colliding particles before the collision is described by the direct arrangement channel Hamiltonian $H_d = K + h$, where K is the kinetic energy operator of the projectile electron and h the internal target Hamiltonian, having eigenenergies w_n and eigenstates $|n\rangle$. The full Hamiltonian of the system is $H = H_d + V_d$, where V_d is the interaction between the electron and the target in the initial (direct) arrangement channel. For simplicity we shall assume here that we are dealing with a neutral target atom of atomic number Z, so that

$$V_d = -\frac{Z}{r} + \sum_{j=1}^{Z} \frac{1}{|\mathbf{r}-\mathbf{r}_j|} \tag{2.14}$$

where $\mathbf{r}$ is the coordinate of the projectile and $\mathbf{r}_j$ those of the target electrons.

The Born series for the direct scattering amplitude is given by

$$f = \sum_{n=1}^{\infty} \bar{f}_{Bn} \tag{2.15}$$

where the nth Born term $\bar{f}_{Bn}$ contains n times the interaction V_d

and (n-1) times the direct Green's operator $G_d^{(+)} = (E-H_d+i\varepsilon)^{-1}$, with $\varepsilon \to o^+$.

The first term $\bar{f}_{Bn}$ is the familiar first Born amplitude, which has been calculated for a large number of electron-atom elastic, excitation and ionization scattering processes[65]. A major goal of the theory is to obtain systematic improvements over the first Born approximation. To this end, we first examine the second Born term $\bar{f}_{B2}$. For the direct transition $|\mathbf{k}_i,o\rangle \to |\mathbf{k}_f,m\rangle$, where $|o\rangle$ and $|m\rangle$ denote respectively the initial and final target states, we have[66]

$$\bar{f}_{B2} = 8\pi^2 \sum_n \int d\mathbf{q} \frac{\langle \mathbf{k}_f,m|V_d|\mathbf{q},n\rangle\langle \mathbf{q},n|V_d|\mathbf{k}_i,o\rangle}{q^2 - k_i^2 + (w_n-w_o) - i\varepsilon}, \varepsilon\to o^+ \quad (2.16)$$

At sufficiently high energies, a good approximation to $\bar{f}_{B2}$ is obtained by replacing the energy differences (w_n-w_o) by an average excitation energy $\bar{w}$, so that the sum on the intermediate target states $|n\rangle$ can be performed by closure. An improvement over this approximation consists in evaluating exactly the first few terms in the sum, while treating the remaining states by closure. This method has been widely used in recent years, along with further improvements.[6] It is worth noting that the sum on intermediate states in (2.16) has been evaluated "exactly" (without using closure) for both elastic scattering[67-69] and 1s-2s excitation[69,70] in e^- - H scattering.

The behaviour of the terms $\bar{f}_{Bn}$ as a function of k_i (the wave number or velocity of the projectile electron) and Δ (the magnitude of the momentum transfer) has been discussed in detail by Byron and Joachain[6]. For direct elastic scattering at small Δ the quantity Re $\bar{f}_{B2}$ (which is governed by polarization effects) gives the dominant correction (of order k_i^{-1}) to the first Born differential cross section. At large Δ the terms $\bar{f}_{Bn}$ $(n \geqslant 2)$ are dominated by processes in which the atom remains in its initial state $|o\rangle$ in all intermediate states; this reflects the fact that at large angles the elastic scattering is governed by the static potential $\langle o|V_d|o\rangle$. We note that for direct elastic scattering the dominant contribution at large k_i is given by the first Born term $\bar{f}_{B1}$ at all momentum transfers.

The situation is different for direct inelastic collisions where for large Δ the first Born term falls off rapidly and the Born series is dominated by the second Born term. The fact that $\bar{f}_{B2}$ falls off more slowly than $\bar{f}_{B1}$ at large Δ is due to the possibility of off-shell elastic scattering in intermediate states, where the projectile can experience the Coulomb potential of the nucleus. Thus, since the values $\Delta < 1$ correspond to angles $\theta < k_i^{-1}$, the angular region in which the first Born approximation is valid shrinks with increasing energy. However, because the dominant contribution to the integrated cross section comes from the region $\Delta < 1$, the first Born values for integrated inelastic cross . sections should be reliable at high energies.

The terms $\bar{g}_{Bn}$ of the Born series for the exchange amplitude are much more difficult to analyze than the direct Born terms $\bar{f}_{Bn}$. We simply remark that for large k_i the term $\bar{g}_{B2}$ falls off more slowly than $\bar{g}_{B1}$, except for elastic exchange scattering at small Δ, where the Ochkur[71] amplitude, which is the leading piece of $\bar{g}_{B1}$, is of order k_i^{-2}. It is also interesting to note that if we know all the direct amplitudes half-off shell, then the exchange amplitude $g(\mathbf{k}_f,m;\mathbf{k}_i,o)$ for a transition $|\mathbf{k}_i,o\rangle \rightarrow |\mathbf{k}_f,m\rangle$ may be obtained by quadratures via the relation[72]

$$g(\mathbf{k}_f,m;\mathbf{k}_i,o) = \bar{g}_{B1}(\mathbf{k}_f,m;\mathbf{k}_i,o)$$

$$+ (2\pi^2)^{-1} \sum_n \int d\mathbf{q} \frac{1}{q^2-k_n^2-i\varepsilon} \bar{g}_{B1}(\mathbf{k}_f,m;\mathbf{q},n) f(\mathbf{q},n;\mathbf{k}_i,o) \ , \quad \varepsilon \rightarrow o^+ \qquad (2.17)$$

where $\bar{g}_{B1}$ $(\mathbf{k}_f,m;\mathbf{k}_i,o)$ is the on-shell first Born exchange amplitude $\bar{g}_{B1}(\mathbf{k}_f,m;q,n)$ is a half-off shell first Born exchange amplitude, $f(\mathbf{q},n;\mathbf{k}_i,o)$ is a half-off shell direct amplitude, and $k_n^2 = k_i^2 - 2(w_n-w_o)$. The equation (2.17) is also interesting because it is non-perturbative in nature, although if one wishes one can of course expand $f(\mathbf{q},n;\mathbf{k}_i,o)$ in a Born series and generate in this way the Born series for $g(\mathbf{k}_f,m;\mathbf{k}_i,o)$.

Let us now turn to ionization collisions. In this case it is difficult to perform calculations of the Born terms $\bar{f}_{Bn}$ beyond first order. Recently, however, the second Born term has been evaluated in the closure approximation for the ionization of atomic hydrogen[73] and helium[74]. By generalizing the arguments given above for inelastic scattering, it is also possible to show[75] that in certain kinematical situations the second Born term $\bar{f}_{B2}$ again falls off more slowly than $\bar{f}_{B1}$ for large k_i.

The effects of exchange in the ionization of atoms by electron impact have been discussed in detail by Rudge and Seaton[76], Rudge[77] and Peterkop[78]. A delicate problem, still unresolved, is the determination of the relative phase of the direct and exchange amplitudes in approximate (for example first Born) calculations. In the case of the ionization of positive ions by electron impact, the equivalent of the first Born approximation is the "Coulomb-Born" approximation[79] which, together with its various exchange derivatives, has been discussed recently by Jacubowicz and Moores[80].

The Glauber approximation

The Glauber[81] scattering amplitude for a direct collision leading from an initial target state $|o\rangle$ to a final state $|m\rangle$ is given by

$$f_G = \frac{k_i}{2\pi i} \int d^2\mathbf{b}\, \exp(i\boldsymbol{\Delta}\cdot\mathbf{b}) \langle m | \{\exp[\frac{i}{k_i}\, \chi_o(\mathbf{b},X)] - 1\} | o \rangle \tag{2.18}$$

where the symbol X denotes the ensemble of the target coordinates and we use a cylindrical coordinate system, with $\mathbf{r} = \mathbf{b} + z\hat{\mathbf{z}}$. The Glauber phase χ_o is given in terms of the direct interaction (2.14) between the projectile and the target by

$$\chi_o = - \int_{-\infty}^{+\infty} V_d(\mathbf{b},z,X)\, dz \tag{2.19}$$

the integration being performed along a z axis perpendicular to $\boldsymbol{\Delta}$ and lying in the scattering plane.

Detailed discussions of the Glauber approximation may be found in the review articles of Joachain and Quigg[82], Bransden and McDowell[5] and Byron and Joachain[6]. We first remark that the Glauber approach may be viewed as an eikonal approximation to a "frozen target" model proposed by Chase[83], in which closure is used with an average excitation energy $\overline{w} = 0$. Interesting insight into the properties of the Glauber method may also be gained by expanding the Glauber amplitude (2.18) in powers of the interaction V_d,

$$f_G = \sum_{n=1}^{\infty} \bar{f}_{Gn} \tag{2.20}$$

and comparing the nth Glauber term $\bar{f}_{Gn}$ with the corresponding n^{th} Born term $\bar{f}_{Bn}$. Because of our choice of z-axis we have $\bar{f}_{B1} = \bar{f}_{G1}$. For $n \geq 2$, the terms $\bar{f}_{Gn}$ are alternately real or purely imaginary, in contrast with the Born terms, which are complex for $n \geq 2$. This special feature of the Glauber amplitude leads to several defects such as i) the absence in the elastic scattering case of the all-important terme Re $\bar{f}_{B2}$ which accounts for long-range polarization effects and ii) identical cross sections for electron- and positron-atom scattering. Other deficiencies of the Glauber amplitude (2.19) include i) a logarithmic divergence for elastic scattering in the forward direction, which is due to the choice $\overline{w} = 0$ made in obtaining the Glauber amplitude and may be traced to the behaviour of the second term $\bar{f}_{G2}$ at $\Delta = 0$ and ii) a poor description of inelastic collisions involving non-spherically symmetric states. Nevertheless, the Glauber amplitude (2.18) has one attractive property : it includes terms from all orders of perturbation theory in such a way as to ensure unitarity. The Glauber approximation has been applied to a variety of atomic scattering processes. Its major role has been to stimulate interest in eikonal methods[6] such as the "eikonal-Born series" (EBS) theory[84] which we shall now examine.

The eikonal-Born series method

The basic idea of the EBS method consists in analyzing the terms of the Born series (2.15) and the Glauber series (2.20) with the aim of obtaining a consistent expansion of the scattering amplitude in powers of k_i^{-1} (i.e. in inverse powers of the projectile velocity). A detailed study of this problem[6,84] shows that for elastic and inelastic (s-s) transitions the Glauber term $\bar{f}_{Gn}$ gives in each order of perturbation theory the leading piece of the corresponding Born term (for large k_i) except in second order where the long range of the Coulomb potential is responsible for the anomalous behaviour of $\bar{f}_{G2}$ at small Δ. Moreover, neither the second Born amplitude $f_{B2} = \bar{f}_{B1} + \bar{f}_{B2}$, nor the Glauber amplitude f_G are correct through order k_i^{-2}. In fact, a consistent calculation of the direct scattering amplitude through that order requires the terms $\bar{f}_{B1}$, $\bar{f}_{B2}$ and Re $\bar{f}_{B3}$ (or $\bar{f}_{G3}$). Since Re $\bar{f}_{B3}$ is very difficult to evaluate, and because $\bar{f}_{G3}$ is a good approximation to Re $\bar{f}_{B3}$ for large enough k_i, it is reasonable to use $\bar{f}_{G3}$ in place of Re $\bar{f}_{B3}$. Thus we obtain in this way the eikonal-Born series direct scattering amplitude[84]

$$f_{EBS} = \bar{f}_{B1} + \bar{f}_{B2} + \bar{f}_{G3} \tag{2.21}$$

We note that the amplitude[85]

$$f_{EBS'} = f_G - \bar{f}_{G2} + \bar{f}_{B2} \tag{2.22}$$

also gives a consistent picture of the direct scattering amplitude through order k_i^{-2}. In addition, exchange effects are taken into account in the EBS theory by keeping the relevant terms in the Born series for the exchange amplitude. For example, in the case of elastic scattering the Ochkur amplitude, which is of order k_i^{-2} for large k_i and fixed Δ, must be taken into account in order to perform a consistent calculation of the differential cross section throuth order k_i^{-2}.

A detailed account of the EBS method and its application to various electron-atom processes at intermediate and high energies may be found in the review article of Byron and Joachain[6]. It is apparent from the foregoing discussion that the EBS theory represents an improvement over the second Born or Glauber approximations. However, the EBS method is a perturbative approach, and since the convergence of the Born series for the direct amplitude is slower at large Δ than at small Δ, it would be desirable to perform an "all order" treatment (especially of the static potentials in the initial and final channels) at large momentum transfers. The target eigenfunction expansion methods discussed previously provide (approximately) such "all-order" treatments. We shall now briefly examine two approaches (the unitarised EBS approximation and the optical potential method) which also present this advantage.

The many-body Wallace amplitude and the unitarized EBS method

In the case of potential scattering, Wallace[86] has obtained improvements over the eikonal approximation by writing down in a systematic way the corrections to the eikonal phase. Detailed studies of the relationships between the terms of the Born, eikonal and Wallace series have been carried out by Byron, Joachain and Mund[87]. In light of these developments, Byron, Joachain and Potvliege[88] and subsequently Franco and Iwinski[89] and Unnikrishnan and Prasad[90] proposed a generalization of the potential scattering Wallace amplitude to the multiparticle case, in the same spirit as that of Glauber's original extension of the potential scattering eikonal amplitude. For the direct transition $|\mathbf{k}_i, o\rangle \rightarrow |\mathbf{k}_f, m\rangle$, the many-body Wallace amplitude reads

$$f_W = \frac{k_i}{2\pi i} \int d^2\mathbf{b} \exp(i\mathbf{\Delta}.\mathbf{b}) \langle m|\{\exp[i(\frac{1}{k_i}\chi_0(\mathbf{b},X) + \frac{1}{k_i^3}\chi_1(\mathbf{b},X)]-1\}|o\rangle \tag{2.23}$$

where χ_0 is the Glauber phase (2.19) while the Wallace correction χ_1, which is of second order in the interaction potential V_d, is given by

$$\chi_1(\mathbf{b},X) = \frac{1}{2}\int_{-\infty}^{+\infty} (\nabla\chi_+).(\nabla\chi_-)dz \tag{2.24}$$

with

$$\chi_+(\mathbf{b},z,X) = -\int_{-\infty}^{z} V_d(\mathbf{b},z',X)dz', \quad \chi_-(\mathbf{b},z,X) = -\int_{z}^{\infty} V_d(\mathbf{b},z',X)dz' \qquad (2.2\ )$$

The terms of the Wallace series, obtained by expanding f_W in powers of V_d, will be called $\bar{f}_{Wn}$.

At this point, we remark that since the excitation energies in both the initial and final channels have been set equal to zero, the long-range polarization effects will be missing from f_W, and

Im $\bar{f}_{W2}$ (= Im $\bar{f}_{G2}$) will diverge at $\Delta = 0$ for elastic scattering. Following Byron, Joachain and Potvliege[88], we therefore construct a new amplitude with $\bar{f}_{B2}$ inserted in the place of $\bar{f}_{W2}$, namely

$$f_{UEBS} = f_W - \bar{f}_{W2} + \bar{f}_{B2} \tag{2.26}$$

At small momentum transfers, where higher-order terms of perturbation theory are rather unimportant, we retrieve the EBS amplitude (2.21) by keeping the terms through order k_i^{-2} in (2.26). On the other hand, the terms $\bar{f}_{B2}$ and $\bar{f}_{W2}$ differ negligibly at large Δ, so that are large angles f_{UEBS} will differ little from f_W, and will provide a more accurate value of the direct scattering amplitude than the EBS' amplitude (2.22). Moreover, the amplitude f_{UEBS} will be nearly unitary at all angles. It is this "all-order" amplitude which is called the "unitarized EBS amplitude". It may be used in eq. (2.17) to obtain a corresponding "all order" exchange amplitude. Applications of this method have been made thus far to elastic scattering[72] and 1s-2s excitation[88] in e^- - H collisions, where excellent results have been obtained for impact energies above 100 eV.

Optical potentials

The basic idea of the optical potential method is to analyze the elastic scattering of a particle from a complex target by replacing the complicated interactions between the projectile and the target particles by an optical potential (or pseudopotential) in which the incident particles moves[91]. Once the optical potential V_{opt} is determined, the original many-body elastic scattering problem reduces to a one-body situation. However, this reduction is in general a difficult task, and approximation methods are necessary.

At intermediate and high energies it is particularly convenient to use the multiple scattering approach developed by Mittleman and Watson[92,93]. We begin by considering direct elastic scattering, for which we write the corresponding direct part V_{opt}^d of the optical potential as a multiple scattering expansion[82] in terms of the projectile-target interaction V_d. That is

$$V_{opt}^d = V^{(1)} + V^{(2)} + V^{(3)} + \ldots \tag{2.27}$$

Here $V^{(1)} = V_{st} = \langle o|V_d|o\rangle$ is the static potential while the second order part reads

$$V^{(2)} = \sum_{n \neq o} \frac{\langle o|V_d|n\rangle\langle n|V_d|o\rangle}{k_i^2/2 - K - (w_n - w_o) + i\varepsilon}, \quad \varepsilon \to o^+ \tag{2.28}$$

The static potential V_{st} is readily evaluated for simple target atoms, or when an independent particle model is used to describe the target state $|o\rangle$, which we assume here to be spherically symmetric. We remark that V_{st} is real and of short range, and hence does not account for polarization and absorption effects which play an important role in the energy range considered here. However, for small values of the projectile coordinate r we note that V_{st} correctly reduces to the Coulomb interaction $- Z/r$ acting between the incident electron and the targer nucleus, and thus should give a good account of large angle direct elastic scattering, as we already remarked in our discussion of the Born series.

Although the second order part $V^{(2)}$ of the direct optical potential is in general a complicated non-local, complex operator, at sufficiently high energies a useful local approximation of $V^{(2)}$ may be found by introducing an average excitation energy $\bar{w}$ and using eikonal methods[94-96]. The resulting $V^{(2)}$ may be written as

$$V^{(2)} = V_{pol} + i\, V_{abs} \tag{2.29}$$

where V_{pol} and V_{abs} are real and central but energy-dependent. The term V_{pol} (which falls off like r^{-4} at large r) accounts for dynamic polarization effects and i V_{abs} for absorption effects due to loss of flux from the incident channel. Recently, the leading, local, contribution of the third order part $V^{(3)}$ has also been evaluated[69] for the case of e^- - H scattering.

Having obtained a local approximation for V^{d}_{opt}, exchange effects may be taken into account by using a local exchange pseudo-potential[93,95,96-98] V^{ex}_{opt}, so that the full optical potential is given by $V_{opt} = V^{d}_{opt} + V^{ex}_{opt}$. An "exact" (partial wave) treatment of this potential is then carried out. It is worth noting that in performing such an exact, full-wave treatment of the optical potential V_{opt}, one generates approximations to all terms of perturbation theory, a feature which is an important advantage for large angle scattering.

Distorted waves

Distorted wave treatments are characterized by the fact that the interaction is broken in two parts, one which is treated exactly and the other which is handled by perturbation theory. This separation is dictated by the physics of the problem, and consequently many kinds of distorted wave methods have been applied to electron-atom (ion) collisions. Detailed discussions of distorted-wave treatments may be found in the reviews of Bransden and McDowell[5] and Henry[4]. We mention in particular the well-known distorted eave Born approximation (DWBA), the distorted wave second Born approximation[99] (DWSBA) which has been very useful in analyzing elastic and inelastic electron-atom collisions at intermediate energies, and the distorted wave impulse approximation[100,101] which has been used with success in (e,2e) spectroscopy studies.

Classical and quasi-classical methods

We shall conclude this review of electron-atom (ion) collisions by a brief discussion of electron collisions with highly excited (Rydberg) atoms. It is in this area that classical and quasi-classical theories, which saw an important resurgence in interest following the work of Gryzinski[102], are very useful. These approaches, which range from purely classical calculations using Monte-Carlo techniques to methods based on correspondence principles, have been reviewed recently by Percival and Richards[103,104].

3. ATOM (ION) - ATOM COLLISIONS

Although the general theory of atom (ion) - atom collisions[105] is formally similar to that of electron-atom collisions, there are important differences. Firstly, since the incident particle is now in general a "composite" object, calculations for atom-atom collisions are more difficult to perform from first principles. Secondly, because the projectile is heavy, at nearly all velocities for which experiments have been performed many channels are open. In particular, excitation or ionization of the target and (or) the projectile can occur, as well as charge transfer (electron transfer).

However, an important simplification can very often be made in the theoretical analysis of atom (ion) - atom collisions, since the wave number k corresponding to the relative motion of the colliding particles is frequently very large (in atomic units). For example, in the case of collisions between protons and atomic hydrogen we have $k > 1$ as soon as the center of mass energy exceeds 1.5×10^{-2}eV, and for a center of mass energy of 1 keV we have $k \simeq 260$ a.u. As a result, the condition $k \gg 1$ is satisfied at all but the lowest colliding energies, so that semi-classical or even classical methods can be used to describe the relative motion, except for very slow collisions which will not be considered here. On the other hand, the motion of the atomic electrons must be described by using quantum mechanics.

In what follows it will be convenient to divide the energy range into "slow" and "fast" collisions. If v denotes the relative velocity of the projectile and the target, and if v_o is a typical velocity of the active electron(s) in the collision, we shall say that the collision is slow when $v < v_o$ and fast when $v > v_o$. For example, if we take for v_o the Bohr velocity of the electron in the target we have $v_o = Z$ a.u. for a K shell electron. Thus, in the case of protons incident on hydrogen atoms in the ground state, we have $v_o = 1$ a.u. and the center of mass energy for which $v = v_o$ is 12.5 keV. For heavy ions, the corresponding energies are much higher, extending into the MeV region; these energies have only been reached recently with the development of heavy-ion accelerators. Of course, as in the case of electron-atom collisions, there is also a loosely defined intermediate energy region around $v = v_o$, for which the collision is neither slow or fast; this region is often the hardest to treat theoretically. We also remark that the fact that $k \gg 1$ does not necessarily imply that the colliding velocity v (expressed in a.u.) is large. For example, if we return to the p-H example considered above, a center of mass energy of 1 keV corresponds to the relative wave number $k \simeq 260$ a.u., but only to the laboratory proton velocity $v \simeq 0.28$ a.u.

For the sake of illustration, we shall essentially consider a one-electron model in which a nucleus A, of mass M_A and charge Z_A, is incident on an atom $(B+e^-)$ containing a nucleus B of mass M_B and charge Z_B, and an electron. We shall denote by $\mathbf{R}$ the position vector of A relative to B, by $\mathbf{r}_A$ and $\mathbf{r}_B$ the position vectors of the electron relative to A and B, respectively and by $\mathbf{r}$ a vector joining an arbitrary origin O located along the internuclear axis to the position of the electron (see Fig. 3). Thus we have

$$\mathbf{r}_A = \mathbf{r} - p\mathbf{R} , \qquad \mathbf{r}_B = \mathbf{r} + q\mathbf{R} \tag{3.1}$$

with p+q = 1. We remark that certain collision processes occuring in many-electron systems can be accurately described in terms of a

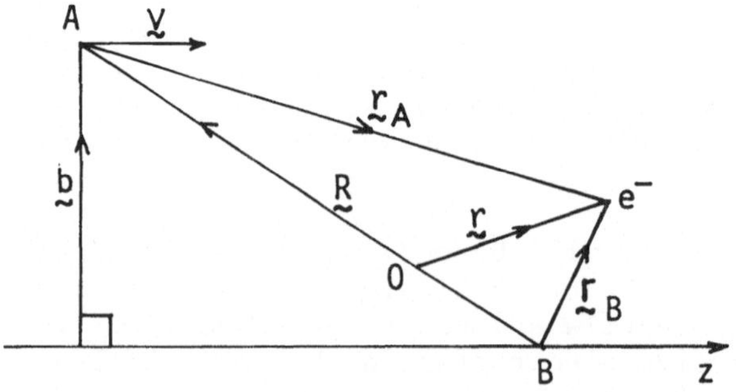

Fig. 3. A coordinate system for a system of one electron and two nuclei A and B.

one-electron model. This is the case for inner-shell processes in heavy atoms, since the strong nuclear Coulomb field experienced by an inner-shell electron is much more important than the field due to other electrons. A review of this topic has been given by Briggs[106].

3.1. Fast atom (ion) - atom collisions

We begin by considering fast collisions, which have recently been reviewed by Bransden[107]. As explained above, the relative motion of the nuclei A and B can then be treated classically, the classical trajectory being determined by assuming an effective internuclear potential $U(\mathbf{R})$ and solving the corresponding classical equations. For each value of the impact parameter vector $\mathbf{b}$, the relative position vector $\mathbf{R}$ of the two nuclei will be some function of the time t, $\mathbf{R} = \mathbf{R}(\mathbf{b},t)$, which can be determined once $U(\mathbf{R})$ is known. The theoretical problem then reduces to solving the time-dependent Schrödinger equation

$$(H_{e\ell} - i \frac{\partial}{\partial t}) \Psi(\mathbf{r},\mathbf{b},t) = 0 \qquad (3.2)$$

where $H_{e\ell}$ is the Hamiltonian of the electron in the field of the two nuclei, namely

$$H_{e\ell} = - \frac{1}{2} \nabla_r^2 - \frac{Z_A}{r_A} - \frac{Z_B}{r_B} + W(\mathbf{R}) \qquad (3.3)$$

It is worth stressing that since $\mathbf{R}$ depends on the time t, $\mathbf{r}_A$ and $\mathbf{r}_B$ also depend in general on t via (3.1). Of course particular cases arise when p = 0 (or q = 0) in which case the origin O coincides with A (or B) and $\mathbf{r}_A$ (or $\mathbf{r}_B$) are independent of t. The operation $\partial/\partial t$ in (3.2) is to be performed with the position vector $\mathbf{r}$ fixed in the laboratory frame. We note that since the effective internuclear potential $U(\mathbf{R})$ has already been taken into account when defining the classical trajectory, the interaction $W(\mathbf{R})$ occuring in (3.3) must be identified with $Z_AZ_B/R - U(\mathbf{R})$. The potential $W(\mathbf{R})$ can be removed by the phase transformation

$$\Psi(\mathbf{r},t) = \Psi'(\mathbf{r},t) \exp[- i \int W(\mathbf{R})\, dt] \qquad (3.4)$$

and for this reason it does not influence the transition probabilities and total cross sections, which only depend on $|\Psi|^2$. On the other hand, since differential cross sections depend on the magnitudes and phases of the probability amplitudes, the proper phase must be restored when calculating angular distributions. Belkič and Salin[108] have recently emphasized this point in calculating differential cross sections for charge exchange. At energies larger than 1 keV per nucleon, it is often sufficient to ignore

$U(\mathbf{R})$, in which case the nuclear motion is rectilinear, and is defined by

$$\mathbf{R}(t) = \mathbf{b} + \mathbf{v}t \; , \qquad \mathbf{b}\cdot\mathbf{v} = 0 \tag{3.5}$$

Expansion in atomic eigenfunctions for excitation

During a fast collision, the electron, for most of the time, will be bound either to nucleus A or nucleus B in an atomic orbital. Therefore an "atomic picture", based on expansions of the scattering wave function in atomic eigenfunctions, is physically reasonable in this case. We shall start by considering direct excitation processes, thus neglecting for the moment charge-exchange (electron capture) reactions which, as we shall see below, have much smaller cross sections when the velocity v is large. Let $\psi_m^B(\mathbf{r}_B)$ be the eigenfunctions of the atom $(B+e^-)$, corresponding to the energies w_m^B and forming a complete orthonormal set. Using the method of the variation of the constants, and setting the origin O at the nucleus B, we write the wave function as

$$\Psi(\mathbf{r}_B,\mathbf{b},t) = \sum_m a_m(\mathbf{b},t)\, \psi_m^B(\mathbf{r}_B)\, \exp(-\, i w_m^B\, t) \tag{3.6}$$

After substitution in the time-dependent Schrödinger equation (3.2), we find that the amplitudes a_m must satisfy the set of coupled first-order differential equations

$$i\, \dot{a}_k(\mathbf{b},t) = \sum_m V_{km}(\mathbf{R})\, \exp[\, i(w_k^B - w_m^B)t]\, a_m(\mathbf{b},t) \tag{3.7}$$

where

$$V_{km}(\mathbf{R}) = \langle \psi_k^B | \frac{Z_A Z_B}{R} - U(\mathbf{R}) - \frac{Z_A}{r_A} | \psi_m^B \rangle \tag{3.8}$$

If the atom is initially in the state $|\psi_o^B\rangle$, with energy w_o^B, the boundary conditions are

$$\lim_{t\to-\infty} a_k(\mathbf{b},t) = \delta_{ko} \tag{3.9}$$

and the total cross section for direct excitation of the level m from the level o is given by

$$\sigma_{mo}^D = \int d^2\mathbf{b}\, |a_m(\mathbf{b}, t=+\infty)|^2 . \tag{3.10}$$

The Born and close-coupling approximations for excitation

At high velocities, $v \gg v_o$, a first order perturbation solution to the equations (3.7) may be found by noting that all the amplitudes a_m apart from a_o remain small as time evolves, and a_o remains close to unity. The coupled equations (3.7) can then be approximated by

$$i\,\dot{a}_k(\underset{\sim}{b},t) = V_{ko}(\underset{\sim}{R})\,\exp[i(w_k^B - w_o^B)t] \qquad (3.11)$$

and the probability amplitude that the target atom $(B+e^-)$ will be left in the level $k \neq o$ after the collision is given to first order by

$$a_k(\underset{\sim}{b},t=+\infty) = -\,i\int_{-\infty}^{+\infty} V_{ko}(\underset{\sim}{R})\,\exp[i(w_k^B - w_o^B)t]dt \qquad (3.12)$$

This result is just the impact parameter version of the first Born approximation for the direct transition $o \rightarrow k$. For sufficiently high energies, the first Born approximation for total direct excitation cross sections becomes accurate (as in the case of electron scattering) and many calculations have been performed for simple systems such as H^++H, H^++He, H+H, H+He^+, etc.[65]. We mention parenthetically that for heavy particle excitation the Born approximation in the impact parameter form is equivalent to that in the wave form. Comparison with the results of higher order calculations shows that, for total cross sections, the Born approximation should be accurate above energies of 200 keV per nucleon for excitation of low-lying strongly coupled states. As in the case of direct electron-atom collisions, these total cross sections decrease like v^{-2} for s-s transitions, while those corresponding to s-p transitions fall off like $v^{-2}\,\ell n\,v$ for large v. We also remark that, just as for electron scattering, first Born differential cross sections for heavy particle excitation are inaccurate at large angles, where it is necessary to allow for the nuclear deflection. The importance of higher order terms in obtaining the high-energy limit of the differential cross section at fixed angle has been discussed by Shakeshaft[109]. Finally, we note that the expression (3.12) also gives a first order approximation to the transition amplitude for ionization, provided that the functions $\psi_k^B(\underset{\sim}{r}_B)$ are taken to be the proper positive energy eigenfunctions of the system $(B+e^-)$.

If we truncate the expansion (3.6) to retain only M discrete terms, the system (3.7) reduces to a close-coupling set of M coupled first order differential equations which must be solved in order to obtain the amplitudes and the cross sections. The close coupling approximation has been used to calculate excitation cross sections for several simple systems[110-113] but in many cases it is not satisfactory because it does not take into account the continuum target states. In addition, it should be remembered that no close-coupling single center expansion can represent charge transfer processes. Since these processes become important at the lower energies, and affect the inelastic cross sections through unitarity, it is clear that close-coupling single center expansions will become inaccurate at the lower energies.

Higher order methods for excitation

Several methods have been used to account approximately for continuum terms in the expansion (3.6). These include second[114] and higher Born[115] calculations, as well as calculations using the Glauber approximation[116-118] and the second order potential method[54]. As in the case of electron scattering, expansions including pseudostates may also be used to take approximately into account the effect of the continuum. A recent example of this approach is the work of Reading and his coworkers[119]. We remark that by including pseudostates it is possible to calculate ionization cross sections as well as those for discrete excitation.

The results of calculations performed on a variety of simple systems indicate that in order to improve on the first Born approximation, it is necessary to use a large basis set, and either to solve the coupled first-order equations directly, or to calculate to at least fourth order of perturbation theory. We also recall that at lower energies it is necessary to take into account the effect of charge transfer processes in calculating inelastic cross sections.

Atomic eigenfunction expansions for charge transfer

In order to calculate charge transfer cross sections, the basis set must be enlarged so that the rearranged system in which the electron is bound to the nucleus A can be represented asymptotically. For the case of fast collisions ($v > v_0$) which we are considering here, and for which an "atomic picture" is sensible, it is reasonable to make an atomic two-center expansion. Let us denote by $\phi_n^A(\mathbf{r}_A)$ the eigenfunctions of the atom $(A+e^-)$, with eigenenergies w_n^A. For a charge transfer reaction $A + (B+e^-) \to (A+e^-) + B$ such that the atom $(A+e^-)$ is left in the state ϕ_n^A after the collision, we have

$$\Psi(\mathbf{r}_B,\mathbf{b},t) \underset{t\to+\infty}{\to} \phi_n^A(\mathbf{r}_A) \exp(-\, i w_n^A t) \exp[\, i(\mathbf{v}\cdot\mathbf{r}_B - \frac{1}{2} v^2 t)] \quad (3.13)$$

The electron translation factor $\exp[\, i(\mathbf{v}\cdot\mathbf{r}_B - v^2 t/2)]$ which appears in (3.13) is required because the nucleus A is moving with respect to the origin of the coordinate system, which we have located here at the nucleus B. The necessity of including such translation factors was first recognized by Bates and McCarroll[120]. Taking into account (3.13) one can replace the single center expansion (3.6) by the two-center expansion

$$\Psi(\mathbf{r}_B,\mathbf{b},t) = \sum_{m=1}^{M} a_m(\mathbf{b},t)\ \psi_m^B(\mathbf{r}_B)\ \exp[-\,i\ w_m^B\ t] + \sum_{n=1}^{N} c_n(\mathbf{b},t)\ \phi_n^A(\mathbf{r}_A)\ \exp[-iw_n^A t]\ \exp[i(\mathbf{v}\cdot\mathbf{r}_B - \tfrac{1}{2}v^2 t)] \quad (3.14)$$

where the sums on m and n run over discrete states. By using the variational method, it may be shown[121] that the amplitudes a_m and c_n must satisfy a system of first order coupled equations. To obtain the total cross section σ_{no}^C for charge transfer from an initial bound state ψ_0^B of the system $(B+e^-)$ to a final bound state ϕ_n^A of the system $(A+e^-)$, these equations must be solved subject to the boundary conditions

$$\lim_{t\to-\infty} a_m(\mathbf{b},t) = \delta_{mo}\ , \qquad \lim_{t\to-\infty} c_n(\mathbf{b},t) = 0 \quad (3.15)$$

and one has

$$\sigma_{no}^C = \int d^2\mathbf{b}\,|c_n(\mathbf{b},t=+\infty)|^2 \quad (3.16)$$

As the velocity increases, the oscillatory factors $\exp(i\mathbf{v}\cdot\mathbf{r}_B)$, which represent the gain in momentum of the captured electron, lead to a rapid decrease of the matrix elements responsible for charge transfers. As a result, charge transfer cross sections fall off much more rapidly at high energies than excitation cross sections.

Charge transfer in the high-velocity limit

The high-energy limit of charge-transfer cross sections has been the subject of many investigations, which have been reviewed by Shakeshaft and Spruch[122]. We simply mention here that this high-energy limit is not given by the first Born approximation, but by the second Born term. For example, in the case of ground state capture of an electron by a nucleus A of charge Z_A from a nucleus B of charge Z_B the leading term of the second Born approximation is given by

$$\sigma_{B2} = \sigma_{BK}\,[0.295 + \frac{5\pi v}{2^{11}(Z_A+Z_B)}] \quad (3.17)$$

where

$$\sigma_{BK} = 2^{18}\pi \frac{(Z_A Z_B)^5}{5v^{12}} \qquad (3.18)$$

is the asymptotic Brinkman-Kramers[123] first order cross section. Thus we see that σ_{B2} decreases like v^{-11} at large v, while σ_{BK} falls off like v^{-12}, a result first obtained by Drisko[124], who also gave arguments to show that the third Born term modifies the coefficient 0.295 of the first term in (3.17) to the value 0.315; this has been confirmed recently by Shakeshaft[125]. It must be stressed, however, that the approach to the high-velocity limit is slow, and that the second term on the right of (3.17) becomes dominant only at energies $E > 40\ (Z_A+Z_B)^2$ MeV/nucleon, whereas for energies higher than 9 MeV/ nucleon the radiative capture process $A + (B+e^-) \rightarrow (A+e^-) + B + h\nu$ is the dominant charge transfer mechanism[126].

Improved treatments of fast charge transfer collisions

Let us consider now the region of intermediate and high (but not asymptotic) velocities. The atomic expansion (3.14) which we wrote above contains truncated summations over a certain number of discrete atomic eigenstates centered about A and B. This truncation of the expansion is often an important limitation, partly because it does not take into account the continuum states, and partly because atomic expansions are slowly convergent at short internuclear distances, where the wave function resembles that to the united atom.

Several methods have been proposed to overcome these difficulties, which have been reviewed recently by Bransden[107], Lin and Richard[127] and McCarroll[128]. Since they will be discussed at length by Bransden during this meeting, I will only mention them very briefly. One possibility, introduced by Gallaher and Wilets[129] and developed by Shakeshaft[130] is to make the expansion in terms of Sturmian functions. These functions have the advantage of forming an infinite discrete set which, in its entirety, is complete. Another approach, proposed by Cheshire et al.[131] is to introduce pseudostates to represent continuum states. An interesting alternative[132] is to use atomic wave functions with variable charges which are determined in a variational way. Another method is to expand the scattering wave function about three, rather than two centers, for example about the two nuclei A and B and their center of charge[133].

Other modifications of the scattering function include the use of "switching factors", which are chosen in order to remove the difficulties associated with the electron translation factors at short internuclear distances. To understand this point, let us return to the two-center atomic expansion (3.14). Choosing the origin O of our coordinate system to be an arbitrary point along the internuclear axis (see Fig. 3) and using (3.1), we may write the expansion (3.14) as

$$\Psi(\mathbf{r},\mathbf{b},t) = \sum_m a_m(\mathbf{b},t)\, \psi_m^B(\mathbf{r}_B)\, \exp[-iw_m^B t]\exp[-i(q\, \mathbf{v}\cdot\mathbf{r} + \frac{1}{2} q^2v^2t)]$$

$$+ \sum_n c_n(\mathbf{b},t)\, \phi_n^A(\mathbf{r}_A)\, \exp[-iw_n^A t]\, \exp[i(p\mathbf{v}\cdot\mathbf{r} - \frac{1}{2} p^2v^2t)] \quad (3.19)$$

The factors $\exp(ip\mathbf{v}\cdot\mathbf{r})$ and $\exp(-iq\mathbf{v}\cdot\mathbf{r})$ are required at large R in order to describe correctly the asymptotic states. However, they have the undesirable property of associating the electron with one or the other nucleus at small R. To avoid this difficulty, Schneiderman and Russek[134] suggested to use more complicated electron translation factors that reduce to the plane wave form as $R \to \infty$. This can be done by replacing in (3.19) the factors $\exp(ip\mathbf{v}\cdot\mathbf{r})$ and $\exp(-iq\mathbf{v}\cdot\mathbf{r})$ by expressions of the form $\exp[if_n(\mathbf{r},\mathbf{R})\mathbf{v}\cdot\mathbf{r}]$, where the switching factors f_n must satisfy the conditions that $f_n \to p$ when $R \to \infty$ with r_A finite and $f_n \to -q$ when $R \to \infty$ with r_B finite. Elsewhere the switching factors are chosen on physical grounds in order to give the best dynamical picture of the electron motion.

The calculations described above are usually very lengthy, but there are some circumstances in which it is possible to obtain accurate results with simpler approaches. For example, except for the case of resonance, the cross section for capture into the ground state has a maximum near incident velocities v which are close to the orbital velocity v_0 of the target electron. At these velocities, capture occurs essentially at large impact parameters, for which the scattering wave function can be approximated by a simple two state expansion involving only the initial and final states. Excellent results for K-K electron transfer at velocities $v \simeq v_0$ have been obtained in this way by Lin et al.[135].

Another example where simplifications can occur is if the charge Z_A of the projectile is much less than the charge Z_B of the target nucleus. In this case the interaction between the incident particle A and the electron can be treated as a perturbation, and the scattering wave function can be expanded in terms of atomic states and pseudostates centered on the target atom. This method has been used by Reading et al.[136] to a number of inner-shell capture processes. In the opposite type of asymmetric charge transfer collision, where a fully stripped, multicharged ion of charge $Z_A \gg Z_B$ captures the electron from the target atom $(B+e^-)$, the basic physical process is different. Except at very high velocities, electron capture from the outer shells takes place mainly into excited states of the system ($A+e^-$). Since many states are populated, the coupled-channel expansion (3.14) is impractical. However, at higher collision velocities, the coupling between final excited channels can be neglected and capture to an individual final state can be approximated by a two-state model. In this spirit, Ryufuku and Watanabe[137] have proposed a simple unitarized distorted wave approximation (UDWA) for calculating total charge transfer cross sections in the case of highly charged, fully stripped ions incident on hydrogen atoms; the agreement with experiment is satisfactory. Finally, we note that an eikonal approximation has been applied by Dewangan[138] and by Tsuji and Narumi[139] to the charge transfer processes $H^+ + H(1s) \rightarrow H(n=1,2) + H^+$. The method has been generalized recently by Chan and Eichler[140] for charge transfer to arbitrary hydrogenic states of the projectile. Despite various shortcomings, it is useful when capture into excites states dominates, as in the case of collisions involving multicharged ions. The classical model of Abrines and Percival[141] has been applied with success in the intermediate velocity range, in particular for the calculation of charge transfer from Rydberg atoms[142-144] and of charge transfer and ionization for fully stripped ions colliding with atomic hydrogen[145]. The charge transfer processes of highly charged ions with atoms have been reviewed by several authors[146-148]. The recent review of Janev and Presnyakov[147] contains a general survey of collision processes of multicharged ions with atoms, including excitation, ionization and charge exchange, for slow as well as fast collisions. Cross sections scaling laws for charge exchange, ionization and electron-loss in collisions of multicharged ions with atoms have been discussed recently by Janev and Hvelplund[149]. We also mention the importance of charge transfer as an ionization process (charge exchange in the continuum), especially for collisions involving highly charged ions[150].

We have seen above that at asymptotically high velocities the second Born term dominates the charge transfer cross section, but that the approach to the high-velocity limit is slow. If we take a finite basic set (even including pseudostates) and solve the corresponding coupled equations, the cross sections will approach the Brinkman-Kramers values at high velocities, which is incorrect. Clearly it is desirable to have other methods which include (approximately) the coupling to all intermediate states, and in particular to continuum states. Among the methods proposed are the impulse approximation[151-154], the continuum intermediate state (CIS) approximation[155] and the continuum distorted wave (CDW) approximation[156-158,108]. Of these the CIS and CDW methods have proved to be the most successful, giving results in excellent agreement with experiment for many systems at high colliding velocities. Detailed discussions of high-energy methods may be found in the reviews of Basu et al.[159] and Belkić et al.[160]

3.2. Slow atom (ion) - atom collisions

When the relative velocity v of the collision is smaller than the classical orbital velocity v_0 of the active electron(s) in the process, it is reasonable to use a "molecular picture" of the collision, in which the scattering wave function is expanded in molecular eigenstates of the "compound system" made of the projectile and the target. This approach, introduced by Massey and Smith[161], is termed the perturbed stationary states (PSS) method and will now be briefly examined.

The perturbed stationary states method

Let us consider again the one-electron model in which a nucleus A is incident on the system $(B+e^-)$. We shall assume that the wave number k associated with the relative motion is large enough so that a classical description of this nuclear motion can be made. The full quantum mechanical formulation of the molecular state expansion can be developed in a similar way and may be found for example in the review of McCarroll[128].

Let us introduce the adiabatic electronic molecular orbitals (MO) $\chi_k(\underset{\sim}{r},\underset{\sim}{R})$, defined by

$$H_{e\ell}\ \chi_k(\underset{\sim}{r},\underset{\sim}{R}) = \varepsilon_k(R)\ \chi_k(\underset{\sim}{r},\underset{\sim}{R}) \tag{3.20}$$

where $H_{e\ell}$ is the electronic Hamiltonian. In its original form, the PSS method consists in expanding the scattering wave function on the basis set of the adiabatic MO wave functions χ_k as

$$\Psi(\underset{\sim}{r},\underset{\sim}{b},t) = \sum_k d_k(\underset{\sim}{b},t)\ \chi_k(\underset{\sim}{r},\underset{\sim}{R})\ \exp[-i\int^t \varepsilon_k\, dt'] \tag{3.21}$$

and in this way one obtains for the coefficients d_k the set of coupled first order differential equations

$$i\,\dot{d}_k(\underset{\sim}{b},t) = \sum_{m\neq k} V_{km}(\underset{\sim}{R})\ \exp[-i\int^t (\varepsilon_k-\varepsilon_m)dt']\,d_m(\underset{\sim}{b},t) \tag{3.22}$$

where

$$V_{km} = \langle \chi_k|\frac{\partial}{\partial t}|\chi_m\rangle \tag{3.23}$$

and the time derivative $\partial/\partial t$ is to be performed with the electron coordinate $\underset{\sim}{r}$ constant with respect to a fixed (laboratory) space axis, chosen as the quantization axis. In general the molecular wave functions are specified with respect to a body-fixed axis, with the internuclear line AB (see Fig. 3) taken as the axis of quantization. Account must be taken of the rotation of the internuclear axis during the collision. Thus, if we denote by $\chi_m(\underset{\sim}{r}',\underset{\sim}{R})$ the MO's quantized along the molecular axis AB we have[106]

$$\frac{\partial}{\partial t} \chi_m(\underset{\sim}{r}',\underset{\sim}{R}) = (v_R \frac{\partial}{\partial R} - i\, \dot{\theta}\, L_{y'})\, \chi_m(\underset{\sim}{r}',\underset{\sim}{R}) \qquad (3.24)$$

where $v_R = dR/dt$ is the radial velocity, $\dot{\theta} = bv/R^2$ is the angular velocity of the internuclear axis and $L_{y'}$ is the y' component of the electronic orbital angular momentum in the rotating coordinate system. The two parts ($v_R\, \partial/\partial R$) and ($- i\theta L_{y'}$) of the coupling are called the radial and rotational coupling, respectively. The radial coupling connects MO's having the same angular symmetry (i.e. $\sigma \leftrightarrow \sigma$, $\pi \leftrightarrow \pi$, etc...) while the rotational coupling connects states of different angular symmetry, e.g. $\sigma \leftrightarrow \pi$. The radial coupling is the dominant contribution near an avoided crossing between two adiabatic molecular potential curves corresponding to states of the same symmetry. On the other hand, rotational coupling can be very important at small internuclear distances, as the σ, π, ... states become degenerate in the united atom limit. When the radial coupling varies too rapidly near an avoided crossing, it is useful to work in a new "diabatic" basis chosen in such a way that the radial coupling remains small, and the system evolves in a gradual and continuous fashion[162].

In order to decide which MO's are important in the coupled-state calculations, it is useful to draw a correlation diagram which relates the molecular energy levels in the separated atom (SA) and united atom (UA) limits, and therefore summarizes the qualitative features of the energy levels as a function of the internuclear distance R. In general, the dominant transitions in slow atom (ion)-atom collisions are those involving degeneracies of molecular energy levels, which fall into three categories : i) those associated with the separated atoms ($R \to \infty$), ii) those associated with the united atom ($R \to 0$) and iii) those occuring at some finite R, for example in the case of curve crossings. As a result, these dominant transitions tend to be highly specific. For example, the cross section for excitation of the 3p level of atomic hydrogen by slow protons is much smaller than the corresponding cross section for 2p excitation. This is due to the fact that there is no mediating degeneracy for the first transition, while the second one is mediatedby a degeneracy between the $2p\sigma_u$ and $2p\pi_u$ states as $R \to 0$. An important application of these ideas is the "promotion" phenomenon, namely the increase in principal quantum number of certain of the MO's during the transition from $R = \infty$ to $R = 0$. This promotion phenomenon is a major ingredient of the Fano-Lichten model which has been applied successfully to estimate the excitation probability for inner-shell electrons in slow homonuclear[163] and heteronuclear[164] heavy-ion-atom collisions. Another example of selectivity occurs in the case of a collision between a multicharged ion A and a neutral atom B. Here the Coulomb repulsion in the final channel creates a number of crossings. For a given velocity, transitions occur in a limited range of internuclear distances which lead to capture of the electron into selected excited states of the $(A+e^-)$ system.

The simplest non-trivial PSS expansion (3.21) is that containing two states; it has been very successful in explaining the measurements of the charge exchange probability for symmetric systems, which oscillates nearly between zero and one as a function of velocity for a fixed scattering angle[9]. In addition, various approximate analytical solutions of the two-state problem have been proposed, such as the Landau-Zener-Stückelberg[165] and the Demkov[166] models. Unfortunately, these approximate treatments are often inaccurate. An even more serious problem is that the original, multistate PSS expansion (3.21) itself suffers from severe deficiencies, to which we now turn our attention.

Modified perturbed stationary states expansions. Travelling molecular orbitals and electron translation factors

The PSS matrix elements V_{km} given by eq. (3.23) exhibit a dependence on the origin of the reference system, and some of them do not even vanish asymptotically as $R \to \infty$. If the complete set of adiabatic MO's χ_k were used in (3.21) there would be no difficulty, but of course truncation of the expansion (3.21) to a finite number of MO's must be made, and it is this truncation which is responsible for the pathological behaviour of the expansion (3.21). A detailed account of the deficiencies of the original PSS method, and of the remedies which have been proposed may be found in the recent review article of Delos[167]. An interesting discussion of the inconsistencies rooted in the description of slow atom (ion) - atom collisions in the "molecular picture" has also been given by Salin[168].

The first task is to ensure translational invariance of the scattering equations even when a finite basis is used. The asymptotic conditions require that the basis functions at large internuclear separation approach the separated atom wave functions times an appropriate electron translation factor (ETF) of the type already discussed above in Section 3.1. This suggests replacing the original PSS expansion (3.21) by an expansion in terms of travelling molecular orbitals,

$$\Psi(\underset{\sim}{r},\underset{\sim}{b},t) = \sum_k d_k(\underset{\sim}{b},t)\ \chi_k(\underset{\sim}{r},\underset{\sim}{R})\ \exp[\,i f_k(\underset{\sim}{r},\underset{\sim}{R})\underset{\sim}{v}\cdot\underset{\sim}{r} - i\int^t \varepsilon_k\, dt'] \qquad (3.25)$$

Each basis function will then have the correct asymptotic form if $f_k \to p$ when $R \to \infty$ with r_A finite, and $f_k \to -q$ when $R \to \infty$ with r_B finite. The form of the electron translation factor $\exp[if_k\, \chi \cdot \underset{\sim}{r}]$ for finite R is in principle arbitrary.

The first type of electron translation factor is the plane wave type proposed by Bates and McCarroll[120] and discussed above for fast collisions. It corresponds to the choice $f_k = p$ if χ_k dissociates to the center A, and $f_k = -q$ if it dissociates to the center B. With the introduction of these plane wave translation factors the coupling matrix elements can be shown to vanish for large R, and there is no dependence on the reference frame[128]. However, it should be remembered that the improvement may be formal, since the choice of translation factor is arbitrary for finite R. Extensive calculations using the Bates-McCarroll plane wave translation factors have been performed recently for He^{2+} - H collisions by Hatton et al.[169] and Winter et al.[170].

The Bates-McCarroll translation factors lead to complicated exchange and overlap integrals, and in addition they present the disadvantage of associating the electron with one or the other nucleus at small R, as we pointed out in Section 3.1. For this reason, Schneiderman and Russek[134] argued that a more flexible form of the switching factors f_k should be chosen, which reduces to the plane wave translation factors as $R \to \infty$, but also takes into account the fact that the electron ceases to be localized on either nucleus at short internuclear separations. The switching factor introduced by Schneiderman and Russek is independent of k, which simplifies the calculations. It has been adopted in several recent calculations for heavy-ion-atom collisions[171-173]. Unfortunately, there is no guarantee that a translation factor chosen in an "ad hoc" way will be optimal. Thus, instead of using arbitrary forms of translation factors, it is important to formulate requirements that would help in their construction. Ideally, a variational principle may be used to determine the switching function. In this way, Riley and Green[174] have obtained general equations for the switching factors, and more recently progress along these lines has also been made by Thorson and Delos[175], Crothers and Hughes[176], Ponce[177] and Schmid[178]. Clearly this is a problem whose time has come.

REFERENCES

1. C.J. Joachain, in Atomic and Molecular Processes in Controlled Thermonuclear Fusion, ed. by M.R.C. McDowell and A.M. Ferendeci (Plenum Press, New York, 1980), p. 147.
2. P.G. Burke and J.F. Williams, Phys. Reports 34 C, 325 (1977).
3. W.D. Robb, in Atomic and Molecular Processes in Controlled Thermonuclear Fusion, ed. by M.R.C. McDowell and A.M. Ferendeci (Plenum Press, New York, 1980), p. 245.

4. R.J.W. Henry, Phys. Reports 68, 1 (1981).
5. B.H. Bransden and M.R.C. McDowell, Phys. Reports 30 C, 207 (1977); 46 C, 249 (1978).
6. F.W. Byron, Jr. and C.J. Joachain, Phys. Reports 34 C, 233 (1977).
7. H.S.W. Massey and C.B.O. Mohr, Proc. Roy. Soc. A 136, 289 (1932).
8. M.J. Seaton, Phil. Trans. Roy. Soc. (London) 245, 469 (1953).
9. See for example B.H. Bransden, Atomic Collision Theory (Benjamin, New York, 1970).
10. L. Castillejo, I. Percival and M.J. Seaton, Proc. Roy. Soc. A 254, 259 (1960).
11. R. Damburg and E. Karule, Proc. Phys. Soc. 90, 637 (1967).
12. A. Temkin, Phys. Rev. 107, 1004 (1957).
13. P.G. Burke and H.M. Schey, Phys. Rev. 126, 147 (1962).
14. Vo Ky Lan, M. Le Dourneuf and P.G. Burke, J. Phys. B 9, 1065 (1976).
15. C. Schwartz, Phys. Rev. 124, 1468 (1961).
16. R.L. Armstead, Phys. Rev. 171, 91 (1968).
17. M.R.H. Rudge, J. Phys. B 8, 940 (1975).
18. M.H. Mittleman, Phys. Rev. 147, 69 (1966).
19. R.K. Nesbet, Phys. Rev. 156, 99 (1967); Adv. Atom. Molec. Phys. 13, 315 (1977).
20. R.K. Nesbet, Variational Methods in Electron-Atom Scattering Theory (Plenum Press, New York, 1980).
21. M.A. Crees, M.J. Seaton and P.M.H. Wilson, Comp. Phys. Comm. 15, 23 (1978).
22. R.K. Nesbet, Comp. Phys. Comm. 6, 275 (1973).
23. R.J.W. Henry, S.P. Rountree and E.R. Smith, Comp. Phys. Comm. (1981).
24. K.A. Berrington, P.G. Burke, M. Le Dourneuf, W.D. Robb, K.T. Taylor and Vo Ky Lan, Comp. Phys. Comm. 14, 367 (1978).
25. J. Hata, L.A. Morgan and M.R.C. McDowell, J. Phys. B 13, 4453 (1980).
26. A. Temkin, in Autoionization, ed. by A. Temkin (Mono Book Corp., 1966), p. 55.
27. A.V. Hazi and H.S. Taylor, Phys. Rev. A 1, 1109 (1970).
28. H.S. Taylor, Adv. Chem. Phys. 18, 91 (1970).
29. H.S. Taylor and L.D. Thomas, Phys. Rev. Letters 28, 1091 (1972).
30. L.D. Thomas, J. Phys. B 7, L 97 (1974).
31. E.J. Heller and H.A. Yamani, Phys. Rev. A 9, 1201 (1974); A 9, 1209 (1974).
32. J.T. Broad and J. Reinhardt, J. Phys. B 9, 1491 (1976); Phys. Rev. A 14, 2159 (1976).
33. E.P. Wigner, Phys. Rev. 70, 15 (1946); 70, 606 (1946); E.P. Wigner and L. Eisenbud, Phys. Rev. 72, 29 (1947).
34. P.G. Burke, A. Hibbert and W.D. Robb, J. Phys. B 4, 153 (1971).
35. P.G. Burke and W.D. Robb, J. Phys. B 5, 44 (1972).
36. P.G. Burke and W.D. Robb, Adv. Atom. Molec. Phys. 11, 143 (1975).

37. P.J.A. Buttle, Phys. Rev. 160, 719 (1967).
38. D.J. Zvijac, E.J. Heller and J.C. Light, J. Phys. B 8, 1016 (1975).
39. C. Bloch, Nucl. Phys. 4, 503 (1957).
40. M.J. Seaton, Mon. Not. Roy. Astron. Soc. 118, 504 (1958); Proc. Phys. Soc. 88, 801 (1966).
41. M. Gailitis, Soviet Phys. JETP 17, 1328 (1963).
42. U. Fano, Phys. Rev. A 2, 353 (1970); A 15, 817 (1977).
43. Gy Csanak, H.S. Taylor and R. Yaris, Adv. Atom. Molec. Phys. 7, 287 (1971); L.D. Thomas, Gy Csanak, H.S. Taylor and B.S. Yarlagadda, J. Phys. B 7, 1719 (1974).
44. P.G. Burke, Adv. Atom. Molec. Phys. 4, 173 (1968).
45. D.E. Golden, Adv. Atom. Molec. Phys. 14, 1 (1978).
46. H. Feshbach, Ann. Phys. (N.Y.) 5, 357 (1958); 19, 287 (1962).
47. T.F. O'Malley and S. Geltman, Phys. Rev. 137, A 1344 (1965); A.K. Bhatia and A. Temkin, Phys. Rev. 182, 15 (1969); Phys. Rev. A 8, 2184 (1973); A 10, 458 (1974); A 11, 2018 (1975).
48. P.G. Burke and T.G. Webb, J. Phys. B 3, L 131 (1970).
49. J. Callaway and J.W. Wooten, Phys. Lett. A 45, 85 (1973); Phys. Rev. A 9, 1924 (1974); Phys. Rev. A 11, 1118 (1975).
50. J. Callaway, M.R.C. McDowell and L.A. Morgan, J. Phys. B 8, 2181 (1976); J. Phys. B 9, 2043 (1976).
51. S.L. Willis and M.R.C. McDowell, J. Phys. B 14, L 453 (1981).
52. W.C. Fon, K.A. Berrington, P.G. Burke and A.E. Kingston, J. Phys. B 12, 1861 (1979); W.C. Fon, K.A. Berrington and A.E. Kingston, J. Phys. B 13, 2309 (1980).
53. P.G. Burke, K.A. Berrington and C.V. Sukumar, J. Phys. B 14, 289 (1981).
54. B.H. Bransden and J.P. Coleman, J. Phys. B 5, 537 (1972); K.H. Winters, C.D. Clark, B.H. Bransden and J.P. Coleman, J. Phys. B 6, L 247 (1973); J. Phys. B 7, 788 (1974). See also ref. 5.
55. T. Scott and B.H. Bransden, J. Phys. B 14, 2277 (1981).
56. W.P. Reinhardt, Comp. Phys. Comm. 17, 1 (1979).
57. L. Schlessinger and C. Schwartz, Phys. Rev. Letters 16, 1173 (1966).
58. L. Schlessinger, Phys. Rev. 171, 1523 (1968).
59. F.A. McDonald and J. Nuttall, Phys. Rev. Letters 23, 361 (1969); Phys. Rev. A 4, 1821 (1971).
60. G. Doolen, G. McCartor, F.A. McDonald and J. Nuttall, Phys. Rev. A 4, 108 (1971).
61. T.N. Rescigno and W.P. Reinhardt, Phys. Rev. A 10, 1584 (1974).
62. P.W. Langhoff, Chem. Phys. Lett. 22, 60 (1973).
63. P.W. Langhoff and W.P. Reinhardt, Chem. Phys. Lett. 24, 495 (1974).
64. J.R. Winick and W.P. Reinhardt, Phys. Rev. A 18, 910 (1978); Phys. Rev. A 18, 925 (1978).
65. K.L. Bell and A.E. Kingston, Adv. Atom. Molec. Phys. 10, 53 (1974).

66. C.J. Joachain, Quantum Collision Theory (North Holland, Amsterdam, 1975), Chapter 19.
67. A.R. Holt, J. Phys. B 5, L 6 (1972).
68. A.M. Ermolaev and H.R.J. Walters, J. Phys. B 12, L 779 (1979).
69. F.W. Byron, Jr. and C.J. Joachain, J. Phys. B 14, 2429 (1981).
70. A.M. Ermolaev and H.R.J. Walters, J. Phys. B 13, L 473 (1980).
71. V.I. Ochkur, Sov. Phys. JETP 18, 503 (1964).
72. F.W. Byron, Jr., C.J. Joachain and R.M. Potvliege, J. Phys. B 15, 3915 (1982).
73. F.W. Byron, Jr., C.J. Joachain and B. Piraux, J. Phys. B 13, L 673 (1980).
74. F.W. Byron, Jr., C.J. Joachain and B. Piraux, J. Phys. B 15, L 293 (1982); H. Ehrhardt, M. Fischer, K. Jung, F.W. Byron, Jr., C.J. Joachain and B. Piraux, Phys. Rev. Letters 48, 1807 (1982).
75. F.W. Byron, Jr., C.J. Joachain and B. Piraux, to be published.
76. M.R.H. Rudge and M.J. Seaton, Proc. Roy. Soc. A 283, 262 (1965).
77. M.R.H. Rudge, Rev. Mod. Phys. 40, 564 (1968).
78. R.K. Peterkop, Theory of Ionization of Atoms by Electron Impact (Colorado Univ. Press, 1977).
79. M.R.H. Rudge and S.B. Schwartz, Proc. Phys. Soc. 88, 563 (1966); 88, 579 (1966).
80. H. Jacubowicz and D.L. Moores, Comm. on Atom. Molec. Phys. 9, 55 (1980); J. Phys. B 14, 3733 (1981).
81. R.J. Glauber, in Lectures in Theoretical Physics, vol. 1, ed. by W.E. Brittin (Interscience, New York, 1959), p. 315.
82. C.J. Joachain and C. Quigg, Rev. Mod. Phys. 46, 279 (1974).
83. D.M. Chase, Phys. Rev. 104, 838 (1956).
84. F.W. Byron, Jr. and C.J. Joachain, Phys. Rev. A 8, 1267 (1973); A 8, 3266 (1973), J. Phys. B 7, L 212 (1974); B 10, 207 (1977); Phys. Reports 34 C, 233 (1977).
85. F.W. Byron, Jr. and C.J. Joachain, J. Phys. B 8, L 284 (1975).
86. S.J. Wallace, Ann. Phys. (N.Y.) 78, 190 (1973).
87. F.W. Byron, Jr., C.J. Joachain and E.H. Mund, Phys. Rev. D 11, 1662 (1975); Phys. Rev. C 20, 2325 (1979).
88. F.W. Byron, Jr., C.J. Joachain and R.M. Potvliege, J. Phys. B 14, L 609 (1981).
89. V. Franco and Z. Iwinski, Phys. Rev. A 25, 1900 (1982).
90. K. Unnikrishnan and M.A. Prasad, J. Phys. B 15, 1549 (1982).
91. C.J. Joachain, ref. 66, Chapter 20.
92. M.H. Mittleman and K.M. Watson, Phys. Rev. 113, 198 (1959).
93. M.H. Mittleman and K.M. Watson, Ann. Phys. (N.Y.) 10, 268 (1960).
94. C.J. Joachain and M.H. Mittleman, Phys. Rev. A 4, 1492 (1971); F.W. Byron, Jr. and C.J. Joachain, Phys. Rev. A 9, 2559 (1974).
95. F.W. Byron, Jr. and C.J. Joachain, Phys. Letters A 49, 306 (1974); Phys. Rev. A 15, 128 (1977); C.J. Joachain, R. Vanderpoorten, K.H. Winters and F.W. Byron, Jr., J. Phys. B 10, 227 (1977).
96. R. Vanderpoorten, J. Phys. B 8, 926 (1975).

97. J.B. Furness and I.E. McCarthy, J. Phys. B 6, 2280 (1973).
98. M.E. Riley and D.G. Truhlar, J. Chem. Phys. 65, 792 (1976).
99. D.P. Dewangan and H.R.J. Walters, J. Phys. B 10, 637 (1977); A.E. Kingston and H.R.J. Walters, J. Phys. B 13, 4633 (1980).
100. I.E. McCarthy and E. Wiegold, Phys. Reports 27 C, 275 (1976); E. Wiegold and I.E. McCarthy, Adv. Atom. Molec. Phys. 14, 127 (1978).
101. A. Giardini-Guidoni, R. Fantoni, R. Camilloni and G. Stefani, Comm. on Atom. Molec. Phys. 10, 107 (1981).
102. M. Gryzinski, Phys. Rev. 115, 374 (1959).
103. I.C. Percival and D. Richards, Adv. Atom. Molec. Phys. 11, 1 (1975).
104. I.C. Percival, in Atomic and Molecular Collision Theory, ed. by F.A. Gianturco (Plenum Press, New York, 1982), p. 431.
105. J.P. Coleman and M.R.C. McDowell, Introduction to the Theory of Ion-Atom Collisions (North-Holland, Amsterdam, 1970); see also ref. 9.
106. J.S. Briggs, Rep. Progr. Phys. 39, 217 (1976).
107. B.H. Bransden, Adv. Atom. Molec. Phys. 15, 263 (1979).
108. Dz. Belkić and A. Salin, J. Phys. B 9, L 397 (1976).
109. R. Shakeshaft, Phys. Rev. A 16, 1458 (1977).
110. M.R. Flannery, J. Phys. B 2, 913 (1969); B 2, 1044 (1969); Phys. Rev. 183, 231 (1969); 183, 241 (1969); J. Phys. B 3, 306 (1969).
111. M.R. Flannery and K.J. McCann, Phys. Rev. A 8, 2915 (1973); J. Phys. B 7, 840 (1974); B 7, 1349 (1974); B 7, 1558 (1974).
112. K.L. Bell, A.E. Kingston and W.A. McIlveen, J. Phys. B 6, 1246 (1973); K.L. Bell, A.E. Kingston and T.G. Winter, J. Phys. B 7, 1339 (1974).
113. K.L. Bell and A.E. Kingston, J. Phys. B 11, 1259 (1978).
114. A.R. Holt and B.L. Moiseiwitsch, J. Phys. B 1, 36 (1968); A.R. Holt, J. Phys. B 2, 1253 (1969).
115. B.H. Bransden and D.P. Dewangan, J. Phys. B 12, 1377 (1979).
116. A.S. Ghosh and N.C. Sil, J. Phys. B 4, 836 (1971).
117. V. Franco and B.K. Thomas, Phys. Rev. A 4, 945 (1971).
118. D.P. Dewangan, J. Phys. B 11, L 37 (1978).
119. J.F. Reading, A.L. Ford and E.O. Fitchard, Phys. Rev. A 16, 133 (1977); E.O. Fitchard, A.L. Ford and J.F. Reading, Phys. Rev. A 16, 1325 (1977); G.L. Swafford, J.F. Reading, A.L. Ford and E. Fitchard, Phys. Rev. A 16, 1329 (1977).
120. D.R. Bates and R. McCarroll, Proc. Roy. Soc. A 245, 175 (1958).
121. B.H. Bransden, Rep. Progr. Phys. 35, 949 (1973).
122. R. Shakeshaft and L. Spruch, Rev. Mod. Physics 51, 369 (1979).
123. H.C. Brinkman and H.A. Kramers, Proc. Acad. Sci. Amst. 33, 973 (1930).
124. R.M. Drisko, Ph. D. Thesis, Carnegie Institute of Technology (1955).
125. R. Shakeshaft, Phys. Rev. A 17, 1011 (1978).
126. J.S. Briggs and K. Dettmann, Phys. Rev. Letters 33, 1123 (1974);

J. Phys. B 10, 1113 (1977).
127. C.D. Lin and P. Richard, Adv. Atom. Molec. Phys. 17, 275 (1981).
128. R. McCarroll, in Atomic and Molecular Collision Theory, ed. by F.A. Gianturco (Plenum Press, New York, 1982), p. 165.
129. D.F. Gallaher and L. Wilets, Phys. Rev. 169, 139 (1968).
130. R. Shakeshaft, Phys. Rev. A 14, 1626 (1976).
131. I.M. Cheshire, D.F. Gallaher and A.J. Taylor, J. Phys. B 3, 813 (1970).
132. I.M. Cheshire, J. Phys. B 1, 428 (1968).
133. D.G.M. Anderson, M.J. Antøl and M.B. McElroy, J. Phys B 7, L 118 (1974); M.J. Antøl, M.B. McElroy and D.G.M. Anderson, J. Phys. B 8, 1513 (1975).
134. S.B. Schneiderman and A. Russek, Phys. Rev. 181, 311 (1969).
135. C.D. Lin, J. Phys. B 11, L 185 (1978); C.D. Lin, S.C. Soong and L.N. Tunnell, Phys. Rev. A 17, 1646 (1978); C.D. Lin and L.N. Tunnell, J. Phys. B 12, L 485 (1979).
136. J.F. Reading and A.L. Ford, J. Phys. B 12, 1367 (1979); A.L. Ford, J.F. Reading and R.L. Becker, J. Phys. B 12, 2905 (1979); A.L. Ford, J.F. Reading and R.L. Becker, Phys. Rev. A 23, 510, (1981).
137. H. Ryufuku and T. Watanabe, Phys. Rev. A 18, 2005 (1978); Phys. Rev. A 19, 1538 (1979).
138. D.P. Dewangan, J. Phys. B 8, L 119 (1975); J. Phys. B 10, 1083 (1977).
139. A. Tsuji and H. Narumi, J. Phys. Soc. Japan 41, 357 (1976).
140. F.T. Chan and J. Eichler, Phys. Rev. Letters 42, 58 (1979); J. Eichler and F.T. Chan, Phys. Rev. A 20, 104 (1979); A 20, 1081 (1979).
141. R. Abrines and I.C. Percival, Proc. Phys. Soc. 88, 861 (1966); 88, 873 (1966). See also ref. 103.
142. D. Banks, K.S. Barnes and J.B. Wilson, J. Phys. B 9, L 141 (1976).
143. A. Salop, J. Phys. B 12, 919 (1979).
144. R.E. Olson, J. Phys. B 13, 483 (1980).
145. R.E. Olson and A. Salop, Phys. Rev. A 16, 531 (1977).
146. L.P. Presnyakov, in Electronic and Atomic Collisions, ed. by G. Watel (North-Holland, Amsterdam, 1978), p. 407.
147. R.K. Janev and L.P. Presnyakov, Phys. Reports 70, 1 (1981).
148. P.T. Greenland, Phys. Reports 81, 131 (1982).
149. R.K. Janev and P. Hvelplund, Comments on Atom. Molec. Phys. 11, 75 (1981).
150. See for example J. Macek, Comments on Atom. Molec. Phys. 6, 169 (1977).
151. B.H. Bransden and I.M. Cheshire, Proc. Phys. Soc. 81, 820 (1963).
152. J.P. Coleman and S. Trelease, J. Phys. B 1, 172 (1968).
153. J.S. Briggs, J. Phys. B 10, 3075 (1977).
154. D.H. Jacubassa-Amundsen and P.A. Amundsen, Z. Phys. A 297, 203 (1980).

155. Dz. Belkić, J. Phys. B 10, 3491 (1977).
156. I.M. Cheshire, Proc. Phys. Soc. 84, 89 (1964).
157. R. Gayet, J. Phys. B 5, 483 (1972).
158. Dz. Belkić and R. Gayet, J. Phys. B 10, 1911 (1977); B 10, 1923 (1977); Dz.Belkić, and R. McCarroll, J. Phys. B 10, 1933 (1977).
159. D. Basu, S.C. Mukherjee and D.P. Sural, Phys. Reports 42 C, 145 (1978).
160. Dz. Belkić, R. Gayet and A. Salin, Phys. Reports 56, 279 (1979).
161. H.S.W. Massey and R.A. Smith, Proc. Roy. Soc. A 142, 142 (1933).
162. For a discussion of diabatic representations, see for example R. McCarroll, ref. 128.
163. U. Fano and W. Lichten, Phys. Rev. Letters 27, 635 (1971); W. Lichten, Phys. Rev. 164, 131 (1967).
164. M. Barat and W. Lichten, Phys. Rev. A 6, 211 (1972).
165. L.D. Landau, Z. Phys. Sowjet. 2, 46 (1932); C. Zener, Proc. Roy. Soc. A 137, 696 (1932); E.C.G. Stückelberg, Helv. Phys. Acta 5, 320 (1932).
166. Yu N. Demkov, Sov. Phys. JETP 18, 138 (1964); see also R.E. Olson, Phys. Rev. A 6, 1822 (1972).
167. J. Delos, Rev. Mod. Phys. 53, 287 (1981).
168. A. Salin, Comments on Atom. Molec. Phys. 9, 165 (1980).
169. G.J. Hatton, N.F. Lane and T.G. Winter, J. Phys. B 12, L 571 (1979).
170. T.G. Winter and G.J. Hatton, Phys. Rev. A 21, 793 (1980); T.G. Winter, G.J. Hatton and N.F. Lane, Phys. Rev. A 22, 930 (1980).
171. J.S. Briggs and J.H. Macek, J. Phys. B 5, 579 (1972).
172. K. Taulbjerg and J.S. Briggs, J. Phys. B 8, 1895 (1975).
173. W. Fritsch and U. Wille, J. Phys. B 11, 4019 (1978).
174. M.E. Riley and T.A. Green, Phys. Rev. A 4, 619 (1971).
175. W.R. Thorson and J.B. Delos, Phys. Rev. A 18, 117 (1978); A 18, 135 (1978).
176. D.S.F. Crothers and J.G. Hughes, Proc. Roy. Soc. A 359, 345 (1978); Phys. Rev. Letters 43, 1584 (1979).
177. V.H. Ponce, J. Phys. B 12, 3741 (1979), J. Phys. B 14, 2823 (1981).
178. G.B. Schmid, J. Phys. B 12, 3909 (1979).

CURRENT THEORETICAL TECHNIQUES FOR ELECTRON-ATOM AND ELECTRON-ION SCATTERING

M. R. C. McDowell

Department of Mathematics
Royal Holloway College
(University of London)
Egham Hill
Egham, Surrey TW20 OEX

1. INTRODUCTION

I do not propose to give a comprehensive discussion of all the theoretical methods available for electron collisions with atoms and ions. There are many excellent text books, including that by Joachain,[1] and a new edition of Bransden[2] is in preparation. A brief general survey was given by Joachain[3] at the first of these summer schools, and I will assume you are all familiar with its contents.

Rather, I will select those methods which I believe to be useful, and of general applicability, and attempt to discuss their success and failure. Where one wants detailed information on a particular collision process the first criterion in a choice of method is the energy range in which results are required. In this lecture I will confine myself to excitation processes from a target in a state $|i\rangle$ to a final state $|f\rangle$. If the initial state $|i\rangle$ is the ground state $|0\rangle$ then provided the energy of the incident electron is less than the ionisation potential of the target I_X, the first choice will be an eigenfunction expansion method, and this will be further discussed in section 2. Recent applications of these methods to excitation of positive ions are discussed in section 3. Conversely if the incident energy is many times I_X, the obvious choice of method is the First Born Approximation (FBA), provided an overall spin change is not involved, or in the case of positive ions the Coulomb Born Approximation (CBA). This will be the topic of section 4. The past decade has seen much effort devoted to developing methods for the intermediate energy range where neither of these criteria obtain. One can hope to push the eigenfunction expansion method to higher energies, with suitable modifications, or to employ higher order terms in the Born series, and to approach the problem from the high energy range.

In the case of a neutral atom target the most important effects appear to be

(i) long-range polarisation type interactions

(ii) exchange

(iii) short-range correlations

though their relative importance varies in different cases. When the target is a positive ion the polarisation interaction is dominated by the Coulomb, even for small values of the residual charge, $q = Z - N$, and for $q \gg 1$, this is the dominant interaction, except close to threshold. One would therefore expect that it would be somewhat easier to obtain results of a given accuracy for highly ionised systems, such as those of interest in controlled thermonuclear reaction devices, and this is indeed the case.

However, one must note that at high Z, relativistic (including spin-orbit) effects become important, both structurally and dynamically, and while progress in dealing with these is being made, there are as yet no general programs available.

It is much more difficult to deal with transitions between excited states. If the energy difference is ΔE_{if} and the two states in question are isolated, then the FBA may be useful if the incident energy $k_i^2 \gg \Delta E_{if}$, or if only a few states lie close to $|i>$ or $|f>$ then an eigenfunction expansion approach looks promising.

For many cases of interest detailed quantal calculations are impractical, but if the principal quantum numbers involved are large ($n_i, n_f > 5$ say), and not too different, total cross sections $\sigma_{n_i n_f}$ may be obtained from the classical methods developed by Percival and his collaborators.[4]

The ionisation problem is much more difficult, especially at low energies. This is because the asymptotic form of the wave function for two outgoing electrons is not well understood. The only available calculational procedures are distorted wave and other variants of the FBA or CBA, possibly with unitarisation via the K-matrix. The theory is outlined, and recent applications reviewed in section 5.

2. EIGENFUNCTION EXPANSION METHODS

We suppose that the incident electron energy is insufficient to further ionise the target, so that only a finite number of states can be excited. It is convenient pedagogically to eliminate target structure problems by considering only a one-electron target at this stage. Let $\phi_n(\underline{r}_1)$ be the target states when electron (1) is bound. Then we want to solve

$$H(1,2)\,\Psi(1,2) = E\,\Psi(1,2) \tag{2.1}$$

subject to the boundary conditions that the wave function is everywhere finite, and behaves asymptotically in the open channels as

$$\Psi(1,2) \underset{r_2\to\infty}{\longrightarrow} e^{i\underline{k}_0\cdot\underline{r}_2} + \sum_{j=0}^{N} f_{oj}(\hat{\underline{r}}_2)\,e^{ik_j r_2}/r_2 \tag{2.2}$$

but vanishes asymptotically in the closed channels. The sum runs over accessible states, and the $f_{oj}(\hat{\underline{r}}_2)$ are the (direct) scattering amplitudes.

Let us expand the total wave function in terms of the $\phi_n(\underline{r}_1)$, remembering to allow for the Pauli principle,

$$\Psi^{\pm}(\underline{x}_1,\underline{x}_2) = \sum_{j=0}^{\infty} (\phi_j(\underline{r}_1)F_j^{\pm}(\underline{r}_2) \pm \phi_j(\underline{r}_2)F_j^{\pm}(\underline{r}_1))\chi(1,2) \tag{2.3}$$

where $\chi(1,2)$ is a two electron spinor, and in the case of this two electron system the + sign refers to the singlet and the − sign to the triplet. Substituting (3) in (1), projecting on each $\phi_j(\underline{r}_1)$ in turn, and integrating over all co-ordinates (including spin) except $\underline{r}_2$, we obtain

$$(\nabla_2^2 + k_j^2)F_j^{\pm}(\underline{r}_2) = 2\sum_{k=0}^{\infty} (V_{jk} + (-1)^S W_{jk})F_k^{\pm}(\underline{r}_2) \tag{2.4}$$

where $S = 0$ or 1 is the total spin.

The direct potentials are

$$V_{jk} = \langle\phi_j\,|\frac{1}{r_{12}}|\,\phi_k\rangle - \frac{Z}{r_2}\delta_{jk} \tag{2.5}$$

while

$$W_{jk}\,F_k^{\pm}(\underline{r}_2) = \phi_k^{*}(\underline{r}_2)\int[\frac{1}{r_{12}} + \Delta E_{12} - E]\phi_j(\underline{r}_1)F_k^{\pm}(\underline{r}_1)d\underline{r}_1 \tag{2.6}$$

and

$$k_j^2 = 2(E - \varepsilon_j) \tag{2.7}$$

in which ε_j is the energy corresponding to ϕ_j. This so far is exact, and intractable. One is forced to truncate the summation to a few states $k = 0, \ldots, N$, say, and this is the close-coupling approximation,

$$(\nabla_2^2 + k_j^2) F_j^{\pm}(\underline{r}_2) = 2 \sum_{k=0}^{N} \{V_{jk} + (-1)^S W_{jk}\} F_k^{\pm}(\underline{r}_2) . \tag{2.8}$$

It is convenient to make a partial wave decomposition,

$$F_j^{\pm}(\underline{r}_2) = \sum_{\ell_2=0}^{\infty} c_{\ell_2} \frac{f^{\pm}(r_2)}{r_2} Y_{\ell_2 m_{\ell_2}}(\hat{\underline{r}}_2) \tag{2.9}$$

and to recouple the angular momentum of the two electrons, so that one considers scattering between channels of fixed total orbital angular momentum L, total spin S, and parity Π. The relevant equations then appear as a set of N_0 coupled integro-differential equations in the radial variable r_2, after integrating over angles, where N_0 is the number of channels. Thus, for example, if only the $n = 1,2$ states are retained, there are four coupled channels $|n = 1, \ell_1 = 0, \ell_2 = L\rangle$, $|n = 2, \ell_1 = 0, \ell_2 = L\rangle$, $|n = 2, \ell_1 = 1, \ell_2 = L - 1\rangle$ and $n_1 = 2, \ell_1 = 1, \ell_2 = L + 1\rangle$ for each L and S, and $\Pi = +1$, the odd parity channel $|n_1 = 2, \ell_1 = 1, \ell_2 = L\rangle$ being isolated.

The close coupling equations (8) may alternatively be derived from the Hulthen or Kohn variational principles by using (3) as a trial function. This suggests that systematic improvement of the results is possible by adding additional terms to the trial function. These may be of two kinds

(i) quadratically integrable functions $\psi_q(\underline{r}_1, \underline{r}_2)$ to represent short range correlations;

(ii) pseudostates, i.e. states $\phi_p(\underline{r}_1)$ which are not eigenfunctions of the target, but which attempt to represent the complement of $\{\phi_j(\underline{r}), j = 0,N\}$ by a finite number of terms.

Indeed, an expansion entirely in terms of a finite number of pseudostates may be made, for example in terms of Sturmian functions, and has been used with considerable success by Shakeshaft[5] in the study of heavy particle collisions. There are of course compensating disadvantages in introducing pseudostates. Such states have thresholds, ε_q which do not form part of the spectrum of the one-electron Hamiltonian, and just as true resonances arise below the real thresholds ε_j

$$H_0 \phi_j = \varepsilon_j \phi_j , \quad H_0 = -\tfrac{1}{2}\nabla^2 - \frac{Z}{r} \tag{2.10}$$

so pseudo-resonances occur below the pseudo-thresholds $\varepsilon_p = \langle\bar{\phi}_p|H_0|\bar{\phi}_p\rangle$. The difficulties may be avoided by carrying out calculations well away from these pseudo-thresholds.

Using both (i) and (ii) above our expansion is now

$$\Psi(1,2)=\mathcal{A}\left\{\sum_{i=0}^{N_0}\phi_i(\underline{r}_1)F_i(\underline{r}_2)+\sum_{p=N_0+1}^{N_1}\phi_p(\underline{r}_1)F_p(\underline{r}_2)\right\}+\sum_{q=1}^{M}c_q\psi_q(\underline{r}_1,\underline{r}_2). \quad (2.11)$$

Burke and collaborators have pointed out[6] that the correlation terms may also give rise to pseudo-resonances and have suggested a technique for dealing with these. If P projects onto the states and pseudostates included and Q onto the L^2 functions, by orthogonalising, we can arrange that $PQ = 0$ so that (1) becomes

$$H'\Psi = P(H - PHQ\frac{1}{Q(H-E)Q}QHP - E)P\Psi = 0 \quad (2.12)$$

or writing the second term in brackets as the optical potential

$$V_{opt} = \sum_{j=1}^{m} |PHQ\,\theta_j\rangle \frac{1}{\bar{\varepsilon}_j - E}\langle\theta_j\,QHP| \quad (2.13)$$

where the θ_j are linear combinations of the ψ_q such that

$$QHQ\,|\theta_j\rangle = \bar{\varepsilon}_j\,|\theta_j\rangle \quad . \quad (2.14)$$

It follows that pseudo-resonances occur at

$$\bar{E}_j = \bar{\varepsilon}_j + \Delta_j - \tfrac{1}{2}i\Gamma_j \quad (2.15)$$

with shift

$$\Delta_j = \langle\theta_j\,QHP\frac{\mathbb{P}}{E-H_j'}QHP\,\theta_j\rangle \quad (2.16)$$

and width

$$\Gamma_j = 2\pi\sum_f |\langle\theta_j\,QH_j'P\psi_f^+\rangle| \quad (2.17)$$

where H_j' is the Hamiltonian of (12) with the pole terms subtracted and $\psi_f^\pm$ is the outgoing wave scattering solution. Burke et al[6] then discuss one method of subtracting the effects of these resonances. They carry out this procedure for the $2s^2\ {}^1S^e \rightarrow 2s\,2p^3\,P^o$ transition in C III and show the effects of (a) averaging over T(E), and (b) averaging the collision strength (Fig. 1).

A similar, but independent, approach has been made by Abu-Salbi and Callaway[7] who, however, find that only the pseudo-resonances associated with pseudo-states are important. They work with the K-

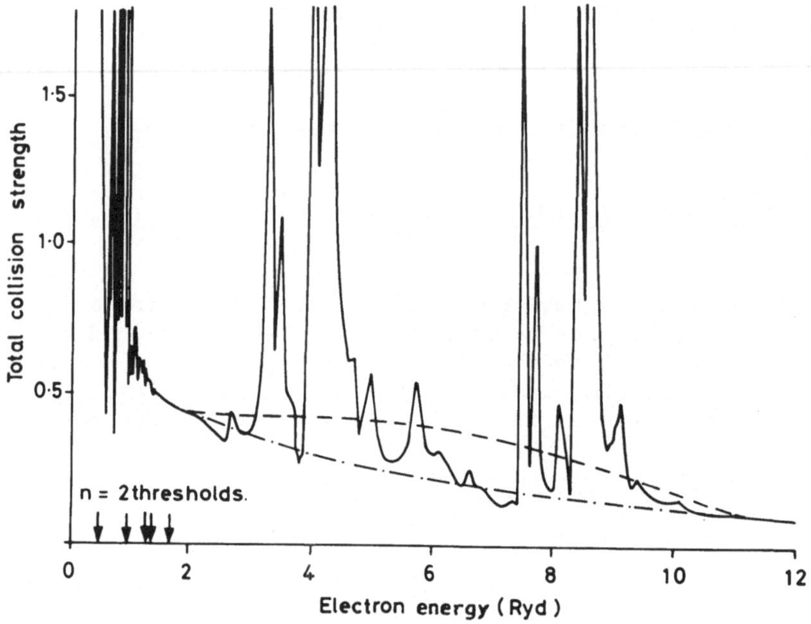

Fig. 1. Total collision strength for the $2s^2\ {}^1S^e - 2s\,2p\ {}^3P^0$ transition in e - C III scattering. Full curve: calculated collision strengths; chain curve: total collision strength obtained by averaging the T matrix over pseudo-resonances; broken curve: total collision strength obtained by averaging the calculated collision strength over pseudo-resonances. From ref (6), by permission of the Institute of Physics.

matrix (see below), locate the pseudo-pole in $\underline{K}$ and subtract it out. They applied their method to $1s \rightarrow 2s$ and $1s \rightarrow 2p$ excitation of C^{5+} and O^{7+} between the $n = 2$ and $n = 3$ thresholds. Their results for the 2p excitation of C^{5+} in the neighbourhood of the pseudo-resonance at $k_i^2 = 38.6$ in the ^1S channel associated with the pseudo-threshold at 47.2 Ry are shown in Fig. 2. The smooth dashed line is obtained by averaging $T(E)$ and the solid unbroken curve by averaging $\underline{K}(E)$.

We now return to the main close-coupling equations (8), and adopt standing wave boundary conditions, in open channel Γ_a

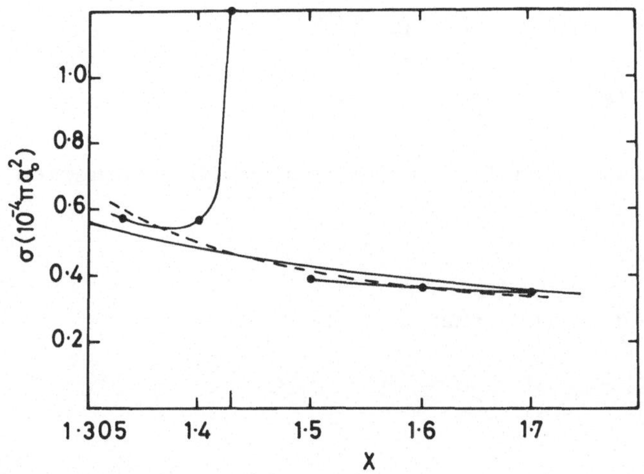

Fig. 2. The 1s - 2p excitation cross section of C^{3+}. The broken curve includes the pseudo-resonance. The dashed curve is obtained by averaging T(E) and the unbroken solid curve by averaging $\underline{K}$(E). From ref. (7) by permission of the American Institute of Physics.

$$\lim_{r_2 \to \infty} \Psi_{\Gamma_a} (\underline{x}_1, \underline{x}_2) = \sum_j G_{ja} (r_2) \tilde{\Psi}_{\Gamma_j} (\overline{r}_2) \tag{2.18}$$

where $\overline{r}_2$ indicates that the coordinate is missing, and the sum is over all open channels. Then we require

$$r\, G_{ja} (0) = 0 \tag{2.19a}$$

$$r\, G_{ja} (r) \underset{r \to \infty}{\sim} k_j^{-\frac{1}{2}} [\delta_{ja} \sin \Theta_j + K_{ja} \cos \Theta_j], \; k_j > 0 \tag{2.19b}$$

with

$$Ⓗ_j = k_j r - \tfrac{1}{2}\, \ell_j \pi + \frac{q}{k_j} \ln 2k_j r + \delta_j \ , \tag{2.20}$$

and δ_j is the usual Coulomb phase

$$\delta_j = \arg \Gamma(\ell_j + 1 - \frac{iq}{k_j}) \ . \tag{2.21}$$

With these boundary conditions the scattering information is contained in the $\underline{K}$-matrix

$$\underline{K} = [K_{ij}] \tag{2.22}$$

which is related to the other formulations by

$$\underline{S} = \frac{\underline{\underline{1}} + i\underline{\underline{K}}}{\underline{1} - i\underline{K}} \tag{2.23}$$

and

$$\underline{T} = \underline{S} - \underline{1} \, . \tag{2.24}$$

Note that if all the elements of $\underline{K}$ are "small" then

$$\underline{T} \simeq 2i\underline{K} \tag{2.25}$$

which will be important later. An important advantage is that since $\underline{K}$ is real symmetric, $\underline{S}$ is unitary.

If we choose a trial function $\Psi_t(\alpha)$ in channel Γ_α then the Kohn variational estimate of $\underline{K}$ is

$$K_{\alpha\beta} = K_{\alpha\beta}(t) - 2\, I_{\alpha\beta}(t) \tag{2.26}$$

with

$$I_{\alpha\beta} = \int \Psi_t\ (\alpha)\ [H - E]\ \Psi_t\ (\alpha)\ d\tau \, . \tag{2.27}$$

The $\underline{K}$-matrix may be determined by numerical solution of the close coupling equations, by an algebraic variational procedure[8] or by the R-matrix method.[9] One advantage of the algebraic method is that other variational estimates e.g. Inverse Kohn, may readily be compared. Although all these variational methods are of second order in the error

$$\delta\Psi = \Psi - \Psi_t \tag{2.28}$$

we do not, in general, know the sign of the error. However, Spruch et al[10] and Gailitis[11] have proved a theorem which is equivalent to stating that if the close coupling equations are solved in a basis (N_0, N_1, M) as in (11) above, and we then add an additional term to any of these sums, then provided we are working at an impact energy below the highest real threshold included, the eigenvalues of $\underline{K}$ increase. Strictly, since the eigenvalues of $\underline{K}$ in a diagonal representation are the tangents of the elastic scattering phase-shifts (the eigenphases) η_i, the theorem says that the eigen-phases are bounded above. Similarly $\sum_i \eta_i \geqq \sum_i \eta_i\,(t)$ and this provides the test on the eigenphase sum used.[12] Thus, within the energy range stated, we can state rigorously which calculation is best, and we can hope to achieve high precision.

The most elaborate calculations to date are those of Callaway[13] on the $1s \rightarrow 2s, 2p$ transitions of atomic hydrogen, below the $n = 3$ threshold. He used the exact $n = 1, 2, 3$ target eigenstates, together with the exact 4f state and eleven pseudo-states (4s type, 3p type, 2d, 1f and 1g). Up to fifteen short range functions were used in each channel, together with additional energy dependent oscillatory functions of the form $r^{-n} \begin{smallmatrix}\cos\\ \sin\end{smallmatrix} (kr)$, $n = 2, 3$. Table 1 gives a comparison for the ^3P channel at four energies. Calculation [12] used three states and three pseudo-states, [14] used three states and twenty Hylleras type correlation terms, [15] is the original 6CC and [16] an eleven state calculation. Clearly the new Callaway calculation is the most accurate. As in the earlier version[17] by Morgan et al, there is excellent agreement with the measurements of Williams[18] except that the new calculations show a strong (but narrow) ^3F resonance at $k^2 = 0.87697$.

This calculation is very much "state of the art". Calculations for other energies and other transitions in atomic hydrogen have recently been reviewed by Callaway and McDowell.[19] The only trans-itions for which there are substantial amounts of data are (1,2), (1,3) and (2,3). Figs. 3 and 4 show comparisons of our best quantal estimates with the semi-empirical formulae of Johnson[20], the FBA[21], and the Classical formulae of Percival and Richards.[4] In neither of

Table 1. Eigenphase sums for ^3P scattering of electrons by atomic hydrogen. The calculations are from references (12), (13), (14), (15) and (16).

k_i	(13)	(12)	(14)	(15)	(16)
0.76	-.417	-.439	-.508	-.476	-.44
0.78	1.482	1.455	1.386	1.410	1.46
0.81	0.682	0.651	0.683	0.584	0.66
0.83	0.436	0.398	0.312	0.320	0.40

these cases do the FBA or the other two methods agree with the best quantum mechanical values to at best 50% below ten times threshold.

Distorted Wave Approximations

In the case of highly charged positive ions ($q > 5$, say) experience has shown that in the near threshold region where only a few L contribute, and couplings are weak, the close coupling equations may be well approximated by a distorted-wave method (DW), high L being treated in the Coulomb-Bethe approximation. The University College and JILA groups calculate their distorted waves using the same scaled Thomas-Fermi-Dirac potential used to obtain accurate target eigen-energies and oscillator strengths, $V(\lambda_i, r)$, so that the radial wave functions satisfy

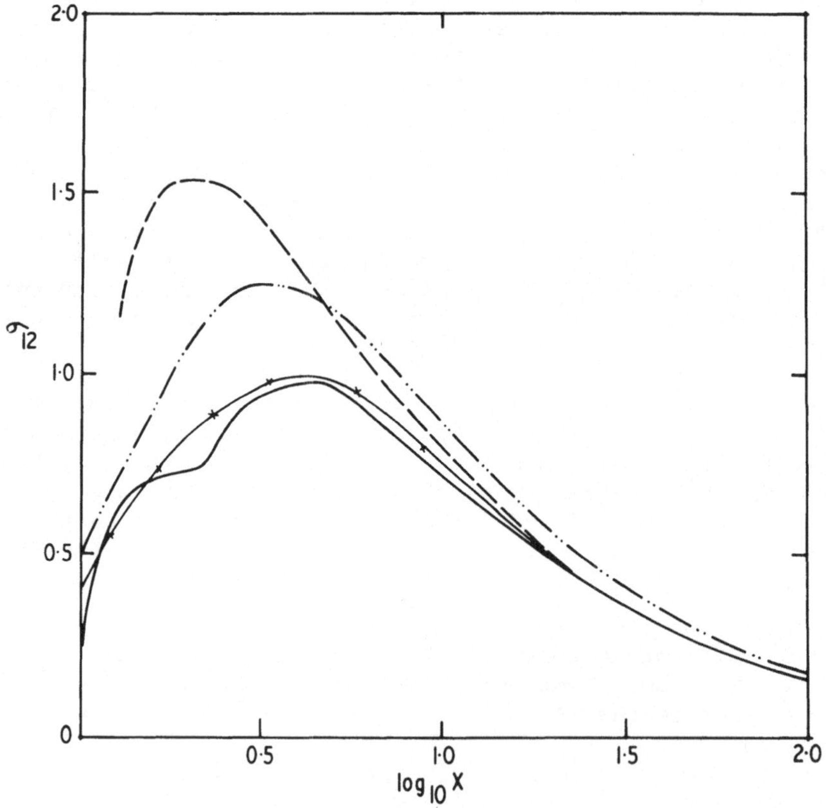

Fig. 3. The $n = 1 \rightarrow n = 2$ cross section in atomic hydrogen. The solid curve is from detailed quantal calculations.[19] The dashed curve is the FBA, the chain curve the classical model and the curve —×—×— is Johnson's semi-empirical formula.

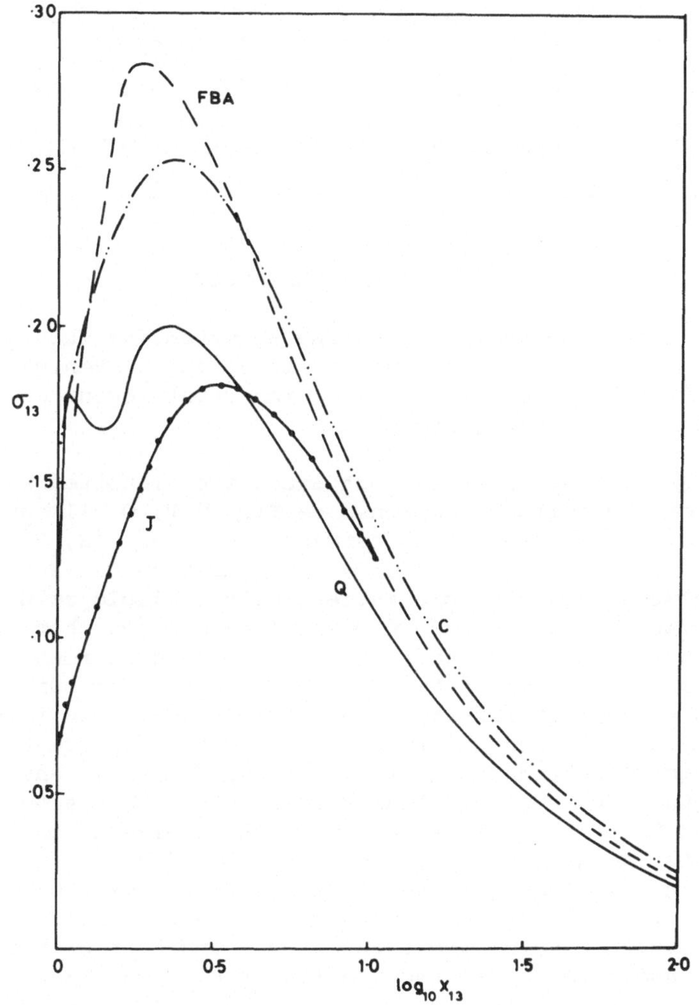

Fig. 4. As for Fig. 3 but $n = 1 \rightarrow n = 3$. The quantal calculations are indicated by Q, the classical by C, while J is the Johnson semi-empirical formula.

$$\left\{\frac{d^2}{dr^2} + k_i^{\,2} - \frac{\ell_2(\ell_2+1)}{r^2} + 2V(\lambda_i, r)\right\} F_i^{DW}(r) = 0 . \qquad (2.29)$$

No exchange effects are explicitly included. Pradhan et al[23] have shown that for He-like ions above the highest threshold included in the model, such a DW calculation gives excellent agreement with full CC calculations. We show in Table 2 a comparison between such a DW calculation (unitarised by using $\underline{K}^{DW}$) and a five state CC calculation of Wyngaarden et al[30] for the $1^1S - 2^1P^0$ transition in O VII. The disagreement is at worst 10%.

Table 2. Collision strengths for O VII ($1^1S \rightarrow 2^1P^0$).

k_i^2 (Ry)	45	50	63	75	126	210
(a) 5DW[23]	0.0208	0.0250	0.0385	0.0480	0.0817	0.1159
(b) 5CC[30]	0.0229	0.0276	0.0389	0.0485	0.0797	0.1131

3. EXCITATION OF POSITIVE IONS NEAR THRESHOLD

There is little point in presenting an extensive survey. Results through 1980 have been reviewed in detail by Henry.[22] We therefore restrict our comments to some important recent calculations of interest in the present context.

Pradhan et al[23] have carried out extensive distorted wave calculations on the helium-like ions Be III, C V, O VII and Fe XXV including all transitions between the 1^1S, 2^3S, 2^1S, 2^3P^0 and 2^1P^0 states, using four or five configuration CI wave functions. These calculations should become increasingly reliable as q increases. Studies by Willis and McDowell[24] have shown that such DW calculations may be superior to five-state CC calculations which are in error by up to a factor of two for He I, due to omission of pseudo-states to correct for the long range polarisation interaction. Pradhan et al investigated the resonance structure in great detail using a quantum-defect method analysis of the K-matrix, and were careful to obtain accurate oscillator strengths. They show that the resonances can in certain cases have very large effects on the rate coefficient γ. An interesting example for $2^1P^0 \rightarrow 2^3S$ in Fe XXV is shown in Fig. 5. They conclude that resonance structure effects will be particularly important in enhancing the rate coefficient in forbidden and inter-combination transitions, and that previous analysis of plasmas using He-like ion line ratios may need to be re-examined.

Similar effects of resonances in the threshold region are found in work on Si III by Baluja et al.[25] Baluja et al use an R-matrix approach to study all transitions within the first twelve states of Si III. There are sixteen allowed transitions and fifty partly-forbidden. The ground state is Mg like, with configuration $(1s)^2\ (2s)^2\ (2p)^6\ (3s)^2\ {}^1S$, and all single and double outer shell excitations up to $3s\,4p\ {}^1P^0$ are included. They give their results as rate coefficients, up to 3×10^4K only. The same authors also consider all transitions among the $2p^2$, $2s\,2p^3$ and $2p^4$ configurations of O III.[26] Since their interest is primarily astrophysical, collision strengths are given up to 2×10^4K only in most cases. The variation of rate coefficient (plotted as effective collision strength γ) where the rate coefficient is defined in terms of effective γ by

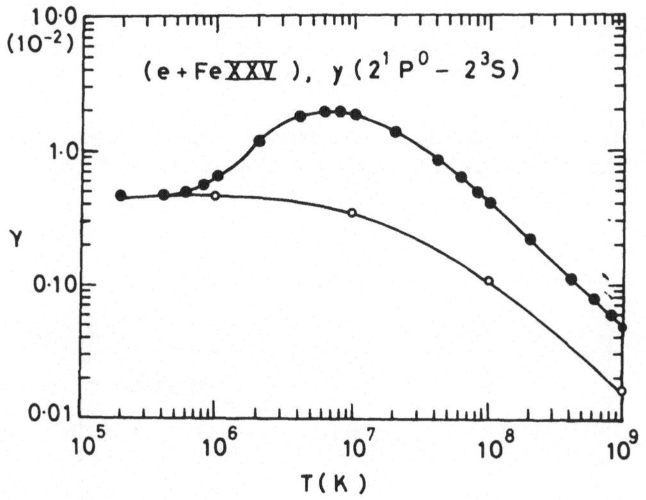

Fig. 5. Rate parameter $\gamma(2^3S - 2^1P^0)$ for Fe XXV. From ref. (23). The upper curve includes the effect of the resonances. By permission of the American Institute of Physics.

$$\alpha_{ij}(T) = \frac{8.63\ 10^{-6}}{w_i T^{\frac{1}{2}}} e^{-E_{ij}/kT} \gamma_{ji}(T) \tag{3.1}$$

with

$$\gamma_{ji}(T) = \int_0^\infty \Omega_{ij}\, e^{-\varepsilon_i/kT}\, d\left(\frac{\varepsilon_i}{kT}\right) \tag{3.2}$$

and

$$Q_{ij} = \Omega_{ij}/w_i\varepsilon_i \tag{3.3}$$

is plotted in Fig. 6 for the transitions within the ground state multiplet. There are significant changes from the earlier nine-state calculation of Eissner and Seaton,[27] largely attributable to the more complete treatment of resonances in the new work. In contrast to the He-like ions these decrease the rate coefficient in this case. It is not expected that these differences would persist to the much higher temperatures of interest in CTR devices.

Mendoza[28] has considered the transitions between levels of the ground state configuration of carbon-like S III, which has an open

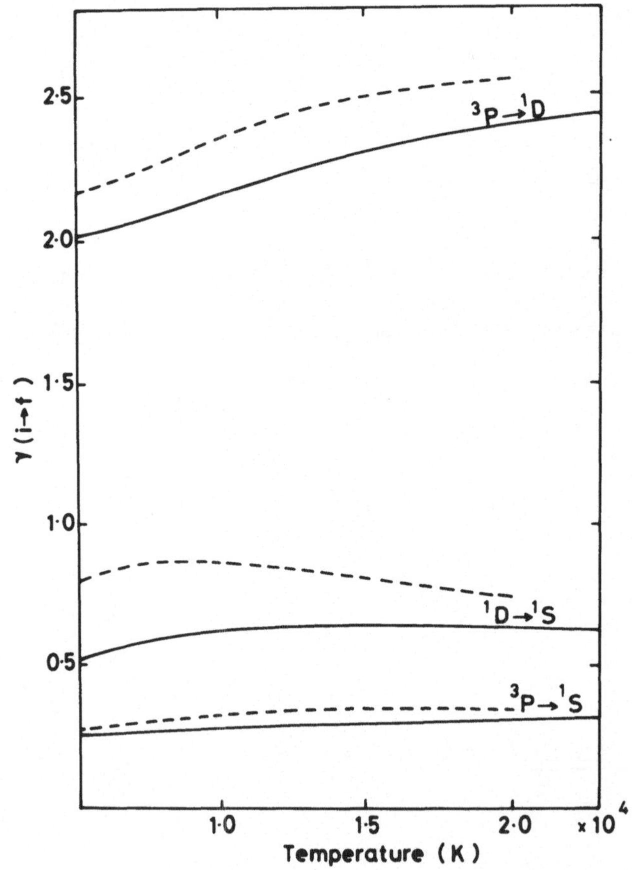

Fig. 6. Effective collision strength for transitions within the ground state configuration of O III as a function of electron temperature. ——, twelve-state R-matrix,[25] ----, nine-state results of Eissner and Seaton.[27]

$(3p)^2$ shell. He paid special attention to configuration interaction effects, convergence of the CC expansion, the effects of high L contributions, primarily as a guide to which might be important in more highly ionised species. The lowest seven target states $3p^2\ {}^3P,\ {}^1D,\ {}^1S;\ 3s\ 3p^3\ {}^5S^0,\ {}^3D^0,\ {}^3P^0,\ {}^1D^0$ were included, and a large number of basis states used in the CI program. Two bases were used, the first including the configurations

$$\text{(I)}\quad \begin{array}{ll} 3s^2\,3p^2,\ 3p^4,\ 3s\,3p^2\,3d,\ 3s^2\,3d^2,\ 3s^2\,3p\,\overline{4f}\,, & \text{even} \\ 3s^2 3p^3\,,\ 3s^2\,3p\,3d^2, & \text{odd} \end{array}$$

and the second, additionally

$$\text{(II)} = \text{(I)} + 3p^3\,3d,\quad 3s\,3p\,3d^2\,, \qquad \text{odd.}$$

Case II gave much improved term energies for the excited odd parity states. Three models were considered

(A) (I), $\ell_2 \geq 4$ in distorted wave (DW)

(B) (I), $\ell_2 \geq 5$ in DW

(C) (II), $\ell_2 > 4$ in DW.

All three models gave rate coefficients in agreement to 15% from 5×10^3 to 2×10^4 K. Because of the known analytic structure of the Coulomb interaction the $\underline{K}$ matrix elements may be fitted, in the region of the resonances E_k' by[29]

$$K_{ij} = K_{ij}(0) - \sum_k C_{ik}(E - E_k)^{-1} C_{kj} \tag{3.4}$$

and the collision strength averaged over the resonances, thus allowing a rate coefficient to be determined. In fact it is more convenient to convert to the $\underline{S}$ matrix, fit above threshold, and extrapolate below using

$$|S_{ij}|^2 = |S_{ij}(0)|^2 + \sum_{p,q} \frac{S_{ip}\,S_{pj}\,S_{iq}^*\,S_{qj}}{1 - S_{pp}\,S_{qq}^*} \tag{3.5}$$

where p,q run over closed channels. Care must be taken if, for example, there is a shape resonance just above the excitation threshold.

Bely-Dubau et al[31] have considered excitation of both the $1^1S \rightarrow 2^1P$ resonance line of O VII and the inner-shell satellite lines excited from $1s^2\,2s$ and $1s^2\,2p$ configurations of O VI. These lines correspond to transitions

$$1s^2\,n\ell \rightarrow 1s\,2\ell'\,n\ell\,, \quad n \geq 2 \tag{3.6}$$

and are important in plasma diagnostics. In O VII the $n = 2$ and 3 lines are most important, those for $n > 3$ tending to blend with the resonance line and increase its apparent intensity and width. Bely-Dubau et al included the $1s^2\,2s$, $1s^2\,2p$, $1s\,2s\,2p$; $1s\,2s^2$ and $1s\,2p^2$ configurations in their target wave function, and assumed LS coupling. The collision strengths were evaluated using the DW K-matrix. They give rate coefficients for five transitions from the ground state and five from $1s^2\,2p\ ^2P$ of O VI for $3 \times 10^5 \leq T \leq 4 \times 10^7$ K, and for excitation of the resonance line of O VII. They then calculate the contribution of the unresolved satellite lines to the resonance line, where the line intensity I_R is $(1 + \alpha)^{-1}\, I_O$, where I_O is the observed intensity. The values of α depends on the spectral resolution of the observations, and their results are shown in Fig. 7. The effect in this case is not significant at temperatures of interest in CTR devices.

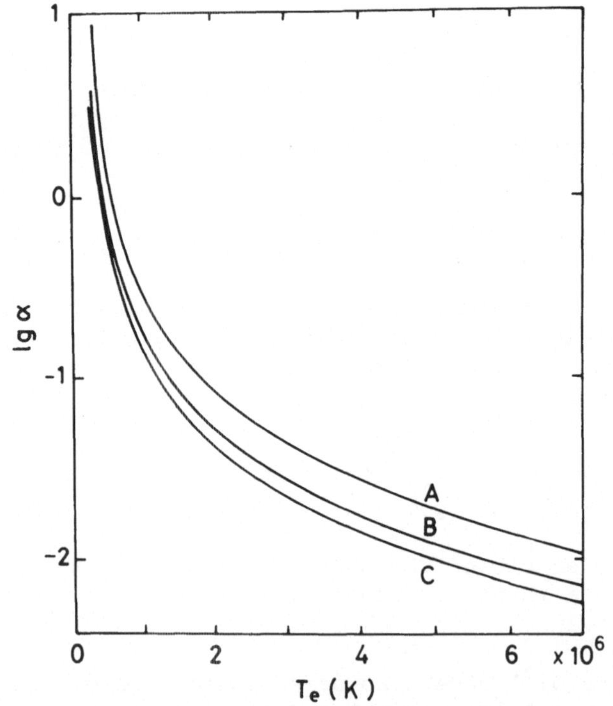

Fig. 7. The contribution $\alpha(T_e)$ of the unresolved satellite lines to the resonance line for three spectral resolutions $\lambda/\Delta\lambda$, A, 500; B, 1000; C, 2000. (From Bely-Dubau et al[31] by permission of the Institute of Physics).

4. EXCITATION OF POSITIVE IONS BY FAST ELECTRONS

At some sufficiently high energy, $k_i^2 \gg \Delta E_{if}$, we would by analogy with potential scattering, expect perturbative methods to be an adequate approach to calculating total (integral) cross sections. The evidence is that such methods are much less useful for differential inelastic cross sections. However, in the context of CTR Fusion, a knowledge of the integral cross sections suffices.

A recent article by Peek and Mann[32] provides a useful connection with the close-coupling methods discussed in section 2, the DW method and the high energy Coulomb-Born (CB) approach, and attempts to include exchange in a consistent manner. They consider scattering from channel $i = |\alpha_i, L_i, S_i, \ell_i, k_i, LS\Pi\rangle$ to channel $f = |\alpha_f, L_f, S_f, \ell_f, k_f, LS\Pi\rangle$ where (L_i, S_i) are the total orbital angular momentum of the target's initial state (L_f, S_f) of its final state, $\alpha_{i,f}$ being the other quantum numbers of those states. Similarly (ℓ_i, k_i) describes an incident electron with orbital angular momentum ℓ_i and energy k_i^2, (ℓ_f, k_f) the scattered electron. The S-matrix is diagonal on this coupled $|LS\Pi\rangle$ representation, and is independent of M_L, M_S, provided spin-orbit coupling is negligible. We drop all terms with $k \neq j$ in (2.8) and display the residual Coulomb potential explicitly, obtaining after performing the angular integrals

$$\left(\frac{d^2}{dr^2} + k_i^2 - \frac{\ell_i(\ell_i+1)}{r^2}\right) f_i = 2(V_{ii} + (-1)^S W_{ii}) f_i \tag{4.1}$$

for the radial scattering functions.

The target orbitals may be treated using CI, but as a further simplification the direct potential is often replaced by its spherical average,

$$\overline{V}_{ii}(r) = \sum_j p_j \int \frac{1}{r_>} P^2(n_j \ell_j, r') dr' \tag{4.2}$$

where p_j is the occupation number of the (n_j, ℓ_j) sub-shell. The Coulomb-Born-Exchange approximation (CBX) is obtained by replacing (4.2) by its asymptotic form for all r,

$$V_{ii}^{CBX}(r) = \frac{q}{r} \tag{4.3}$$

and the CB approximation by dropping exchange (W_{ii}) in (4.1) and the corresponding exchange matrix element.

At intermediate and high energies it is sufficient to replace W_{ii} by a local approximation to it. American authors tend to favour some variant of the Free Electron Gas approximation which has been carefully examined by Truhlar and collaborators,[33] while U.K. authors tend to use the Furness-McCarthy approach of expanding the integrand of W_{ii} in Taylor series and replacing ∇_i^2 by the channel momentum $K_i^2 = k_i^2 - 2V_{ii} - 2|\varepsilon_1|$. Again this has been examined in detail by Bransden et al[34] for the 3CC model for H and He^+. A simple form is to put

$$\tilde{W}_{ii} f_i = W_{ii} f_i$$

and

$$\tilde{W}_{ii} = \frac{4\pi |\phi_i(r)|^2}{K_i^2} . \tag{4.4}$$

We note that as $k_i^2 \to \infty$ this gives the Bonham-Ochkur approximation to the FBA exchange term. The full form is

$$\tilde{W}_{ii}^{(F)} = \frac{1}{4}\left[k_i^2 - 2V_{ii}(r) - \left[(k_i^2 - 2V_{ii}(r))^2 - 32\pi|\phi_i|^2\right]^{\frac{1}{2}}\right] \tag{4.5}$$

which reduces to (4.4) when $|W_{ii}| << |V_{ii}|$. Note that the expression (their eqn. (7)) of Mann and Peek contains misprints. Mann and Peek then calculate S-matrix elements, though in my view it would be preferable to calculate the K-matrix and ensure unitarity. They do this in some cases, these approximations being denoted by DWX II, CBX II, but in their approach they assume all S-matrix elements other than S_{if} are zero.

Data from many sources, using these methods, has been compiled by the Los Alamos group.[35,36] For one-electron targets both CBX and DWX results agree well (± 10%) with 3CC, for 1s → 2p, 3p, 3d, but there are severe discrepancies for s-s transitions for $X = (\Delta E_{if}/k_i^2) < 2$, as is evidenced by the well known case of He^+. For two-electron targets exchange is important even for $1^1S \to 2^1P$, as is shown in Fig. 8, but DWX agrees reasonably closely with 5CC. For $1^1S \to 2^1S$, CBX breaks down below $X \simeq 10$, but DWX is in good accord with the 5CC unpublished calculations of Robb (for CV). Somewhat surprisingly CBX and DBX also agree well for $1^1S \to 2^3P$. It should be borne in mind however that

(a) there is no evidence that 5CC is itself accurate for small q. Indeed it is known to be poor for $q = 0$.[24]

(b) There are no available CC cross sections for excitation of the $n = 3$ levels, and none for excited-excited transitions.

Peek and Robb also present results for Li- and Be-like targets. They conclude, and I concur, that DWX II is the best rapid approximation to many-state close coupling in these few electron targets. An exception is for transitions in Li-like ions when the final state is nearly degenerate with two other levels, i.e. $2^2S \to 3^2P$.

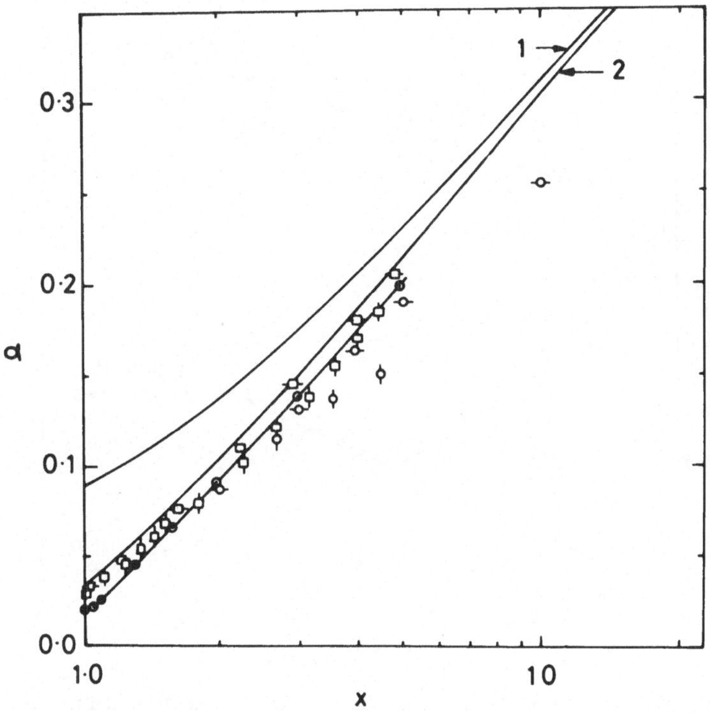

Fig. 8. Collision strengths for the 1^1S-2^1P transition in C^{4+} are shown as a function of x. Curves 1 and 2 represent the DW and DWX results respectively from ref. (32). (From ref. (32) by permission of the American Institute of Physics.)

A similar situation is well known for H (1s → 3d).[19] A simplified close-coupling calculation retaining the long range couplings between the nearly degenerate final states is desirable. It is also important to note that both in optically forbidden transitions ($2^2S \rightarrow 3^2S$) and in transitions between states of large principal quantum number very many ($\geqq$ 400) values of L make important contributions at $X \gg 1$. The final conclusion must be that when many-state close coupling calculations are not available or are prohibitively expensive, DWX II should always be used rather than CBX II or simpler approximations.

Tully and Baluja[38] have compared the CBI with twelve-state

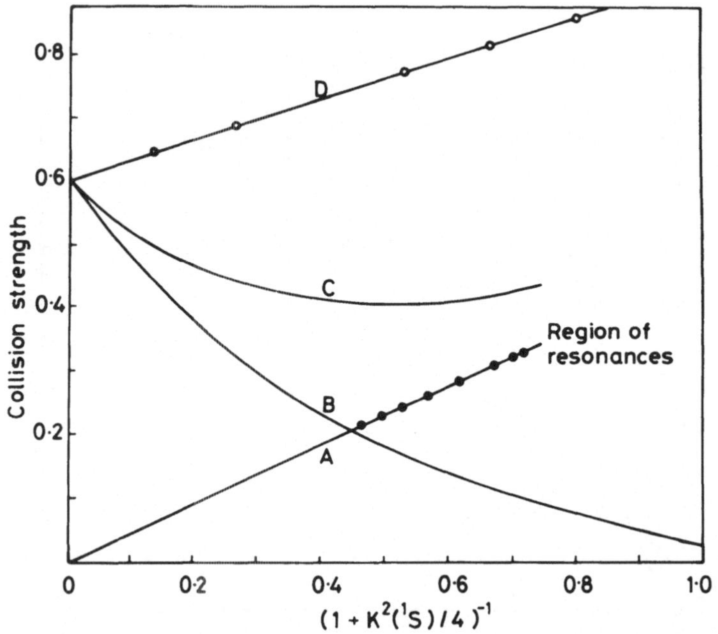

Fig. 9. Collision strengths for electron impact excitation of $O^{2+}(2p^2\ ^1D \to 2p^2\ ^1S)$. A, linear fit to the R-matrix results ($L \leq 3$) of Baluja et al (1981); B, Coulomb-Born ($L \geq 4$); C, A + B; D, Coulomb-Born ($L \geq 0$). (From ref. (38) by permission of the Institute of Physics.)

R-matrix results[37] for $2p^2\ ^1D \to 2p^2\ ^1S$ in O III (4363 Å) but both calculations use CBI for $L \geq 4$. Their results show that (Fig. 9) the CB is a factor of 2.17 higher at threshold.

5. ELECTRON IMPACT IONISATION OF POSITIVE IONS

We consider, following Moores and Jakubowicz,[39] ionisation of an $(N + 1)$-electron target ion of nuclear charge Z by an incident electron of momentum $\underline{k}_0$, energy k_0^2. After ionisation it is scattered with momentum $\underline{k}_f$, energy k_f^2, and a target electron is ejected with energy κ^2, momentum $\underline{\kappa}$. The process may be expressed in terms of a direct amplitude $f(\underline{k}_f,\underline{\kappa})$, an exchange amplitude $g(\underline{\kappa}_f,\underline{\kappa})$ and a capture amplitude $h(\underline{k}_0,\underline{\kappa},\underline{k}_f)$. Phillips and McDowell have shown[40] that capture may normally be neglected. The direct amplitude is

$$f(\underline{k}_f,\underline{\kappa}) = -(2\pi)^{-5/2}\, e^{i\Delta(\underline{k}_f,\underline{\kappa})} \int \Phi_f^*(\underline{x},\cdots,\underline{x}_{N+2})[H-E]\Psi(\underline{x},\cdots,\underline{x}_{N+2})d\tau \quad (5.1)$$

where Φ_f is asymptotic to

$$\Phi_f(\underline{x}_1,\cdots,\underline{x}_{N+2}) \underset{\substack{r_{N+1}\to\infty \\ r_{N+2}\to\infty}}{\longrightarrow} \chi(\underline{x}_1,\cdots,\underline{x}_{N-1})\times\phi(q_1,-\underline{\kappa}|\underline{x}_{N+1})\phi(q_2,-\underline{k}_f|\underline{x}_{N+2}) \quad (5.2)$$

Here the ϕ functions are regular Coulomb Functions of charge q_i and momentum $\underline{k}_i$, $\phi(q_i,-\underline{k}_i|\underline{\kappa})$, including spin. The asymptotic charges q_1, q_2 satisfy

$$\frac{q_1}{\kappa} + \frac{q_2}{k_f} = \frac{q}{\kappa} + \frac{q}{k_f} - \frac{1}{|\underline{k}_f - \underline{\chi}|} \quad (5.3)$$

while

$$\Delta(\underline{k}_f,\underline{\kappa}) = 2\left[\frac{q_1}{\kappa}\ln\left(\frac{\kappa}{k_0}\right) + \frac{q_2}{k_f}\ln\left(\frac{k_f}{k_0}\right)\right]. \quad (5.4)$$

The exchange amplitude is related to the direct by[41]

$$g(\underline{\kappa},\underline{k}_f) = f(\underline{k}_f,\underline{\kappa}) \quad (5.5)$$

so that the total ionisation amplitude is

$$f^{\underline{s}}(\underline{k}_f,\underline{\kappa}) = f(\underline{k}_f,\underline{\kappa}) + (-1)^s g(\underline{k}_f,\underline{\kappa}) \quad (5.6)$$

where $\underline{s}$ is the total spin of the two outgoing electrons. Thus the total ionisation cross section is

$$\sigma_{ion}(k_0^2) = \frac{1}{8\pi}\frac{k_f}{k_0}\frac{1}{(2L_0+1)(2S_0+1)} \sum_{\substack{M_{S_0}M_{L_0} \\ M_{S_1}M_{L_1}}}$$

$$\int_0^{E/2} d\kappa^2 \int d\hat{\underline{k}}_f\, d\hat{\underline{\kappa}}\left[|f(\underline{k}_f,\underline{\kappa})|^2 + |g(\underline{k}_f,\underline{\kappa})|^2 - \mathrm{Re}\, f^*(\underline{k}_f,\underline{\kappa})g(\underline{k}_f,\underline{\kappa})\right] \quad (5.7)$$

with

$$E = k_f^2 + \kappa^2 \quad (5.8)$$

and

$$k_i^2 = I_x + k_f^2 + \kappa^2 \tag{5.9}$$

where $(L_O, M_{L_O}, S_O, M_{S_O})$ describe the initial target state, and similarly for the final target state.

This, so far, is exact, and in view of the many-body nature of the problem approximations must be made. The main difficulty is with (3), and with the unknown relative phase of f and g, which determines the third term in (7). This is usually chosen to maximise Re f^*g, and minimise σ_{ion}, at least in Born approximation, in which all practical calculations are carried out.

The usual procedure is to choose

$$q_1 = q \ , \quad q_2 = q - 1 \tag{5.10}$$

where q_2 refers to the slower of the outgoing electrons.

The final state is then

$$\Phi_f(\underline{x}_1, \cdots, \underline{x}_{N+2}) = \Psi_f(\underline{x}_1, \cdots, \underline{x}_{N+1})\phi(q, -\underline{k}_f, \underline{x}_{N+2}) \tag{5.11}$$

and the initial state

$$\Phi_i(\underline{x}_1, \cdots, \underline{x}_{N+2}) = \Psi_i(\underline{x}_1, \cdots, \underline{x}_{N+1})\phi(q, \underline{k}_0, \underline{x}_{N+2}) . \tag{5.12}$$

The $(N + 1)$-electron target functions can be obtained in a variety of approximations. Clearly Ψ_f corresponds to scattering of an electron by an N-electron target, and Ψ_i is a bound state of an $(N + 1)$-electron target. Moores and Jakubowicz[43] obtain these by using a close-coupling expansion as in section 2.

We can write (7) as

$$\sigma_{ion}(k_0^2) = \sigma_d + \sigma_e - \sigma_{int} \tag{5.13}$$

in an obvious notation. This is the CBX class of approximation. Dropping σ_{int} gives the CB approximation, and putting $\sigma_e = 0$ the CBNX set. In practice this works well since σ_e and σ_{int} tend to cancel out. The three approximations, using $1s\ 2s^2\ {}^2S$, $1s\ 2s\ 2p\ {}^4P$ and $1s\ 2s\ 2p\ {}^2P$ basis states are compared for Li-like N V in Fig.10. There is excellent agreement with the experimental measurements of Crandall et al[42] up to the onset of autoionisation at $X = 4$. The models can be further improved (at additional expense) by replacing the Coulomb waves of (11) and (12) by distorted waves, leading to the CBNXD etc. class of models. Autoionisation contributions are readily included in principle by increasing the bases used in the target structure calculations and allowing for excitation to these states. The assumption is made that radiative decay can be neglected. Detailed studies of Li-like and Be-like ions have been reported. In all cases the agreement with the Lotz semi-empirical formulae[44] is within 25% for $X < 5$, but it gives a rate coefficient 40% higher for Fe XXIV, Fe XXV for $T > 10^8$ K.

Such elaborate calculations are extremely expensive and time consuming, and a much simpler model has been developed by Younger,[45] with the direct aim of producing moderately accurate results for many ions. Younger uses the CBX or CBXD method in all cases with maximum interference terms. Radial functions for incident, scattered and ejected electrons are computed using the potential

$$V(r) = -\frac{Z}{r} + \sum_{i=1}^{N} J_i(r) \tag{5.14}$$

with the spherically averaged terms

$$J_i(r) = \int_0^\infty \frac{1}{r_>} P_i^2(\rho)\,d\rho \tag{5.15}$$

for light elements, and the Riley and Truhlar[33] semi-classical exchange potential for the Ne-like sequence. Here i runs over occupied orbitals. For both incident and scattered electron, the potential is that of the target ion V_T, but that of the residual ion V_F for the ejected electron. The treatment of $g(\underline{k},\underline{K})$ is peculiar. It is not set equal to $f(\underline{K},\underline{k})$ but instead the ejected electron is computed in V_T, the scattered in V_F. Younger remarks that "such an unphysical choice is tolerated in order to ensure orthogonality of overlapping orbitals in the matrix element and has been found to yield theoretical data in reasonable agreement (better than 25%) with available experimental data on light ions." This seems an inadequate reason!

For He^+ (1s), O VI (2s), C IV (2s), N V (2s) his results agree well with the CBX values. The agreement with experiment for these light ions is excellent except for C IV (2s).

Younger has also considered Ne-like,[46] Na-like[47] and Be-like[48] sequences. For the Ne-like sequence experimental data are available for Na II, Mg III. In both cases his DWX results tend to overestimate at all energies, but to be most accurate when $X \to 0$. His results (solid line) are compared with the Lotz formulae and experiment in Fig. 11. The scaled cross section $\bar{\sigma} = k_0^2 I_X \sigma_{ion}$ is plotted. Clearly the Lotz result is worse than DWX, but neither is satisfactory. Younger remarks that inclusion of W_{ii} in the distorted wave equation reduces (improves) the calculated cross section

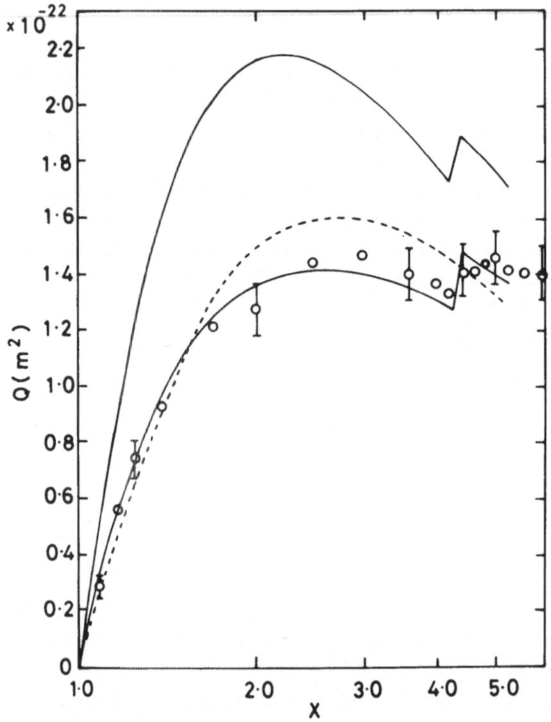

Fig. 10. Ionisation cross section of N^{4+}. ---, CBNX; —— (upper), CB; —— (lower), CBX. The experimental points are the crossed-beam data of Crandall.[42] (From ref. (39) by permission of the Institute of Physics.)

by a factor of three for Na II at X = 1.125, but is much less significant for Mg III. What is surprising is that the DWX results are in best agreement with experiment (both for Na II and Mg III) for small X, i.e. near threshold, and diverge increasingly from it as the impact energy increases.

In two further papers Younger considers Na-like[47] and Be-like[48] ions. In the first case results are given for Mg II, Aℓ II, P V and Ar VIII, as well as the $Z = \infty$ limit for transitions to $2p^6$, $2p^5\,3s$ and $2s\,2p^6\,3s$ configurations of the final ion, for $1.125 \leq X \leq 5.00$. The results for Mg II are in excellent agreement with experiment. The contributions from inner shell

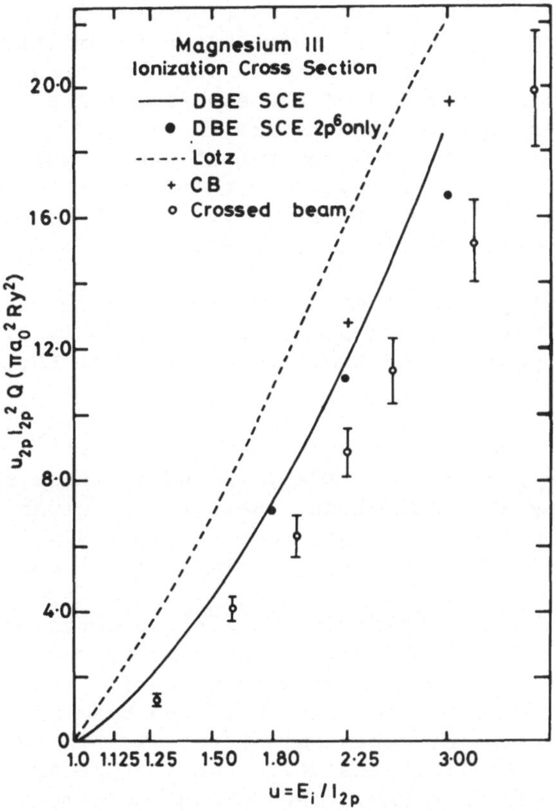

Fig. 11. Fano plot of the scaled electron-impact-ionisation cross section uI^2Q for Mg III; —— distorted-wave Born exchange with semi-classical exchange in the distorted potentials; ● same as solid line but omitting inner-shell ionisation from $2s^2$; + Coulomb-Born no exchange, O crossed-beam experiment, ref. (47); --- Lotz semi-empirical formula, ref. (42).

ionisation grow rapidly with $q = Z - N$, as can be seen in Fig. 11. The calculated cross sections may be expressed as

$$X \, I_X^2 \sigma = A\left(1 - \frac{1}{X}\right) + B\left(1 - \frac{1}{X}\right)^2 + C \ln X + D X^{-1} \ln X \qquad (5.16)$$

where A, B, C, D are tabulated and C is the Coulomb-Bethe constant. An analytic expression for the rate coefficient is then obtained.

For Be-like ions, Younger includes contributions from the ground $1s^2\,2s^2\,{}^1S$ state and the $1s^2\,2s\,2p\,{}^3P$ metastable. The ground state contribution scales classically, but is dominated by that from the metastable state, by a factor of two for C III to a factor of 1.35

for Ar XV. The calculations are in fair accord with experiment[50,51] for C III and N IV if the ^3P contribution is omitted, but with this in, overshoot by ≃ 20%. There is much better agreement in the case of O V, though the slope of the experimental curve on a Bethe plot is larger. Again calculations are restricted to the region $X \leq 5.0$, so there must be some doubt that the calculated cross sections have in fact "rolled over" to their final Coulomb-Bethe form. Results are given for C III, N IV, O V, F VI, Ar XV and the $Z = \infty$ limit, as are coefficients of (13).

This series of papers gives extremely valuable data, and in spite of the simplicity of the approach, is probably of sufficient accuracy for CTR purposes for the ions considered, though unsatisfactory for $Z > 20$.

An even simpler model has been used by McGuire,[52] namely the Plane Wave Born Approximation (PWBA), including inner shell ionisation followed by autoionisation. Results are given for C^{p+} $(1 \leq p \leq 4)$, N^{p+} $(1 \leq p \leq 3)$, O^{p+} $(1 \leq p \leq 4)$, F^{p+} $(1 \leq p \leq 5)$. Atomic orbitals were obtained using a simple model potential.

Good agreement with experiment (± 25%) is obtained for C II,[53] C III,[50] C IV and C V,[51] as well as for some ions of O and N, allowing for autoionisation contributions and for the fact that in some cases (e.g. O^{4+}) the composition of the beam in the experiments was uncertain (in terms of metastables) to 50%. The worst agreement is obtained for the Be-like ions N IV and O V, near threshold. The accuracy of these calculations, considering the simple wave functions involved is remarkable, but not entirely unexpected: many years ago the near equivalence of CB and PWBA for Li^+ above the peak of the cross section was demonstrated by Economides and me.[54]

Similar, scaled, PWBA calculations, as well as a modification of the Lotz formulae have been used in analysing recent ionisation measurements in the heavy ions Zn II, Ga II.[55] Here autoionisation, both via inner shell processes and excitation, enhances the cross section by as much as a factor of 2.5 for $X < 2.5$. The inner shell contributions were calculated in a 2CC model, and CI wave functions.

The PWBA is accurate for Zn II close to threshold (I_{Zn} + 2 eV) and at 100 eV or more above it, and for Ga II at 150 + eV above threshold. In both cases there is a major contribution from autoionisation of doubly excited states. In Ga II, these are probably $3d^9\,4s^2\,n\ell$ levels, since the deserved detailed structure is in good agreement with the fine structure splittings calculated by Pindzola et al[56] for the $j = 1$ levels of $4s^2\,4p$.

For Zn II contributions of the order of $0.14\ \pi a_0$ arise from $3d^9\,4s\ (^1P)\ 5p\ ^2P^o$ and perhaps as much as $0.2\ \pi a_0^2$ from the $3d^9\,4s\,5\ell$ levels, compared with a total cross section of

75.1 (± 7.7%) 10^{-18} cm^2 (0.85 ± 0.06 πa_o^2) at 87 eV.

Griffin et al[57] have extended such calculations of the auto-ionisation contribution to ionisation from $np^6\, nd^m \rightarrow np^5\, nd^{m+1}$ transitions in Ti IV, Zr IV and Hf IV. They write

$$\sigma_{ion}(k_i^2) = \sigma_{ion}^{d} + \sum_j \sigma_{excit,j}\, B_j^a \tag{5.17}$$

where the branching ratio B_j^a is

$$B_j^a = \sum_m A_{jm}^a \Big/ \left(\sum_m A_{jm}^a + \sum_k A_{jk}^r\right) \tag{5.18}$$

where A_{jm}^a is the autoionisation rate to channel m, A_{jk}^r the radiative rate to lower bound state k.

They calculate a unitarised excitation cross section using eqn (2.25). Simple target wave functions were used (relativistic HF), exchange neglected, and r_{12}^{-1} replaced by $r_>^{-1}$. The incident and scattered electrons were represented by distorted waves, calculated using a localised exchange potential. This dipole approximation may be expressed in terms of the optical oscillator strengths $f_{J_iJ_f}$ and an effective Gaunt factor g,

$$\sigma_{i,f}(k_i^2) = \frac{8\pi}{\sqrt{3}} \frac{1}{k_i^2\, \Delta E_{if}} f_{J_iJ_f}\, g \quad (\pi a_o^2) \tag{5.19}$$

with

$$g = \frac{\sqrt{3}}{2\pi} \sum_{\ell_i \ell_f} \frac{\ell_>\, R^{(1)}(\ell_i,\ell_f)^2}{r_{if}^{\;2}} \tag{5.20}$$

where

$$R^{(1)}(\ell_i,\ell_f) = \int_0^\infty\!\!\int_0^\infty \frac{1}{r_>} P_{n_i\ell_i}\, P_{n_f\ell_f}\, dr\, F_{k_i\ell_2} F_{k_f\ell_2'}\, dr^1 \tag{5.21}$$

$$r_{if} = \int_0^\infty r\, P_{n_i\ell_i}\, P_{n_f\ell_f}\, dr \tag{5.22}$$

so that all quantities except $F_{k_i\ell_2}(r')$ and $F_{k_f\ell_2'}(r')$, the distorted waves, can be obtained from an atomic structure code. When these results are added to the Lotz cross section for direct ionisation, they require scaling by a factor of 0.4 to bring them into agreement with experiment[58] for Ti IV. Part of this is due to neglecting exchange matrix elements in the DW approximation. A calculation of ^2F excitation at one energy including exchange reduced the contribution by 22%. Similar agreement given the 0.4 scaling is found for Zr IV (Fig. 12) but not for Hf IV, where the discrepancy is attributed to the thirty-four levels of the $5p^5\,4f^{14}\,5d^2$ configuration being omitted. If the Lotz result is taken as an upper bound on direct ionisation, experiment indicates an enhancement due to indirect processes of a factor of 10 for Ti IV, Hf IV and of 20 for Zr IV. Similar calculations, it is suggested, are required for Fe-ions, especially Fe III, IV, V, to establish ionisation rates near the wall in CTR devices. A rough estimate of a factor of 2.5 enhancement for Fe IV, based on the decrease in importance of $3p \rightarrow 3d$ as the d-shell fills.

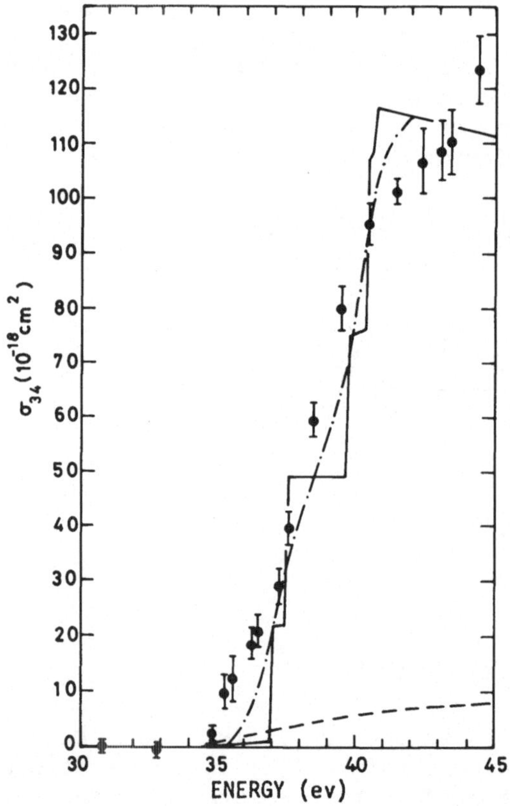

Fig. 12. Electron-impact ionisation cross section for Zr^{3+}. From ref. (57) by permission of the American Institute of Physics.

6. FURTHER REMARKS

A few weeks prior to this summer school, a NATO ASI meeting on "Atomic Physics of Highly Ionised Atoms" was held at Cargese[59], where D. H. Crandall reviewed the same topics as I have covered here, but from a different point of view. This was followed by a meeting of the International Atomic Energy Agency Co-ordinated Research Project (CRP) on "Atomic Data for Diagnostics of Magnetic Fusion Plasmas", which discussed, among other things, the available data and its evaluation. A bibliography for 1978 - April 1982 has been prepared by Itikawa[60].

Ionisation

The theoretical and experimental data for ionisation of atoms and ions with $Z \leq 10$ have been critically reviewed and evaluated by Bell et al[61]. In each case the recommended cross section has been fitted by an expression of the form

$$\sigma(E) = \frac{1}{IE} \{ A \ln (E/I) + \sum_{i=1}^{N} B_i (1 - \frac{I}{E})^i \}$$

where E is the incident energy (rydbergs), I is the ionisation potential, and the cross section is in πa_0^2's. The form is chosen to ensure the correct high energy limit, and scaling

$$I^2 \sigma(E) = \sigma_c (E/I)$$

where $\sigma_c(X)$ is the same for all memebrs of the same isoelectronic sequence. Rate coefficients are computed and given as analytic fits.

During the CRP it became clear that there was no experimental data for fluorine ions (F^{q+}, $q = 0, \ldots 9$), which is an important sequence for diagnostics in fusion plasmas. There is an urgent need for such work, to confirm or otherwise the Belfast group's recommended values.

In a note for the CRP meeting Dunn[62] comments that there are good beam measurements on Na II, Mg II, Mg III, Si II, Si III, Si IV, Ar II, III, IV, V. VI. The Na isoelectronic sequence shows strong excitation-autoionisation effects which increase rapidly with Z. Effects of resonant inner-shell dielectronic capture-double autoionisation may also be important. Work is urgently needed on $C\ell^{q+}$ (all q).

There are very few beam measurements for species in the range $19 \leq Z \leq 36$ except for K II, Ca II, Ti IV, Zn II, Ga II, Xe II, III, IV and VII, and for $Z > 36$, for Rb II, Sr II, Zr IV, Cd II, Xe II, Xe III, Xe IV, Cs II, Ba II, Hf IV,

Ta IV, Hg II, III, Ti III. Note that there are no measurements on any iron ion. Our lack of understanding of the contributions of autoionisation and inner shell excitation makes predictions for any ion with $Z > 19$ hazardous.

Excitation

An extensive programme of evaluation of the data for carbon and oxygen ions is being undertaken by the Japanese and American data centres. The available data on the main species of interest in fusion have been looked at by the IAEA CRP group. It is convenient to finish by quoting extensively from their report.

"It is all too often current practice to obtain excitation rate coefficients for plasma interpretation from Gaunt factor formulas and oscillator strengths. It is now clear that such results are not reliable and that significantly improved reliability is obtained from detailed calculations of collision strengths for individual transitions using close coupling or distorted wave approximations. The number of calculations required is very large (at least hundreds of thousands) but a beginning has been made and the techniques are being refined which will allow the production of the extensive data desired for fusion plasma applications.

Lorenz of IAEA in consultation with active members of the Fusion Community has suggested that the three highest priority groups of targets are[63]

Priority 1	C, O, Ti, Fe
Priority 2	Al, Cr, Ni
Priority 3	H (and D, T), He ,

though other targets, especially Li and the rare gases (Priority 4) may be important for diagnostics. In each case, all stages of ionisation need to be considered.

Experiment

The role of experiment in electron impact excitation is foreseen to be principally in testing theory. All experiments to date have relied on measurements of the radiation emitted by the excited system. Such measurements become progressively more difficult as the charge state of the ion increases, both because the cross sections decrease rapidly with increasing q, and because the photon wave lengths become shorter. The crossed beams technique has produced a few accurate tests of collision theoretical models, notably for the resonance line 2s - 2p $^2P_{1/2,3/2}$ of Li-like ions from Be II[64] to NV.[65] More recently Dunn et al[66] have obtained results on the resonance lines of Na-like systems Al III and Mg II which add to the previous measurements of Zapesochnyi[67] for Mg II. Apart from resonance transitions there are a few measurements on, for example, intercombination lines of Li II[68] and on several transitions in systems of less interest to fusion. There are of course two classic measurements on the 1s - 2s transition in He II,[69] but unfortunately no measurement of the 1s - 2p transition which extends to high energies. (See, however, the results of Zapesochnyi.[69]) An absolute measurement of the total n = 1 to n = 2 excitation cross section (σ_{12}) in He II is of high priority, particularly near threshold.

Theory

Theoretical calculations of excitation cross sections of positive ions are generally carried out in few state close-coupling, or a simplification of that known as distorted wave exchange (DWX). The results are often unitarised (DWXII). For low q and neutral atoms (q = 0) it is essential to include short range correlations, and also to ensure that the leading non-Coulomb term of the long-range interaction in each open channel considered is correctly represented, if accurate results are to be obtained. Unfortunately there are no published calculations satisfying these criteria.

For all species (other than hydrogen-like) accurate results require the use of accurate target wave functions. These should reproduce the observed energy levels, though all matrix diagonalisation methods have difficulty obtaining this accuracy for the ground state. In addition it is desirable to obtain accurate oscillator strengths, in the sense that the dipole length and velocity results should be in agreement to better than 10%. These properties do not suffice to ensure accuracy and, as we have mentioned, it is also important to obtain accurate static dipole polarisabilities of each level. These can be systematically improved variationally.

Further, for high Z ions, ($Z \geq 20$) relativistic effects should be included, at least to the level of the Breit interaction, and full account taken of intermediate and j - j coupling.

Few, if any, published calculations meet these criteria fully, so it would be dangerous to suppose that the accuracy achieved so far is better than 20% except for hydrogenic systems. Readers are invited to consider the comparison of theory with experiment for the benchmark case of Be II 2s - 2p. (Ref. 69, Fig. 5.7).

Targets of Priority Class 1

Carbon and Oxygen

There are few reliable calculations or measurements on the neutral or singly ionised species. The available data on the neutrals are summarised by Bransden and McDowell.[70] For C II there are 5CC (five-state close coupling) and CB XII (Unitarised Coulomb-Born, with exchange) calculations by Robb and Mann,[71] but it is difficult to assess the unreliability for such open shell systems. Further accurate calculations are required for C I and C II especially, in view of their special interest in Tokamaks with carbon inner liners.

The Japanese and US data centers are currently preparing a compilation and selection of data on ions of C and O, which will be part of the publication programme of this CRP. There is a significant number of detailed calculations for excitation from the ground states of each of these ions, which in most cases should allow a recommendation to be made.

Ti and Fe

There is a substantial amount of data (theoretical) for ions of Fe.[72,73] The current situation for excitation of Fe ions illustrates both what can be accomplished by detailed calculations and that significant work remains to be done before a complete set of reasonably reliable rate coefficients is obtained. A recent compilation[72] selects calculated cross sections for about 1150 transitions of $\Delta n = 0$ type for Fe IX-Fe XXVI and $\Delta n = 1$ type for Fe XVII-Fe XXVI from low lying states. All of the selected data are from DWXII or few state calculations. It is desirable to represent collision strengths and rate coefficients for each transition by a four parameter fit in a standard form involving powers of transition energies such as is given on p.206, ref. 73. For the 1150 transitions mentioned, about half of the calculations provide a cross section at only one energy, whereas about seven values well spread in energy are needed to allow a reasonably accurate fit. Thus the work is far from finished for the 1150 transitions considered so far. Sufficient detail has been provided primarily for Be-like through H-like cases, but refinements may significantly change these results. It seems clear that many of these calculations needed to be carried out in intermediate rather than L-S coupling[74] and further that the effects of resonances are not yet correctly accounted for in these highly charged ions.[75] From the perspective of plasma fusion diagnostics it is a matter of highest priority to improve and complete the detailed calculations for specific J to J' transitions for all $\Delta n = 0$ and most $\Delta n = 1$ transitions from low lying states of Fe ions. Some of this work is currently in progress at Los Alamos Scientific Laboratory (J. B. Mann and A. L. Merts).

There is as yet little data of any sort for Ti-ions. Bhatia et al[76] have reported DWX values for various transitions among $2s^m\ 2p^n$ configurations for Ti^{q+} ($q = 13, \ldots, 19$). In view of the general considerations discussed above, their reliability must be open to question.

Al, Cr, Ni

There are DWX II (unitarised, distorted wave exchange) calculations available for some transitions in Al II to Al VI. Mann[65] gives coefficients from which cross sections and rate coefficients may be obtained for a few transitions. It is again difficult to assess the reliability of these calculations at this stage, especially for the lower stages of ionisation. Various approximations for H-like and He-like ions of both Al and Ni are discussed by Henry.[69] There have been no calculations on Cr ions since 1970. In view of the priority attached to these species a major research programme needs to be initiated.

There are very extensive DWX calculations for He-like ions by Pradhan et al[77] which include resonance effects by a backwards Gailitis extrapolation, and these calculations also provide collision strengths and rate parameters.[78]

However, it recently became apparent that the reliability of this work decreases with increasing atomic number beyond $Z = 6$ and for $Z = 26$ may be in error by more than an order of magnitude (see the discussion of Fe^{q+} ions above).

Lower Priority Systems

There are reasonably accurate results available for H-like and He-like ions of B, N, F, Si, Ne, Ar and Kr. Again many charge states of N, Ne and Ar have been considered. Readers are referred to Itikawa's bibliography for details.[60] No attempt can yet be made to assess the reliability of the calculations, but the uncertainties will increase with increasing Z and decreasing q."

REFERENCES

1. C. J. Joachain, "Quantum Collision Theory," North-Holland Publishing Co. Ltd., Amsterdam (1975).
2. B. H. Bransden, "Atomic Collision Theory," Benjamin, New York

(1970) Second edition 1983.
3. C. J. Joachain, Theoretical methods for atomic collisions, in "Atomic and Milecular Processes in Controlled Thermonuclear Fusion," M. R. C. McDowell and A. Ferendeci, ed., Plenum Press, New York (1980).
4. I. C. Percival and D. Richards, Adv. Atom. Mol. Phys. 11, (1975).
5. R. Shakeshaft, Phys. Rev. A 17, 1011 (1978).
6. P. G. Burke, K. A. Berrington and C. V. Sukumar, J. Phys. B. 14, 289 (1981).
7. N. Abu-Salbi and J. Callaway, Phys. Rev. 24, 2372 (1981).
8. J. Callaway, Phys. Rep. 45, 89 (1978).
9. P. G. Burke and W. D. Robb, Adv. Atom. Mol. Phys. 11, 143 (1975).
10. Y. K. Hahn, T. F. O'Malley and L. Spruch, Phys. Rev. 134, 397 (1964), and 134, 911 (1964).
11. M. K. Gailitis, J.E.T.P. 20, 107 (1964).
12. S. Geltman and P. G. Burke, J. Phys. B. 3, 1062 (1970).
13. J. Callaway, Phys. Rev. (1982) in press.
14. A. J. Taylor and P. G. Burke, Proc. Phys. Soc. A 92, 366 (1967).
15. P. G. Burke, S. Ormonde and W. Whittaker, Proc. Phys. Soc. A 92, 319 (1967).
16. J. Callaway and J. W. Wooton, Phys. Rev. A 9, 1294 (1974).
17. L. A. Morgan, J. Callaway and M. R. C. McDowell, J. Phys. B. 10, 3297 (1977).
18. J. F. Williams, J. Phys. B. 9, 1519 (1976).
19. J. Callaway and M. R. C. McDowell, to be presented at the SPIG Conference, Dubrovnik, August 1982.
20. L. C. Johnson, Astrophys. J. 174, 227 (1972).
21. K. Omidvar, Phys. Rev. A 140, 38 (1965).
22. R. J. W. Henry, Phys. Rep. 68, 1 (1981).
23. A. K. Pradhan, D. W. Norcross and D. G. Hummer, Phys. Rev. A 23, 619 (1981).
24. S. L. Willis and M. R. C. McDowell, J. Phys. B. 14, L453 (1981).
25. K. L. Baluja, P. G. Burke and A. E. Kingston, J. Phys. B. 14, 1333 (1981).
26. K. L. Baluja, P. G. Burke and A. E. Kingston, J. Phys. B. 14, 119 (1981).
27. W. Eissner and M. J. Seaton, J. Phys. B. 7, 2533 (1974).
28. C. Mendoza, J. Phys. B. 15, 867 (1982).
29. M. J. Seaton, J. Phys. B. 2, 1817 (1974).
30. W. L. Wyngaarden, K. Bhadra and R. J. W. Henry, Phys. Rev. A 20, 1409 (1979).
31. F. Bely-Dubau, J. Dubau, P. Faucker and L. Steenman-Clark, J. Phys. B. 14, 3313 (1981).
32. J. M. Peek and J. B. Mann, Phys. Rev. A 25, 749 (1982).
33. M. E. Riley and D. G. Truhlar, J. Chem. Phys. 65, 792 (1976).
34. B. H. Bransden, M. Crocker, I. E. McCarthy, M. R. C. McDowell and L. A. Morgan, J. Phys. B. 11, 3411 (1978).
35. N. H. Magee, Jr., J. B. Mann, A. L. Merts and W. D. Robb, Los Alamos National Laboratory Report LA-6691-MS.

36. A. L. Merts, J. B. Mann, W. D. Robb and N. H. Magee, Jr., Los Alamos National Laboratory Report LA-8267-MS.
37. J. Callaway and M. R. C. McDowell, to be presented at SPIG Conference, Dubrovnik, August 1982. (Proceedings published by Yugoslav. Institute of Physics.)
38. J. A. Tully and K. L. Baluja, J. Phys. B. 14, L831 (1981).
39. H. Jakubowicz and D. Moores, Comm. Atom. Mol. Phys. 9, 55 (1980).
40. D. H. Phillips and M. R. C. McDowell, J. Phys. B. 6, L165 (1973).
41. R. K. Peterkop, "Theory of ionization of atoms by electron impact," Colorado Assoc. Univ. Press, Boulder, Colorado (1977).
42. D. H. Crandall, R. A. Paneuf, B. E. Hasselquist and D. C. Gregory, J. Phys. B. 12, L249 (1979).
43. H. Jakubowicz and D. Moores, J. Phys. B. 14, 3733 (1981).
44. W. Lotz, Z. Physik, 216, 241 (1968).
45. S. M. Younger, Phys. Rev. A 22, 111 (1980).
46. S. M. Younger, Phys. Rev. A 23, 1139 (1981).
47. S. M. Younger, Phys. Rev. A 24, 1272 (1981).
48. S. M. Younger, Phys. Rev. A 24, 1278 (1981).
49. S. O. Martin, B. Peart and K. T. Dolder, J. Phys. B. 1, 537 (1968).
50. D. R. Woodruff, M.-C. Hublet, M. F. A. Harrison and E. Brook, J. Phys. B. 11, L679 (1978).
51. D. H. Crandall, R. A. Paneuf and D. C. Gregory, ORNL-TM-7020, unpublished.
52. E. J. McGuire, Phys. Rev. 25, 192 (1982).
53. K. L. Aitken, M. F. A. Harrison and R. D. Rundle, J. Phys. B. 4, 1189 (1971).
54. D. C. Economides and M. R. C. McDowell see also G. Peach, J. Phys. B. 4, 1670 (1971).
55. W. T. Rogers, C. Stefani, R. Camilloni, G. H. Dunn, A. Z. Msezane and R. J. W. Henry, Phys. Rev. A 25, 737 (1982).
56. M. S. Pindzola, D. C. Griffin and C. Bottcher, Phys. Rev. A 25, 211 (1982).
57. D. C. Griffin, C. Bottcher and M. S. Pindzola, Phys. Rev. A 25, 1374 (1982).
58. R. A. Falk, G. H. Dunn, D. C. Griffin, C. Bottcher, D. C. Gregory, D. H. Crandall and M. S. Pindzola, Phys. Rev. Lett. 47, 494 (1981).
59. D. H. Crandall, Atomic Physics of Highly Ionised Atoms, Proc. NATO Summer School, Cargèse, to be published by Plenum Press (1982).
60. Y. Itikawa, Supplement to Institute of Plasma Physics, Nagoya, Report No. IPPJ - AM - 7, to be published.
61. K. L. Bell, H. B. Gilbody, J. G. Hughes, A. E. Kingston and F. J. Smith, Culham Laboratory Report CLM - R216 (1982).

62. G. H. Dunn, *in* Report of the IAEA CRP on Atomic Data for Diagnostics of Magnetic Fusion Plasmas, IAEA Vienna, to be published.
63. A. Lorenz, *in* Report of the IAEA CRP on Atomic Data for Diagnostics of Magnetic Fusion Plasmas, IAEA Vienna, to be published.
64. P. O. Taylor, R. A. Phaneuf and G. H. Dunn, Phys. Rev. A 22, 435 (1980).
65. D. C. Gregory, G. H. Dunn, R. A. Phaneuf and D. H. Crandall, Phys. Rev. A 20, 410 (1979).
66. G. H. Dunn, R. A. Falk, D. S. Belic, D. H. Crandall, D. C. Gregory and C. Cisneros, Private communication.
67. I. P. Zapesochny, V. A. Kel'man, A. I. Imre, A. I. Daschenko and F. F. Danch, Sov. Phys. JETP 42, 989 (1976).
68. W. T. Rogers, J. Ø. Olsen and G. H. Dunn, Phys. Rev. A 18, 1353 (1978).
69. R. J. W. Henry, Phys. Rep. 68, 1 (1981).
70. B. H. Bransden and M. R. C. McDowell, Phys. Rep. 46, 249 (1978).
71. N. H. Magee, Jr., J. B. Mann, A. L. Merts and W. D. Robb, LA-6691-MS (1977).
72. M. S. Pindzola and D. H. Crandall, ORNL/TM-7957 (1981).
73. A. L. Merts, J. B. Mann, W. D. Robb and N. H. Magee, Jr., LA-8267-MS (1980).
74. R. E. H. Clark, N. H. Magee, Jr., J. B. Mann and A. L. Merts, Astrophys. J. 254, 412 (1982).
75. A. K. Pradhan, Phys. Rev. Lett. 47, 79 (1981).
76. A. K. Bhatia, U. Feldman and G. A. Doschek, J. Appl. Phys. 51, 1464 (1980).
77. A. K. Pradhan, D. W. Norcross and D. G. Hummer, Phys. Rev. A 23, 619 (1981).
78. A. K. Pradhan, D. W. Norcross and D. G. Hummer, Astrophys. J. 246, 1031 (1981).

EXPERIMENTAL ASPECTS OF ELECTRON IMPACT IONIZATION AND EXCITATION OF POSITIVE IONS

Kenneth T. Dolder

School of Physics
The University
Newcastle upon Tyne, NE1 7RU, U.K.

1. INTRODUCTION

Plasmas of greatest interest in fusion and astrophysics are not in thermodynamic equilibrium so that details of each significant atomic process are needed to construct theoretical models and interpret observations.

Two of the more important types of reaction are electron impact ionization,

$$e + A^{n+} \rightarrow A^{m+} + (m-n)e \qquad 1.1$$

and excitation,

$$e + A^{n+}(n,\ell) \rightarrow e + A^{n+}(n',\ell'). \qquad 1.2$$

In plasmas which are "thin" to electromagnetic radiation, the ionization equilibrium (and hence almost all important plasma properties is mainly determined by the balance between the rates of ionization and recombination. Electron impact is usually the dominant ionization process and when molecular ions are absent, this is primarily balanced by dielectronic and radiative recombination.

For pure, hot, hydrogenic plasmas, bremsstrahlung is the major energy loss mechanism, but when impurities occur, line radiation may predominate. Energy losses can then be calculated only when excitation cross sections are known for each constituent.

A great deal of theoretical and experimental effort has therefore been directed towards the study of ionization and excitation. We will be concerned only with experimental aspects but references to some theoretical reviews are included in the selected bibliography in section 5.

Three experimental approaches to the study of inelastic electron-ion collisions have been developed. The most prolific, accurate and versatile technique employs crossed beams. Pure, well-collimated beams of ions and electrons are made to intersect and the flux of ionization products, or light produced by a specific excitation process, is measured absolutely. These methods have been extensively developed and reviewed, and a newcomer to this field could do worse than study articles by Dolder and Peart[1], Dunn[2], Salzborn[3] and Crandall[4], in that order.

Crossed beam techniques have now been extended to multiply-charged ions such as N^{4+} and O^{5+}, but this is still a far cry from the very highly-charged species encountered in fusion plasmas and hot stellar atmospheres. These may, however, be studied with ion traps or by plasma spectroscopy.

Plasma experiments give reaction rates, rather than cross sections, and they require time-resolved measurements of spectral emission from a well-diagnosed, transient plasma. Ionization rate coefficients can then be deduced if the temperature, density and any change of volume of the plasma in the direction of observation are known. Various rate coefficients can then be inserted into a computer programme of a model plasma until a rate is found for which the calculated temporal variation of line intensity coincides with the measurements.

This method gives good agreement with crossed beam experiments for hydrogen-like ions but, for helium-like ions the results are consistently about 15% too low, and for lithium-like ions the discrepancy increases to about 40%.

Plasma experiments can also be used to obtain rate coefficients for excitation (α_{if}). The intensity of radiation per unit volume, per steradian from a transition of frequency ν is given by,

$$I_{fi}(t) = \frac{h\nu}{4\pi} \cdot \frac{A_{fj}}{\sum_r A_{fr}} \cdot N_e(t)\, N_i(t)\, \alpha_{if} \qquad 1.3$$

where N_e is the electron density and N_i is the density of ions in the initial state. A_{fj} represents the transition probability for radiation from state f to j and A_{fr} is the transition probability from f to any state, r.

Equation 1.3 assumes that:

(1) State f is populated only by transitions from state i.

(2) No process other than direct excitation (e.g. electron capture or cascade) contributes to the population of f.

(3) The transition probabilities are not modified by fields or collisions within the plasma.

(4) The plasma temperatures and densities remain constant during excitation and radiation.

The first twc conditions may be the most elusive. Metastable states and cascade can provide other routes to the population of final states. Gabriel and Jordan[7] reviewed the interpretation of spectral intensities from plasmas, and descriptions of the determination of excitation rate coefficients are given by Henry[8] and Kunze[5]. More recent measurements for the ionization of Fe^{9+} were described by Datla et al[6].

The third approach applied to ionization (but not excitation) involves trapped ions. Donets et al[9] have developed a technique derived from experiments by Baker and Hasted[10] in which the space charge of an electron beam, constrained by an axial magnetic field, is used to trap ions which are progressively ionized to higher states by successive electron impacts. The distribution of charge states can at any time be monitored with a mass spectrometer and the results compared with a simple model in which ionization cross sections are fitted parameters. In this way ionization cross sections can be determined at chosen electron beam energies. This method is the only one which yields cross sections as functions of electron energy for very highly-charged species, e.g. Ar^{17+}. The results are in reasonable agreement with theory and crossed beam measurements, but the method neglects the influences of processes such as multiple ionization, charge exchange, recombination and ion losses from the trap. It is therefore inherently less accurate than good crossed beam experiments. Moreover, measurements have so far only been made with electron energies above 2 keV, whereas in applications to fusion plasmas and astrophysics, it is the nearer-threshold cross sections that are more important.

This article will concentrate on beam techniques, but some data obtained by other methods will be included in discussions of results.

2. IONIZATION OF POSITIVE IONS BY CROSSED BEAM TECHNIQUES

Details of crossed beam experiments have been reviewed many times and will not be repeated here. Instead, we will draw attention to the scope of the measurements and some more interesting recent advances.

The most important development has been the application of crossed beams to highly-charged ions, notably by Crandall and his colleagues at ORNL. Results were obtained for isoelectronic series (e.g. Mg^+, Al^{2+} and Si^{3+}) and these offer clues to the behaviour of very highly-charged species beyond the range of crossed beam experiments. They also provide much more systematic tests for theory and the empirical formulae used by plasma physicists. Measurements also revealed further evidence of the important role sometimes played by autoionization and they suggest that the relative contribution of autoionization tends to increase with ionic charge. It has also been possible to obtain results for multiple ionization of ions with more than a single charge, e.g.

$$Ar^{2+} + e \rightarrow Ar^{5+} + 4e \qquad 2.1$$

This type of process must clearly be included in any detailed description of ionization equilibrium.

Table 1 lists measurements of ionization made with crossed beams and a discussion of some of the more interesting conclusions forms the basis of the following section.

3. DISCUSSION OF SOME IONIZATION MEASUREMENTS

3.1 Introduction

This section will primarily be concerned with the impact which recent measurements have made upon our understanding of autoionization and the validity of empirical formulae.

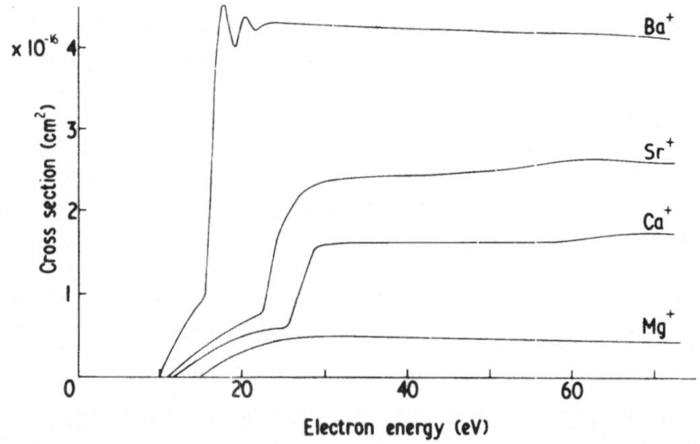

Fig. 1. Measured[11] ionization cross sections of Mg^+, Ca^+, Sr^+ and Ba^+. Large autoionization contributions occur mainly from np→nd transitions, where n = 3, 4 and 5 for Ca^+, Sr^+ and Ba^+, respectively.

Table 1. Measurements of Ionization with Crossed Beams

REACTION	REFERENCE	COMMENT
$He^{+} \rightarrow He^{2+}$	Dolder et al (1961)[60]	Results to 1 keV
$He^{+} \rightarrow He^{2+}$	Peart et al (1969)[61]	Results to 10 keV
$Li^{+} \rightarrow Li^{2+}$	Lineberger et al (1966)[62]	
$Li^{+} \rightarrow Li^{2+}$	Peart & Dolder (1968)[63]	
$B^{3+} \rightarrow B^{4+}$		He-like sequence
$C^{4+} \rightarrow C^{5+}$	Crandall et al (1979)[64]	
$N^{5+} \rightarrow N^{6+}$		
$Be^{+} \rightarrow Be^{2+}$	Falk & Dunn (1982)[65]	
$C^{3+} \rightarrow C^{4+}$	Crandall et al (1978)[66]	
$N^{4+} \rightarrow N^{5+}$		Li-like sequence
$N^{4+} \rightarrow N^{5+}$	Müller et al (1982)[68]	
$O^{5+} \rightarrow O^{6+}$	Crandall et al (1979)[36]	
$B^{+} \rightarrow B^{2+}$		Be-like sequence
$C^{2+} \rightarrow C^{3+}$	Crandall et al (1979)[69]	(large metastable
$N^{3+} \rightarrow N^{4+}$		contamination)
$O^{4+} \rightarrow O^{5+}$		
$Mg^{+} \rightarrow Mg^{2+}$	Martin et al (1968)[67]	
$Mg^{+} \rightarrow Mg^{2+}$		Na-like sequence
$Al^{2+} \rightarrow Al^{3+}$	Crandall et al (1982)[70]	
$Si^{3+} \rightarrow Si^{4+}$		
$Na^{+} \rightarrow Na^{2+}$	Hooper et al (1966)[71]	
$Na^{+} \rightarrow Na^{2+}$	Peart & Dolder (1968)[72]	
$K^{+} \rightarrow K^{2+}$	Hooper et al (1966)[71]	
$K^{+} \rightarrow K^{2+}$	Peart & Dolder (1968)[72]	
$Rb^{+} \rightarrow Rb^{2+}$	Peart & Dolder (1975)[73]	
$Cs^{+} \rightarrow Cs^{2+}$		
$Rb^{+} \rightarrow Rb^{3+}$		
$Rb^{+} \rightarrow Rb^{4+}$	Hughes & Feeney (1981)[74]	
$Rb^{+} \rightarrow Rb^{5+}$		
$Ca^{+} \rightarrow Ca^{2+}$	Peart & Dolder (1975)[73]	Very large auto-
$Sr^{+} \rightarrow Sr^{2+}$		ionization
$Ba^{+} \rightarrow Ba^{2+}$	Peart & Dolder (1968)[75]	
$Ba^{+} \rightarrow Ba^{2+}$	Peart et al (1973)[76]	Energy-resolved electrons
$Ba^{+} \rightarrow Ba^{2+}$	Feeney et al (1972)[77]	

/cont'd

Table 1. (continued)

REACTION	REFERENCE	COMMENT
$C^+ \rightarrow C^{2+}$ $N^+ \rightarrow N^{2+}$ $N^{2+} \rightarrow N^{3+}$	Aitken et al (1971)[78]	
$O^+ \rightarrow O^{2+}$ $O^{2+} \rightarrow O^{3+}$	Aitken & Harrison (1971)[79]	
$Ne^+ \rightarrow Ne^{2+}$	Dolder et al (1963)[80]	
$Ar^{2+} \rightarrow Ar^{4+}$ $Ar^{2+} \rightarrow Ar^{5+}$ $Ar^{3+} \rightarrow Ar^{5+}$	Müller & Frodl (1980)[81]	
$Ar^{4+} \rightarrow Ar^{5+}$ $Ar^{4+} \rightarrow Ar^{6+}$ $Ar^{5+} \rightarrow Ar^{6+}$	Müller et al (1980)[82]	
$Xe^{3+} \rightarrow Xe^{4+}$	Gregory et al (1981)[83]	
$Mg^{2+} \rightarrow Mg^{3+}$	Peart et al (1969)[84]	
$Tl^+ \rightarrow Tl^{2+}$	Divine et al (1976)[85]	
$Zn^+ \rightarrow Zn^{2+}$ $Ga^+ \rightarrow Ga^{2+}$	Rogers et al (1982)[86]	
$Ti^{3+} \rightarrow Ti^{4+}$ $Zr^{3+} \rightarrow Zr^{4+}$ $Hf^{3+} \rightarrow Hf^{4+}$	Falk et al (1982)[87]	
$Li^+ \rightarrow Li^{3+}$	Peart & Dolder (1969)[110]	

3.2 Autoionization

When an electron strikes an ion it may either eject an outer-shell electron directly, or excite an inner shell electron. If the excitation energy (χ_e) exceeds the outershell ionization energy (χ_i), autoionization may occur. The dramatic effects which this can produce are illustrated in figure 1 by measurements[11] of electron impact ionization of Mg^+, Ca^+, Sr^+ and Ba^+. With the exception of Mg^+, the cross sections all rise abruptly due to the onset of autoionization. Clear evidence of autoionization has been observed for a number of other ions and results[12] for N^{4+} and O^{5+} are illustrated by figures 2 and 3.

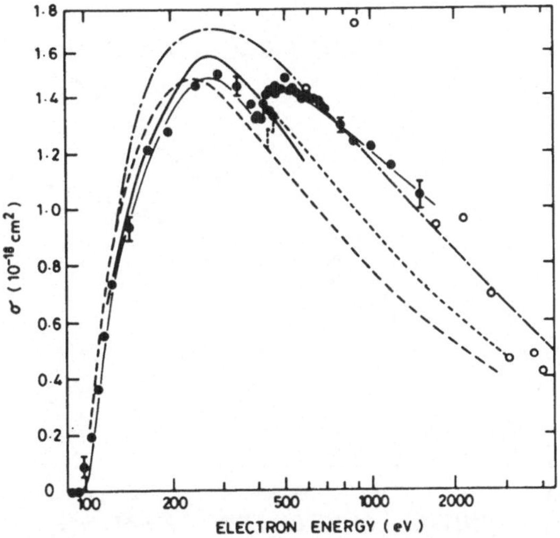

Fig. 2. Ionization cross sections for N^{4+} showing secondary peak due to autoionization. The solid and open circles denote results with crossed beams[36,66] and an ion trap.[88] Solid curves and dashed curves are Coulomb-Born calculations by Moores[89] and Sampson and Golden.[90] The chain curve is based on the Lotz formula and the short-dashed curve is Henry's[91] calculation of autoionization.

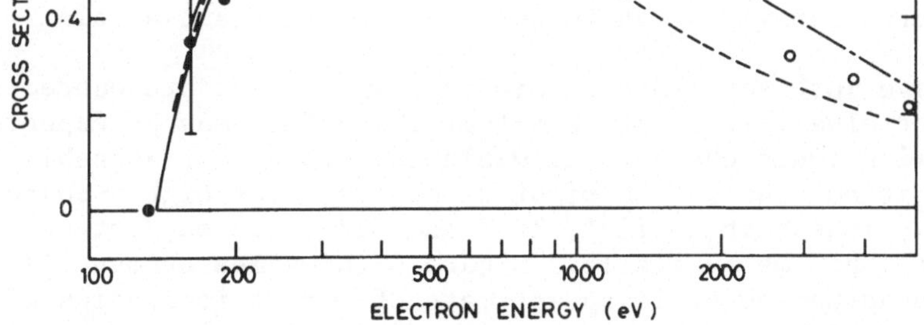

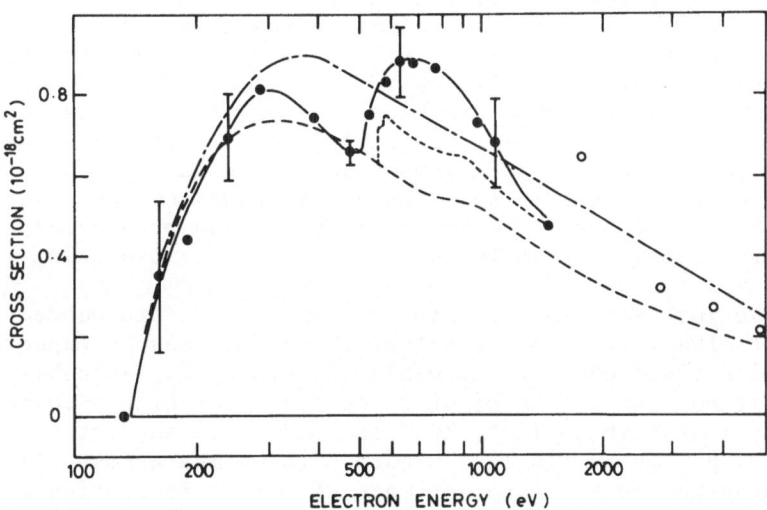

Fig. 3. Ionization cross sections for O^{5+} showing secondary peaks due to autoionization. Measurements[36,66] are compared with the Lotz formula (chain curve), Coulomb-Born[90] (long dashes) and close-coupling calculation of excitation[91] (short dashes).

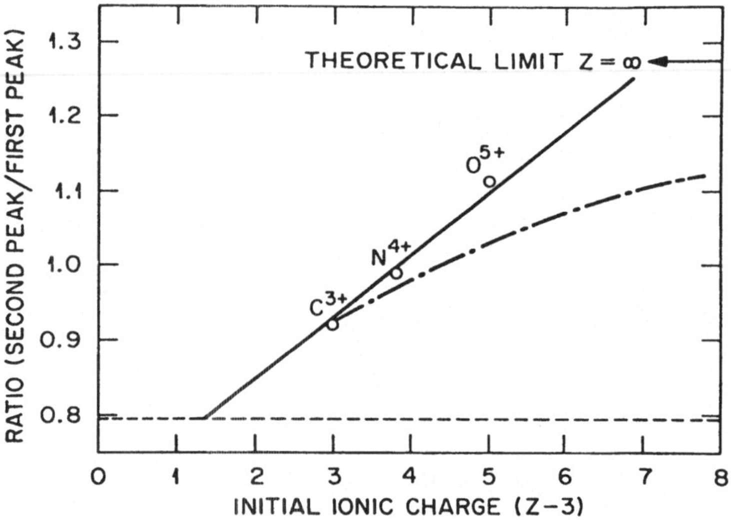

Fig. 4. Ratio of heights of second and first peaks in ionization cross functions of the Li-like ions, C^{3+}, N^{4+} and O^{5+}. The second peak contains autoionization contribution and its relative magnitude increases with ionic charge. The broken curve is a theoretical prediction based on the Coulomb-Born approximation.

For an isoelectronic series (e.g. C^{3+}, N^{4+} and O^{5+}) the ratio χ_e/χ_i tends to increase with ionic charge and this might lead one to guess that autoionization will be relatively more important for highly-charged ions. There is now some evidence to support this view and figure 4 shows the ratio of the second and first peaks in the ionization cross section for three Li-like ions. The first peak is due solely to direct ionization, whereas the second contains an autoionization contribution. A solid curve is drawn through experimental values[12] of the ratio and the broken curve is the prediction of a Coulomb-Born theory[13] which includes both direct and autoionization.

We have just seen that autoionization is very pronounced for the heavier alkali-like ions and that its effect may be expected to increase with ionic charge. It would therefore be reasonable to suppose that autoionization might be enormous for highly-charged, alkali-like ions such as Ti^{3+}, Zr^{3+} and Hf^{3+}, and supporting evidence has recently become available. Figure 5 shows measurements[14] for Ti^{3+}. The dashed curve is an estimate of direct ionization given by the empirical formula of Lotz (see section 3.3) whilst the solid curve was obtained by adding this estimate to the result of a distorted wave calculation of $3p^6 3d \rightarrow 3p^5 3d^2$ excitation, which had been divided (empirically) by a factor 2.5. The chain curve shows the result of folding a 2 eV energy spread (characteristic of the experiment) into the continuous curve. The theory is far from rigorous,

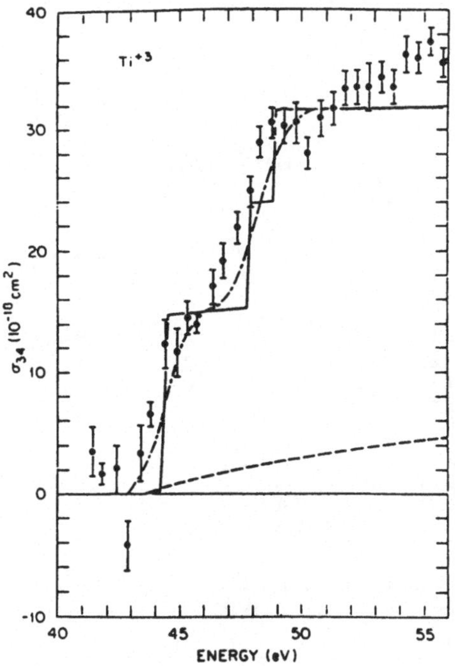

Fig. 5. Measured[14] cross sections for Ti^{3+} which indicate direct ionization whilst the solid curve is a semi-empirical calculation which includes auto-ionization. The chain curve was obtained by folding the experimental electron energy distribution into the theoretical curve.

but it is not hard to believe that, for Ti^{3+}, autoionization can be an order of magnitude larger than direct ionization.

There is also evidence of strong inner shell effects in cross sections for multiple ionization by single collisions. Figure 6 illustrates measurements by Müller and Frodl[15] for multiple ionization of Ar^{2+} and Ar^{3+}. Thresholds for outer shell ionization and direct ionization from the L-shells are marked in the figure, and it can be seen that cross sections increase by as much as an order of magnitude due to the onset of L-shell ionization. The dashed curves in figure 6 represent normalised Bethe calculations for direct double ionization.

It will have been noticed that it was necessary to divide a theoretical estimate for auto-ionization of Ti^{3+} by a factor 2.5 to obtain a fit with experiment and, in general, the theoretical results for autoionization are not very accurate. This was illustrated by Crandall[16] who compared measured inner shell excitation cross sections with theory for the Li-like series, Be^+, C^{3+}, N^{4+} and O^{5+}. His results, reproduced in Table 2, indicate that even six-state

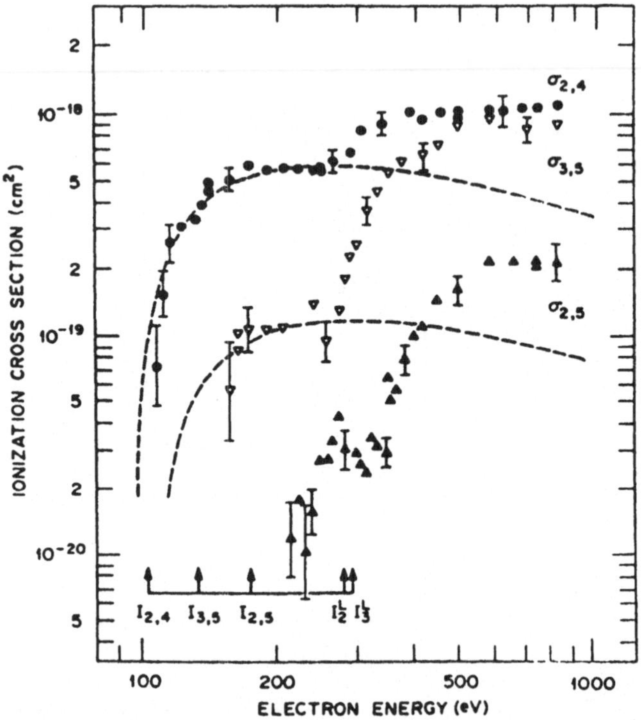

Fig. 6. Measured[15] multiple ionization cross sections $\sigma_{n,m}$ for Ar^{2+} and Ar^{3+} ions. The figure includes thresholds for L-shell ionization which coincide with very abrupt increases in $\sigma_{n,m}$ due to inner shell effects. The dashed curves are theoretical estimates only of direct ionization.

close-coupling theory[8] may be in error by 40%, or much more in the case of O^{5+}. The excitation cross sections refer to electron energies about 1.1 times threshold but it must be emphasized that the data in Table 2 is somewhat tentative. For example, the measurements for O^{5+} are subject to considerable error and the Coulomb-Born calculations are not strictly applicable to ions with small ionic charge (e.g. Be^{+}). Nevertheless, there is reason to doubt the accuracy of most autoionization calculations and, in view of complicated electron interactions which must occur, especially in larger ions, this is hardly surprising. The matter should be of some concern to fusion physicists because they deal with highly-charged species for which autoionization might be very important. Some idea of its magnitude for Fe^{15+} and uncertainties in the theory can be obtained from calculations by Cowan and Mann[32] and by Kim and Cheng.[33] The former (which is probably more reliable) predicts a four-fold increase due to autoionization, whilst the latter estimate is an order of magnitube larger.

Table 2. Comparisons of theoretical and experimental cross sections for inner shell excitation of Li-like ions, at energies about 1.1 times the threshold for exciting $1s^22s \rightarrow 1s2s2\ell$.

Ion	Energy (eV)	Experiment	Scaled Coulomb-Born	Six-state close-coupling	$\frac{\text{Expt. result}}{\text{6 c-c result}}$
B^+	130	17	23	12.2	1.4
C^{3+}	340	3.2	3.7	2.15	1.5
N^{4+}	460	1.8	2.0	1.27	1.4
O^{5+}	612	2.1 ± 0.7	1.1	0.74	2.8

3.3 Empirical Formulae for Ionization

In the absence of reliable measurements or theory, plasma physicists resort to one of the empirical formulae which can be used to predict ionization cross sections. The most detailed appraisals of these formulae have been given by Itikawa and Kato[17], and Crandall.[18]

Expressions most widely used are based on the work of Lotz,[19,20] Golden and Sampson,[21,22,23,24] and Burgess.[25]

For ions with initial charge greater than +3, Lotz suggested

$$\sigma(E) = 4.5 \times 10^{-14} \sum_j \frac{r_j}{I_j E} \ln\left(\frac{E}{I_j}\right) \text{ cm}^2 \qquad 3.1$$

for the cross section at incident electron energy, E. Here, r_j represents the number of electrons in the sublevel of ionization energy I_j. For less highly-charged ions he proposed the more complicated expression,

$$\sigma(E) = \sum_j \frac{a_j r_j}{I_j^2} \frac{\ln X_j}{X_j} \{1 - b_j \exp\left[-c_j (X_j - 1)\right]\} \qquad 3.2$$

where $X_j \equiv E/I_j$ and a_j, b_j and c_j represent adjustable parameters. Numerical values of these parameters were given for $(\text{He-Ga})^+$, $(\text{Li-Zn})^{2+}$ and $(\text{Be-Ga})^{3+}$. Equation 3.2 is therefore restricted to ions for which parameters were evaluated and it is only likely to be valid for the limited range of energies over which they have been fitted.

Golden and Sampson suggested an expression which contains four adjustable parameters which depend only on the subshell of the ejected electron. Numerical values have been determined for j = 1s to 4f and a choice is made to ensure the correct Bethe asymptote for a hydrogenic ion. Their expression can be written,

$$\sigma(E) = \pi a_o^2 \sum_j \left(\frac{n}{Z_{eff}(j)}\right) \frac{I_H}{I_j} \cdot \frac{r_j}{u_j} \Big[A_j \ln u_j + D_j (1 - u_j^{-1})^2$$

$$+ \left(\frac{c_j}{u_j} + \frac{d_j}{u_j^2}\right) (1 - u_j^{-1}) \Big] \qquad 3.3$$

where n is the principal quantum number of level j, I_H = 13.6 eV, $u_j = E/I_j$ and Z_{eff} is the effective charge binding an electron in level j of the initial ion. A_j, D_j, c_j and d_j represent the adjustable parameters.

Burgess derived the "ECIP" (exchange classical result added to a long-range impact parameter contribution) formula widely used for highly-charged ions in astrophysics, but it is more complicated to apply and will not be discussed here.

Itikawa and Kato[17] compared the Lotz, and Golden and Sampson formulae with experimental data for no fewer than 27 ions. The agreement was frequently within 25%. In many cases agreement was much better, but larger deviations were noted for Na^{+}, O^{+} and C^{3+}.

Crandall also discussed the validity of the Lotz formula for a selection of highly-charged ions. He found (with the exception of C^{3+}) that it agreed with experiment within a few percent but that the agreement was partly fortuitous. The simple Lotz formula (equation 3.1) usually overestimates direct ionization and omits autoionization, but it happens that for many ions the autoionization contribution compensates fortuitously for the overestimate of direct ionization and so there is often remarkably good agreement above the autoionization thresholds. Unfortunately, this is far from being universally true, and we conclude this section by drawing attention to three examples in which the omission of inner shell effects leads to serious inaccuracies in the Lotz formula.

Figure 7 compares measurements[26] for the ionization of Ga^{+} with predictions of Lotz and a scaled plane wave Born calculation.[27] The substantial autoionization below 25 eV causes the measured cross section to rise much more steeply than these theoretical predictions.

Measurements[28] for Xe^{3+} are compared in figure 8 with the Lotz formula to which estimates have been added for contributions for

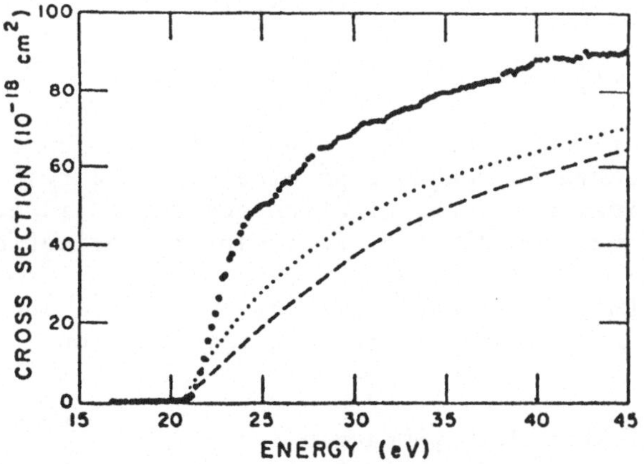

Fig. 7. Measured[26] cross sections for the ionization of Ga^{+} compared with a scaled Born calculation[92,93] (dashed curve) and the Lotz formula (dotted curve).

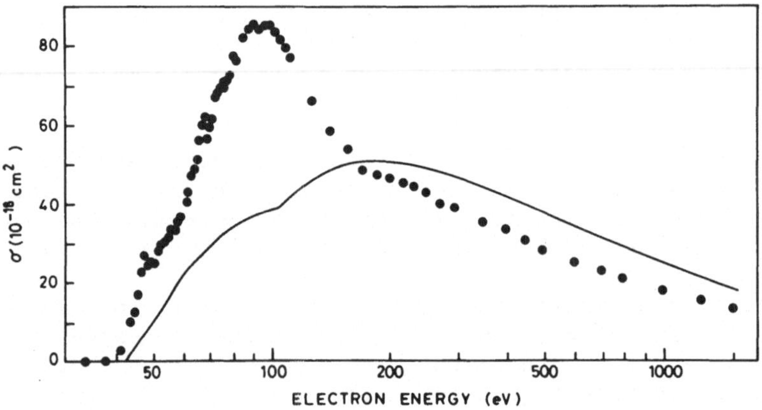

Fig. 8. Measured[28] ionization cross sections for Xe^{3+}. The continuous curve illustrates predictions of the Lotz formula for direct ionization from 5s, 5p and 5d subshells. The disparity is attributed to further large contributions from autoionization.

ejection of 4d, 5s and 5p electrons. This result still falls far below the near-threshold measurements and illustrates the large magnitude and complex nature of inner-shell ionization, especially in multiply-charged ions.

A further complication may arise from a capture-autoionization process which has recently been suggested.[29] It is possible that the capture of an incident electron by an ion might form a doubly-excited, short-lived, resonant, intermediate state which could decay with the Auger emission of two electrons, i.e.

$$A^{n+} + e \rightarrow {}^{**}A^{(n-1)+} \rightarrow A^{(n+1)+} + 2e. \qquad 3.4$$

Results of calculations[29] for Fe^{15+} are illustrated by figure 9 in which the dotted curve is the Lotz prediction for direct ionization, the dashed curve includes autoionization and the solid curve illustrates the additional contribution from capture-ionization.

We conclude that inner-shell effects are subtle and various. They cannot be described by any simple generalization.

3.4 Ionization of Very Highly-Charged Ions

It was pointed out in section 1 that results for the ionization of very highly-charged ions have only been obtained by plasma spectroscopy and trapped ion techniques. The former method gives reaction rates whilst the latter provides cross sections.

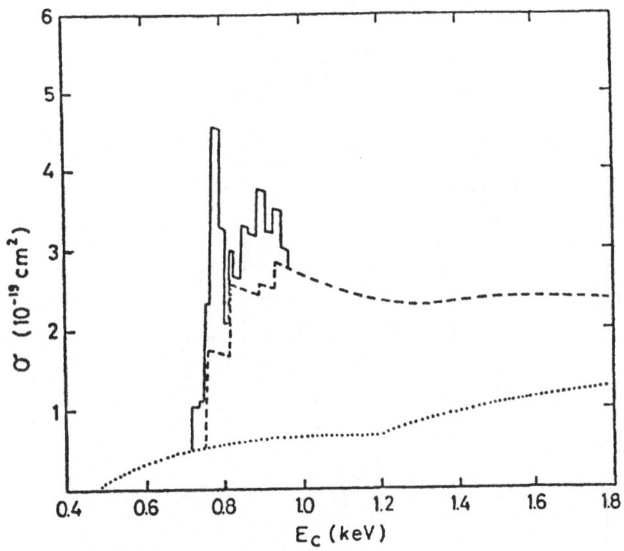

Fig. 9. Calculated cross sections for the ionization of Fe^{15+}. The dotted curve is the Lotz prediction for direct ionization, the dashed curve includes excitation-autoionisation and the solid curve includes the further contribution from capture-autoionization.

Figure 10 illustrates measurements[30] of the ionization of Ar^{n+} ions (n = 1 to 5) obtained with crossed beams. The results fit quite well to the curves.

$$\sigma(E) = A \cdot \frac{\ln E/\chi_n}{E \cdot \chi_n} \qquad 3.5$$

where χ_n represents the respective ionization energies and A is an empirical constant. Donets[9] has studied more highly-charged Ar^{n+} ions ($8 \leq n \leq 17$) with the "electron beam ion source" (EBIS) mentioned in section 1. His measurements are illustrated by figure 11 in which the dashed curves show results of a classical binary-encounter theory by Salop.[31] We note that the measurements do not extend to the near-threshold regions which are of particular interest to plasma- and astro-physics.

4. EXCITATION OF IONS BY ELECTRON IMPACT

4.1 Introduction

Line radiation is often the dominant energy loss from thin

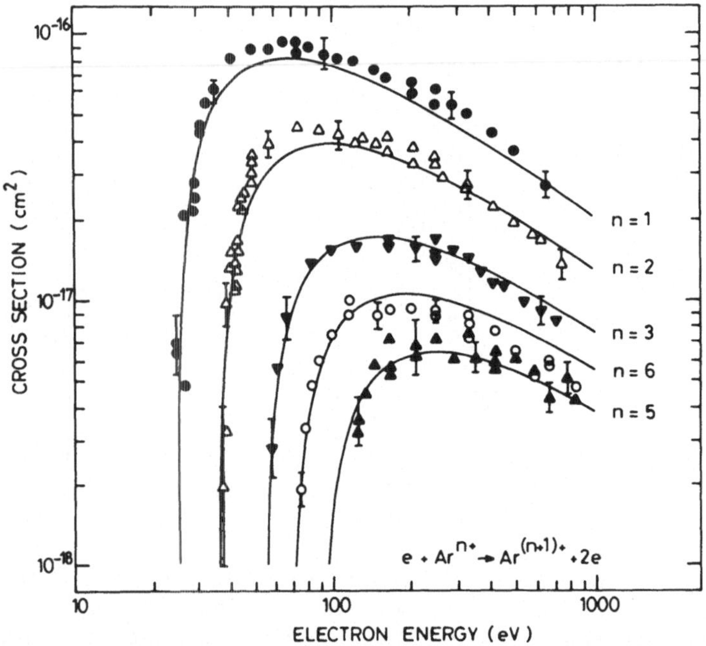

Fig. 10. Measured cross sections for Ar^{n+} (n = 1 to 5). The solid curves are given by $\sigma = A\ell n\varepsilon/\chi_n{}^2\varepsilon$ where ε is the electron energy in units of the ionization energy χ_n, and $A = 1.4 \times 10^{-13}$ $cm^2.(eV)^2$.

plasmas and this implies that a knowledge of excitation cross sections forms an essential part of fusion physics. Most reliable experimental information has come from crossed beam experiments, and the meticulous, time-consuming development of absolute radiometric calibration techniques at JILA made the decisive contribution. In principle, the experimental method is simple. Beams of ions and electrons cross at right angles and their intersection is viewed with a photomultiplier prefaced by an interference filter, to select the required radiation. The main problem is absolute calibration of the optical system by comparison with a standard black body. In crossed beam experiments the photomultiplier typically records a few hundred photons per second, but the intensity in the same bandwidth from a black body (often operating at the melting point of copper) is many orders of magnitude larger. Apart from the obvious problem of relative intensity, one must take account of variation in response of the optical system to polarization of the radiation and the position from which it is emitted. Remember that ions move as they radiate, and the lifetime of a particular transition will determine the point of origin of a photon, and hence the efficiency with which it is detected. Details of the calibration techniques were given by Taylor[34] and Taylor and Dunn.[35] The accuracy of these calibrations is accepted to be ±10%, or better, which is an impressive achievement.

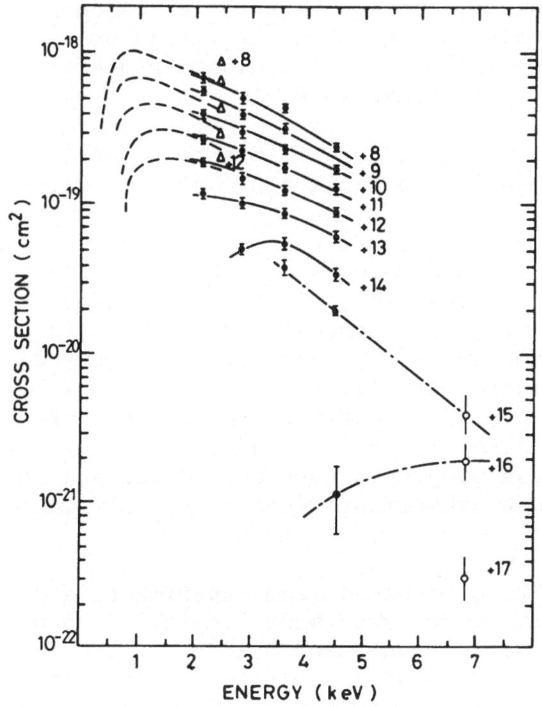

Fig. 11.
Cross sections for ionization of highly-charged Ar^{n+} ions (n = 8 to 17) compared with classical theory.

The next major step was an extension of these measurements to multiply-charged ions by the group at ORNL. Optical calibration was extended to the ultraviolet and a massive ion source was used to provide adequate ($\sim 10^{-6}$ A) ion beams. The ratio of signal to background was often much smaller when working with highly-charged ions (e.g. N^{4+}) and counting times of 10^5 sec. were needed to obtain results with random errors of a few percent.

Reaction rates for electron excitation (α_{if}) have been deduced from observations of well-characterized plasmas and the articles by Kunze[5] and Gabriel and Jordan[7] are standard references. The intensity (I_{fj}) of radiation from a given transition per steradian, per unit volume, is

$$I_{fj}(t) = \frac{h\nu}{4\pi} \cdot \frac{A_{fj}}{\sum_r A_{fr}} \cdot N_e(t)\; N_i(t)\; \alpha_{if} \qquad 4.1$$

where N_e and N_i represent the electron and ion densities, A_{fj} is the transition probability of the observed radiation ($f \rightarrow j$) and A_{fr} is the probability for transitions to any state, r. Equation 4.1 assumes that f is populated only by transitions from state i, and that no other processes (e.g. cascade electron capture, etc.) contribute to the population of f. It also assumes that no fields or collisions modify the transition probabilities and that plasma properties are known and constant throughout the transitions.

These assumptions cannot fully be realized in practice but this is the only technique which has provided experimental data for very highly-charged ions.

There is general agreement that new methods must be developed. One approach would be to collect and analyse electrons which have experienced inelastic collisions with ion targets. This could (in principle) give absolute excitation cross sections without recourse to optical calibration, and the number of electron-ion collisions might be enhanced by merging the beams. Successful accomplishment of such an experiment offers a glittering prize to an ambitious young experimentalist. But beware, these proposals have been discussed for more than twenty years and, if such experiments were easy, they would have been completed long ago.

We will concentrate on results of crossed beam experiments and conclusions which have been drawn. More experimental details can be found in reviews by Dolder and Peart[1], Dunn[2], and Crandall.[16,36,4] Crandall[36] also prepared a lucid outline of theoretical methods which is especially recommended to experimentalists. More profound reviews of theory have been given by Seaton[37], Henry[38], and Robb.[39] Theoretical results for more than 100 transitions have been listed by Merts et al[40] whilst calculated cross sections for about 800 transitions of Fe ions have been compiled by Pindzola and Crandall.[41] Applications of excitation cross sections to astrophysics and fusion have been discussed by many authors including Gabriel and Jordan[7], Lorenz et al[42], and Post.[43]

Table 3 lists a number (but probably not all) of measurements of excitation cross sections for atomic ions which have been obtained with intersecting beams.

4.2 Brief Comments on Theory

There is a striking qualitative difference between the electron excitation of positive ions and neutral atoms. In the case of atoms the cross section is zero at threshold, whereas for ions it is finite and often has almost its maximum value. This is due to the attractive ionic field which causes a number of partial waves to contribute even at threshold. Low energy behaviour is often of greatest importance in plasma physics and the finite threshold cannot be explained classically or semi-classically.

At high energies (i.e. one of two orders above theshold) the

Table 3. Crossed Beam Measurements of Excitation of Positive Ions by Electron Impact. The table excludes excitation-autoionization measurements which are included in Table 1 and the symbol A, N or R in the third column indicates whether the measurements are absolute, normalized to theory or in arbitrary units.

TRANSITION	REFERENCE		COMMENT
He^+ (1S-2S)	Dance et al (1966)[45]	N	H-like
He^+ (1S-2S)	Dolder & Peart (1973)[46]	N	H-like
He^+ (1S-2P)	Daschencko et al (1975)[44]	N	H-like
Li^+ (2^1S-2^3P)	Rogers et al (1978)[48]		A, He-like, $\Delta s \neq 0$
Be^+ (2S-2P)	Taylor et al (1980)[50]	A	Li-like
Be^+ (2S-2P)	Gregory et al (1979)[47]	A	Li-like
C^{3+} (2S-2P)	Taylor et al (1977)[51]	A	Li-like
N^{4+} (2S-2P)	Gregory et al (1979)[16]	A	Li-like
Mg^+ (3^2S-3^2P)	Kel'man et al (1975)[94]	A	
Ca^+ (4^2S-4^2P)	Kel'man & Imre (1975)[95]	A	
Ca^+ (4S4P; 4P4D)	Zapesochnyi et al (1976)[59]	A	
Ca^+ ($4S-4P_{\frac{1}{2}},4P_{\frac{3}{2}}$)	Taylor & Dunn (1973)[35]	A	Also measured polarization
Sr^+ (5S-5P)	Zapesochnyi et al (1976)[59]	A	Also measured polarization
Ba^+ ($6^2S-6^2P_{\frac{1}{2},\frac{3}{2}}$)	Bacon & Hooper (1969)[96]	A	
	Pace & Hooper (1973)[98]	A	
Ba^+ ($6\ S-6\ P_{\frac{1}{2},\frac{3}{2}}$; $6^2P_{\frac{3}{2}}-7^2S,6^2D$)	Crandall et al (1974)[97]	A	
	Zapesochnyi et al (1976)[59]	A	
Ar^+ (4S-4P)	Zapesochnyi et al (1973)[99]	R	
Kr^+ (5S-5P)		R	
Hg^+ ($6^2S-6^2P_{\frac{3}{2}}$)	Crandall et al (1975)[100]	A	
Hg^+ (6S-7S)	Phaneuf et al (1976)[101]	A	
Zn^+ ($4p^2P$, $5s^2S$)	Rogers et al (1982)[102]	A	

Bethe-type energy dependence,

$$\sigma = \frac{A}{E} \ln(BE) \qquad 4.2$$

is followed for $\Delta \ell = 1$, $\Delta s = 0$ transitions, but when $\Delta \ell \neq 1$ and $\Delta s = 0$,

$$\sigma = \frac{C}{E} \qquad 4.3$$

and when $\Delta s \neq 0$,

$$\sigma = \frac{D}{E^3}. \qquad 4.4$$

In these expressions A, B, C and D are constants.

Two valuable checks on theory, which do not rely on radiometric calibration, are provided by measuring the polarization of emitted radiation. The constant A in equation 4.2 is given by,

$$\sigma = \frac{4\pi a_o^2}{E} \cdot \frac{f}{\Delta E} \cdot \ell n\ BE \qquad 4.5$$

where f is the oscillator strength, whilst ΔE and E represent energies of the transition and the incident electron (Rydbergs). McFarlane[58] showed that B can be evaluated in the Bethe approximation from the energy at which the polarization passes through zero, and since A is defined by the oscillator strength (which is known for many transitions), the high-energy cross section is defined absolutely by the polarization.

A more obvious check is provided simply by comparing the measured and calculated energy-dependence of polarization.

Plasma physicists often speak of a "collision strength" (Ω) defined by,

$$\Omega \equiv \omega\sigma E \qquad 4.6$$

where ω represents the statistical weight of the lower (initial) state of the target ion. This scales approximately as Z_{eff}^{-2} where Z_{eff} is the effective charge on the commuting electron.

For ionization we paid considerable attention to empirical formulae which provide rough estimates of cross sections. Astrophysicists often use the "g bar formula" for excitation. This predicts that,

$$\sigma = \frac{2\pi}{\sqrt{3}} \cdot \frac{f}{E} \cdot \frac{\bar{g}}{\Delta E} \cdot \pi a_o^2 \qquad 4.7$$

where f is the oscillator strength for the particular transition, E is the incident electron energy (Rydbergs), ΔE is the energy of the transition (Rydbergs), $\bar{g}$ is the "effective Gaunt factor", and a_o is the radius of the first Bohr orbit. Reliable oscillator strengths for very many transitions can be obtained from tables and so the prime uncertainty is usually associated with $\bar{g}$. Crandall[4] critically reviewed the choice of values for $\bar{g}$. For resonant transitions of single-charged ions, reasonable agreement with experiment is obtained by assuming $\bar{g} \approx 0.2$, but larger values are needed for more highly-charged ions. Crandall quotes Younger and Wiese[57] who suggested $\bar{g} \approx 1.0$ for $\Delta n = 0$ transitions of more highly-charged ions whereas, when $\Delta n \neq 0$, the best values ranged from 0.05 to 0.7. Generally, it is assumed that equation 4.7 is valid within a factor 2 or 3, but clearly there can be exceptions (see figure 15). Just as autoionization could play havoc with empirical formulae for ionization, one might expect that resonances and other correlation effects will upset simple predictions for excitation.

4.3 Results for H-like Ions

Two transitions (1S-2S) and (1S-2P) have been studied for the simplest ion, namely He^+. Dashchenko et al[44] obtained non-absolute results for He^+ (1S-2P) which, when normalized, agreed quite well with theory. The situation is, however, dramatically different for He^+ (1S-2S). Independent measurements by Dance et al[45] and Dolder and Peart[46] both indicated a much slower variation of cross section with energy than predicted by even the most sophisticated theory. If the measurements are normalized to theory at high energies (≳500 eV) they indicate a threshold cross section less than half that predicted by elaborate close-coupling calculations. This discrepancy is especially worrying in view of its magnitude and the unique simplicity of the target ion. It has been discussed several times, notably by Seaton[37], Henry[38] and Crandall.[36]

Crandall[36] speculated that theory might have overestimated the monopole term in the Hamiltonian. Since monopole radiation dominates the (1S-2S) transition, and dipole radiation dominates (1S-2P), it would explain why theory is satisfactory in one case, but not in the other. The writer is tempted to look more critically at the experiments. Although there were substantial differences in technique between the two measurements, both used very similar electron guns and collectors and these were probably the least satisfactory components. It is possible that these guns might have produced molecular ions which became trapped in the He^+ beams. Perhaps these ions might then have reacted very mysteriously with He^+ (2S) to produce spurious results. It should be emphasized that neither experiment revealed anything to support this hypothesis and it could be dismissed as wild speculation if something equally strange had not been observed by Gregory et al[47] during their study of N^{4+}. Near threshold the measured cross section was (as one would expect) independent of electron current, but at higher energies (specifically at 52 eV) it apparently decreased with electron current. As the current increased from 50 to 250 μA, the apparent cross section halved. It was suggested that this might be due to trapped molecular ions in the beam but no plausible, quantitative explanation is apparent.

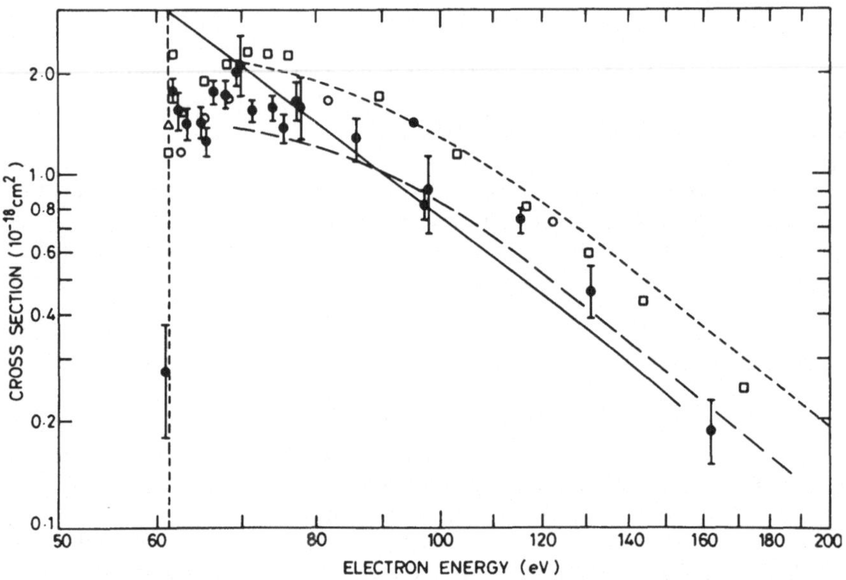

Fig. 12. Solid circles show measurements of Li^+ (λ5485 Å) radiation compared with distorted wave theories (short and long dashes respectively). The crosses, hollow squares and circles are five-state close-coupling calculations (see ref. 2).

4.4 Results for He-like Ions

Rogers et al[48] have made the only study of a He-like ion by observing λ5485 Å radiation from Li^+, caused primarily by 1^1S-2^3P excitation. The experiment was also notable because it provided the only data for a spin-changing transition. The results are illustrated by figure 12 where it can be seen that there is structure associated with resonance effects near threshold but at high energies the measurements follow the predicted E^{-3} energy dependence. Close-coupling calculations[38,48] reproduce structure but no theory agrees closer than about ±50% with the magnitude of the measurements at energies between 60 and 160 eV.

More experimental and theoretical work is clearly needed on spin-changing collisions of two-electron ions because they are believed to persist over a broad range of temperatures in fusion plasmas, and it has been predicted[49] that in highly-charged ions there are extremely large resonances in intercombination transitions which could enhance energy-averaged rate coefficients by factors of about six.

4.5 Results for Li-like Ions

Measurements for Be^+, C^{3+} and N^{4+} provide the best available set of data for an isoelectronic sequence and the painstaking investigation[50] of Be^+ (2S-2P) resonance radiation is often cited as a benchmark experiment". These results are compared in figure 13 with eight theoretical predictions and it can be seen (as one would hope) that although even five-state, close-coupling calculations do not quite fall within the estimated experimental error ($\sim \pm 10\%$). It was thought that this disparity might have been due to errors in the radiometric calibration but protracted checks and corroborating evidence from measurements of polarization (see section 4.2) discount this.

The results[51,52] for C^{3+} and C^{4+} offer some crumbs of comfort for fusion physicists. In both cases there is excellent agreement with two-state[53] close-coupling theory. This is presumably due to the increasing dominance of the ionic Coulomb field for more highly-charged ions so that <u>some</u> correlation effects become relatively unimportant and their influence is further reduced by the increased spacing between fine structure levels. This simple conclusion is unfortunately not entirely supported by preliminary results for O^{5+} but the experiment is of such difficulty that the results are still open to question. Moreover, inner-shell effects may well complicate the excitation just as they influenced the ionization of highly-charged ions.

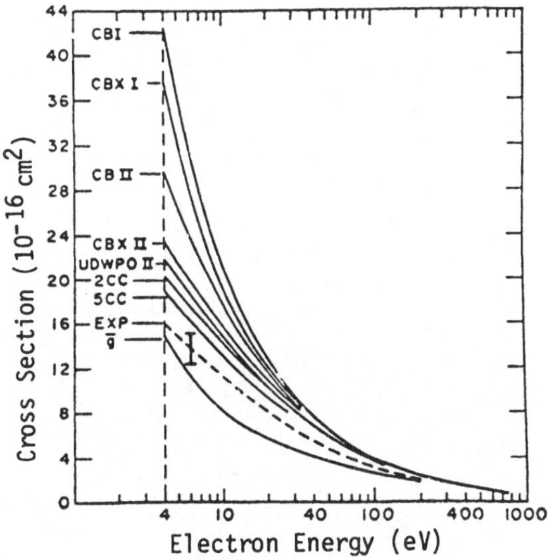

Fig. 13. Measured cross sections for Be^+ (2S-2P) compared with various theories including Coulomb-Born (CBI), Coulomb and Born Exchange (CBXI and CBXII), two state (2 cc) and five state (5 cc) close-coupling calculations and predictions of the $\bar{g}$ formula.

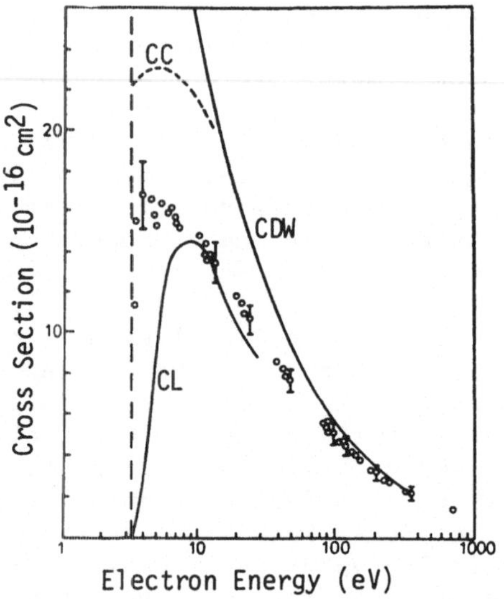

Fig. 14. Measured cross sections for Ca^+ K line excitation compared with three-state close-coupling (cc), Coulomb distorted wave (CDW) and classical binary encounter (CL) calculations.

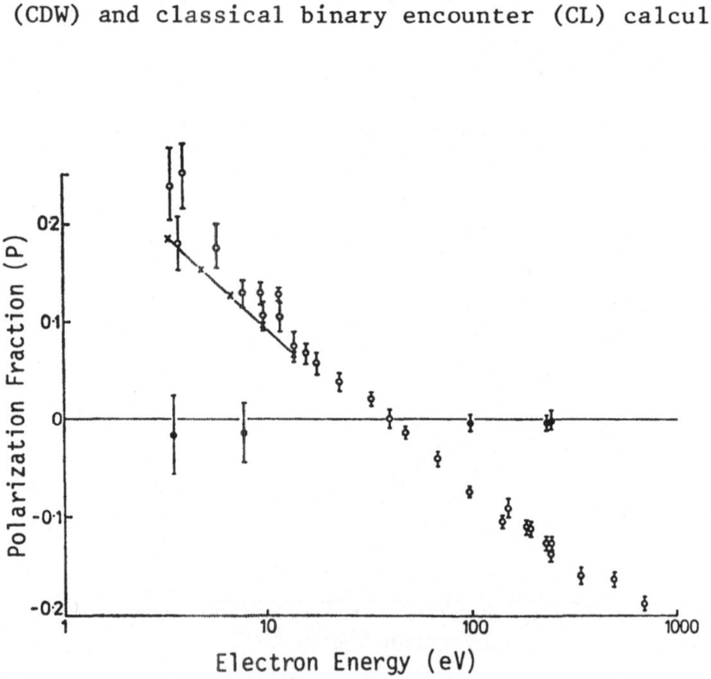

Fig. 15. Measured polarization of the H and K lines of Ca^+ denoted by closed and open circles, respectively. Continuous lines indicate theoretical values which predict zero polarization for the H line.

4.6 Results for Complex Ions

Results have been reported for transitions Mg^+, Ca^+, Sr^+, Ba^+, Ar^+, Kr^+, Zn^+ and Hg^+. The study[35] of the H and K transitions in Ca^+ ($4S-4P_{\frac{3}{2},\frac{1}{2}}$) is highly relevant to astrophysics and many of the optical calibration techniques were originally developed for this experiment.

Figure 14 compares measurements for the K line (λ3934 Å) with results of calculations based on classical binary encounter[56] (CL), three-state close-coupling[54] (CC) and Coulomb distorted wave[55] (CDW) approximations.

The quantum calculations are not sufficiently sophisticated to explain the result and the agreement with classical theory is probably partly fortuitous. It can also be seen, as we would expect, that classical theory fails to explain the finite cross section at threshold.

Evidence that the experiment is more accurate than theory was provided by measuring the polarization of radiation from the two transitions. The solid and open circles in figure 15 illustrate measurements for the H and K lines plotted against electron energy. They agree well with calculated polarizations indicated by the continuous lines.

The availability of cross sections for a range of transitions provides an opportunity to test the $\bar{g}$ formula (equation 4.7). If $\bar{g}$ were constant for all transitions, the cross sections, multiplied by $(\Delta E)^2/f$, and plotted against $\Delta E/E$ should fall about a universal curve. Figure 16 illustrates results for a number of transitions with $\Delta n = 0$, $\Delta \ell = 1$ and $s = 0$ and compares them with predicted result assuming $\bar{g} = 0.2$. At energies greater than ten times threshold, the agreement is quite good but, at lower energies, the measurements for multiply-charged ions are several times too large (see section 4.2).

We have also noted that the $\bar{g}$ formula could not be expected to accommodate large contributions to cross sections from resonances. A striking example of the influence of resonances near threshold was manifested by the non-absolute measurements of resonance radiation from K^+ by Zapesochnyi et al.[59] These are illustrated by figure 17. An even more dramatic example was the dominance of autoionization (inner-shell excitation) on the ionization function of Ti^{3+} which was illustrated by figure 5. One is inclined to speculate that just as inner-shell effects contributed relatively more to the ionization of highly-charged ions, a similar trend might be found for excitation. This would obviously be relevant to fusion.

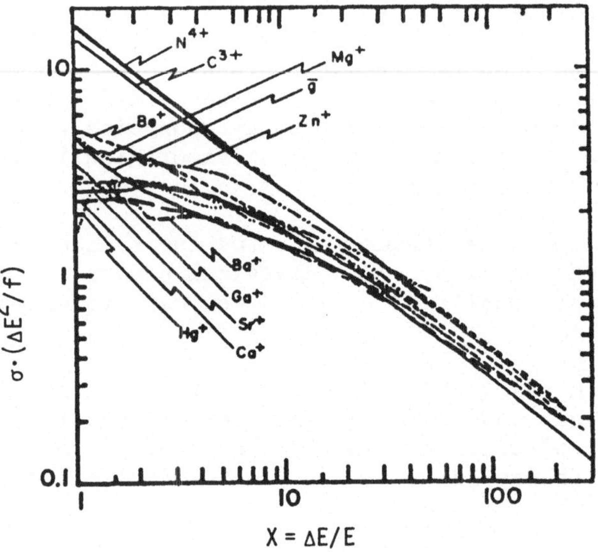

Fig. 16. Scaled cross sections $\sigma(\Delta E)^2/f$ plotted against electron energy in units of threshold energy for transitions with $\Delta n = 0$, $\Delta \ell = 1$ and $\Delta s = 0$. Prediction of the $\bar{g}$ formula is also shown.

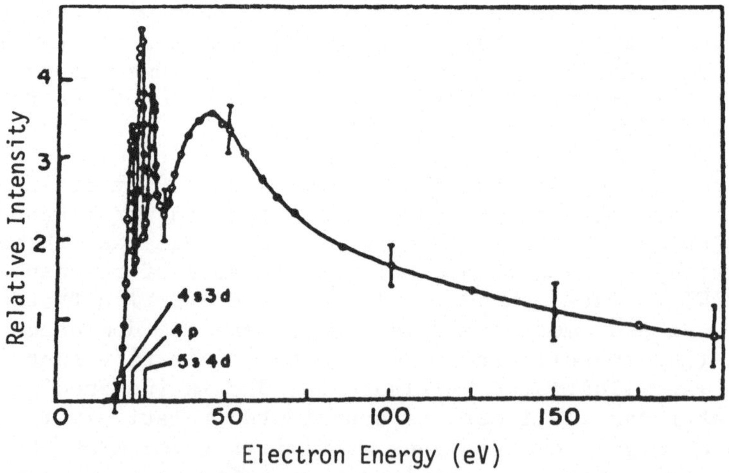

Fig. 17. Measured excitation functions of K^+ resonance radiation showing effects of resonances near threshold.

5. BIBLIOGRAPHIES AND REVIEWS

This article has necessarily been brief and incomplete. It merely attempted to indicate some active, current developments in crossed beam experiments with electrons and atomic ions. One omission was any discussion of the formation or destruction of H_2^+ or H^-. It might seem that these are far too fragile to be of concern in thermonuclear physics, but they are, in fact, very relevant to ancillary devices such as divertors and neutral beam injectors.

Fortunately there are a number of excellent reviews and bibliographies from which the literature can be explored and this section will draw attention to a few of the articles which the author has found particularly helpful.

There could hardly be a more useful series of lectures than those that were delivered last September at a NATO ASI at Baddeck, in Canada. These are about to be published by Plenum. Crandall[4] and Salzborn[3] respectively discussed experimental aspects of the excitation and ionization of ions and their accounts amplify the present article. Applications of results to fusion and astrophysics are reviewed by Post[43] and Dalgarno[103] and there are articles by Claeys[104] and Dolder[105] on ion-ion collisions, many of which also relate to fusion. The introduction to the theory of electron-atom and electron-ion collisions by Kim[106] is very warmly recommended both to experimentalists and budding theoreticians, whilst those interested in collisions of molecular ions will find several excellent reviews of theoretical and experimental techniques. These articles naturally draw attention to earlier reviews and so they will not be listed here.

A valuable bibliography of publications between 1940 and 1977 on collisions between electrons and positive atomic ions has been prepared by Takayanagi and Iwai,[107] whilst a more wide-ranging and comprehensive bibliography of atomic and molecular data relevant to fusion was prepared by the IAEA in Vienna.[108] Recommended cross sections and rates for reactions relevant to fusion are being collected and evaluated under the auspices of Culham Laboratory, and the first volume,[109] which deals with electron impact ionization of light ions, has just appeared. Empirical formulae for ionization were critically evaluated by Itikawa and Kato.[17]

A newcomer to this field will discover that much as been done but each year the rate of progress tends to increase. Clearly there is still plenty of scope for originality and enterprise.

REFERENCES

1. K. Dolder and B. Peart, Repts. Prog. Phys., 39:693 (1976).
2. G.H. Dunn, "Electron Ion Collisions" in "Physics of Ionized

Gases", H. Matic and B. Kidric, eds., Inst. Nucl. Science, Belgrade (1980).
3. E. Salzborn, "Electron Impact Ionization of Ions" in "Physics of Ion-Ion and Electron-Ion Collisions", NATO ASI, Baddeck, Canada: to be published by Plenum Press (1982).
4. D.H. Crandall, "Electron Impact Excitation of Ions" in "Physics of Ion-Ion and Electron-Ion Collisions", NATO ASI, Baddeck, Canada: to be published by Plenum Press (1982).
5. H-J. Kunze, Space Sci. Rev., 13:565 (1972).
6. R.U. Datla, M. Blaha and H.J. Kunze, Phys. Rev. A 12:1076 (1975).
7. A.H. Gabriel and C. Jordan, Case Studies in Atomic Collision Physics, M.R.C. McDowell and E.W. McDaniel, eds., North Holland, Amsterdam, 2:211 (1972).
8. R.J.W. Henry, J. Phys. B, 12:L309 (1979).
9. E.D. Donets and V.P. Ovsyannikov, Joint Inst. Nuc. Res. Reprint P7-10780, Dubna (1977).
10. F.A. Baker and J.B. Hasted, Phil. Trans. Roy. Soc. A 261:33 (1966).
11. B. Peart and K. Dolder, J. Phys. B, 8:56 (1975).
12. D.H. Crandall, R.A. Phaneuf, B.E. Hasselquist and D.C. Gregory, J. Phys. B, 12:L249 (1979).
13. D.H. Sampson and L.B. Golden, J. Phys. B, 12:L785 (1979).
14. R.A. Falk, G.H. Dunn, D.C. Griffin, C. Bottcher, D.C. Gregory, D.H. Crandall and M.S. Pindzola, Phys. Rev. Letts., 47:494 (1981).
15. A. Müller and R. Frodl, Phys. Rev. Letts., 44:29 (1980).
16. D.H. Crandall (invited paper) Proc. 12th ICPEAC, Gatlinburg, S. Datz, ed., North Holland Press, Amsterdam (1981).
17. Y. Itikawa and T. Kato, Inst. of Plasma Physics, Nagoya Univ., Japan, Rept. No. IPPJ-AM-17 (1981).
18. D.H. Crandall, Physica Scripta, 23:153 (1981).
19. W. Lotz, Z. Phys., 216:241 (1968).
20. W. Lotz, Z. Phys., 220:466 (1969).
21. L.B. Golden and D.H. Sampson, J. Phys. B, 10:2229 (1977).
22. L.B. Golden and D.H. Sampson, J. Phys. B, 13:2645 (1980).
23. D.H. Sampson and L.B. Golden, J. Phys. B, 11:541 (1978).
24. D.L. Moores, L.B. Golden and D.H. Sampson, J. Phys. B, 13:385 (1980).
25. A. Burgess and I.C. Percival, Adv. At. & Mol. Phys., 4:109 (1968).
26. W.I. Rogers, G. Stefani, R. Camilloni, G.H. Dunn, A.Z. Msezane and R.J.W. Henry, Phys. Rev. A (in press, 1982).
27. E.J. McGuire, Phys. Rev. A, 16:62 (1977).
28. D.C. Gregory, P.F. Dittner and D.H. Crandall, Proc. 12th ICPEAC, Gatlinburg, S. Datz, ed., North Holland Press, Amsterdam (1981).
29. K.J. LaGattuta and Y. Hahn, ibid.
30. A. Müller, E. Satzborn, R. Frodl, R. Becker, H. Klein and H. Winter, J. Phys. B, 13:1877 (1980).
31. A. Salop, Phys. Rev. A, 14:2095 (1976).
32. R.D. Cowan and J.B. Mann, Astrophys. J., 232:940 (1979).
33. Y-K. Kim and K-T. Cheng, Phys. Rev. A, 18:36 (1978).

34. P.O. Taylor, PhD Thesis, University of Colorado (1972). Available from University Microfilms, Ann Arbor, Michigan.
35. P.O. Taylor and G.H. Dunn, Phys. Rev. A, 8:2304 (1973).
36. D.H. Crandall, R.A. Phaneuf, B.E. Hasselquist and D.C. Gregory, J. Phys. B, 12:L249 (1979).
37. M.J. Seaton, Adv. in At. & Mol. Phys., D.R. Bates & I. Estermann, eds., Academic Press, New York, 11:83 (1975).
38. R.J.W. Henry, Phys. Repts., 68:1 (1981).
39. W.D. Robb, Atomic & Molecular Processes in Controlled Thermonuclear Fusion, NATO ASI Series B, Vol. 53, M.R.C. McDowell and A.M. Ferendeci, Plenum Press, New York (1980).
40. A.L. Mertz, J.B. Mann, W.D. Robb and N.H. Magee Jr., Los Alamos Rept. LA-8267MS (1980).
41. M.S. Pindzola and D.H. Crandall, ORNL Report (to be published).
42. A. Lorenz, Phys. Repts. 37c:56 (1978).
43. D.E. Post, "Atomic Collision Processes in Thermonuclear Fusion Plasmas", Proc. NATO ASI, Baddeck, Canada: to be published by Plenum Press (1982).
44. A.I. Dashchenko, I.P. Zapesochnyi, A.I. Imre, V.S. Bukstich, F.F. Danch and V.A. Kel'man, Sov. Phys. JETP, 40:249 (1975).
45. D.F. Dance, M.F.A. Harrison and A.C.H. Smith, Proc. Roy. Soc. A, 290:73 (1966).
46. K. Dolder and B. Peart, J. Phys. B, 6:2415 (1973).
47. D. Gregory, G.H. Dunn, R.A. Phaneuf and D.H. Crandall, Phys. Rev. A, 20:410 (1979).
48. W.T. Rogers, J. Olsen and G.H. Dunn, Phys. Rev. A, 18:1353 (1978).
49. A.K. Pradhan, D.W. Norcross and D.G. Hummer, Phys. Rev. A, 23:619 (1981).
50. P.O. Taylor, R.A. Phaneuf and G.H. Dunn, ibid A, 22:435 (1980).
51. P.O. Taylor, D.C. Gregory, G.H. Dunn, R.A. Phaneuf and D.H. Crandall, Phys. Rev. Letts., 39:1256 (1977).
52. D.C. Gregory, G.H. Dunn, R.A. Phaneuf and D.H. Crandall, Phys. Rev. A, 20:410 (1979).
53. N.H. Magee Jr., J.B. Mann, A.L. Mertz and W.D. Robb, Los Alamos Rept., LA-6691-MS (1977).
54. P.G. Burke and D.L. Moores, J. Phys. B, 1:575 (1968).
55. A. Burgess and V.B. Sheorey, ibid, 7:2403 (1974).
56. A.N. Tripathi, K.C. Mathur and S.K. Joshi, Phys. Rev. A, 1:337 (1970).
57. S.M. Younger and W. Wiese, J. Quant. Spec. & Rad. Trans., 22:161 (1979).
58. S.C. McFarlane, J. Phys. B, 7:1756 (1974).
59. I.P. Zaposochnyi, V.A. Kel'man, A.I. Imre and F.F.Danch, Sov. Phys. ZETP, 42:989 (1976).
60. K. Dolder, M.F.A. Harrison and P.C. Thoneman, Proc. Roy. Soc. A, 264:367 (1961).
61. B. Peart, D.S. Walton and K. Dolder, J. Phys. B, 2:1347 (1969).
62. W.C. Lineberger, J.W. Hooper and E.W. McDaniel, Phys. Rev., 141:151 (1966).
63. B. Peart and K. Dolder, J. Phys. B, 1:872 (1968).

64. D.H. Crandall, Physica Scripta, 23:153 (1979); see also ref. 69.
65. R.A. Falk and G.H. Dunn, Phys. Rev. A (to be published; see ref. 16).
66. D.H. Crandall, R.A. Phaneuf and P.O. Taylor, ibid., 18:1911 (1978).
67. S.O. Martin, B. Peart and K. Dolder, J. Phys. B, 1:537 (1968).
68. A. Müller, E. Salzborn, R. Becker, R. Frodl, H. Klein and H. Winter, J. Phys. B, 13:1877 (1980).
69. D.H. Crandall, R.A. Phaneuf and D.C. Gregory, Oak Ridge Rept., ORNL/TM 7020 (1979).
70. D.H. Crandall, R.A. Phaneuf, R.A. Falk, D.S. Belic and G.H. Dunn, Phys. Rev. A, 25:143 (1982).
71. J.W. Hooper, W.C. Lineberger and F.M. Bacon, Phys. Rev., 141:165 (1966).
72. B. Peart and K. Dolder, J. Phys. B, 1:240 (1968).
73. B. Peart and K. Dolder, ibid, 8:56 (1975).
74. D.W. Hughes and R.K. Feeney, Phys. Rev. A, 23:2241 (1981).
75. B. Peart and K. Dolder, J. Phys. B, 1:872 (1968).
76. B. Peart, J. Stevenson and K. Dolder, ibid, 6:146 (1973).
77. R.K. Feeney, J.W. Hooper and M.T. Eleford, Phys. Rev. A, 6:1469 (1972).
78. K.L. Aitken, M.F.A. Harrison and R.D. Rundel, J. Phys. B, 4:1189 (1971).
79. K.L. Aitken and M.F.A. Harrison, ibid, 4:1176 (1971).
80. K. Dolder, M.F.A. Harrison and P.C. Thoneman, Proc. Roy. Soc. A, 274:546 (1963).
81. A. Müller and R. Frodl, Phys. Rev. Lett., 44:29 (1980).
82. A. Müller, E. Salzborn, R. Frodl, R. Becker, H. Klein and H. Winter, J. Phys. B, 13:1877 (1980).
83. D.C. Gregory, P.F. Dittner and D.H. Crandall, Proc. XII ICPEAC, Gatlinburg (1981).
84. B. Peart, S.O. Martin and K. Dolder, J. Phys. B, 2:1176 (1969).
85. T.F. Divine, R.K. Feeney, W.E. Sayle and J.W. Hooper, Phys. Rev. A, 13:54 (1976).
86. W.T. Rogers, G. Stefani, R. Camilloni, G.H. Dunn and R. Henry, Phys. Rev. A, 25:737 (1982).
87. R.A. Falk, G.H. Dunn, D.C. Griffin, C. Bottcher, D.C. Gregory, D.H. Crandall and M.S. Pindzola, Phys. Rev. Letts., 47:494 (1981).
88. E.D. Donets and A.I. Pikin, Sov. Phys. JETP, 43:1057 (1976).
89. D.L. Moores, J. Phys. B, 11:403 (1978).
90. L.B. Golden and D.H. Sampson, J. Phys. B, 13:2645 (1980).
91. R.J.W. Henry, ibid, 12:L309 (1979).
92. E.J. McGuire, Phys. Rev. A, 16:62 (1977).
93. E.J. McGuire, ibid, 16:73 (1977).
94. V.A. Kel'man, A.I. Dashchencko, I.P. Zapesochnyi and A.I. Imre, Sov. Phys. Dokl., 20:38 (1975).
95. V.A. Kel'man and A.I. Imre, Opt. Spect., 38:709 (1975).
96. F.M. Bacon and J.W. Hooper, Phys. Rev., 178:182 (1969).

97. D.H. Crandall, P.O. Taylor and G.H. Dunn, Phys. Rev. A, 10:141 (1974).
98. M.O. Pace and J.W. Hooper, Phys. Rev. A, 7:2033 (1973).
99. I.P. Zapesochnyi, A.I. Imre and A.I. Dashchencko, Sov. Phys. JETP, 36:1056 (1973).
100. D.H. Crandall, R.A. Phaneuf and G.H. Dunn, Phys. Rev. A, 11:1223 (1975).
101. R.A. Phaneuf, P.O. Taylor and G.H. Dunn, ibid, 14:2021 (1976).
102. W.T. Rogers, G.H. Dunn, J. Olsen, M. Reading and G. Stefani, Phys. Rev. A (in course of publication).
103. A. Dalgarno, "Electron-Ion and Ion-Ion Collisions in Planetary Atmospheres and Interstellar Space", Proc. NATO ASI, Baddeck, Canada: to be published by Plenum Press (1982).
104. W. Claeys, "The Measurement of Ion(Atom)-Ion(Atom) Charge Exchange", ibid., (1982).
105. K. Dolder, "The Measurement of Inelastic Ion-Ion and Electron-Ion Collisions", ibid., (1982).
106. Y-K. Kim, "Theory of Electron-Ion and Electron-Atom Collisions", ibid., (1982).
107. K. Takayanagi and T. Iwai, Inst. Plasma Phys., Nagoya Univ., Japan, Rept. IPPJ-AM-7 (1978).
108. CIAMDA 80, IAEA, Vienna (1980).
109. K.L. Bell, H.B. Gilbody, J.G. Hughes, A.E. Kingston and F.J. Smith, Culham Rept., CLM-R216 (1982). Available from HMSO, London.
110. B. Peart and K. Dolder, J. Phys. B, 2:1169 (1969).

THE THEORY OF CHARGE EXCHANGE AND IONIZATION BY HEAVY PARTICLES

B. H. Bransden*

Joint Institute for Laboratory Astrophysics
University of Colorado and National Bureau of Standards
Boulder, Colorado 80309 U.S.A.

1. INTRODUCTION AND COUPLED CHANNEL METHODS

In the 1979 ASI on atomic processes of interest in fusion research, I surveyed the basic principles of the theory of charge exchange,[1] concentrating on theoretical models in which the wave function of the system was expanded in some suitable set of functions.† At low to intermediate relative velocities of the colliding ions ($v \lesssim 1$ a.u.) the expansion functions are frequently taken to be adiabatic, or diabatic, molecular orbitals, while over a region extending to somewhat higher velocities ($0.1 \lesssim v \lesssim 3$ a.u.), the expansion functions can be taken to be atomic orbitals, supplemented by pseudostate functions. Some higher order perturbation methods, like the continuum distorted wave approximation, which are useful at higher velocities ($v \gtrsim 2.5$ a.u.), were also discussed briefly in I. At this School, I shall not repeat this general material, but I shall concentrate on the special approximations which have been developed to describe charge exchange and ionization in the collision of fully stripped ions of charge Z (in atomic units) with atomic hydrogen in the ground state

$$A^{Z+} + H(1s) \rightarrow A^{(Z-1)+} + H^{+} \tag{1}$$

$$A^{Z+} + H(1s) \rightarrow A^{(Z-1)+} + H^{+} + e^{-} \tag{2}$$

---

*JILA Visiting Fellow, 1981-82; permanent address: The University of Durham, Durham, England.

†These lectures will be referred to as I.

In general, the charge exchange reaction (1) is the most important process at low energies, the cross section being larger than those for excitation or ionization. The cross section for charge exchange rises to a maximum if Z is small (except for the symmetrical case with Z = 1) at some velocity in the region $v <$ 1 a.u. Above this velocity the cross section diminishes rapidly ($\propto v^{-12}$) and excitation and ionization become the most important reactions, the latter cross sections ultimately diminishing like $\log v/v^2$. For large Z, the capture cross sections are rather flat for $v \lesssim 1$ a.u., and decrease rapidly for large v. The capture cross sections are large (for example, at the maximum, $\sigma = 5 \times 10^{-15}$ cm^2 for Z = 6), and scale roughly like $Z^{\alpha(v)}$, where $\alpha \approx 1$ at small v, $\alpha \approx 2$ at $v \approx 1$ and $\alpha \approx 3$ at higher velocities. The states into which the electron is captured are rather specific. For example, for Z = 2 and for velocities below 1 a.u. capture is almost entirely into the n = 2 levels of He^+, while for Z = 3, capture is mainly into the n = 2 and n = 3 levels, of Li^{2+}. For Z = 6 capture is mainly into the n = 4 and n = 5 levels of C^{5+} and for large Z the most likely value of n is ~Z/2, with the distribution in n broadening as Z increases. Taking into account that (for the case in which $A^{(Z-1)+}$ is a one-electron ion) the level n is n^2 degenerate, one sees that the coupled channel method faces severe difficulties as Z increases, irrespective of what basis is used, since it is necessary to represent all the important asymptotic channels in the wave function. However, if only total, and not partial cross sections, are required the basis set can be economized. For example, a minimum basis containing just the H(1s) orbital together with the most important final state functions has four terms for Z = 2, 10 for Z = 3 and 26 for Z = 6. Above the cross section maximum, the distribution in n becomes broader and, in addition, transitions via the continuum become important, and continuum levels have to be represented in some way in the expansion of the wave function.

Some of these problems have been overcome in a new expansion method, called the "one and a half centered" expansion (OHCE), due to Reading and his coworkers,[12] who write the wave function in the form:

$$\Psi = \sum_{m=1}^{N} A_m(b,t)\, \phi_m^B(\vec{r}_B)\, e^{-i\varepsilon_m t} + \sum_{j=1}^{m} C_j(b,\infty)\gamma_j(t)\phi_j^A(\vec{r}_A)\, e^{i(\vec{v}\cdot\vec{r}_B - \frac{1}{2} v^2 t)} \quad . \tag{3}$$

The first sum is about the nucleus B which may be taken to be the nucleus of charge Z and is an expansion in terms of a pseudostate set, large enough to represent the wave function well in the interaction region. The second sum is required to satisfy the asymptotic boundary conditions and is necessary to represent the wave functon well at large values of t. It contains the wave functions about the center A. The functions $\gamma_j(t)$ are predetermined and satisfy $\gamma_j(t) \to 0$, $t \to -\infty$; $\gamma_j(t) \to 1$, $t \to +\infty$. The constants $C_j(b,\infty)$ are the charge exchange amplitudes. By projecting the Schrödinger equation with the functions ψ_m and ϕ_j, coupled differential equations are found for the amplitudes $a_m(b,t)$, together with algebraic equations for the constants $C_j(b,\infty)$. An important advantage of the method is that, because only single center integrals are required to construct the potential matrix coupling the amplitudes a_m, a very large basis set can be used, and the expansion can be pushed to convergence.

The most detailed coupled channel calculations have been carried out for Z = 2, 3, 6, and 8 and details of some recent calculations, restricted to those which are translationally invariant, are shown in Table 1. For Z = 2 and Z = 3 the agreement with the experimental data is good, up to velocities of the order of 2-3 a.u., while for Z = 6 the agreement is good up to the largest velocity for which calculations were made ~1 a.u. (see Figs. 1-3). For Z = 8, the experimental data not sufficient at the time of writing for a detailed comparison to be made; but there is good reason to believe that the predicted cross sections are accurate. Calculations using both linear and non-linear heavy particle trajectories suggest that the linear trajectory approximation is adequate for $v > 0.1\ Z^{1/2}$ a.u. (for total cross section calculations). It is interesting to note that for velocities down to at least v = 0.1 a.u. there appears to be little advantage between a molecular orbital and two-center atomic orbital basis, although, doubtless, at low velocities the molecular orbital approach would seem to be more attractive.

Because of the difficulties expansion methods face as Z increases, other approximations (which are not in general expected to be so accurate), have been introduced and we shall discuss some of these, starting with those expected to be useful at intermediate velocities.

2. INTERMEDIATE VELOCITIES

The Unitarized Distorted Wave Approximation (UDWA)

As discussed in I, in many circumstances the cross section for energies at, and in a region above, the cross section maximum can be found by summing the results of a series of two-state coupled channel calculations. However, below the cross section maximum, the cross sections found in this way are too large and the total probability for charge exchange, summed over all final states, can violate unitarity for some range of the impact parameter b. At sufficiently high energies, the solution of the two-state equations

Table 1. Some recent coupled channel calculations for $A^{Z+} + H(1s) \rightarrow A^{(Z-1)+} + H^{+}$ which allow for the change in momentum of the captured electron, and where A^{Z+} is a fully stripped ion

A^{Z+}	Type of Basis	No. of Terms in Basis	Type of Translational Factor	Energy Range (keV/amu)	Reference
He^{2+}	2 center - atomic	16 (8 on each center*)	plane wave	2-200	2
	2 center - Sturmian	up to 24	plane wave	5-50	3
	molecular orbital	up to 12	switching factor	1-20	4
	"1 1/2 center" - pseudostate	54 about one center + predetermined second center function	plane wave	25-130	5
Li^{3+}	2 center - atomic	20 (10 on each center)	plane wave	1.4-200	6
	molecular orbital	6	switching factor	2-10	7
	"1 1/2 center" - pseudostate	54 about one center	plane wave	50-200	8
	2 center - atomic + pseudostates	up to 34	plane wave	0.2-20	9
C^{6+}	molecular orbital	10 and 33	optimized translational factors	0.013-27	10
	2 center - atomic	11 (10 about A, 1 about H)	plane wave	0.1-1.0	11
O^{8+}	molecular orbital	33	optimized translational factors	0.02-34	59

*Unpublished results with 11 on each center.

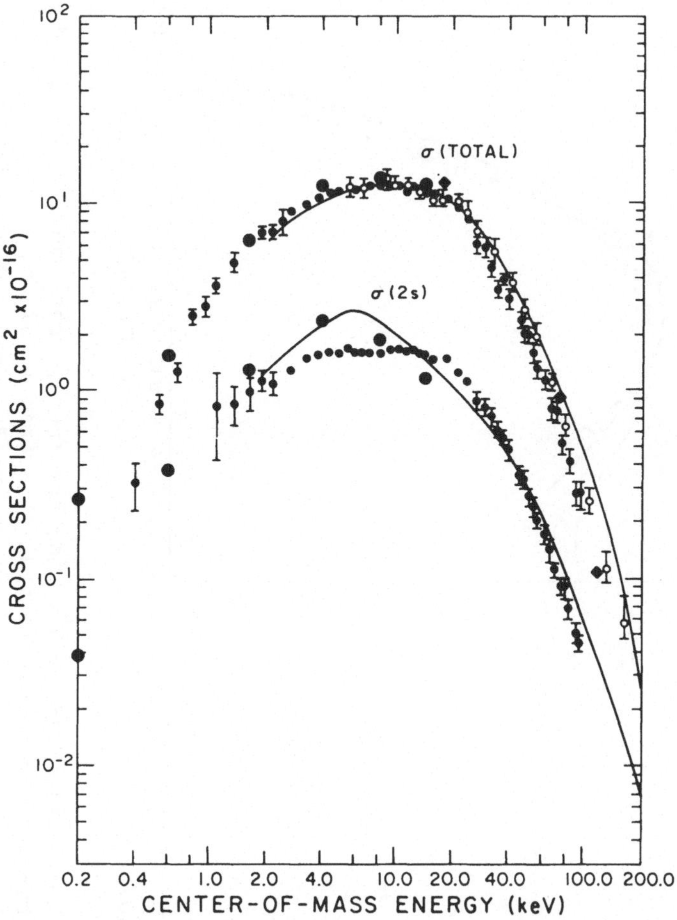

Fig. 1. Cross sections for $He^{2+} + H(1s) \rightarrow He^{+} + H^{+}$. The upper curves and data points refer to the total charge exchange cross section and the lower curves and data points to capture into the 2s level of He^{+}. Experimental data: • • – Nutt et al.,[48] Shah and Gilbody.[49] Theoretical cross sections: —— – atomic expansion (Bransden and Noble[2]); ● ● ● – molecular orbital expansion (Hatton et al.[50]); ◆ ◆ 'one and a half center' expansion (Reading et al.[12]). Note: The Sturmian expansion of Winter[3] produces cross sections similar to those of Bransden and Noble,[2] while the molecular orbital expansion calculations of Kimura and Thorsen[4] and Crothers and Todd[51] produce similar cross sections to those of Hatton et al.

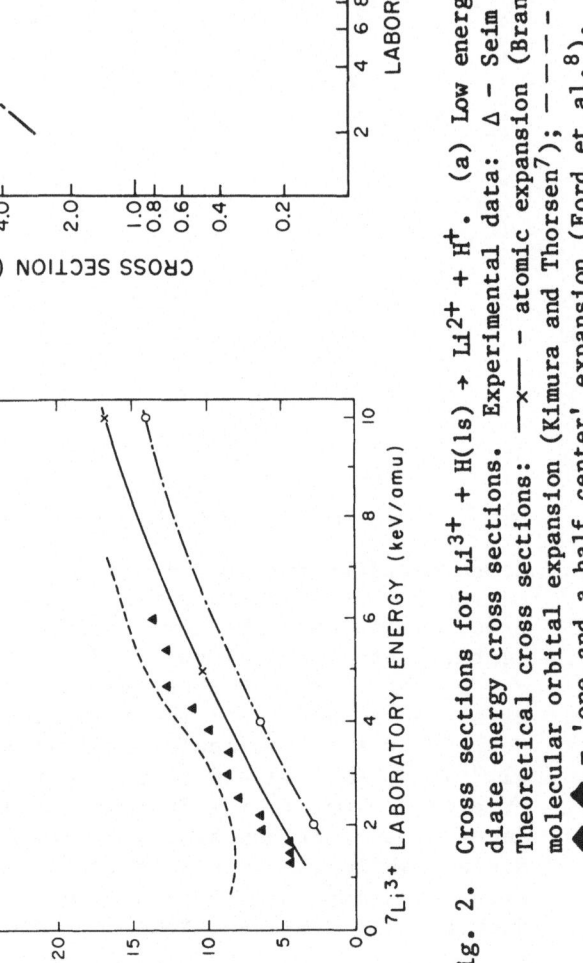

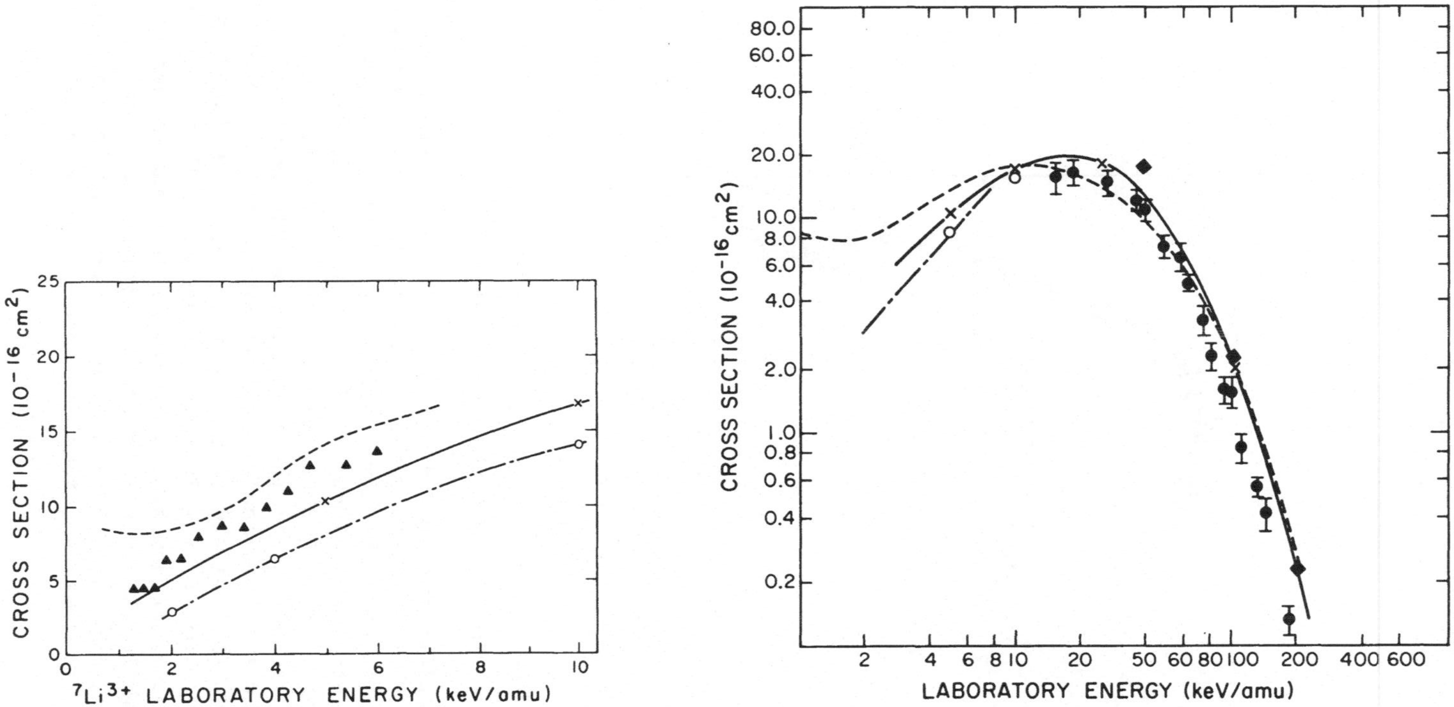

Fig. 2. Cross sections for $Li^{3+} + H(1s) \rightarrow Li^{2+} + H^+$. (a) Low energy cross sections; (b) intermediate energy cross sections. Experimental data: Δ - Seim et al.[52]; ● - Shah et al.[53] Theoretical cross sections: —×— - atomic expansion (Bransden and Noble[6]); — — — — molecular orbital expansion (Kimura and Thorsen[7]); – – – – UDWA (Ryufuku and Watanabe[14]); ◆ ◆ - 'one and a half center' expansion (Ford et al.[8]).

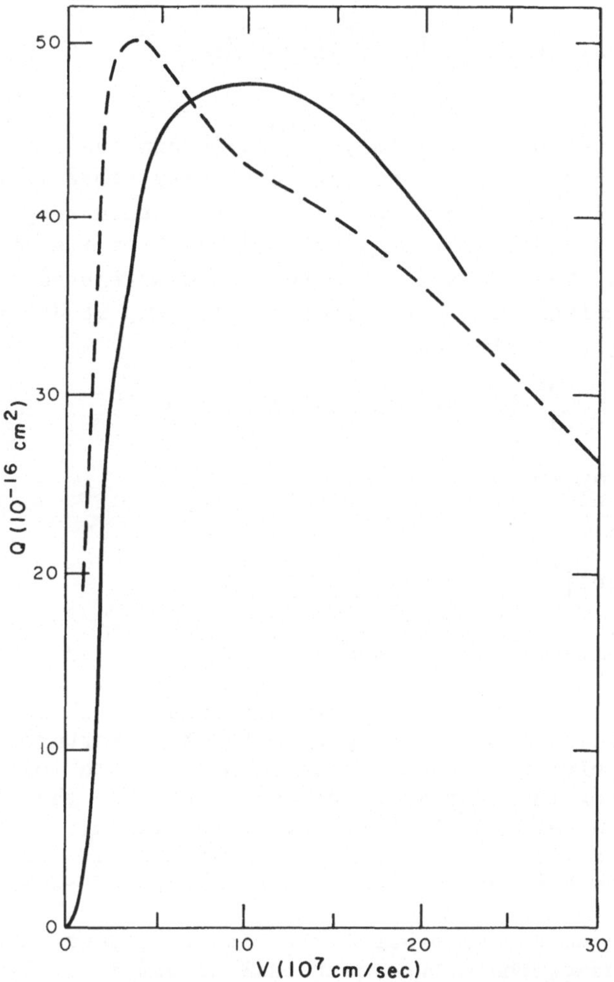

Fig. 3. Cross sections for $C^{6+} + H(1s) \rightarrow C^{5+} + H^+$. Theoretical cross sections: ——— - molecular orbital expansion (Green et al.[10]); — — — - UDWA (Ryufuku and Watanabe[14]). Note: The atomic orbital expansion of Fritsch[11] is in good agreement with the molecular orbital calculations.

(obtained from eq. (19) of I) can be approximated by the distorted wave approximation of Bates[13]

$$C_j^{DW}(b,t=\infty) = \int_{-\infty}^{\infty} \frac{M_{21} - S^* M_{11}}{1 - |S|^2} \, e^{i\delta(t)} \, dt \qquad (4)$$

where $S = S_{12}$ and $S_{11} = S_{22} = 1$ and

$$\delta(t) = \int_{-\infty}^{t} dt' \frac{M_{22} - S^* M_{12} - M_{11} + SM_{21}}{1 - |S|^2} . \quad (5)$$

Both the individual probabilities, $|C_j^{DW}(b)|^2$ and the sum of the probabilities over j, can violate unitarity (that is either $|C_j^{DW}(b)|^2$ or $\Sigma_j |C_j^{DW}(b)|^2$ can exceed 1, for certain ranges of b), if the velocity v is too low or if the charge Z of the incident ion is too high. Ryufuku and Watanabe[14] have attempted to cure this defect by unitarizing the distorted wave amplitudes according to the following prescription

$$C_j^{UDW}(b) = -i\, C_j^{DW}(b) \left[\sin \sqrt{p}/\sqrt{p}\right] \quad ; \quad j \neq 0$$

$$C_o^{UDW}(b) = \cos \sqrt{p}$$

where

$$p = \sum_j |C_j^{DW}(b)|^2 \quad (6)$$

The following comments can be made:

1. Methods of unitarization are not unique. This particular method can be obtained by omitting from the full perturbation series all matrix elements except those connecting the initial state with some state of the rearranged system. This has the effect of omitting all even terms in the series. In addition, the time ordering within each term of the series is neglected. The latter approximation has been estimated to cause errors of ~30%.

2. Because no second-order terms with continuum intermediate states are included, the method must fail at sufficiently high energies because direct computation of second-order terms shows these to be increasingly important. This may not be serious until $E \gtrsim 200$ or 300 keV/amu. Equally at low velocities, second-order terms involving discrete states are certainly important, and the method cannot be expected to be accurate much below the cross section maximum, although unlike the usual distorted wave approximation it cannot fail through a crude violation of unitarity.

3. The Stark mixing between the degenerate levels with different ℓ for a given n, is not taken into account properly so that while total cross sections may be adequate, the distribution in ℓ of the final state cannot be.

The UDW model has been applied over a wide energy range ($0.01 \lesssim E \lesssim 5000$ keV/amu) for various Z in the interval $1 \lesssim Z \lesssim 20$. At low velocities, as expected, the computed cross sections do not agree well with the results of coupled channel calculations (or with the experimental data), but above $E \simeq 10$ keV/amu the predicted cross sections show a similar energy variation to the data, but are rather large in absolute value. It is interesting to note that for $Z = 2$ and $Z = 3$ and above ~25 keV/amu the UDWA cross sections agree very closely with the coupled channel cross sections (with or without pseudostates), which also tend to be larger than the experimental cross sections (see Fig. 2(b)). For larger Z, the UDWA cross sections appear to exceed the data by a factor of up to 3.

In an attempt to improve the UDWA Ryufuku[15] has included the excitation and ionization channels in the calculation of p, which has the effect of reducing the cross sections at high energies and which improves the agreement with the experimental data. The results at low energy are little altered. The ionized states are represented by Coulomb waves centered about the heavy ion, so that the effect calculated is "charge exchange into the continuum." For small Z, one would certainly expect direct ionization to be of greater importance than charge exchange into the continuum in the energy region above 80 keV/amu, but as Z increases it is perhaps reasonable to suppose that projectile centered ionization dominates until much higher energies are reached.

The UDWA cross sections exhibit a scaling behavior for $E \gtrsim 20 \cdot Z^{0.35}$ keV/amu. Writing

$$\sigma(E) = \alpha\, \tilde{\sigma}(\tilde{E}) \quad \text{with} \quad E = \beta \tilde{E} \tag{7}$$

it is found that the cross sections can be fitted by a universal function $\tilde{\sigma}(\tilde{E})$ if $\alpha = Z^{1.07}$, $\beta = Z^{0.35}$. If the universal function is determined from the calculations for $Z = 14$, the predicted cross sections for the partially stripped ions Fe^{q+}, Xe^{q+} etc. $12 \lesssim q < 25$ appear to agree reasonably well with the data.

The Presnyakov and Ulantsev Model

Presnyakov and Ulantsev[16] have proposed a model based on a single centered atomic expansion about the projectile

$$\psi(\vec{r},t) = \sum_m C_m(b,t)\, \chi_m^A(\vec{r}_A)\, e^{-i\varepsilon_m t} \quad . \tag{8}$$

To satisfy the boundary condition, we must have

$$\psi(\vec{r},t) \underset{t\to-\infty}{\sim} \phi_0^B(\vec{r}_A - \vec{R}(t))\, e^{i[\vec{v}\cdot\vec{r}_A - \varepsilon_0^B t + (v^2/2)t]} \tag{9}$$

where ψ_0^B is the H(1s) wave function centered about the target proton and the notation is the same as in I. This condition determines the coefficients C_m in the limit $t \to -\infty$. Note that the retention of the continuum in the expansion (8) is vital. Solving the infinite set of equations for the coefficients $C_m(b,t=+\infty)$ is, in lowest order, equivalent to the UDWA model and although a systematic improvement is possible, no such calculations have been reported.

The Classical Trajectory Monte-Carlo Method

Both the UDWA and the Presnyakov and Ulantsev model are based on approximations to the coupled impact parameter equations. In contrast, a purely classical treatment can be carried out based on a model developed by Abrines and Percival.[17] The hydrogen atom in the initial state is represented by an electron moving in a classical elliptical orbit about the proton, such that the energy of the system is equal to ε, the binding energy of ground state hydrogen. In addition to the energy five parameters are required to specify the orbit completely. These can be taken to be (a) the polar angles of the major axis (θ,ϕ), (b) an angle χ specifying the orientation of the plane of the orbit, (c) the position of the electron on the orbit at some particular time t_0, (d) the square of the orbital angular momentum L^2, which can vary from 0 to L^2_{max}, where $L^2_{max} = |\varepsilon|/2$ (in atomic units).

The Newtonian (or Hamiltonian) equations of motion for a system composed initially of a hydrogen atom specified in this way, together with an incident fully stripped ion of energy E specified by an impact parameter b, can be solved numerically. The calculations are repeated for different values of b, and for a random selection of the five parameters specifying the initial hydrogen atom. This distribution over the hydrogenic parameters can be shown to reproduce the actual velocity distribution of an electron in ground state hydrogen which is a special feature of the Coulomb potential. To obtain statistically significant results from one to two thousand orbits must be computed at each energy. The final configurations in which the electron remains in orbit about the proton contribute to the excitation cross section; those in which the electron is free contribute to ionization; and those in which the electron is finally in an orbit about the fully stripped ion to charge exchange.

The total energy E and angular momentum J can be calculated for the electron in an orbit about the ion after the collision. This provides the principal and angular momentum quantum numbers of the final state through the equations $E = -1/(2n^2)$, $J = \sqrt{j(j+1)}$. The distributions in n and j are, of course, continuous, but by assigning all n between 0.5 and 1.5 to n = 1, those between 1.5 and 2.5 to n = 2, and so on, the distribution of the final hydrogenic levels can be predicted.[18] Extensive applications[19-23] of the classical trajectory method have been made to electron capture by both fully and partially stripped ions from ground state atomic hydrogen (see Fig. 4). In the case of partially stripped ions, the

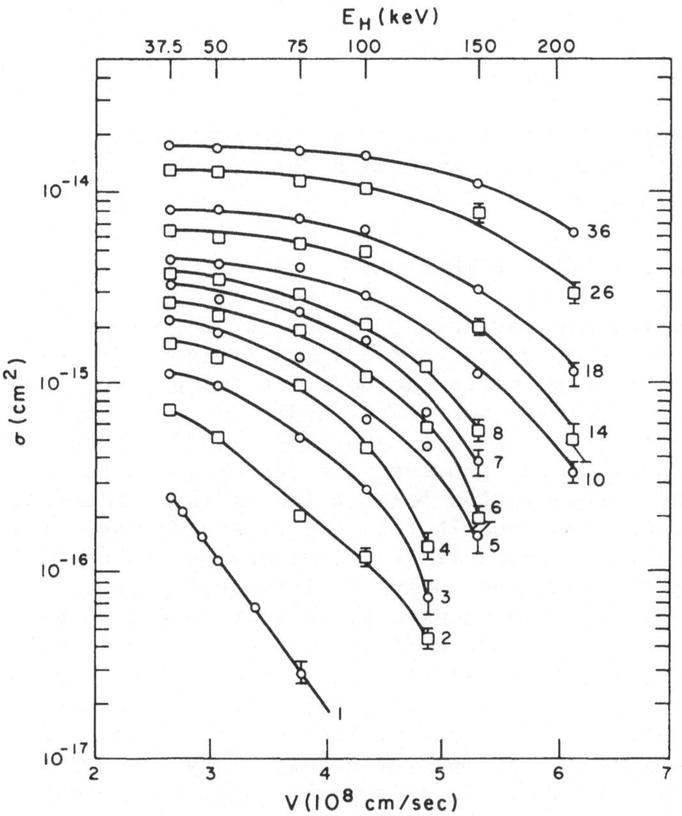

Fig. 4. Classical trajectory Monte-Carlo cross sections for electron capture by fully stripped ions of charge Z from H(1s).

system has been treated as a one-electron problem, the screening of the bound electrons in the projectile being taken into account through an effective charge. Over the energy range 40 to ~150 keV/amu the agreement with experiment appears satisfactory (~20%) except for small Z, for which the cross sections computed by the classical method appear to be of the correct order of magnitude but exhibit a somewhat different energy variation.[24] The upper limit of the energy range for which this method is useful is dictated by the difficulty of obtaining a statistically significant result, because the rate of charge exchange becomes very small compared with ionization. Below ~40 keV/amu, specifically quantum effects due to the pseudo-molecular formation are expected to be significant. At the lower energies the cross sections show little velocity variation and scale like

$$\sigma_{clas.} = 4.6\ Z \times 10^{-16}\ cm^2 \quad . \tag{10}$$

At higher energies the classical cross sections exhibit an approximate Z^3 behavior, in common with the continuum distorted wave and other high energy models.

3. INTERMEDIATE TO HIGH VELOCITIES

At velocities greater than 3 or 4 a.u. (E greater than 200 to 400 keV/amu), the theoretical methods available are rather limited, because low order perturbation methods are inapplicable, in particular neither the first nor the second-order Born (or Brinkman-Kramers) approximations are accurate, and indeed although the second Born approximation provides the correct asymptotic cross section, in the energy region below a few MeV/amu it actually provides[25] a less accurate result than the first Born approximation! Potentially the one-and-a-half center expansion method of Reading and his coworkers should be useful in this energy region because it allows the use of an extensive pseudostate basis chosen so that the important continuum intermediate states can be properly represented. As briefly discussed in I, the most satisfactory method available at present is the continuum distorted wave method (CDW). This has not been extensively exploited for collisions between fully stripped ions and atomic hydrogen for large Z, but where it has been tested, for small Z, it appears to give accurate results for $v \gtrsim 2.5$ a.u. For example, the experimental data for the Li^{3+} + H reaction are very well fitted by the CDW results of Crothers and Todd.[26] A detailed critical account of the CDW method, and its range of validity, has been given by Belkić, Gayet and Salin[27] in a review of high energy theories.

Eikonal (Glauber) Models

A different approach which is designed to be useful in the intermediate to high velocity region was derived by Mittleman and Quong[28] and independently by Dewangan.[29] The exact amplitude for charge exchange to a final level f can be represented as

$$C(b) = -i \int_{-\infty}^{\infty} dt \int d\vec{r}\ X_f^*(\vec{r},t)\ \frac{Z_B}{r_B}\ \Psi(\vec{r},t) \tag{11}$$

where, in the notation of I, X_f is the unperturbed wave function in the final state and $\Psi(\vec{r},t)$ is the exact wave function. In the eikonal approximation Ψ is approximated by

$$\Psi(\vec{r},t) = \Phi_i(\vec{r},t)\ \exp\ i \int_t^{\infty} dt'\ \frac{Z_A}{r_A} \tag{12}$$

where

$$\exp i \int_t^\infty dt' \frac{Z_A}{r_A} = \exp i\eta\, Z_A \log(r_A - z_A) \tag{13}$$

and $\eta = 1/v$. If the eikonal phase factor (13) is set equal to unity, the amplitude reduces to the Brinkman-Kramers first-order amplitude. Remarkably enough, Chen and Eichler[30] and Dewangan[29] have shown that the eikonal amplitude can be evaluated analytically. The cross section takes the form of the product of the Brinkman-Kramers cross section and a slowly varying factor. For capture into all levels with principal quantum number n, it is found that

$$\sigma_{eik}(v) = \frac{\pi\eta}{\sinh(\pi\eta)} \exp\left[-2\eta \tan^{-1}\left(\frac{v}{2} - \eta\omega\right)\right] \times \left\{\frac{23}{48} + \left(\frac{1}{6} + \frac{5\omega}{6}\right)\eta^2 + \frac{5\omega^2\eta^4}{12}\right\} \sigma_{BK}(n,v) \tag{14}$$

where $\omega = \frac{1}{2}\,[1 - (Z^2/n^2)]$.

The Brinkman-Kramers cross section, σ_{BK}, for capture into any hydrogenic final state has been evaluated by Sil in closed form.[31] Because the factor multiplying σ_{BK} varies only slowly with n, the ratios of cross sections for capture into various excited states are approximately given by the Brinkman-Kramers cross section, although the absolute magnitude is very different. As is well known, the eikonal approximation, although contributing to all orders of perturbation theory, omits completely the real part of the second term in the Born series. For this reason, it cannot reproduce the correct high energy limit of the charge exchange cross section, and the range of validity of the model is difficult to estimate. By comparison with experiment, the eikonal cross section appears to be accurate[32] for capture by fully stripped ions from H(1s) for $Z \lesssim 3$ in an energy range from ~50 to ~500 keV/amu. For larger Z ($Z = 5,6$) the theoretical cross sections are too large, but as Z increases further (for example, for $Z = 20$ and $Z = 25$), agreement is again good at high energies 1-3 MeV/amu. Although more work needs to be done to establish accuracy limits, the eikonal model is clearly useful in predicting high energy total cross sections and perhaps the distribution in final states.

Although the Brinkman-Kramers cross section for the reaction $A^{Z+} + H(1s) \rightarrow A^{(Z-1)+} + H^+$, scales like Z^5 at very large values of v, at moderate values of v ($2 < v < 10$), it has been shown[33] to scale like Z^3. Because the extra factor in σ_{eik} varies slowly with Z, this Z^3 behavior is also shown by the eikonal cross sections.

4. CHARGE EXCHANGE AT LOWER VELOCITIES

Although the coupled channel molecular orbital method is expected to be accurate at velocities in the region v < 1 a.u., as we have noted the large number of important channels often makes complete calculations impracticable and makes it important to look for simple models. The coupled equations for the amplitudes $C_\lambda(b,t)$ of the adiabatic molecular orbital ψ_λ are [see I, eq. (26)]

$$i\,\dot{C}_\lambda(t) = \sum_{\mu\neq\lambda} \langle\psi_\lambda|-i\frac{\partial}{\partial t}|\psi_\mu\rangle$$

$$\times \exp i\int_{-\infty}^{t} dt'\{\varepsilon_\lambda[R(t')] - \varepsilon_\mu[R(t')]\}\; C_\mu(t) \qquad (15)$$

where $\varepsilon_\lambda(R)$ is the molecular energy corresponding to the level λ. In these equations the translational factors describing the change in the momentum of the capture electron have been omitted, which is a poor approximation except at very small values of v. As explained in I, the interaction is large when there is a crossing or pseudo-crossing of two potential energy curves. To see how these crossings occur, consider a diabatic basis (which does not diagonalize the electronic Hamiltonian) formed from the perturbed asymptotic atomic levels for the system A^{Z+} + H(1s). To lowest order the energy of the original arrangement is (atomic units)

$$V_o(R) = -\frac{1}{2} + O(1/R^4) \quad . \qquad (16)$$

In the charge exchange channel corresponding to capture into a level n, the corresponding energy is

$$V_n(R) = -\frac{Z^2}{2n^2} + \frac{Z-1}{R} + O(1/R^2) \quad . \qquad (17)$$

The crossings of these diabatic curves are at (see Fig. 5)

$$R_c(n) = (Z-1)\left[\frac{Z^2}{2n^2} - \frac{1}{2}\right]^{-1} \quad . \qquad (18)$$

The number of crossings, in this case, is limited. However, for many electron systems, the number of crossings can be enormous.

The Multi-Crossing Landau-Zener Model

If it is assumed that (1) only in the vicinity of the crossing points is the interaction significant, (2) the region of R about each crossing in which the interaction takes place is so restricted that these regions do not overlap , then at each crossing the probability of transfer from one diabatic curve to another is given by the solution of a two-state coupled channel problem. Let

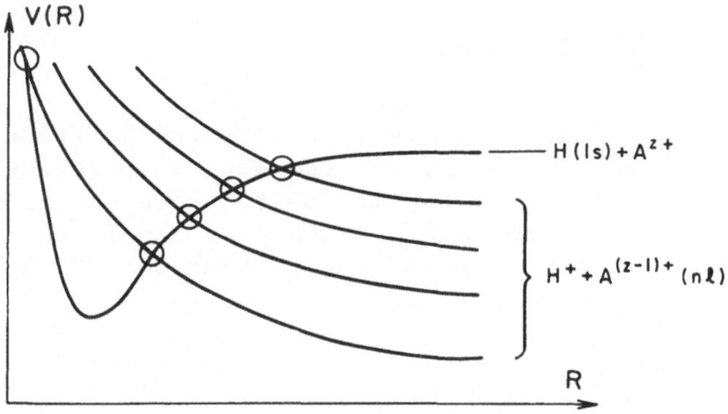

Fig. 5. Schematic diagram of diabatic potentials for the A^{Z+} + H(1s) system. The circles show the crossing points where the coupling is strong.

us label the crossing points by n = 1,2,3...N, with $R_c(1) > R_c(2) > \ldots > R_c(N)$ and let the probability of a transfer from the initial potential curve to a rearranged curve at $R = R_c(n)$ be p_n. Consider the case in which there is only one crossing point $R_c(1)$, then as the crossing point is traversed at $t < 0$ (with $t = 0$ as the time of closest approach), the probability of the system remaining in the original arrangement is $(1-p_1)$ and the probability is p_1. After passing the point of closest approach R will increase again and at some $t > 0$, the crossing point will be traversed again, so that the final probability of finding the system in the rearranged state is $p_1(1-p_1) + (1-p_1)p_1 = 2p_1(1-p_1)$. By keeping track of the probabilities at the n crossing points the generalization of this result to obtain the probability P_n of finding the system in a rearranged level n, is[34]

$$P_n = (p_1p_2\ldots p_n)(1-p_n)\{1 + p_{n+1}p_{n+2}\ldots p_N\}^2 + (1-p_{n+1})^2$$
$$+ [p_{n+1}(1-p_{n+2})]^2 + [p_{n+1}p_{n+2}(1-p_{n+3})]^2$$
$$+ \ldots + [p_{n+1}p_{n+2}\ldots p_{N-1}(1-p_N)]^2\} \quad \text{for} \quad n < N-1$$

$$P_{N-1} = (p_1p_2\ldots p_{N-1})(1-p_{N-1})[1 + p_N^2 + (1-p_N)^2]$$

$$P_N = 2p_1p_2\ldots p_N(1-p_N) \quad . \tag{19}$$

The cross section is given by:

$$\sigma_n(v) = 2\pi \int_0^{R_n} P_n(b,v)b\, db \quad .$$

At each curve crossing p_n could be computed in principle from the two-state coupled equations, but the general practice has been to employ the Landau-Zener approximation to these equations. For a description of this approximation reference should be made to the texts by McDowell and Coleman[35] or by Bransden.[36] It is found that

$$p_{LZ} = \exp[-\frac{\pi G^2}{2v_R F}] \tag{20}$$

where v_R is the radial velocity, and G and F are determined by the potential matrix H_{ij} coupling the two channels:

$$G = 2\, H_{12}(R_c) - S(R_c)H_{11}(R_c)$$

$$F = \left|\frac{\partial}{\partial R}(H_{11} - H_{22})\right|_{R=R_c} \tag{21}$$

and S is the overlap integral between the two diabatic wave functions. In applying (20) and (21) to the present problem, we have $H_{11} = V_0$ and $H_{22} = V_n$ while G can be calculated from the perturbed atomic wave functions. A particular approximation, valid for $4Z^{1/2} << R_c << 2Z$ and $\ell^2 << Z$ is

$$G_{n\ell} = 2\, [\frac{2(2\ell+1)}{\pi n^3}]^{1/2}\, e^{-\ell(\ell+1)/2Z}\, R_c\, e^{-R_c^2/3Z} \quad . \tag{22}$$

The quantity G can also be expressed in terms of the difference in energy between the two corresponding non-crossing adiabatic potential curves at $R = R_c$. This can be calculated exactly for a one-electron system and, to within 17%, the exact numerical results are fitted by $(Z > 4)$

$$G = 18.26\, Z^{-1/2} \exp(-1.324\, R_c\, Z^{-1/2}) \quad . \tag{23}$$

This expression has been used in most numerical work, together with the approximations $F \approx (Z-1)/R^2$ and $v_R = v(1 - b^2/R^2)^{1/2}$. Further details about the calculation of F and G can be found in the reviews of Janev and Presnyakov[37] and of Greenland.[38] Because of the exponential dependence of p_n on $R_c(n)$ only a restricted number of channels are populated.

The multichannel Landau-Zener method has been used by Salop and Olson[34] for Z = 6, 7, 8, 10, 14 and 18, and in a model[39] allowing for rotational coupling between the degenerate levels with different angular momentum, but the same principal quantum number, it has been used by Janev and his coworkers[40] for 5 < Z < 76. The calculated cross sections can be represented by

$$\sigma = 2.25 \times 10^{-16} Z \log(15/v) \text{ cm}^2$$

for $Z > 16$ and $0.04 < v < 2$. Some representative results are shown in Fig. 6.

The Absorbing Sphere Model

The same physical ideas form the basis of several other models. For example, in the interaction of a partially stripped ion with atomic hydrogen, the number of crossings is too large to apply the multichannel Landau-Zener method effectively. If the number of crossings is very large, then a characteristic distance can be introduced R so that for $b \leq R$ the transition probability is effectively unity. For example R can be defined as the distance for which the Landau-Zener single crossing formula is at a maximum. This condition is

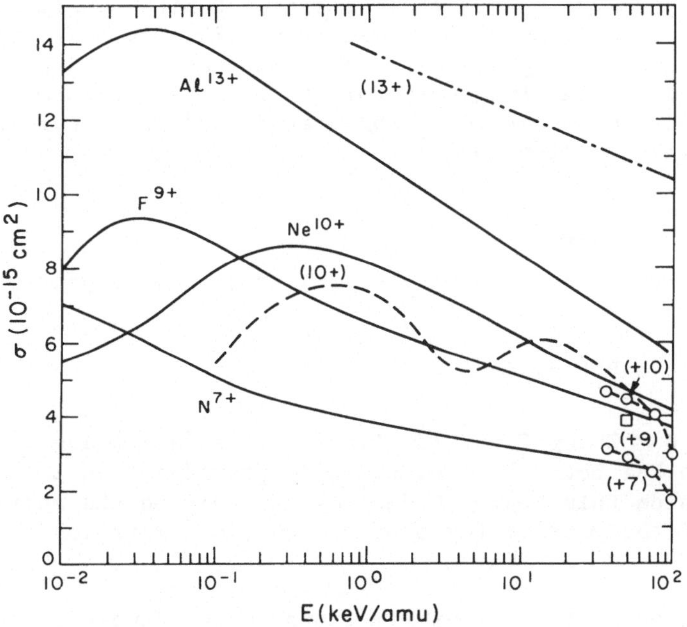

Fig. 6. Total electron capture cross sections for $A^{Z+} + H(1s)$. —— - Multichannel Landau-Zener calculations with allowance for rotational coupling[40]; o-o-o - classical Monte-Carlo model[19,55]; — — — - UDWA (Ryufuku and Watanabe[14]); —•—• - decay model.[55]

$$\frac{\pi G^2(\bar{R})}{2v_R F(\bar{R})} = 0.15 \quad . \tag{24}$$

The cross section is then given by

$$\sigma = \pi \bar{R}^2 \quad . \tag{25}$$

Using (23) it is found,[41] that $\bar{R} \sim Z^{1/2}$ and $\sigma \propto Z$. Further developments of this absorbing sphere model have been reported by Duman et al.[42]

The Decay Model

For large Z, and particularly for partially stripped ions incident on H(1s), the number of final states for charge exchange is so large as to form an effective continuum of states. If this is the case, the process can be viewed as the tunneling of the bound electron through the potential barrier between the atomic well and the quasi-continuum of ionic states, in a manner exactly similar to field ionization.[43] A mathematical model of this process can be made by solving the Schrödinger equation for a hydrogen atom in the field of a charge Z at a fixed position R with purely outgoing wave boundary conditions for the electron. This is an eigenvalue problem with complex eigenvalues $E = \varepsilon + i\Gamma$ where ε and Γ are functions of Z and R. The imaginary part of E, Γ, determines the decay probability of the perturbed hydrogen atom, and the total transition probability is given by

$$P = 1 - \exp - \int_{-\infty}^{\infty} \Gamma(R)\, dt \tag{26}$$

where $R^2 = b^2+v^2t^2$ in the straight line trajectory approximation and as usual

$$\sigma = 2\pi \int_0^{\infty} b\, P(b,v)\, db \quad . \tag{27}$$

For fixed v, $\sigma \propto Z \log Z$ and for fixed Z, $\sigma \propto (a - b \log v)$. This approximation has been shown to be valid provided $v \ll Z^{1/2}$ and $R \gg 2Z$; outside this region P depends not only on the static parameters of the barrier but also on the time variation of its height.

Clearly, none of the approximations to the coupled channel molecular orbital expansion method can be very accurate. In particular all of them neglect the momentum transfer of the electron which is certainly important for $v > 0.1$ a.u. However, these methods do provide a guide to the order of magnitude of the cross section and to the η and ℓ distributions.

5. IONIZATION

The construction of theoretical models for the ionization of H(1s) by a fully stripped ion carrying a charge Z,

$$A^{Z+} + H(1s) \rightarrow A^{Z+} + H^{+} + e^{-} \quad , \qquad (28)$$

is not easy because if Z is large the perturbation is too strong for first-order methods to be accurate, except at very high velocities outside the range of interest. If the first-order impact parameter method is used,[35,36,44] the probability of ionization, $P_I(b)$, exceeds unity for impact parameters $b < b_{min}$, where $b_{min} \approx 2Z/v$. It is possible to introduce a cut-off procedure

$$\sigma_I(v) \approx 2\pi \int_{b_{min}}^{\infty} P_I(b)\, b\, db \qquad (29)$$

but this is clearly not satisfactory because the impact parameter below which $P_I(b)$ is undetermined is large for large Z. There are a number of higher order methods, which have been used with more or less success:

(a) The indirect calculation of ionization using a pseudostate expansion with a coupled channel formalism. This technique has been widely and successfully used for small Z, and also for inner shell ionization.

(b) The classical trajectory Monte-Carlo method. This is also successful, but over a limited velocity region.

(c) The generalization of the continuum distorted wave method[45] and similar methods such as the impulse approximation. These methods are suitable for high velocities.

(d) The direct calculation of ionization with a special choice of pseudostates using a coupled channel formalism.

(e) A coupled channel molecular orbital formalism in which as $R \rightarrow \infty$ the molecular orbitals describing the final state lead to autoionizing atomic levels, or to ionized atomic levels.

In the preceding paragraphs, methods (a) and (b) have been described briefly. Method (c) has not been applied widely, although it is potentially useful. In general, for the fully stripped ion — hydrogen atom system, ionization is unimportant compared with charge exchange at low energies, so that method (e), although in principle an accurate method is not so interesting, although in other contexts it is an important and successful approach. This leaves method (d), which will now be described briefly. As implemented by Janev and Presnyakov,[46] a system of close coupled equations based on an atomic expansion is written down in which the continuum is represented by one (or more) p pseudostates. This pseudostate is chosen (following an idea introduced by van Regemorter), so that it has the same oscillator strength as the hydrogenic continuum. To allow for multipole transitions other than the dipole transition, the effective oscillator strength f_{eff} is written as

$$f_{eff} = K\, f_{cont} \tag{30}$$

where for hydrogen K = 1.5 and f_{cont} = 0.4350. Although the coupled equations could be solved directly, a simplified procedure has been used by Janev and Presynakov. Consider the interaction of an ion carrying charge Z with atomic hydrogen. A three-channel approximation will be made. In the first channel the hydrogen atom is in the ground state and in this channel the amplitude of the wave function, using the straight line impact parameter formulation is a(b,t). In the second and third channels the hydrogen atom is in the p_0 and $p_{\pm 1}$ sublevels of an effective continuum state and in these channels the amplitude of the wave function is $c_0(b,t)$ and $c_1(b,t)$ respectively. In the dipole approximation, the interaction between the incident ion and the hydrogen atom is

$$V(\vec{r},\vec{R}) = -Z\, \vec{\mu}\cdot\vec{R}/R^3 \tag{31}$$

where $\vec{\mu}$ is the dipole moment. The potential matrix coupling the three channels is given by the matrix elements of V with respect to the wave functions of the hydrogen atom. Using these, the following coupled equations are found

$$\begin{aligned} i\,\dot{a} &= \frac{Z\lambda}{R^2}(\cos\theta\; c_0 + \sin\theta\; c_1)\, e^{-i\omega_i t} \\ i\,\dot{c}_0 &= \frac{Z\lambda}{R^2}\cos\theta\; e^{i\omega_i t}\, a \\ i\,\dot{c}_1 &= \frac{Z\lambda}{R^2}\sin\theta\; e^{i\omega_i t}\, a \end{aligned} \tag{32}$$

where

$$\cos\theta = \frac{vt}{R}\,, \quad \sin\theta = \frac{b}{R}\,, \quad \lambda^2 = \left(\frac{f_{eff}}{2\omega_i}\right) \tag{33}$$

and ω_i is the ionization potential. (In the corresponding approximation for excitation of an (np) discrete level, ω_i is replaced by $(\varepsilon_n-\varepsilon_i)$ and $\lambda^2 = f_{on}/2\omega$.) Further approximations were made by Janev and Presnyakov to obtain the result

$$\sigma_I = 2\pi(Z\,\lambda_{eff}/\omega_i)\,D(\beta_i) \tag{34}$$

where $\beta_i = Z\,\lambda_{eff}\omega_i/v^2$ and D(x) is a numerical function which has been tabulated. For $x > 2$

$$D(x) \approx \tfrac{1}{2}\,x(1 - 0.125\,x^{-3/2})\,e^{-(2x)^{1/2}} \tag{35}$$

and for small x, $x \ll 0.01$

$$D(x) = 4x\log(1.4/x) \tag{36}$$

In further calculations, the 2p level of hydrogen was included in the basis, and this provides an important contribution to the cross section via the $1s \to 2p \to ks$ sequence. In Fig. 7, the results of the test case of $H^+ + H(1s) \to H^+ + H^+ + e^-$ are shown, which are seen to be very satisfactory. An application to a

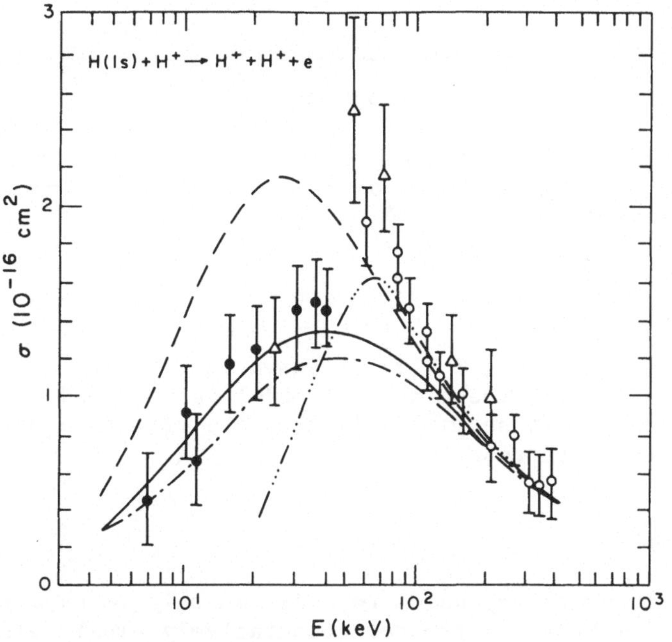

Fig. 7. Cross section for $H^+ + H(1s) \to H^+ + H^+ + e^-$. Experimental data: ● - Fite et al.[56]; ○ - Gilbody and Ireland[57]; Δ - Park et al.[58] Theoretical cross sections: – – – - first Born approximation; —••—••— - classical Monte-Carlo model[19]; ——— - dipole close coupling model with (2p) intermediate level[46]; —•—•— - same without 2p intermediate level.[46]

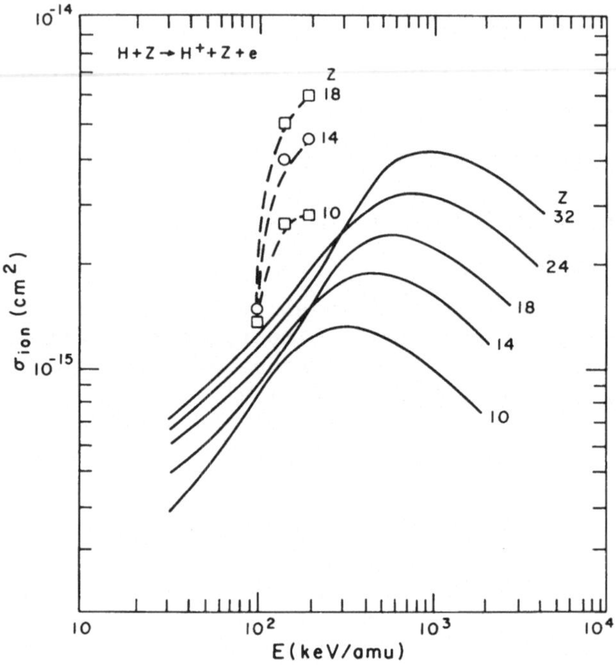

Fig. 8. Cross sections for $A^{Z+} + H(1s) \rightarrow A^{Z+} + H^+ + e^-$.
—— - Dipole close coupling model[46]; □, ○ - classical Monte-Carlo method.[19]

number of cases with $10 \lesssim Z \lesssim 32$ is shown in Fig. 8. Unfortunately, the agreement with the classical model is not good at energies below 1 MeV/amu and it seems likely that the effective oscillator model should be preferred.

6. CONCLUSION

Because accurate expansion methods can only be used to describe the $A^{Z+} + H(1s)$ system for comparatively small values of Z, a number of simple approximations have been developed, some, but not all, of which have been described in these lectures. Further details can be found in the recent reviews of Greenland[38] and Janev and Presnyakov.[37] The accuracy of these methods for large Z is not known with any certainty and there is room for further theoretical and experimental work, even on the one-electron system. For several, or many, electron systems the uncertainties are naturally greater, and many interesting problems remain to be solved, in particular, the calculation of multiple ionization and charge transfer.

REFERENCES

1. B. H. Bransden, in: "Atomic and Molecular Processes in Controlled Thermonuclear Fusion," M. R. C. McDowell and A. M. Ferendeci, eds., Plenum, London (1979), p. 185.
2. B. H. Bransden and C. J. Noble, J. Phys. B 14:1849 (1981).
3. T. G. Winter, Phys. Rev. A 25:697 (1982).
4. M. Kimura and W. Thorsen, Phys. Rev. A 24:3019 (1981).
5. J. F. Reading, A. L. Ford, and R. L. Becker, J. Phys. B 15:625 (1982).
6. B. H. Bransden and C. J. Noble, J. Phys. B 15:451 (1982).
7. M. Kimura and W. Thorsen, XII ICPEAC Abstracts, North-Holland, Amsterdam (1981), p. 638.
8. A. L. Ford, J. F. Reading, and R. L. Becker, J. Phys. B 15:in press (1982).
9. W. Fritsch and C. D. Lin, J. Phys. B 15:L281 (1982).
10. T. A. Green, E. J. Shipsey, and J. C. Browne, Phys. Rev. A 25:546, 1346 (1982).
11. W. Fritsch, J. Phys. B 15:in press (1982).
12. J. F. Reading, A. L. Ford, and R. L. Becker, J. Phys. B 14:1995 (1981).
13. D. R. Bates, Proc. Roy. Soc. A 247:294 (1958).
14. H. Ryufuku and T. Watanabe, Phys. Rev. A 18:2005 (1978); Phys. Rev. A 19:1538 (1979); Phys. Rev. A 20:1828 (1979).
15. H. Ryufuku, Phys. Rev. A 25:720 (1982).
16. L. P. Presnyakov and A. D. Ulantsev, Sov. J. Quant. Electron. 4:1320 (1975).
17. R. Abrines and I. C. Percival, Proc. Phys. Soc. 88:861, 873 (1966).
18. A. Salop, J. Phys. B 12:919 (1979).
19. R. E. Olson and A. Salop, Phys. Rev. A 16:531 (1977).
20. A. Salop and R. E. Olson, Phys. Lett. 71A:407 (1979).
21. R. E. Olson, A. Salop, R. A. Phaneuf, and F. W. Meyer, Phys. Rev. A 16:1867 (1977).
22. R. E. Olson, Phys. Rev. A 18:2464 (1978).
23. K. H. Berkner, W. G. Graham, R. V. Pyle, A. S. Schlachter, J. W. Stearns, and R. E. Olson, J. Phys. B 11:875 (1978).
24. H. B. Gilbody, Physica Scripta 23:143 (1981).
25. P. R. Simony and J. M. McGuire, J. Phys. B 14:L737 (1981).
26. D. S. F. Crothers and N. R. Todd, J. Phys. B 13:2277 (1980).
27. Dz. Belkić, R. Gayet, and A. Salin, Phys. Repts. 56c:279 (1979).
28. M. H. Mittleman and J. Quong, Phys. Rev. 167:76 (1968).
29. D. P. Dewangan, J. Phys. B 8:L119 (1975); J. Phys. B 10:1083 (1977).
30. F. T. Chen and J. Eichler, Phys. Rev. Lett. 42:58 (1979); Phys. Rev. A 20:1841 (1979); J. Eichler and F. T. Chen, Phys. Rev. A 20:104 (1979).
31. N. C. Sil, Ind. J. Phys. 28:232 (1954).
32. J. Eichler, Phys. Rev. A 23:498 (1981).

33. D. S. F. Crothers and N. R. Todd, J. Phys. B 13:547 (1980).
34. A. Salop and R. E. Olson, Phys. Rev. 13:1312 (1976).
35. M. R. C. McDowell and J. P. Coleman, "An Introduction to Ion-Atom Collisions," North-Holland, Amsterdam (1970).
36. B. H. Bransden, "Atomic Collision Theory," 2nd Ed., Benjamin, New York (1982).
37. R. K. Janev and L. P. Presnyakov, Phys. Repts. 70:1 (1981).
38. P. T. Greenland, Phys. Repts. 81:131 (1982).
39. V. A. Abramov, F. F. Baryshnikov, and V. S. Lisitsa, Sov. Phys.-JETP 47:469 (1978).
40. R. K. Janev and D. S. Belić, JILA preprint (1982); D. S. Belić, B. H. Bransden, and R. K. Janev, JILA preprint (1982).
41. R. E. Olson and A. Salop, Phys. Rev. A 14:576 (1976).
42. E. L. Duman, L. I. Men'shikov, and B. M. Smirov, Zh. Eksp. Teor. Fiz. 76:516 (1979).
43. T. P. Grozdanov and R. K. Janev, Phys. Rev. A 17:880 (1978).
44. J. Bang and J. M. Hansteen, K. Dan. Vidensk. Selsk. Mat-Fys. Medd. 31:13 (1959).
45. Dz. Belkić, J. Phys. B 11:3529 (1978).
46. R. K. Janev and L. P. Presnyakov, J. Phys. B 13:4233 (1980).
47. H. van Regemorter, Astrophys. J. 132:906 (1962).
48. W. L. Nutt, R. W. McCullough, K. Brady, M. B. Shah, and H. B. Gilbody, J. Phys. B 11:1457 (1978).
49. M. B. Shah and H. B. Gilbody, J. Phys. B 11:121 (1978).
50. G. T. Hatton, N. F. Lane, and T. G. Winter, J. Phys. B 12:L571 (1979).
51. D. S. F. Crothers and N. R. Todd, J. Phys. B 14:3251 (1981).
52. W. Seim, A. Müller and E. Salzborn, Phys. Letts. 80A:20 (1980).
53. M. B. Shah, T. V. Goffe, and H. B. Gilbody, J. Phys. B 11:L233 (1978).
54. R. E. Olson, Phys. Rev. A 24:1726 (1981).
55. E. L. Duman and B. M. Smirnov, Sov. J. Plasma Phys. 4:651 (1978).
56. W. L. Fite, R. J. Stebbings, D. G. Hummer, and R. T. Brackman, Phys. Rev. 119:663 (1960).
57. H. B. Gilbody and J. V. Ireland, Proc. Roy. Soc. A 277:137 (1964).
58. J. T. Park, J. E. Alday, J. M. George, and J. L. Peacher, Phys. Rev. A 15:508 (1977).
59. E. J. Shipsey, T. A. Green, and J. C. Browne, preprint (1982).

EXPERIMENTS ON ELECTRON CAPTURE AND IONIZATION BY MULTIPLY CHARGED IONS

F.J. de Heer

FOM-Institute for Atomic and Molecular Physics
Kruislaan 407
1098 SJ Amsterdam, The Netherlands

1. INTRODUCTION

A great number of review articles have been devoted to the capture and ionization processes in collisions between multiply charged ions and atoms which play an important role in controlled thermonuclear fusion, astrophysics, and the development of short wavelength lasers, see for instance Invited Papers and Progress Reports at ICPEAC XI and ICPEAC XII[1,2], the IAEA Technical Committee Meeting on Atomic and Molecular Data for Fusion[3], H.B. Gilbody[4], A. Dalgarno[5], Janev and Hvelplund[6], A.V. Vinogradov and I.I. Sobel'man[7] and R.W. Waynant and R.C. Elton[8]. In this paper we shall follow the scheme of a review article, which we have written three years ago[9]. There we discussed the experimental methods which have been used for the determination of electron capture and ionization cross sections and experimental results were given, which were compared with theory. The different behaviour of singly and multiply charged ions was explained and attention was given to the formation of excited states, decaying by photon or electron emission. Separate attention was given to experiments with atomic and molecular hydrogen as a target and to the scaling of electron loss and electron capture cross sections. In this paper we shall discuss the progress in the different fields of our previous paper[9], further indicated by I. Therefore, it may be useful to use I as an introduction. As new important activities we mention electron capture experiments in the low energy (eV) impact region with multiply charged ions produced by recoil-ion technique [10,11,12] and by the use of laser radiation impinging on a surface[13]. In many cases these ions are fully stripped. New experiments have also been performed in the low energy keV impact region with fully stripped ions coming from an Electron Cyclotron Resonance Ion Source (ECRIS)[14] and from an Electron Beam Ionization Source (EBIS)[15,16].

Part of this work is concerned with the measurement of capture of electrons into excited projectile states by observation of the emitted soft X-rays[17,18]. In connection with Li-beam probing diagnostics in controlled thermonuclear fusion a series of experiments has been started with Li as a target[19,20,21].

In connection with the fast increasing interest of this field, it has not been possible to include all newest results in this article and for some recent developments the reader is referred to Proceedings of the Symposium on Production and Physics of Highly Charged Ions[22] in Stockholm and to Atomic Physics of Highly Ionized Atoms of the Nato Summerschool at Cargèse[23].

In this paper we shall extend the description of I[9] on experimental methods (section 2). A few remarks are given about ion sources for the production of multiply charged ions (section 3). The role of potential curves crossing in electron capture processes is illustrated with quantal and classical models (sections 4 and 5). Symmetric resonance multiple charge transfer and transfer ionization are mentioned in sections 6 and 7. Experimental results with H and H_2 targets are shown in section 8, partly in the frame-work of the so-called scaling laws. The recent work with Li targets is considered in section 9. Section 10 in particular deals with the distribution over n,ℓ-states of the projectile formed in the electron capture process.

2. EXPERIMENTAL METHODS TO DETERMINE ELECTRON CAPTURE AND IONIZATION CROSS SECTIONS

We shall mention the main methods, but for a more complete description the reader is referred to I[9]. In addition new developments and improvements on existing methods will be discussed.

A. Condensor Plates Method

We consider the processes:

$$A^{q+} + B \rightarrow A^{(q-1)+} + B^{+} \quad \text{(electron capture)} \qquad (1)$$

$$A^{q+} + B \rightarrow A^{q+} + B^{+} + e \quad \text{(ionization)} \qquad (2)$$

A beam of ions A^{q+} enters a chamber filled with a target gas of atoms B. The slow ions B^+ and electrons e, formed by reactions (1) and (2) are collected on the two opposite plates of a condensor positioned along the beam line as illustrated in Fig. 1 of I. The cross sections for production of slow ions and electrons can be evaluated according to

$$\sigma_{i;e} = q\iota_{i;e}/LI_q N \qquad (3)$$

where I_i respectively I_e are the currents of slow ions and electrons formed in the target gas of density N. I_q is the current of projectile ions of charge q and L is the length of the condensor plates. From equations (1) and (2) it follows that

$$\sigma_e = \sigma_I \quad \text{and} \quad \sigma_{q,q-1} = \sigma_i - \sigma_e \tag{4}$$

where σ_I is the ionization cross section and $\sigma_{q,q-1}$ the electron capture cross section. The equations (4) are only a good approximation if electron stripping from the projectile, multiple ionization and multiple electron capture are negligible, so in particular for atomic hydrogen as a target at low impact energies. The condensor plates method has been applied in many experiments. Here we quote the ionization cross section measurements of Berkner et al. for Fe^{q+} incident on H_2 (q = 11-22) at 1.1 MeV amu^{-1} [24] and electron capture cross section measurements by Winter and coworkers[20] for 5q-30q keV $Ne^{1,2+}$-Li collisions.

B. Charge State Selection of the Projectile

When A^{q+} collides with atom B a variety of processes are possible presented by

$$A^{q+} + B \rightarrow A^{(q-n)+} + B^{m+} + (m-n)e \tag{5}$$

where $m-n \geq 0$. In I[9] we discussed a set-up used by Crandall et al.[25], who shot a beam of multiply charged ions through an oven of high temperature (~2800 K) filled with H_2. At this temperature about 90% of the gas in the oven is dissociated into atomic hydrogen. The charge of the beam particles, after having passed the hydrogen oven is determined by electrostatic deflection. In the case of H as a target only one electron can be captured and the process is uniquely defined. When the target contains many electrons, $\sigma_{q,q-n}$ includes all possible electron capture processes together with simultaneous ejection of electrons where $(m-n) > 0$. This is called transfer ionization (or capture ionization).

B'. Charge State Selection of the Target Ion

The slow target ions formed in the gas can also be analyzed by electrostatic deflection (or magnetic deflection) methods. In that case a hole has to be made in one of the condensor plates in the collision chamber introduced in section A and a voltage has to be applied across the plates to extract the target ions for charge analysis. Recently in many experiments with multiply charged ions, the time of flight method has been used. For explanation we show a scheme of the apparatus used by Hvelplund et al.[26] (see Fig. 1) in collisions between H^+, He^{q+}, O^{q+} and Au^{q+} colliding with He at velocities from v_0 to $10v_0$ ($v_0 = 2.19 \times 10^8$ cm/s, the electron velocity in the

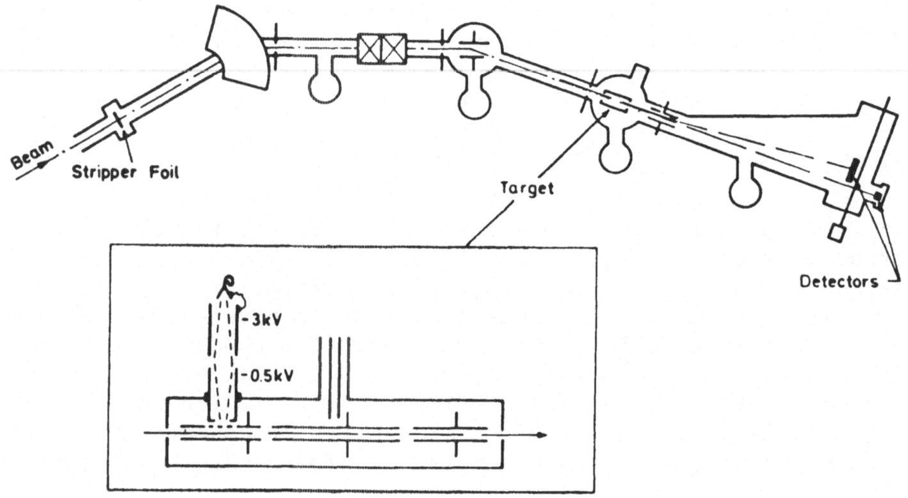

Fig. 1. Schematic diagram of the apparatus of Hvelplund et al.[26]

first Bohr orbit of H), q varying from 1 to 21. In their collision cell, which is 22 cm long, two condensor plates are present in the center, surrounded by guard plates. According to the method discussed under A it was possible to collect the slow ions (and electrons). Operating with He as a target (see Eq.(5)), this leads to the (apparent) cross section for slow ion production, σ_i, according to

$$\sigma_i = \sigma_i(m=1) + 2\sigma_i(m=2) \qquad (6)$$

where the two terms on the right side respectively present cross sections for production of singly and doubly charged He ions. In order to discriminate between He^+ and He^{2+} formation, a guard plate in the collision cell was used as the first electrode in a time of flight spectrometer. A hole in this plate was covered with nickel mesh of high transparency. Two other electrodes were applied to accelerate the extracted He ions to a channel-electron multiplier where they were counted. In this experiment the accelerator was operated in a pulsed mode with a repetition period of 1 μs and a pulse width of 10 ns. These pulses of ions were detected either by a channel-electron multiplier or a solid-state detector to start a time-to amplitude converter for registrating the slow He ions as a function of time. This is the so called delayed coincidence method, where in the time of flight spectrum two peaks are seen corresponding to He^+ and He^{2+}, with the largest delay time for He^+.

A different (coincidence) time of flight method has been applied by Shah and Gilbody[27,28] to measure the ionization (see Eq.(2)) in the case of a crossed beam experiment of keV protons or multiply charged ions and a thermal neutral atomic hydrogen beam (see Fig. 2). In such a crossed beam experiment it is possible to collect the slow

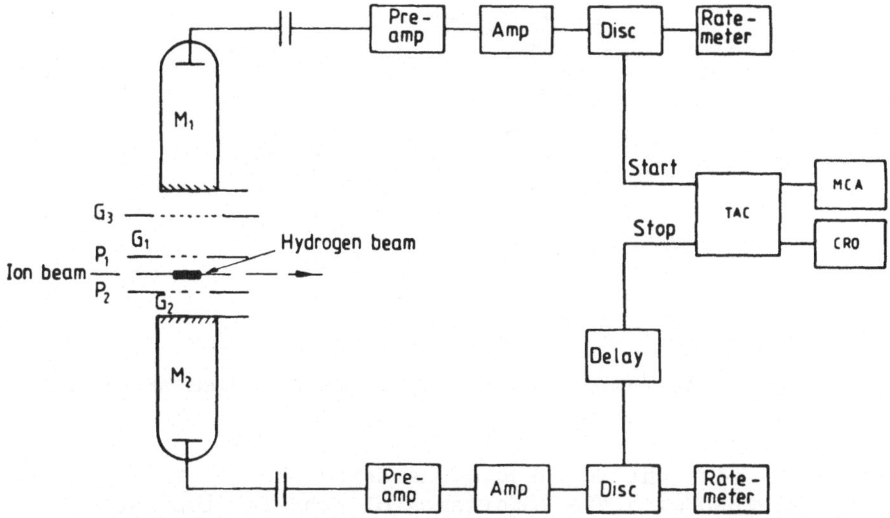

Fig. 2. Electrode arrangement and electronics for signal recovery from crossed-beam intersection region (Shah and Gilbody[27])

H^+ ions and electrons by means of the condensor plate method (see for instance ref. 27), but a problem is that slow ions and electrons are also formed in the background gas. Further, the atomic hydrogen beam, coming from a tungsten oven, contains a small fraction molecular hydrogen, which can also be ionized. Shah and Gilbody[27,28] identified the slow H^+ ions formed in the ionization process by applying a time of flight analysis and counting these ions in coincidence with the electrons from the same events. As shown in Fig. 2 an electric field was applied across the beam intersection region by two plane grids G_1 and G_2 of high transparency. Slow protons are formed both by electron capture and ionization (see Eqs.(1) and (2)) but due to the coincidence between electrons and protons, the processes are distinguished. The electric field pulls the protons to G_1 and the electrons to G_2. Then they enter particle multipliers M_1 and M_2 respectively and are counted in coincidence. By determination of the time of flight of the slow ion collected with respect to that of the electron, product ions can be selected with respect to their charge-to-mass ratio and H^+ target ions are easily distinguished from H_2^+ ions and ions from the background gas.

B''. Charge State Selection of the Projectile and Target Ion in Coincidence

Justiano et al.[30] and Groh et al.[31] investigated 100-1100 q eV Ar^{q+}-Ne ($q \leq 10$) and 10q keV Xe^{q+}-Xe ($q = 2$-12) collisions respectively by measuring the end products of projectile ions and target ions in coincidence. Justiano et al.[30] used a target chamber filled with gas and Groh et al.[31] used a thermal atomic beam target intersected

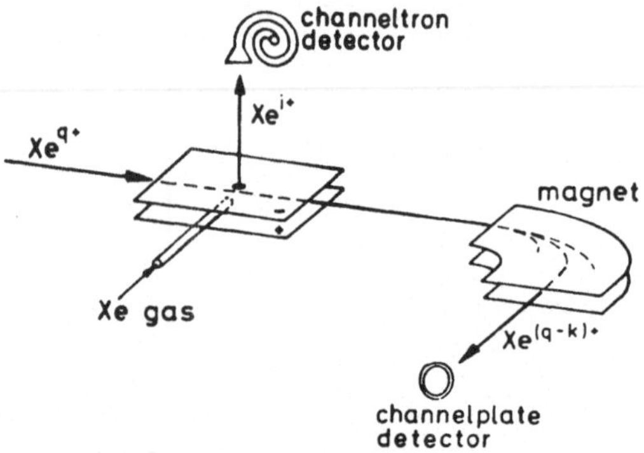

Fig. 3. Schematic diagram of the experimental arrangement of Groh et al. [31]

at right angles with the ion beam. Considering Eq.(5) these experiments are concerned with the importance of transfer ionization (electron production, $m > n$) in electron capture processes for which $n > 0$. By the coincidence measurement m and n are determined simultaneously. In Fig. 3 we show the schematic diagram of the set up of Groh et al.[31]. The target ions produced in the interaction region are extracted through a small hole in one of two opposite plates, post accelerated and detected with a channeltron. This detector gives a start pulse to a time-to-amplitude convertor. The fast ion beam is analyzed by a magnetic field on charge and incident on a channelplate detector, which provides the stop pulse for the time-to-amplitude convertor. The fast ions have the same flight time from the collision region whereas the flight time of the recoil ions is determined by their charge-to-mass ratio as in the time of flight experiments discussed before. When the primary beam current is not too large, pure ionization processes with $n = 0$ and $m > 0$ (see Eq.(5)) can also be investigated (see ref. 30).

C. Charge State, Energy and Angle Selection of Projectile and Target Ions

We have shown in I[9] that this type of experiment gives complete information about the kinematics of the collision, including the potential change, ΔE, involved in the collision. Thus one gets information whether the particles after the reaction have been formed into an excited state or the ground state, which is not the case in the methods described in the previous sections. The collision is governed by conservation of energy and momentum as indicated in I[9] by the three equations (12-14) with 6 unknown quantities, E_0, E_1, E_2, Θ, ϕ and ΔE. E_0 and E_1 are the projectile energies before and after the collision respectively, E_2 is the target ion energy after the collision and Θ and ϕ are the scattering angles of projectile and target respectively. If 3 quantities are measured, for instance E_0, E_1 and Θ, we get all information about the reaction. This has been done by Afrosimov et al.[32] as discussed in I[9]. They investigated the reaction

$$He^{++} + He \rightarrow He^{+}(n_1) + He^{+}(n_2) \qquad (6)$$

where n_1 and n_2 are the principal quantum numbers of the He^+ states. E_0 at $\Theta = 0$ and E_1 at various angles Θ were measured by means of a rotating electrostatic parallel-plate analyzer, with variation of E_0 between 2 and 60 keV. Two problems are encountered. First because of a good energy resolution needed to determine ΔE, it is not possible to use very high projectile energies. In some of their experiments Afrosimov and coworkers[37] reach $\Delta E/E$ values of about 5×10^{-5}. In order to overcome this difficulty, retardation technique has been applied for instance by Sato and Moore[34]. Secondly in many experiments atomic hydrogen is bombarded by heavy projectile ions and the scattering is limited to very small angles. In this case it is not possible to measure the relevant differential scattering cross section, but still useful information is obtained about the formation of excited states in the collision process. For more information about this kind of experiments we refer to the progress report of Panov[35] and to Huber[36,37].

For reasons of completeness we mention the introduction of the channelplate as a position sensitive detector in scattering experiments, also combined with coincidence technique, thus reducing the acquisition time for differential scattering data considerably (see refs. 38, 39 and 40).

D. Measurement of Photon Emission

Photons will be emitted when short living excited states are formed in the reaction, for instance

$$A^{q+} + B \rightarrow A^{(q-1)+*} + B^{+} \qquad (7)$$

$$A^{q+} + B \rightarrow A^{q+} + B^{*} \qquad (8)$$

where the excited states are indicated by an asterisk. Usually the photons are observed by monochromators which collect photons emitted in a direction perpendicular to or at the so called magic angle of $54^o44'$ with respect to the ion beam. These measurements provide σ_{exc}, the cross sections for formation of projectile or target ions and atoms into excited states. For more technical details on optical measurements see for instance F.J. de Heer[41,42] and references therein. Most measurements cover the wavelength range between about 10 and 600 nm. Recently, because relatively strong beams of completely stripped ions became available, shorter wavelengths have also been observed, i.e. in the soft X-ray region using filtering detectors[18], proportional counters[17] and crystal spectrometers[43]. Technical information on detection in this wavelength region can be found in references 44-47.

Two details are mentioned: First Matsumoto et al.[48] applied an optical attenuation method to distinguish between electron capture by ground state and metastable Ar^{2+} ions in Na vapour, observing photon emission from Ar^{+*} along the beam path as a function of position. The method has some similarity with that introduced by Gilbody[49] to measure and control excited state populations of beams in charge changing collisions and has also been applied by Brazuk and Winter[50]. Secondly to increase the speed of observation, multi-channel techniques are applied. In the vacuum ultraviolet region this is done by putting a channelplate detector on the Rowland circle at the position of the exit slit of the vacuum monochromator. In fact one uses the monochromator as a spectrograph, but instead of photographic recording photoelectric recording can be applied. Johnson et al.[51] and Kadota et al.[21,52] have performed measurements in this way. For details of operation of channelplates we refer to Wijnaendts et al.[38,39] and to Martin et al.[53]. Multichannel detection in the wavelength region above 200 nm is possible and monochromators equipped with diode-array detectors are advertised by Princeton Applied Research and by Tracor. Recently position sensitive photomultipliers are manufactured by Galileo Electro-Optics Corporation and by Surface Sciences Laboratories in the USA. The development of these detectors is described by Rees et al.[54]. Previously image intensifiers have also been used for multichannel detection[55]. In the (soft) X-ray region position sensitive proportional counters have been developped[56,57,58].

D'. Measurement of Electron Emission

We have already described the measurement of the cross section for electron production by means of the condensor plates method under A, but here we consider experiments in which the energy spectrum of the electrons at a certain angle of emission with respect to the projectile beam is determined. The measurements give information about the role of excited atomic and quasi-molecular states in the transfer ionization and pure ionization processes. Mostly the electron spectrum is determined by means of an electrostatic analyzer with sufficient energy resolution and in the case of collisions between multiply charged ions and multi-electron targets it consists of a series of autoionization or Auger-electron peaks, sometimes superposed on a continuum. Let us illustrate the origin of the autoionization peaks formed in the following processes:

$$A^{q+} + B \rightarrow A^{(q-2)+**} + B^{2+} \quad (9)$$

$$A^{q+} + B \rightarrow A^{q+} + B^{**} \quad (10)$$

In Eq.(9) two electrons from B are captured into a doubly excited state of $A^{(q-2)+**}$, which autoionizes to $A^{(q-1)+} + e$ and thus leads to an electron of a specific energy determined by the autoionization level. Eq.(9) corresponds to Eq.(5) with n = 2 and m = 2. Using the charge analyzing method under section B", where the projectile and target ion end-products are measured in coincidence, one finds n = 1 and m = 2 and the origin of the reaction is not identified. Similarly in Eq.(10) the target is doubly excited and autoionizes to $B^+ + e$, thus transforming Eq.(10) with n = m = 0 to Eq.(5) with n = 0 and m = 1.

Mann et al.[43] have identified Auger-electrons from 1s 2s nℓ and 1s 2p nℓ configurations of Ne^{3+} formed in collisions between 1.4 MeV/ amu A^{12+} up to U^{40+} and Ne. Their electrostatic analyzer had a resolution of 1 eV. Morgenstern et al.[59] investigated the autoionization states formed by electron capture in collisions of Ne^{q+} ions (q = 1-3) and He, H_2 and Xe between 1 and 10 keV impact energy. A schematic drawing of their apparatus is given in Fig. 4. A good energy resolution of the spectra is obtained by detecting the electrons at $\Theta = 180^o$ with respect to the beam direction, so that Doppler broadening is minimized. Because in this example the energies of the (autoionization) electrons are much smaller than those of the Auger electrons of Mann et al.[43], much better energy resolution had to be used (~60 meV). The electrons result mainly from autoionizing Ne^{**} and Ne^{+**} states formed in single or double (see Eq.(9)) charge transfer.

E. Crossing and Merging Beams Techniques

In I[9] we have shown different set ups using this kind of techniques and we shall not repeat that discussion here. We showed a set up of Fite et al.[29] in which a proton beam is crossed by a neutral

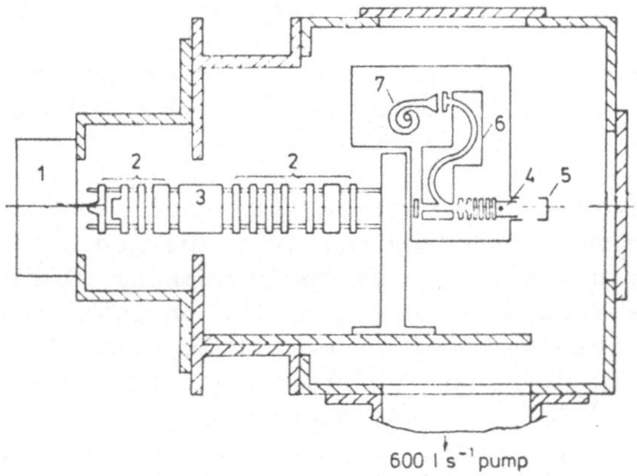

Fig. 4. Schematic drawing of the apparatus of Morgenstern et al.[59]: 1. ion source; 2. ion focusing optics; 3. Wien filter; 4. scattering centre; 5. Faraday cup; 6. double hemispherical electron energy analyser; 7. channel electron multiplier.

H beam and electron capture and ionization processes are investigated. Stebbings et al.[60] have used the same set up to observe Ly-α photons from excited states of H. The H beam comes from a hydrogen furnace. In section B' we discussed the time of flight-coincidence method applied by Shah and Gilbody[27,28] to measure the ionization process more accurately. In addition the furnace technique was improved by them to produce H beams of density $\sim 3 \times 10^{10}$ atoms/cm^3 in the interaction region. To increase the density of the hydrogen beam direct current[37], radio frequency[61] and microwave[62] discharges have been applied for its production and densities of $\sim 3 \times 10^{12}$ atoms/cm^3 have been reached, but mostly with a lower dissociation degree than in the furnace technique.

An experiment of crossed beams of highly excited states of H and of N^{3+} has been carried out by Kim and Meyer[63] to measure the electron loss cross section at 40 keV/amu collision energy according to

$$H(n) + N^{3+} \begin{array}{l} \nearrow H^+ + e + N^{3+} \\ \searrow H^+ + N^{2+} \end{array} \tag{11}$$

The value of the principal quantum number, n, of the excited atoms varied between 9 and 24. The N^{3+} particles come from one of the two keV accelerators and are crossed at 90° by a keV H beam produced by passing keV protons through a neutralizer cell containing H_2O. The selection of the n quantum numbers for H(n) particles participating in the reaction has been done by application of appropriate ionizing electric fields in the interaction region. The reactions in Eq.(11) are detected by recording the protons formed with a charge analyzer. For more details see ref. 63. Some problems in the experiment are discussed by Olson[64]. Experiments of a similar kind, also using merging beams technique (see I[9]) have been carried out by Burniaux et al.[65], Koch and Bayfield [66]and Bayfield et al.[67]. Much of the work with H beams is reviewed by Gilbody[68].

Next to atomic hydrogen beams a series of experiments has been started with Li beams crossed by multiply charged ions, measuring electron capture cross sections by the condensor plates method and the excited states formed by detecting the photons emitted (see refs. 20,21,52). The Li beams are coming out of an oven. In general with crossed beams, it is difficult to obtain accurate absolute cross sections, because the density of the atomic beam has to be known and besides in optical experiments the density profile is needed[20,21,52]. For the density determination often use is made of cross sections which have been measured previously with a Li target[20,21,52]. In Fig. 5 we show the apparatus of Kadota et al.[52]. They observe the LiI resonance line at 670.8 nm produced by impact of 500 eV electrons from a gun positioned at the axis of the ion beam normally used. Cross section data of Leep and Gallagher[69] are used to determine the

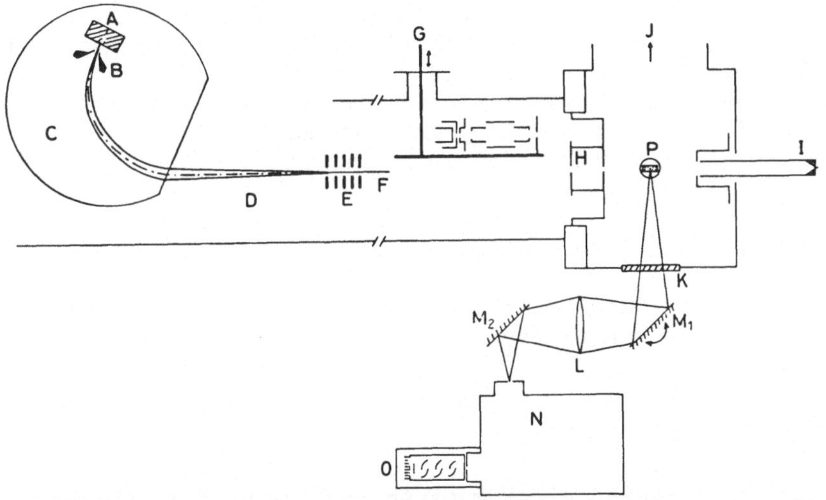

Fig. 5. Schematic diagram of the apparatus of Kadota et al.[52]

density of the Li beam. The density profile is obtained by means of a turnable mirror and lenses as shown in Fig. 5, where a one to one image is made of the intersecting region to the entrance slit of the monochromator. In the set up of Winter et al.[20] the interaction region is scanned by a plate with a rectangular hole, which can be moved in the direction of the beam and interrupt the radiation from that region at any position.

In order to investigate reactions between two charged particles crossed beams techniques are unavoidable. For this topic we refer to a progress report of Gilbody[70] and a comment by Dolder[71]. Reactions of importance are of the following type

$$H^+ + X^{n+} \rightarrow H + X^{(n+1)+} \qquad \text{(electron capture)} \qquad (12)$$

$$H^+ + X^{n+} \rightarrow H^+ + X^{(n+1)+} + e \qquad \text{(ionization)} \qquad (13)$$

3. ION SOURCES FOR MULTIPLY CHARGED IONS

In I[9] we have made some short notes about ion sources with high and low arc discharge currents for multiply charged ions. Further we have remarked in the introduction of this paper (section 1) that succesfull progress has been made in the development and application of ion sources for completely stripped ions in the acceleration region of about 0.5 - 20 kV. Recoil-ion and laser-beam techniques have been mentioned to obtain multiply charged ions (including fully stripped) in the eV region. Progress with small type ion sources has been made by Huber and coworkers in the low energy keV region, obtaining beams of ions with charges up to about $q = 8$ in the case of a miniature type EBIS source and a Redhead source[73,74] with relatively low energy spread. Okuno et al.[75,76] applied the so-called injected-ion drift-tube technique to determine the cross sections for symmetric single and multiple electron charge transfer from 0.04 eV up to 20 eV for Kr^{2+}-Kr and Xe^{2+}-Xe systems. A recent review about this topic of ion sources is given by Winter[77].

4. ELECTRON CAPTURE INVOLVING POTENTIAL CURVES CROSSING

Here we repeat some of the considerations given in section 4.3 of I[9]. Let us take a non-resonance ($\Delta E \neq 0$) electron capture reaction of the kind

$$A^{2+} + B \rightarrow A^{+} + B^{+} \quad (\Delta E \neq 0) \tag{14}$$

and the potential-curve diagram given in Fig. 6. Here ΔE represents the internal energy difference of the reactants and products of the reaction at infinite separation. This diagram shows that for multiply charged ions the energy difference of the potential curves varies strongly as a function of nuclear distance, mainly as a consequence of the Coulomb repulsion of the reaction products A^+ and B^+. As a consequence a so-called crossing occurs at a distance R_c approximated by

$$R_c = (q-1)/\Delta E \text{ in a.u.} \tag{15}$$

where q is the charge of the projectile, i.e. 2. In this equation we disregard polarization effects between the collision partners. If the molecular states corresponding with $A^{2+} + B$ and $A^+ + B^+$ have the same symmetry, the curves representing eigenvalues of the electronic Hamiltonian as a function of R cannot cross each other. The dashed curves correspond to these "adiabatic" states of the molecule and these curves are followed in a slow collision. However, when the heavy particles have a moderate or large velocity, a jump is possible from one adiabatic state to another. The system then follows the solid curves which correspond to "diabatic" states of the molecule. At low velocities, $v < 1$ a.u., the molecular aspect of the collision is important and when the relevant molecular levels have the same symmetry, one can apply the theory of Landau-Zener[78-80]. In analogy

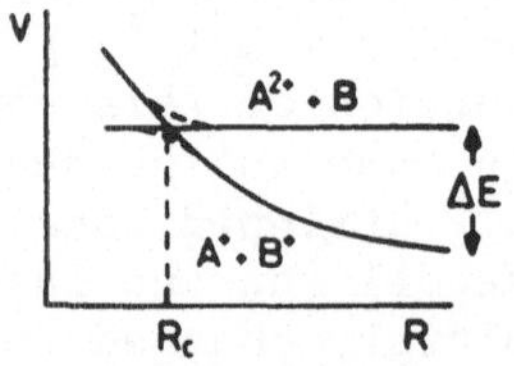

Fig. 6. Potential diagram related to reaction (14). Dashed curves correspond to adiabatic states, solid curves to diabatic states in the molecule.

with the adiabatic criterion of Massey[81,82] discussed in I[9], a maximum in the cross section is obtained at v_{max} given by

$$\frac{a'\Delta E(R_c)}{h\,v_{max}} = 1$$

where (16)

$$a' = \frac{\Delta E(R_c)}{\frac{d}{dR}(V_1 - V_2)_{R=R_c}} \qquad \text{and} \qquad \Delta E(R_c) = 2H_{12}$$

H_{12} is the matrix element for coupling of diabatic states. The denominator in (16) is equal to the difference of the slopes of the diabatic potential curves at $R = R_c$. On the basis of geometry we can understand that the maximum cross section according to the Landau-Zener theory is of the order of $\frac{1}{2}\pi R_c^2$. As an example for empirical application of the Landau-Zener theory we consider the measurements of Huber [36] on single electron capture in Ar^{2+}-Ne collisions below 2 keV using the method of determination of E_0, E_1 and Θ as discussed in section 2C. By integrating over Θ a total electron capture cross section can be obtained for the different exit channels. It is found that the dominant reaction, being exothermic, has a ΔE value of +6.1 eV and the potential diagram similar as in Fig. 6 leads to $R_c = 2.7$ Å. In Figs. 7 and 8 the experimental data are given and Huber[36] has fitted a Landau-Zener curve to it by taking the relevant R_c value from above and $H_{12} = 0.5$ eV. In the maximum the cross section is about 10^{-15} cm^2, which is roughly half the value πR_c^2. It is known that the Landau-Zener theory has severe limitations, but it is very useful for explanation of the physical phenomena.

In general when we work with highly charged ions more than one crossing plays a role in the reaction. This is demonstrated in Fig. 9 for the reaction

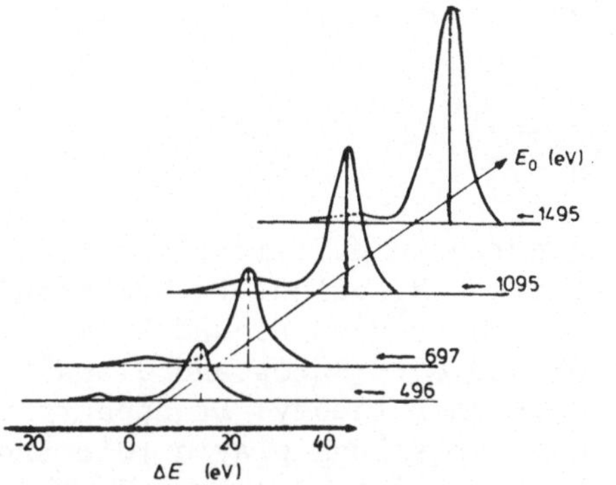

Fig. 7. Kinetic energy spectra of Ar^+ ions formed in Ar^{2+}-Ne collisions in dependence on the collision energy ($\Theta = 0^o$) (Huber[36])

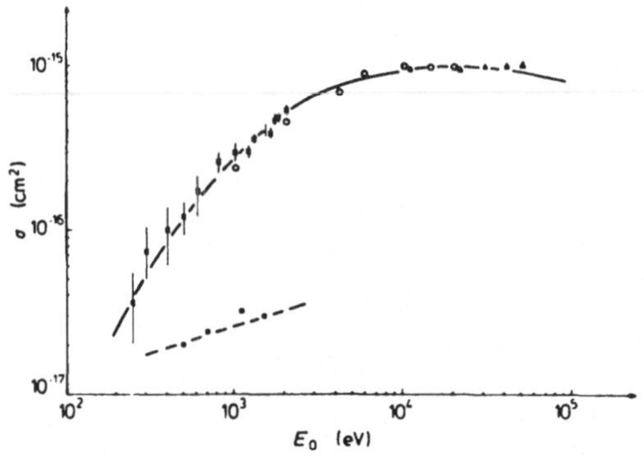

Fig. 8. Cross sections for single charge exchange in Ar^{2+}-Ne collisions. ×, exothermic reaction channel with $\Delta E = +6.1$ eV; ■, sum over the endothermic reactions. ○, ▲ total cross sections from Salzborn[83] and Fedorenko[84]. The full curve corresponds to a Landau-Zener fit (Huber[36]).

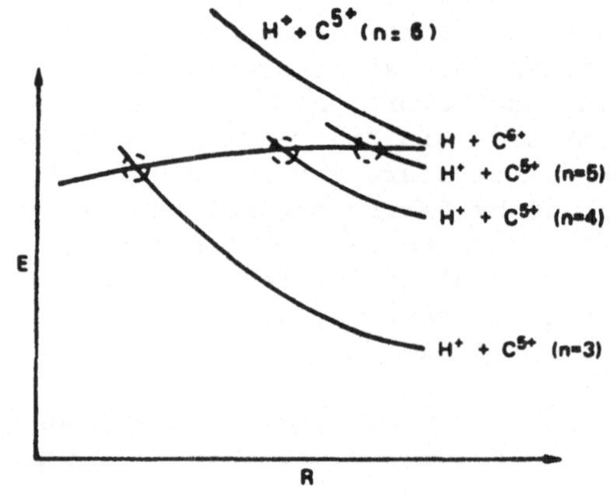

Fig. 9. Diagram of diabatic potential curves for reactions between C^{6+} and H.

$$C^{6+} + H \rightarrow C^{5+}(n) + H^{+} \qquad (17)$$

The electron can be captured in different excited states of C^{5+}, each corresponding to a different R_c value. It has been found experimentally that for many cases the region around $3 \lesssim R_c \lesssim 10$ au is the most important (see refs. 85 and 86). Salop and Olson[87] introduced a kind of multi-crossing Landau-Zener theory. It appears that for higher ion charges more and more crossings play a role and lead to a flattening of the curve of σ_c vs. impact energy E. This is illustrated in Fig. 10 and has also been verified by experiment (see section 4.5 and Fig. 21 of I[9]). So far we have only considered coupling between

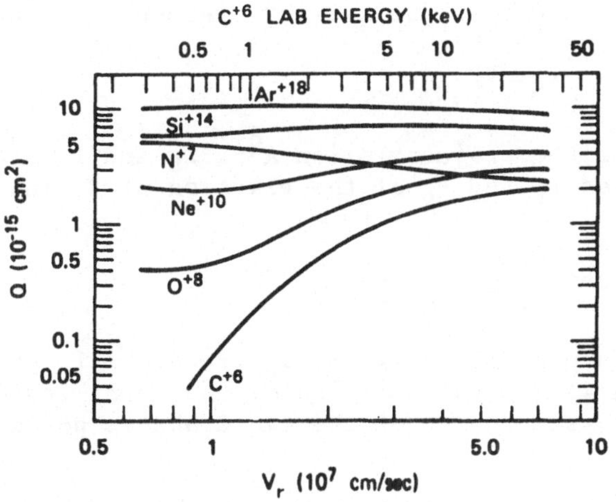

Fig. 10. Total Landau-Zener cross sections for electron transfer between H(1s) and various stripped ions (Salop and Olson[87])

molecular states of the same symmetry or radial coupling of adiabatic states. It has also been shown that rotational coupling, which connects states of different symmetry, has to be considered (see for instance Russek[88]).

5. CLASSICAL MODEL FOR ELECTRON CAPTURE

Ryufuku et al.[89] introduced a classical model which may be considered as an extension of the theory of Bohr and Lindhard[90]. The same treatment is given by Mann et al.[43]. Because of its importance for scaling and explanation of experimental data we shall discuss this model. In Fig. 11 we show a schematic situation for the potential of an electron initially on target atom B (H in Fig. 11), which is transferred to the projectile ion A^{q+} ($q = Z = 4$ in Fig. 11). Charge transfer from the target to the projectile ion occurs if the electron has sufficient energy to overcome the potential barrier. The potential at internuclear position r ($r = x$ in Fig. 11) is written as

$$V(r) = -\frac{1}{R} - \frac{q}{R-r} \qquad (18)$$

where r is the distance of the electron from B^+ and R the internuclear distance. The maximum value of V(r) for $0 < r < R$ is given by

$$V_{max} = -\frac{1}{r_m} - \frac{q}{R-r_m} = -(q^{\frac{1}{2}}+1)^2/R \qquad (19)$$

which corresponds to the height of the potential barrier. The electron will be transferred if the quasi-resonance condition is fulfilled and the electron passes the barrier, respectively according to

$$I_B^* = I_p^* \quad \text{and} \quad I_B^* \geq V_{max} \tag{20}$$

where p stands for the projectile ion A^{q+}. I_B^* and I_p^* are the decreased binding energies I_B and I_p of the electron at B^+ and A^{q+} respectively:

$$I_B^* = I_B - \frac{q}{R} \quad \text{and} \quad I_p^* = I_p - \frac{1}{R} \tag{21}$$

According to Ryufuku et al.[89] I_B^* and I_p^* are diabatic potentials (see Fig. 6 in section 4) and the solution of the quasi-resonance condition gives the crossing point of the two diabatic potential curves:

$$R_c = \frac{q-1}{I_B - I_p} = \frac{q-1}{\Delta E} \tag{22}$$

equivalent to Eq.(15). The electron can be captured into different states with principal quantum number n at the projectile, corresponding to different R_c values indicated by R_n:

$$R_n = \frac{q-1}{q^2/2n^2 - |I_B|} \tag{23}$$

Because of the second equation in (20) we have an extra limitation for R_n:

$$R_n \leq (2q^{\frac{1}{2}} + 1)/|I_p| \tag{24}$$

From the last two equations we get the conditions which the integer n has to fulfil, dependent on I_B and q:

$$n \leq n_p \equiv q\left[2|I_B|\left(1 + \frac{q-1}{2q^{\frac{1}{2}}+1}\right)\right]^{-\frac{1}{2}} \tag{25}$$

where I_B is given in atomic units. The most probable n corresponds to the largest possible R_n and the classical cross section is

$$\sigma_{cl} = A\pi R_n^2 \tag{26}$$

where A is assumed to be ½ (see refs. 92 and 43) as the transfer probability of one electron at the crossing point of the relevant potential curves. For projectile ions with many electrons A is taken 1, because of the many curve crossings in the region around R_n. This result of the classical model has its equivalence in the electron capture cross section predicted by the Landau-Zener theory introduced in section 4.

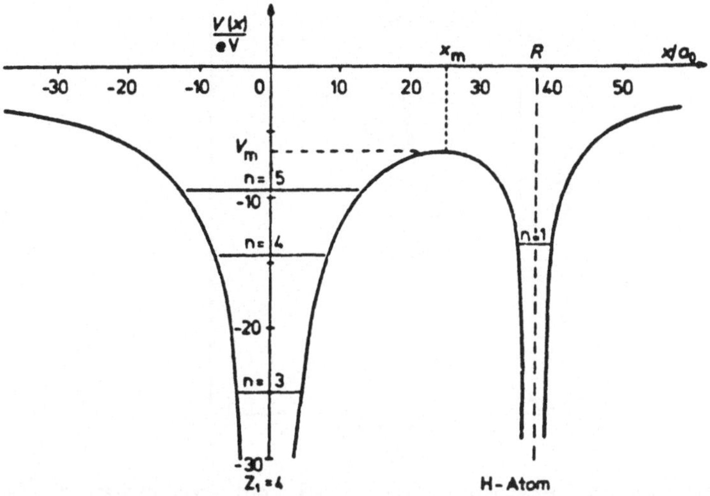

Fig. 11. Schematical potential diagram in a collision of a nucleus of charge $Z_1 = q = 4$ and a H-atom ($x \equiv r$ in Eq.(19))(Huber[91])

Equations (23) and (25) lead to strong variations in σ_{cl} of (26) as a function of q or I_B: For example when we increase q continuously in Eq.(25), n_p will also vary continuously, but n increases stepwise only taking integer values. A fixed value of n in (23) leads to a decrease of R_n when q increases in (23), but when n jumps by one an increase occurs in R_n. As a consequence σ_{cl} in (26) oscillates as a function of q as was first illustrated by Ryufuku et al.[89] (and later by Mann et al.[43]) who showed that their unitary distorted wave approximation (UDWA) leads to similar results as the classical model. In Fig. 12 Ryufuku et al.[89] show that the oscillations in their UDWA theory are consistent with experimental data on electron capture from a H target obtained by the projectile charge analyzing method. For the partially stripped ions, indicated by Z^{q+}, they took the effective charge as the average value of $Z_n = nI_p^{\frac{1}{2}}$ over orbitals occupied dominantly by the captured electron. Further experimental agreement with the classical model has been found by Mann et al.[43] (see table 5 of their article). They used the recoil ion technique to produce eV Ne^{10+}, Ne^{8+} and N^{5+} ions and investigated the n-state selective electron capture in different target gases by observing the emission of X-rays and Auger electrons (see sections 2D and 2D'). Bliman and coworkers[14,96,97] found the oscillations for electron capture, investigated by the projectile-charge analysis, in the 2q-10q keV energy region for $Ar^{q+}(2 \le q \le 12)$, $C^{q+}(2 \le q \le 6)$, $N^{q+}(2 \le q \le 7)$ and $O^{q+}(2 \le q \le 8)$ incident on Ar and D_2. Ryufuku et al.[89] show that these oscillations are typically only present at low impact energies and no more at higher energies (see their Fig. 3 at 25 keV/amu).

Recently Kaneko et al.[16] have studied these oscillations experi-

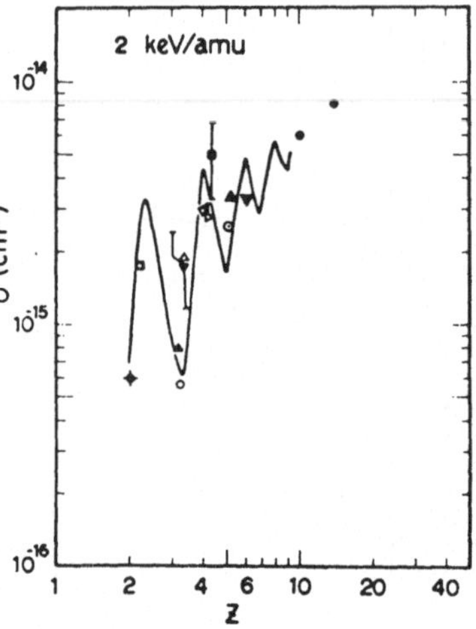

Fig. 12. Comparison between theoretical and experimental cross sections for charge transfer in collisions of completely and partially stripped ions having effective charge Z with atomic hydrogen at ion impact energy of 2 keV/amu. —— and ● denote the UDWA results. Experimental data: ▲(B^{3+}), ▼ (C^{3+}), and ■(O^{4+}) Bayfield et al.[93]; □(B^{2+}), ○(B^{3+}), Δ(N^{3+}), ∇(B^{4+}), ◁ (C^{4+}), ▷(N^{4+}), ⊙(N^{5+}), ▲(O^{5+}), and ∇(O^{6+}) Crandall et al.[25]; and ◆(He^{2+}) Nutt et al.[95] (Ryufuku et al.[89])

mentally as a function of (effective) charge for fully stripped up to lithium-like ions into helium target gas in the energy range of 0.5 - 0.4 q keV. The data come close together near a single curve when they are plotted as a function of the effective core charge of the projectile ion. For more details about the equations used for effective screening in the projectile and the target one should consider ref. 16.

Kim et al.[90] and Meyer et al.[99] have observed oscillations in the electron capture as a function of q for ions of Ta, W and Au colliding with H_2 and H at energies of 25-102 keV/amu. These oscillations have a different origin as discussed before and have been attributed to interference between the scattering amplitudes arising from the long range Coulomb and short-range (screened Coulomb) forces acting on the transferred electron during the collision.

6. SYMMETRIC RESONANCE MULTIPLE CHARGE TRANSFER

Relatively little attention has been paid to symmetric resonance multiple charge transfer processes, indicated by

$$A^{q+} + A \rightarrow A + A^{q+} \quad (q > 1) \qquad (27)$$

We mentioned already the work of Okuno et al.[75,76] at eV energies in section 3. More work has been done by Kaneko et al.[100] for 2-49 keV Ne^{q+} and Ar^{q+} ions (q = 1-4) in their own gas. Kaneko et al.[100] refer to most of the previous important work in this field. The experiment has been carried out by detecting the fast neutrals formed after the projectile particles passed through a gas cell, deflecting the charged particles out of the beam. For detection of the neutrals, secondary emission of electrons in an electron multiplier was applied. Kaneko et al.[100] show that in the interpretation of the results the effect of curve crossing is very important for sufficiently large q, for instance in Ar^{3+} on Ar. For more details see their article.

7. TRANSFER IONIZATION

In I[9] and ref. 41 we discussed results on electron production due to transfer ionization. This process is characterized by Eq.(5) with $n > 0$ and $(m-n) > 0$, but can have its origin due to autoionization of the reaction products according to Eqs.(9) and (10). It can only occur with multi-electron targets and not with atomic hydrogen. The potential-curves crossing is again important. Theoretically the description of the direct transfer ionization process is complicated, when the electron is formed via a coupling of a discrete quasi-molecular state in the entrance channel and a continuum state with a free electron in the exit channel. Transfer ionization is considered in the review of Niehaus[101] and before by Kishinevskii and Parilis[102]

Winter et al.[103] have determined the total cross section, σ_e, for electron production by the condensor plates method of section 2A for Ne^{q+} (q = 1-4) on He, Ne and Ar, and for Ar^{q+} (q = 1-8) on He, Ne, Ar, Kr and Xe at impact energies around $v \sim 0.5$ au. Their results showed that σ_e values become very large for sufficiently high q values, when the relevant transfer ionization processes become exothermic.

Recently more important work about transfer ionization has been performed by Justiano et al.[30] and Groh et al.[31] (see section 2B") applying coincidence in charge state selection of both the projectile- and target-ion. In this way they were able to distinguish between transfer ionization (Eq.(5) with $n > 0$, $(m-n) > 0$) and direct ionization ($n = 0$, $m > o$), but the experiments did not give information about the role of the excited states in the collision process or its origin as discussed in section 2D' (see Eqs.(9) and (10)).

In Fig. 13 we show the normalized charge state distributions for recoil ions from two-electron capture in 10q keV Xe^{q+}-Xe collisions as measured by Groh et al.[31]. For increasing q, target ions with larger m values are formed in the transfer ionization process. Autoionizing states of the projectile formed after transfer of two electrons or capture of inner-shell electrons from the target atom with

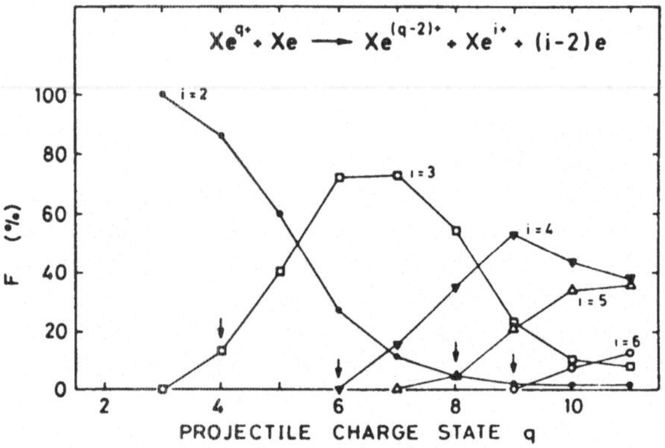

Fig. 13. Normalized charge-state distributions: fractions F_i of recoil ions in charge states i created in double-electron-capture collisions of Xe^{q+} ions with Xe atoms. The arrows indicate the lowest projectile charge state q for which the reaction creating a recoil ion in charge state i (i = 3,4, 5,6) is exothermic (Groh et al.[31])

subsequent Auger ionization can become quite important when q is so large, that these processes are exothermic.

Woerlee et al.[104], as discussed in I[9] ref. 41, measured the energy spectrum of electrons and showed that for 100 keV Ne^{q+} on Ar the peaks in their spectrum were due to

$$Ne^{q+} + Ar \rightarrow Ne^{(q-2)+**} + Ar^{2+} \quad (q = 3,4) \tag{28}$$

where the doubly excited states of $Ne^{(q-2)+**}$ autoionize (see Eq.(9)) and lead to an electron of specific energy determined by the autoionization level. It appeared that the double electron capture process was important when potential curve crossings of the relevant entrance and exit channels occurred at $3 \lesssim R_c \lesssim 10$ au (see Winter et al. [85]). Similar results were obtained by Ogurtsov et al.[105] for 4-50 keV He^{2+}, Ne^{2+} and Ne^{3+} on Ar and Xe.

Morgenstern et al.[50] recently measured the energy spectrum of the electrons from Ne^{q+} (q = 1-3) on He, H_2 and Xe at 1-10 keV impact energy with 60 meV energy resolution (see also section 2D'). They could identify many autoionizing Ne^{**} and Ne^{+**} states, which, similarly as in the experiment of Woerlee et al.[104], were formed by single or double charge transfer.

Many of the processes measured by Mann et al.[43] (see also section 2D') leading to the emission of Auger electrons can be classified as transfer ionization, both in the primary collision at MeV energy and in the secondary collision of the highly stripped recoil-ion at eV energy. In many collisions one or more innershell target electrons are captured followed by Auger-electron ejection from the excited target ion.

In I[9] and ref. 41 we also described an experiment of Niehaus and Ruf [106] for He^{2+} on Hg at velocities as low as 0.1 au and we indicated that in their case electrons are formed by autoionization of the quasimolecule formed during the collision.

8. COLLISIONS WITH H AND H_2 AND SCALING LAWS

A series of experiments with atomic and molecular hydrogen as a target has been described in I[9] and in a review of Gilbody [68]. Berkner et al.[107] give a biliography of the measurements up to 1981. It is not possible for us to summarize all the work that has been published after that time.

Let us next consider electron capture data from charge analysis experiments starting with H as a target. In general the more recent data do not change the general picture of electron capture which we illustrated in I[9]. There in Fig. 19 we showed the data of Gardner et al.[108] who bombarded atomic hydrogen with Fe^{q+} (q = 4-13) of energy 27-290 keV/amu. In their velocity range

$$\sigma_{q,q-1} \sim q^{\alpha(v)}$$

where $\alpha(v)$ increases from 1.5 to 3 with increasing velocity. More recently in their studies about scaling laws Janev and Hvelplund[6] (see also Knudsen[109]) have found similarly that

$$\sigma_{q,q-1} = \sigma_o(\tilde{v})q^{\alpha(\tilde{v})} \tag{29}$$

where $\tilde{v}$ is a reduced velocity equal to $v/q^{1/4}$. The result of their work is given in Fig. 14 in the velocity region from $v \simeq 0.1$ to 100 au. Data for He as a target are included as well. For $0.1 \lesssim v \lesssim 1$, α is weakly dependent on $\tilde{v}$ and has a value close to 1, which is in agreement with the classical theory introduced in section 5. At very high velocities, $\tilde{v} \gg 1$, α becomes equal to 5 and in a restricted region of intermediate values of $\tilde{v} \simeq 2-4$, $\alpha = 3$ just as in the Brinkman-Kramers[110] approximation.

In Fig. 21 of I[9] we illustrated for C^{q+} and B^{q+} on H that at low velocities (0.3×10^8 cm/s < v < 1×10^8 cm/s) $\sigma_{q,q-1}$ is little dependent on v for sufficiently large q ($\gtrsim 4$). As we explained in section 4 this flat behaviour is caused by the occurrence of many potential-curve crossings for large q.

In Fig. 22 of I[9] we showed that $\sigma_{q,q-1}$ is little dependent on the structure of the projectile for q $\gtrsim 4$, for O^{q+}, N^{q+}, C^{q+} and B^{q+} colliding with atomic hydrogen. Experimental data, similarly as in

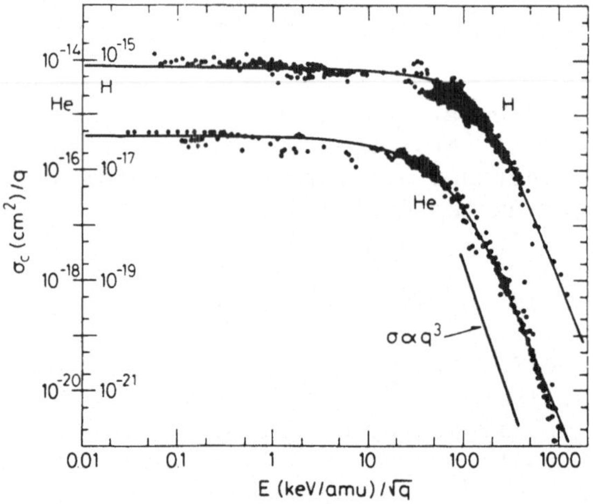

Fig. 14. Reduced electron-capture cross sections $\tilde{\sigma}_c = \sigma_c/q$ vs. reduced energy $\tilde{E} = E(\text{keV/amu})/\sqrt{q}$ for H and He targets. The solid lines are fits to the experimental data (only data with $q \geq 5$). The solid line marked $\sigma \propto q^3$ indicates the functional dependence of reduced cross section versus reduced energy for a cross section which is proportional to q^3 at fixed energy (Janev and Hvelplund[6]).

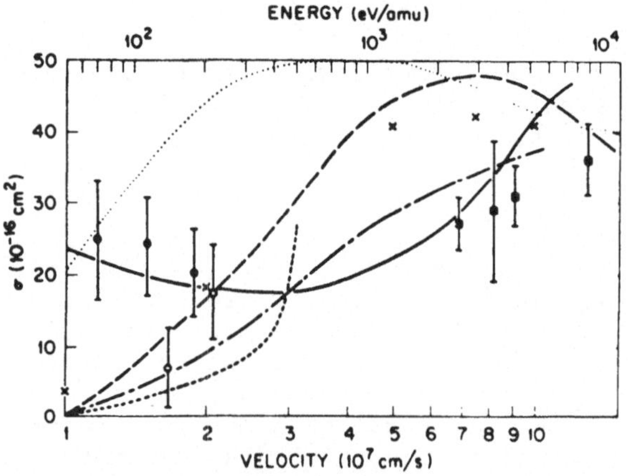

Fig. 15. Electron-capture cross sections for C^{5+} + H (solid points and solid curve) and C^{6+} + H (open points, crosses and broken curves). Circles are present experimental results, and squares are data of Crandall et al.[25]. The solid curve is the 5-PSS calculation of Shipsey et al.[111], the long-dashed curve is the 6-PSS of Salop and Olson[112], the dash-dot curve is the 11-PSS of Vaaben and Briggs[113], the short-dashed curve is the 3-PSS of Bottcher and Heil[114], the dotted curve is the distorted-wave calculation of Ryufuku and Watanabe[115] and the crosses are the 8-PSS and 10-PSS of Green et al.[116]

Fig. 21 were taken from Crandall et al.[25] and Bayfield et al.[93, 94]

In section 3 we showed that Phaneuf[13] introduced a technique to measure electron capture cross sections for C^{6+} ($1 \leq q \leq 6$) colliding on H at very low energies (11-387 eV/amu). Data for C^{6+} and C^{5+} are given in Fig. 15 and compared with different theoretical calculations. The velocity is so small, that the cross section is dependent on the velocity, different from the behaviour in Fig. 21 of I[9] discussed before. More results have been obtained by Phaneuf and coworkers and have been submitted to the Physical Review.

We mention recent work of Seim et al.[117] for Li^{q+}, N^{q+} and Ne^{q+} on H ($q = 2$-5) in the energy region of 3q to 12q keV.

Data for collisions between multiply charged ions and highly excited states of hydrogen are also needed in fusion research. In section 2E we introduced the crossed beams experiment of Kim and Meyer[63] on electron loss (sum of electron capture and ionization) from H(n) (n = 9-24) in collisions with N^{3+} at 40 keV/amu. As found before by other groups (see refs. 63-65) these cross sections are very large. Kim and Meyer[63] find that the cross section for energy loss for H(n) is equal to $\sigma_n = 1.19 \times 10^{-15}\ n^{3.12}$ cm^2 whereas different theories (including CTMC) predict a proportionality of σ_n with n^2. Olson[64] has pointed out later on that the interpretation of the experiment is complicated, because some of the H(n) levels may be excited to an H(n') level in the collision with N^{3+}. This gives rise to extra Stark ionization of highly excited hydrogen atoms (see section 2E) in the set up of Kim and Meyer[63] and apparent electron loss cross sections which are larger than the true ones.

For H_2 as a target many electron capture measurements have been performed as well (see for instance the review of Gilbody[68], the bibliography of Berkner et al.[107], Wirkner-Bott et al.[118] and Schrey and Huber[72]. Bliman and coworkers[14,97] succeeded to include fully stripped ions of oxygen, nitrogen and carbon on D_2 in their experiments at 2q-10q keV impact energy (see also section 3). Similarly Afrosimov et al.[15] used C^{6+}, N^{7+}, O^{8+}, Ne^{10+} and Ar^{18+} colliding with H_2 (and He) at 0.5 - 8 keV/amu. Just as for H, the electron capture cross section in general is found to be little dependent on v at relatively low velocities. For Ar^{18+}-H_2 very large cross sections are obtained ($>10^{-14}$ cm^2).

Instead of going in further detail for H_2, just as in I[9] we want to compare the electron capture cross sections for H and H_2. In Fig. 23 of I we illustrated that

$$\sigma_{q,q-1}(H) > \sigma_{q,q-1}(H_2) \quad \text{at small v} \tag{30}$$

$$\sigma_{q,q-1}(H) < \tfrac{1}{2}\sigma_{q,q-1}(H_2) \quad \text{at large v} \tag{31}$$

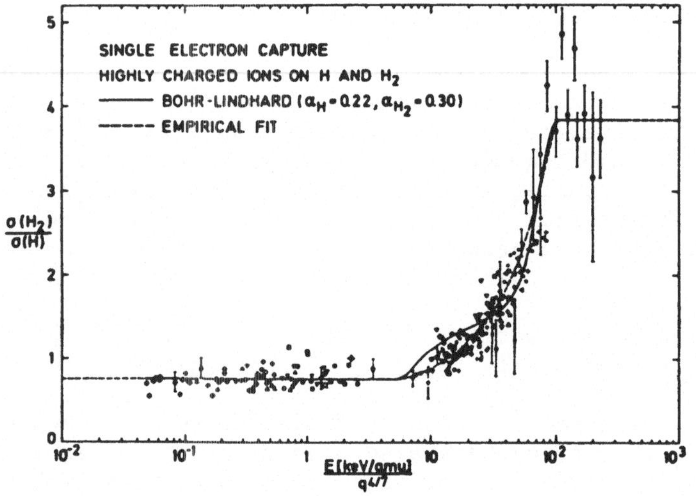

Fig. 16. Comparison of electron capture cross sections for H_2 and H targets (Knudsen[109]).

This is also shown by Knudsen[109] (see Fig. 16) who plotted the ratio $\sigma(H_2)/\sigma(H)$ versus $E(keV/amu)/q^{4/7}$ and compared it with the Bohr-Lindhard[90] theory. Olson and Salop[119] have explained that two factors lead to a smaller coupling matrix element in H_2 at low energies, its greater ionization potential relative to that of H and the Franck-Condon factors in H_2. Thus, electron capture cannot take place at as large internuclear separation for the molecular target as for the atomic target. At high energies a qualitative interpretation (see Hvelplund[120]) may be possible with the semi-classical theory of Thomas[121] (see also Vriens[122]). In that theory an atomic electron is captured by the passing nucleus as a result of two successive binary encounters. The atomic electron which is to be captured has to gain momentum in the first binary encounter of the incident nucleus with this electron. In H_2 we have two electrons and thus the probability for this binary encounter is two times larger than in H. Next in order to be captured this atomic electron has to undergo a second, elastic, binary encounter with the atomic nucleus in such a way that its final velocity is approximately parallel to the incident nucleus velocity. Because H_2 has two nuclei, this chance is again two times larger than in H. This altogether explains the factor 4 at high energies in Fig. 16. Recent experiments in which electron capture for H and H_2 has been compared are those of McCullough et al.[123] for Ba^{2+}, Ti^{2+}, Mg^{2+}, Zn^{2+}, Kr^{2+} and Br^{2+} ions between 0.8 and 40 keV and of Huber[37] for Ar^{q+} $(2 \leq q \leq 6)$ and Ne^{q+} $(2 \leq q \leq 4)$ below 15 keV impact energy.

Next we start with measurements on ionization of atomic and molecular hydrogen. Shah and Gilbody[27,28] (see also ref. 68) succeeded to perform high precision cross section measurements for ionization of H by applying crossed beams technique and identifying the target ions in the intersection region by means of a time of flight analysis with respect to the electrons from the same (ionization) event (see section 2B'). In Fig. 17 we show cross sections determined in this way for ionization of H(1s) by multiply charged ions of oxygen, nitrogen and carbon. Shah and Gilbody[28] indicate that at high velocities the cross sections in Fig. 17 can be described by a simple scaling relation

$$\sigma_{el} = \sigma_o q^n \quad (32)$$

where σ_o and n are constants for a given impact velocity. At 145 keV amu^{-1}, they obtain $n = 1.46 \pm 0.5$, which is close to the values found by Berkner et al.[24] for ionization of molecular hydrogen by very fast Fe^{q+} (see also I[9] section 4.5, Fig. 20), namely $n = 1.43$ at 1100 keV amu^{-1} and $n = 1.40$ at 277 keV amu^{-1}. Berkner et al.[24] used the condensor plates method (see section 2A). In Fig. 17 comparison is made with theoretical data of Olson and Salop[124] using the classical trajectory Monte Carlo (CTMC) method. We see that with increasing charge the maximum in the ionization cross section shifts to higher im-

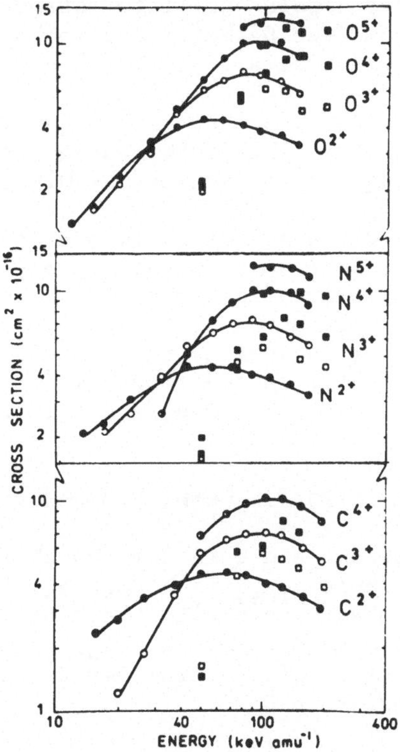

Fig. 17. Cross sections for ionization of H by C^{q+}, N^{q+} and O^{q+} ions ($q = 2-5$). Circular points refer to the experiment of Shah and Gilbody[28], square points to the theory of Olson and Salop[124].

pact energies. This is due to the fact that when the charge of the projectile increases, the probability of electron capture with respect to ionization increases because of the larger recombination (binding) energy and the electron can easier adjust itself to the translational motion of the projectile. Recently Shah and Gilbody[125] measured ionization of H by Li^{q+} ions ($q = 1$-3). They found that n varied from 1.24 - 1.42 for 57-387 keV amu^{-1} impact energy.

In I[9] we compared the ionization cross sections for H and H_2. From the experimental data of Berkner et al.[24] for Fe^{q+} on H_2, which were compared with theoretical CTMC σ_I values for Fe^{q+} on H, it appeared that in the intermediate velocity range

$$\sigma_I(H) \approx \tfrac{1}{2}\,\sigma_I(H_2) \tag{33}$$

similar as in the case of ionization by electrons. However, Eq.(33) will be no more valid at low impact velocities, where the molecular aspect in the collision is important and electron capture is dominating and at very high velocities where the Born approximation becomes valid and the ionization cross section becomes proportional to the integral of the weighted dipole oscillator strength of the target in the ionization continuum (see Bethe[126] and Inokuti[127]). The work on ionization of H_2 by Berkner et al.[24], who also measured charge changing cross sections, was extended to 3400 keV/amu Fe^{q+} ($q = 3$-25) ions and C^{q+} ($q = 4$-6), Nb^{q+} ($q = 28$-36) and Pb^{q+} ($q = 52$-59) between 0.31 and 4.65 MeV/amu (ref. 128). This work can be seen as a continuation on studies about scaling laws as started by Olson et al.[129].

As discussed in section 4.6 of I Olson et al.[129] derived a scaling rule for electron loss from a hydrogen atom in collision with a multiply charged ion. Electron loss from H is determined by the sum of charge exchange and ionization cross sections. Olson et al.[129] have calculated σ_{loss} for the energy range of 50 to 5000 keV/amu (1.4 au $< v < 14$ au) and for q in the range of 1-50, using the classical trajectory Monte Carlo method. The result is represented approximately by

$$\sigma_{loss} = 4.69\times10^{-16}\times\{(32q/E)[1-\exp(-E/32q)]\}\,\mathrm{cm}^2 \tag{34}$$

where E is given in keV/amu and q is the ion charge state. We see that at low energies, where σ_{loss} is almost equal to $\sigma_{q,q-1}$, σ_{loss} is proportional to q in agreement with the classically model introduced in section 5. At high velocities σ_{loss} is almost equal to σ_I and becomes proportional to q^2. For further discussion see section 4.6 of I[9] and Janev and Hvelplund[6].

A few remarks will be made about the scaling of σ_I cross sections separately. This has been considered in great detail by Janev and Hvelplund[6] and by Knudsen[109]. In Fig. 18 we show reduced ionization cross sections versus reduced energy for hydrogen and helium targets compared with different theoretical results. According to Knudsen[109] the data for He scale accurately around and above the maximum in σ_I in agreement with the Bohr[130] theory. At lower E/q values the scaling breaks down. For H the available data support the scaling, but more data are needed for further confirmation. None of the theoretical data shows agreement with experiment in the (entire) considered energy region. This may be due to the fact that capture-to-continuum, measured as ionization, may give an extra contribution (see ref. 136).

So far we have discussed experiments on charge state analysis of the projectile and ionization measurements with the condensor plate method. In addition to that we are interested to the formation of excited states in the electron capture process. We have seen in sections 3C and 3D, that so called energy gain or loss (determination of the initial E_0 and final energy E_1 of the projectile-ion at small angles Θ) and optical measurements are very well suited for this purpose.

In the case of H as a target little work has been done so far, although in many groups preparations are made to do these energy state selective measurements, which have already been performed with stable target gases (H_2, He etc.). For atomic hydrogen only optical measurements have been carried out for proton impact (see the pioneers work of Stebbings et al.[60]) and for He^{2+} impact (see Shah and Gilbody[127] and Bayfield and Khayrallah[138]), observing $n = 2$ states of the projectile ion produced by electron capture. For more details consider the review of Gilbody[68].

In molecular hydrogen more work has been done. One of the most complete experiments has been carried out by Afrosimov et al.[139] who determined elementary reactions in He^{2+}-H_2 collisions between 1.2 and 100 keV. A time of flight coincidence measurement was performed of the target ions H_2^+ and H^+ with respect to the projectile ion formed (see section 2B"). But in addition, the E_0, E_1 and Θ values of the projectiles were determined (see section 2C) by means of a moveable parallel plate analyser, so that the electronic states of the particles after the collision could be evaluated. Some results are summarized in Fig. 19. At energies $E > 10$ keV the most important process is

$$He^{2+} + H_2 \rightarrow He^+ + H_2^+ \tag{35}$$

It has been found that 90-98% of the He^+ ions are formed in excited states with principal quantum number $n = 2$. At $E < 10$ keV the probability for this reaction falls off sharply at the cost of predomination of capture of an electron to the 1s state of He^+ accompanied by simultaneous dissociation of the H_2^+ ion:

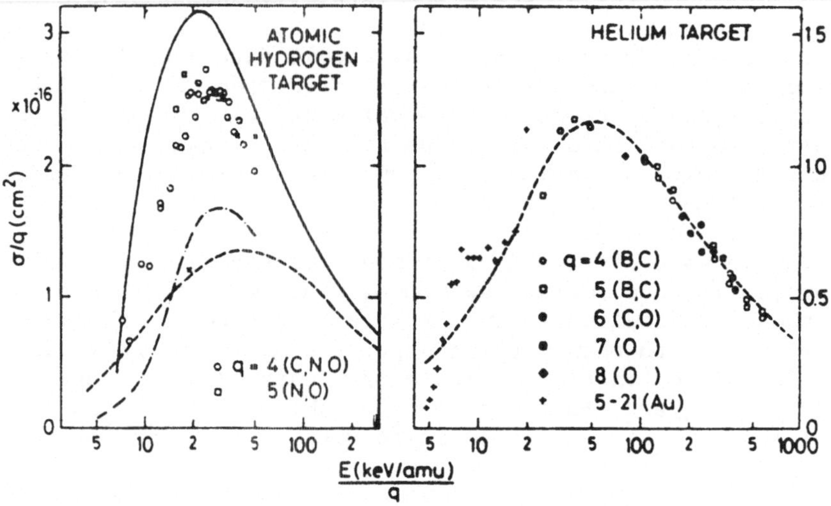

Fig. 18. Single-ionization cross sections. Points are experimental data. The theoretical results are: — Bohr[130] (q=4); --- close coupling[131,132] (all q); -·-·- UDWA[133,134] (q=5); × CTMC[124,135] (q=4).

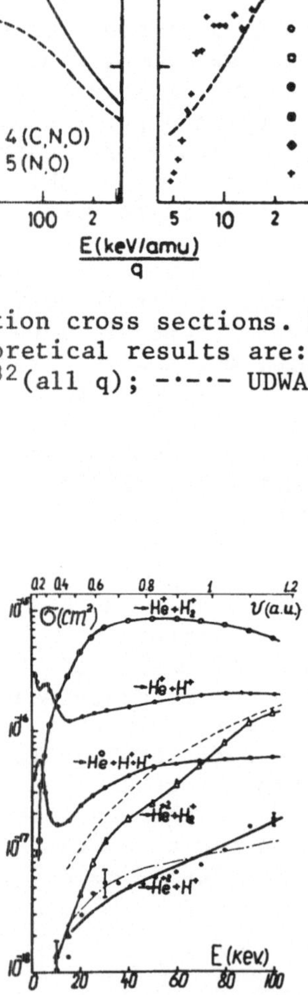

Fig. 19. Experimental cross sections for processes in He^+-H_2 collisions (Afrosimov et al.[139]). Dashed and dot-dashed curves represent similar cross sections in H^+-H_2 collisions at the same velocity from Afrosimov et al.[140].

$$He^{2+} + H_2 \rightarrow He^{+}(1s) \begin{cases} + H^{+} + H \\ + H^{+} + H^{+} + e^{-} \end{cases} \qquad (36)$$

Both ways given in Eq.(36) contribute about equally to this dissociation cross section. According to Afrosimov et al.[139], at low energies, autoionization of the $(HeH_2)^{++}$ quasi-molecule plays a role, leading to He^{+} ions in the ground state and to the released energy being used for ionization and dissociation of the H_2^{+} ion. Other processes are:

$$He^{2+} + H_2 \rightarrow He + 2H^{+} \qquad (37)$$

$$He^{2+} + H_2 \rightarrow He^{2+} + H_2^{+} + e^{-} \quad \text{direct ionization} \qquad (38)$$

$$He^{2+} + H_2 \rightarrow He^{2+} + H^{+} + \ldots \quad \text{ionization with dissociation} \qquad (39)$$

The processes (38) and (39) are endothermic and become important at high velocities. Comparison has been made with earlier data on H^{+}-H_2 (see ref. 140). For more detailed analysis of this experiment see ref. 139. The importance of dissociative charge transfer at relatively low energies has also been found in collisions with singly charged ions (see refs. 141 and 142).

The role of excited states in electron capture from H_2 has also been investigated in optical experiments (section 3D). Details of the results of such an experiment for instance have been shown by El-Sherbini et al.[143] for Ar^{6+} on H_2 at 100-1200 keV and are partly similar as in collisions with target atoms (see I[9] section 5). Hvelplund et al.[144] studied electron capture for Au^{q+} in H_2 (q = 12-18) at 20 MeV. Both experiments are concerned with the population of specific n,ℓ levels in the projectile, which aspect of the collision will be treated in section 10. Under the same category fall the experiments of Khayrallah and Bayfield[145] and Shah and Gilbody[146] for He^{2+} on H_2 at about 10-70 keV. The four groups mentioned before detected the radiation in the vuv, uv and visible spectral region. Important are the recent experiments of Bliman[17] and Afrosimov et al.[18] (see also section 2D) who bombarded H_2 with fully stripped ions and looked to radiation in the soft X-ray region. Afrosimov et al.[18] measured 2p-1s and the sum of nℓ-1s emissions from O^{7+*} and C^{5+*} particles formed after electron capture in collisions of O^{8+} and C^{6+} with H_2 (and He). Cross sections for Lyman-α and the total line emission are given in Fig. 20 together with the relative yield for Lyman-α emission. In these collisions with fully stripped ions in general relatively high n states of the projectile will be populated and some part of the Lyman-α emission is formed via cascading from higher levels in the projectile.

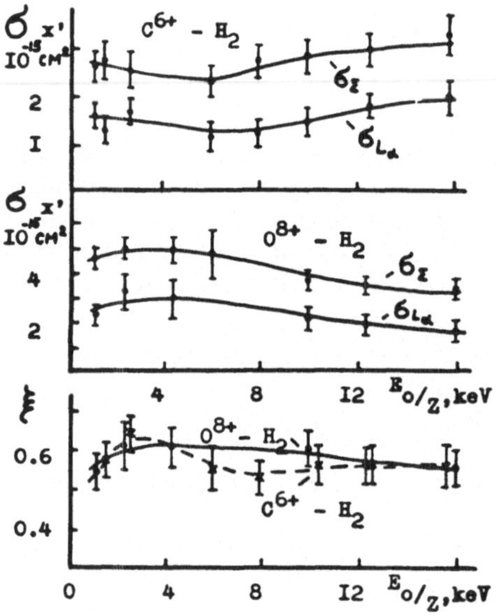

Fig. 20. Cross sections $\sigma_{L\alpha}$ for Lα-line emission and total cross sections σ_{Σ} for X-ray emission in C^{6+}, O^{8+}-H_2 collisions and the relative yield $\xi = \sigma_{L\alpha}/\sigma_{\Sigma}$ (Afrosimov et al.[18])

9. COLLISIONS WITH Li

The investigations with Li have become very important in connection with fusion in Tokamak-experiments for the diagnostics by means of Li-beam probing of α-particles and impurities of oxygen and carbon in the outer region of the magnetically confined hot plasma. These problems are discussed by Winter[147]. Winter has also reviewed the different experiments and we shall give a few considerations here. McCullough et al.[19] have measured one- and two-electron capture cross sections for He^{2+} and He^{+} between 6.7 and 800 keV passing though a collision cell filled with Li vapour. For He^{2+} both single and double electron capture cross sections were determined by means of analyzing the projectile charge. The neutral particles formed in the process had to be detected by means of secondary emission of electrons on a surface. Similar measurements were performed by Murray et al.[148]. Optical measurements in a crossed beams experiment have been carried out for He^{2+} on Li (15-150 keV) by Kadota et al.[21,52] observing the photons emitted from He^{+*} formed after one-electron capture. Kadota et al.[21,52] measured Lyman-α and Lyman-β radiation and showed that their "total capture" cross section, σ_T, given by

$$\sigma_T = \sigma_{em}(2p\rightarrow 1s) + \sum_{n\geq 3} \left(1 + \frac{A_{np,2s}}{A_{np,1s}}\right) \sigma_{em}(np\rightarrow 1s) \tag{40}$$

should be approximately equal to σ_{21} as measured by McCullough et al.[19] and Murray et al.[148]. The A's in this equation refer to transition probabilities. Except for capture into the 1s- and 2s-state, capture into all other states results in np→1s or np→2s emissions ($n \geq 3$) in the last step. The right hand side of Eq.(39) represents the sum of all these emissions. Direct electron capture into 2s- and 1s-states, which do not radiate, according to Bransden and Ermolaev [149] is expected to be unimportant and therefore $\sigma_T \approx \sigma_{21}$. Experimental σ_T and σ_{21} data are compared with each other in Fig. 21 and also with theory. We see that the agreement between the charge analysis data (σ_{21}) and the optical data (σ_T) is very good, both deviating from theory at higher velocities. The data of Kadota et al.[21,52] indicate that at relatively low energies the $n = 3$ level of He^+ is dominantly excited, as predicted by theory and that the $n = 3$-2 transition at 1640 Å is a good candidate for diagnostics. One of the reasons that Li-beam probing is attractive has been stated by Winter[147]. Because of its low ionization potential, some reactions which are exothermic for Li will be endothermic for atomic hydrogen and thus emission for diagnostics of impurities in the fusion reactor can be induced with a relative low background from the H containing plasma itself. Secondly, at low impact energies the electron capture cross section increases with decreasing target ionization potential (see Eq.(24) in section 5 and the empirical scaling law of Müller and Salzborn[152]).

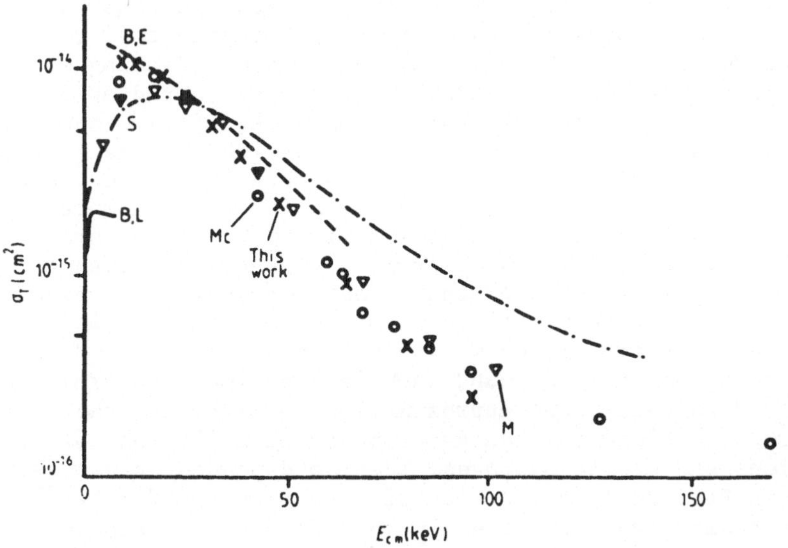

Fig. 21. Total one-electron capture cross sections for He^{2+}-Li: B, L, optical data of Barrett and Leventhal[150]; ×, optical data of Kadota et al.[21,52]; Mc o and M ∇, charge detection data of McCullough et al.[19] and of Murray et al.[148] respectively. Theoretical data: ---, B, E, Bransden and Ermolaev[149]; -·-·-, S, Shipsey et al.[151].

Other experiments with Li as a target are reported by Winter et al.[20,147] for Ne^{q+} ($q = 1,2$) and by Pan Guang Yan et al.[153] for C^{q+} and O^{q+} ($q = 1,2,3$) in the keV region. It has been shown by Winter[147] that some empirical rules govern the state selection in the electron capture process:

a) The relevant curve crossing distance R_n can be derived according to the classical theory introduced in section 5, in particular Eq.(24).

b) In the electron capture, the core remains unchanged. For instance in the case of C^{2+}, the projectile may enter the collision chamber in the ground $1s^2 2s^2\ {}^1S$ or the metastable $1s^2 2s2p\ {}^3P$ state. It appears that for not too low velocities the electron is captured without changing the core configuration. This had led to methods to determine the fraction of metastables in the primary ion beam (see Brazuk and Winter[50] and Matsumoto et al.[48] and section 3D).

c) Multiplet states for which the energy defect in the reaction is almost the same are populated according to statistical weight $(2J + 1)$.

d) In the case of systems with L-S coupling, the spin conservation rule of Wigner[154] holds for the collision process.

10. FORMATION OF EXCITED STATES

In the previous sections we have discussed the behaviour of total cross sections for electron capture, ionization and transfer ionization. We have also treated some experiments dealing with the formation of excited states, either giving rise to radiation or to emission of electrons. In particular we have summarized some recent work in H_2, H and Li. Some experiments on excited states have been summarized in section 5 of I[9] and in reference 36. A great deal of this work is part of the thesis of Bloemen[155], where a more extensive description is given. In this article we shall put emphasis on the behaviour of formation of excited projectile states as a function of principal and azimuthal quantum number. The recent reviews of Janev[156] on this topic are not yet incorporated in this paper, but should be certainly read. We start with the theory and then look how the predictions can be found back in experiment. Ryufuku and Watanabe[133] used their UDWA approximation to calculate the total and partial cross sections for charge transfer in collisions of fully stripped ions and atomic hydrogen. Their n dependent cross sections are given in Fig. 22 as a function of impact energy for Ne^{10+} and Si^{14+} with atomic hydrogen. The n number of the most probable state is denoted by n_m, m meaning maximum.

At low impact energies (<10 keV/amu), charge transfer occurs through the level crossing mechanism. The value n_m is determined from the effective outer level-crossing point and the distribution for $n > n_m$ becomes vanishingly small. The n_m value for this case is found by Ryufuku and Watanabe[133] to be

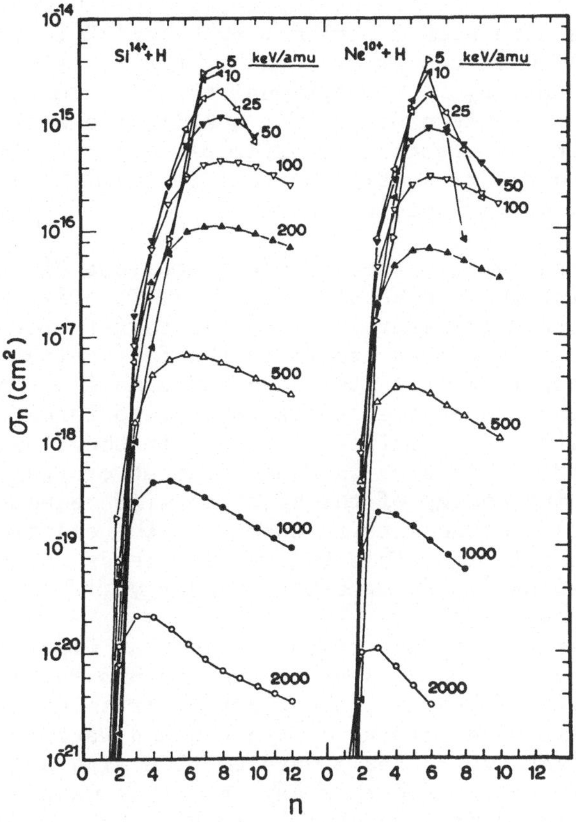

Fig. 22. Dependence of the partial cross sections, σ_n, for charge transfer in the $Ne^{10+}+H(1s)$ and $Si^{14+}+H(1s)$ collision systems on the principal quantum number, n, of the final state, for the collision energies of 5 to 2000 keV/amu (Ruyfuku and Watanabe[133])

$$n_m \le n_p \propto q^{0.768} \tag{41}$$

where n_p can take non integer values and n_m is the largest possible integer. Eq.(41) is approximately equivalent to Eq.(25) in section 5 where n_m has been derived by the classical method using the resonance condition for electron transfer or curve crossing for diabatic potentials.

Between about 10 and 100 keV/amu, the most probable principal quantum number remains unchanged, but the distribution broadens towards 100 keV/amu including n numbers larger than n_m. The broadening is due to the fact that at higher impact velocity or energy the transition is no more localized at curve crossings, qualitatively understandable by the adiabatic criterion of Massey[81,82] (see also

Dinterman and Delos[157]). In the energy region 50-100 keV/amu n,ℓ distributions have also been calculated by Olson[158] for fully stripped ions with q = 1-20 in atomic hydrogen by the classical CTMC method. It is found that n_m is proportional to $q^{3/4}$ similarly as in Eq.(41) by Ryufuku and Watanabe[133]. The qualitative explanation of Olson[157] is that in the electron transfer process the electron chooses an orbit on the projectile so that it tries to preserve its orbital energy and orbital radius.

With increasing energy, above 100 keV/amu (see Fig. 22), n_m decreases and the distribution becomes narrower. This is considered to be due to the momentum transfer, because the target electron in the transfer process has to adjust itself to the fast velocity of the projectile. For that reason it has to be bound more strongly for higher speeds of the projectile, corresponding to larger dominance of transfer to shells with smaller n number. In the classical theory of Bohr and Linhard[90] (see also Knudsen[109]) that orbital is dominating where the binding energy of the electron with respect to the projectile ion is equal to the kinetic energy of the electron in the frame of the projectile ion. This theory also leads to smaller n_m values with increasing impact velocity or energy.

In Fig. 23 the ℓ-dependent cross sections at impact energies of 1,25 and 500 keV/amu, as calculated by Ryufuku and Watanabe[133], are given. At low energy, 1 keV/amu, the electron transfer is localized at the level crossings. According to Ryufuku and Watanabe the transfer must go to a state which has a wavefunction with a large amplitude at the crossing point and they say that this wavefunction usually has $\ell < n-1$, as found for the example in Fig. 23. In the case of 25 keV/amu for $n = 3-6$ a maximum is present at $\ell < n-1$ and for $n \geq 7$ a maximum is found at $\ell = n-1$. So level-crossing still appears effective for low n states (smaller R_c values), but not for high n states (larger R_c values). At higher energies, as shown for 500 keV/amu, the linear momentum transfer to the electron becomes important and the angular momentum ℓ of the electron is suppressed.

Olson[158] has determined the ℓ-value distribution at 50 and 100 keV/amu collisions by the CTMC method. Olson concludes that for low n values the largest possible ℓ values dominate and for large n values there are maxima at $\ell_{max} = n_{max}$. His conclusion differs from that of Ryufuku and Watanabe[133] presented before. Koike has solved this discrepancy, by studying the data of Olson[158] more carefully and comparing them with recent theoretical data of Ryufuku[160] at 50 keV/amu. His analysis leads to the result in Fig. 24, that both Olson's and Ryufuku's calculations around 50 keV/amu lead to a value of

$$\ell_{max} + 1 \approx q^{\alpha} \qquad (42)$$

where α is equal to 0.9 and 1.0 respectively. The origin of this rule

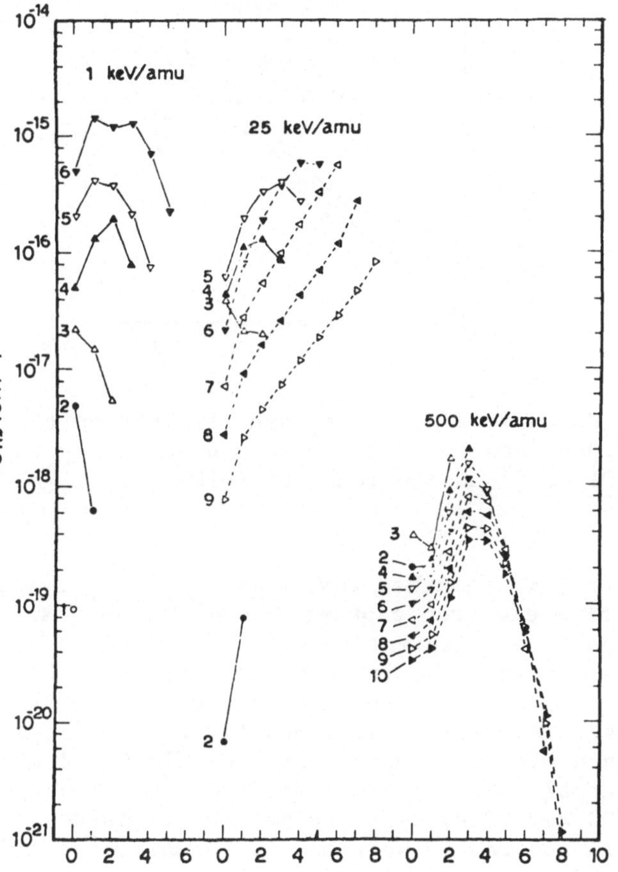

Fig. 23. Dependence of the partial cross sections $\sigma_{n\ell}$ for charge transfer in the $Ne^{10+} + H(1s)$ collision system on the angular quantum number ℓ of the final state, at collision energies of 1, 25 and 500 keV/amu, where the numbers indicated express the principal quantum number n (Ryufuku and Watanabe[133])

has still to be persued. In the case that the orbital considered has an n value smaller than q-1, ℓ_{max} cannot be larger than n-1, the maximum ℓ value following from the calculations of Ryufuku and Watanabe[133] illustrated before (see Fig. 23) when n is not too small ($\gtrsim 5$).

Next we want to see whether these theoretical results are in agreement with experiment. This comparison can only be qualitative, because almost no relevant experimental data have been obtained for completely stripped ions on atomic hydrogen (see section 8). Therefore we consider some experiments in other targets.

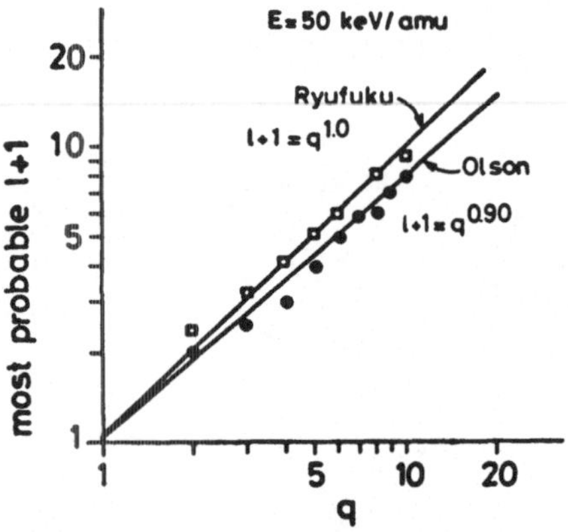

Fig. 24. Most probable $\ell+1$ in the case of electron capture of fully stripped ions on atomic hydrogen according to Ryufuku[160] and Olson[158] as concluded by Koike[159].

In section 5 we remarked that Mann et al.[43] investigated the n-state selective electron capture for Ne^{10+}, Ne^{8+} and N^{5+} in the case of eV-impact on different targets. The most probable n in electron capture appeared to be in agreement with the classical model in section 5 (see Eq.(25)) and we have seen that Eq.(41) of Ryufuku and Watanabe leads to similar predictions for n_m. In I[9] we illustrated the ℓ-dependence of some measurements. We discussed the experiment of Afrosimov et al.[161] (see also Panov[35]) for Ar^{6+}-He collisions between 6 and 100 keV. They used the method of charge, state and angle selection as introduced in section 2C. Formation of Ar^{5+*} occurred mainly in the 3d, 4s and 4p levels with R_c values between 3 and 10 au. In this region of relatively low impact energy per amu the population of levels appears to be determined by the level crossing mechanism as predicted by Ryufuku and Watanabe[133]. At higher energies (200-1200 keV) optical measurements in the vuv region have been carried out by El-Sherbini et al.[143] for Ar^{6+} in H_2. In Fig. 25 we show their data. The $n_{max} \approx q^{3/4}$ rule leads to $n_{max} \lesssim 3.8$. We see that $n=4$ and $n=5$ levels are dominantly excited and that in the higher energy region f and g levels ($\ell = 3,4$) become relatively important in qualitative agreement with Koike's[159] rule that for $n \leq q-1$ the maximum possible ℓ value is important. Hvelplund et al.[144] studied the electron capture for highly charged Au^{q+} ions ($q = 10-18$) at 20 MeV ($v \sim v_o$, 100 keV/amu) into molecular hydrogen and observed the capture to high-lying Rydberg states in the projectile by using a monochromator for the emitted photons in the wavelength region of about 200-600 nm. It appeared that the $\Delta n = 1$ transitions dominated, which means that the highest possible ℓ values are important in the electron capture process. Their results in Fig. 26 for Au^{13+} on H_2 show that the lines corresponding to transitions between two n values present structure, which indicates a non-degeneracy according to different ℓ values.

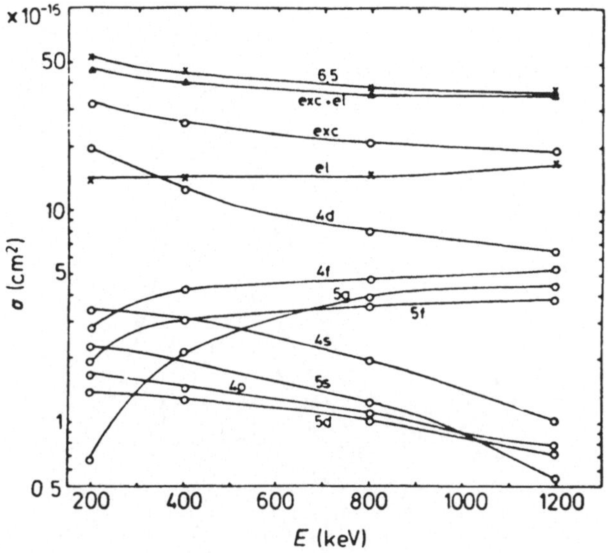

Fig. 25. Cross sections for capture into excited states ($n\ell$), for total capture into excited states (exc), for electron production (el) and for total single electron capture (6,5) plotted against the projectile energy in $Ar^{6+} + H_2$ collisions (El-Sherbini et al.[143])

The ℓ splitting is assumed to be caused by core polarization. The spectrum corresponds to $n = 11$ and $\Delta n = 1$. For Au^{13+} $\ell_{max} = 12$, larger than the n value of 11 considered in Fig. 26. So according to Koike [159] the dominating ℓ value for this case should be $n-1 = 10$, in agreement with the illustrated results.

11. CONCLUSION

We have summarized the experimental activities in the field of electron capture and ionization by multiply charged ions incident on different target gases, with the emphasis on H, H_2 and Li. Many measurements have been carried out on total electron capture and ionization cross sections without any information about the excited states formed in the collision process. Many of these results are fitted very well be empirical and theoretical scaling laws. A smaller amount of experiments has been carried out regarding the excited states, detecting photons and electrons or measuring the energy of the projectile ions before and after the collision. So far little of this work has been done with atomic hydrogen as a target. Efforts in this direction are under the way and results may be expected very soon.

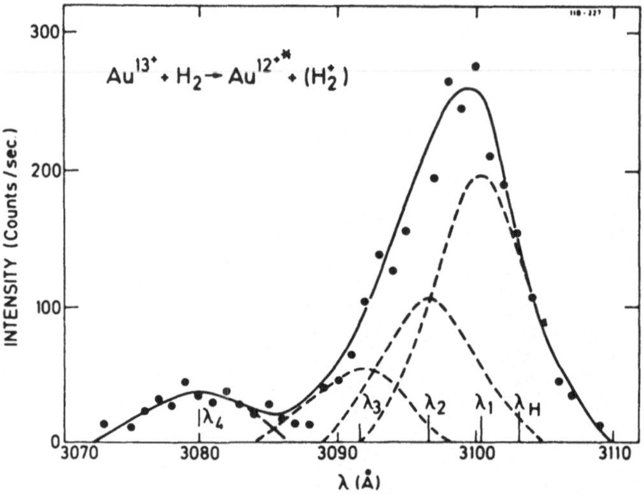

Fig. 26. The spectrum of the $n_i = 11$, $\Delta n = 1$ transition. The dashed curves represent a deconvolution of the spectrum by assuming the satellite line at 3080 Å to be due to a $\ell = 7$ to $\ell = 6$ transition and the splitting caused by induced dipole polarizability of the core. $_H$ is the wavelength found by neglecting core polarization, and $\lambda_1 \ldots \lambda_4$ are the wavelengths for the $\ell \rightarrow \ell - 1$ ($\ell = 10,9,8,7$) transitions, taking core polarization into account (Hvelplund et al.[144])

The author is indebted to Dr. F. Koike and Prof.dr. H. Winter for their critical remarks on the manuscript and to Mrs. A. Smit for typing it. This work is part of the research program of the Stichting voor Fundamenteel Onderzoek der Materie (Foundation for Fundamental Research on Matter) and was made possible by financial support from the Nederlandse Organisatie voor Zuiver-Wetenschappelijk Onderzoek (Netherlands Organization for the Advancement of Pure Research).

REFERENCES

1. *Symposium on electron capture by multiply charged ions, Proceedings ICPEAC XI, in Kyoto 1979, Invited Papers and Progress Reports,* eds. N. Oda and K. Takayanagi, p.387 (North Holland, Amsterdam, 1980).
2. *Symposium on collisions of multiply charged ions with atoms (dedicated to Jim McDonald)* and other articles in *Proceedings ICPAEC XII in Gatlinburg 1981, Invited Papers,* ed. S. Datz (North Holland, Amsterdam, 1982).
3. *Atomic and Molecular Data for Fusion* in Physica Scripta, *23 no.2* (1981), edited by H.W. Drawin and K. Katsonis.

4. H.B. Gilbody in *Advances in Atomic and Molecular Physics, Volume XV*, eds. D.R. Bates and B. Bederson, p.293 (Academic Press, New York, 1979).
5. A. *Dalgarno in Advances in Atomic and Molecular Physics, Volume XV*, eds. D.R. Bates and B. Bederson, p.37 (Academic Press, New York, 1979).
6. R.K. Janev and P. Hvelplund, *Comments in Atomic and Molecular Physics, XI*, 75 (1981).
7. A.V. Vinogradov and I.I. Sobel'man, Soviet Physics, JETP *36*, 1115 (1973).
8. R.W. Waynant and R.C. Elton, Proceedings IEEE *64*, 1059 (1976).
9. F.J. de Heer in *Atomic and Molecular Processes in Controlled Thermonuclear Fusion*, eds. M.R.C. McDowell and A.M. Ferendici, p.351 (Plenum Press, New York, 1979).
10. H.F. Beyer, K-H. Schartner, F. Folkmann and P.H. Mokler, J.Phys. B: Atom.Molec.Phys. *11*, L 363 (1978).
11. C.L. Cocke, T.J. Gray and E. Justiano, *Abstracts of ICPEAC XI in Kyoto 1979*, eds. K. Takayanagi and N. Oda, p.600 (North Holland, Amsterdam, 1979).
12. C.R. Vane, M.H. Prior and R. Marrus, Phys.Rev.Letters *46*, 107 (1981).
13. R.A. Phaneuf, *Abstracts of ICPEAC XII in Gatlinburg 1981*, ed. S. Datz, p.688 (North Holland, Amsterdam, 1981) and Phys.Rev. *A24*, 1138 (1981).
14. S. Bliman, J. Aubert, R. Geller, B. Jacquot and D. van Houtte, Phys.Rev. *A23*, 1703 (1981).
15. V.V. Afrosimov, A.A. Basalaev, E.D. Donets, K.O. Lozhkin and M.N. Panov, J.E.T.P. Letters *34*, 171 (1981).
16. Y. Kaneko, T. Iwai, S. Ohtani, K. Okuno, N. Kobayashi, S. Tsurubuchi, M. Kimura, H. Tawara and S. Takagi, *Abstracts of ICPEAC XII in Gatlinburg 1981*, ed. S. Datz, p.696 (North Holland, Amsterdam, 1981) and ref. 2 p.697.
17. S. Bliman, private communication.
18. V.V. Afrosimov, A.A. Basalaev, Yu.S. Gordeev, E.D. Donets, S.Yu. Ovchinnikov, M.N. Panov and A.N. Zinoviev, *Abstracts of ICPEAC XII in Gatlinburg 1981*, ed. S. Datz, p.692 (North Holland, Amsterdam, 1981).
19. R.W. McCullough, T.V. Goffe, M.B. Shah, M. Lennon and H.B. Gilbody, J.Phys.B: Atom.Molec.Phys. *15*, 111 (1982).
20. H. Winter and E. Rille and E. Rille, J. Weiser and H. Winter, *Abstracts of ICPEAC XII in Gatlinburg 1981*, ed. S. Datz, p.740 and 742 (North Holland, Amsterdam, 1982).
21. K. Kadota, D. Dijkkamp, R.L. van der Woude, Pan Guang Yan and F.J. de Heer, Phys.Letters *88A*, 135, 1982.
22. *Production and Physics of Highly Charged Ions in Stockholm 1982*, to be published in Physica Scripta, edited by L. Liljeby.
23. *Atomic Physics of Highly Ionized Atoms, Cargèse 1982*, edited by R. Marrus (to be published by Plenum Press, New York).
24. K.H. Berkner, W.G. Graham, R.V. Pyle, A.S. Schlachter, J.W. Stearns and R.E. Olson, J.Phys.B: Atom.Molec.Phys. *11*, 875 (1978).

25. D.H. Crandall, R.A. Phaneuf and F.W. Meyer, Phys.Rev. *A19*, 504 (1979).
26. P. Hvelplund, H.K. Haugen and H. Knudsen, Phys.Rev. *A22*, 1930 (1980).
27. M.B. Shah and H.B. Gilbody, J.Phys.B: Atom.Molec.Phys. *14*, 2361 (1981).
28. M.B. Shah and H.B. Gilbody, J.Phys.B: Atom.Molec.Phys. *14*, 2831 (1981).
29. W.L. Fite, R.F. Stebbings, D. Hummer and R.T. Brackmann, Phys. Rev. *119*, 663 (1960).
30. E. Justiano, C.L. Cocke, T.J. Gray, R.D. DuBois and C. Can, Phys.Rev. *A24*, 2953 (1981).
31. W. Groh, A. Müller, C. Achenbach, A. Schlachter and E. Salzborn, Physics Letters *85A*, 77 (1981).
32. V.V. Afrosimov, A.A. Basalaev, G.A. Leiko and M.N. Panov, Soviet Physics - JETP *47*, 837 (1979).
33. V.V. Afrosimov, Yu.S. Gordeev, A.M. Polyanskii and A.P. Shergin, Soviet Physics - Techn.Phys. *17*, 96 (1972).
34. Y. Sato and J.H. Moore, Phys.Rev. *A19*, 495 (1979).
35. M.N. Panov, ref.1, p.437.
36. B.A. Huber, J.Phys.B: Atom.Molec.Phys. *13*, 809 (1980).
37. B.A. Huber, Z.Phys.A - Atoms and Nuclei *299*, 307 (1981).
38. R.W. Wijnaendts van Resandt, H.C. den Harink and J. Los, J.Phys E: Sc.Instrum. *9*, 503 (1976).
39. R.W. Wijnaendts van Resandt and J. Los, *Proceedings ICPEAC XI in Kyoto 1979, Invited Papers and Progress Reports*, eds. N. Oda and K. Takayanagi, p.831 (North Holland, Amsterdam 1980).
40. J.A. Fayeton, J.C. Brenot and J.C. Houver, *Proceedings ICPEAC XII in Gatlinburg 1981, Invited Papers*, ed. S. Datz, p.525 (North Holland, Amsterdam 1982).
41. F.J. de Heer, ref.1, p.427.
42. F.J. de Heer in *Advances in Atomic and Molecular Physics, Volume II*, p.327-384, eds. D.R. Bates and I. Esterman (Academic Press, New York, 1966).
43. R. Mann, F. Folkmann and H.F. Beyer, J.Phys.B: Atom.Molec.Phys. *14*, 1161 (1981).
44. P. Bogen in *Plasma Diagnostics*, ed. W. Lochte-Holtgreven, p.424 (North Holland, Amsterdam, 1968).
45. M.G. Hobby, N.J. Peacock and J.E. Bateman, CLM-R203 (Culham Laboratory 1980).
46. J.A.R. Samson, *Techniques of Vacuum Ultraviolet Spectroscopy* (John Wiley and Sons, New York, 1967).
47. A.E. Sandström in *Handbuch der Physik, Volume XXX*, ed. S. Flügge, p.78 (Springer-Verlag, Berlin, 1957).
48. A. Matsumoto, H. Maezawa, S. Ohtani and T. Iwai, *Abstracts of XII ICPEAC in Gatlinburg*, ed. S. Datz, p.644 (North Holland, Amsterdam 1981) and A. Matsumoto, S. Ohtani and T. Iwai, J.Phys.B: Atom.Molec.Phys. *15*, 1871 (1982).
49. H.B. Gilbody, Inst.Phys.Conf.Ser. *no.38*, 156 (1978).

50. A. Brazuk and H. Winter, J.Phys.B: Atom.Molec.Phys. *15*, 2233 (1982).
51. B.M. Johnson, K.W. Jones, D. Gregory, T.H. Kruse and E. Träbert, Physics Letters *86A*, 285 (1981).
52. K. Kadota, D. Dijkkamp, R.L. van der Woude, A. de Boer, Pan Guang Yan and F.J. de Heer, FOM report nr. 52553 (1981) and in press in J.Phys.B: Atom.Molec.Phys.
53. C. Martin, P. Jelinsky, M. Lampton, R.F. Malina and H.O. Anger, Rev.Sci.Instr. *52*, 1067 (1981).
54. D. Rees, I. McWhirter, P.A. Rounce, F.E. Barlow and S.J. Kellock, J.Phys.E: Sci.Instrum. *13*, 763 (1980).
55. R. Iredale, C.W. Hinder and D.W.S. Smout in *Advances in Electronics and Electron Physics 28B* 965 (1969).
56. A. Gabriel, Rev.Sci.Instr. *48*, 1303 (1977).
57. A. Gabriel, F. Dauvergne and G. Rosenbaum, Nuclear Instr. and Meth. *152*, 191 (1978).
58. P. Platz, J. Ramette, E. Belin, C. Bonnelle and A. Gabriel, J.Phys.E: Sc.Instrum. *14*, 448 (1981) and Eur-CEA-FC-1057, Fontenay-aux-Roses (1980).
59. R. Morgenstern, A. Niehaus and G. Zimmerman, J.Phys.B: Atom. Molec.Phys. *13*, 4811 (1980).
60. R.F. Stebbings, W.L. Fite, D.G. Hummer and R.T. Brackmann, Phys. Rev. *119*, 1939 (1960).
61. J.P. Toennies, W. Welz and G. Wolf, J.Chem.Phys. *71*, 614 (1979).
62. H. Morgner and A. Niehaus, J.Phys.B: Atom.Molec.Phys. *12*, 1805 (1979).
63. H.J. Kim and F.W. Meyer, Phys.Rev.Letters *44*, 1047(1980).
64. R.E. Olson, Phys.Rev. *A23*, 3338, 1981.
65. M. Burniaux, F. Brouillard, A. Jognaux, T.R. Govers and S. Szucs, J.Phys.B: Atom.Molec.Phys. *10*, 2421 (1977).
66. P.M. Koch and J.E. Bayfield, Phys.Rev.Lett. *34*, 448 (1975).
67. J.E. Bayfield, G.A. Khayrallah and P.M. Koch, Phys.Rev. *A9*, 209 (1974).
68. H.B. Gilbody, Physica Scripta *24*, 712 (1981).
69. D. Leep and A. Gallagher, Phys.Rev. *A10*, 1082 (1974).
70. H.B. Gilbody in ref.2, p.223.
71. K. Dolder, *Comments on Atomic and Molecular Physics, XI*, 211 (1982).
72. H. Schrey and B.A. Huber, J.Phys.B: Atom.and Molec.Phys. *14*, 3197 (1981).
73. B. Huber, Z.Phys. *A275*, 95 (1975).
74. P.A. Redhead, Can.J.Phys. *45*, 1791 (1967).
75. K. Okuno, T. Koizumi and Y. Kaneko, Phys.Rev.Letters *40*, 1708 (1978).
76. K. Okuno, T. Koizumi and Y. Kaneko, *Abstracts of ICPEAC XI in Kyoto 1979*, eds. K. Takayanagi and N. Oda, p.594 (North Holland, Amsterdam 1979).
77. H. Winter in ref. 23.
78. L. Landau, Z.Phys.Sowjet *2*, 46 (1932).
79. C. Zener, Proc.Roy.Soc. *A137*, 696 (1932).

80. J.B. Hasted and A.Y.J. Chong, Proc.Phys.Soc. *80*, 441 (1962).
81. H.S.W. Massey, Rep.Progr.Phys. *12*, 248 (1949).
82. J.B. Hasted in *Advances in Electronics and Electron Physics, Volume XIII,* ed. L. Martin, p.1-78 (Academic Press, New York, 1960).
83. E. Salzborn, Habilitationsschrift, Giessen (1976).
84. N.V. Fedorenko, Sov.Phys. - Tech.Phys. *24*, 769 (1954).
85. H. Winter, E. Bloemen and F.J. de Heer, J.Phys.B: Atom.Molec. Phys. *10*, L453 and L599 (1977).
86. H.J. Zwally and D.W. Koopman, Phys.Rev. *A2*, 1851 (1970).
87. A. Salop and R.E. Olson, Phys.Rev. *A13*, 1312 (1976).
88. A. Russek, Phys.Rev. *A4*, 1918 (1971).
89. H. Ryufuku, K. Sasaki and T. Watanabe, Phys.Rev. *A21*, 745 (1980).
90. N. Bohr and J. Linhard, K.Dan.Vidensk.Selsk.Mat. - Fys.Medd. *28*, 1 (1954).
91. B.A. Huber, Habilitationsschrift, Bochum (1981).
92. R.A. Mapleton, *Theory of Charge Exchange* (Wiley, New York, 1972).
93. J.E. Bayfield, P.M. Koch, L.D. Gardner, I.A. Sellin, D.J. Pegg, R.S. Peterson and D.H. Crandall, *Contributed Papers of ICAP5 in Berkeley,* 1976, eds. M.H. Prior and H.A. Shugart, p.126 (University of California, Berkeley, 1976)
94. L.D. Gardner, Ph.D. Thesis (Yale University, 1978).
95. W.L. Nutt, R.W. McCullough, K. Brady, M.B. Shah and H.B. Gilbody, J.Phys.B: Atom.Molec.Phys. *11*, 1457 (1978).
96. S. Bliman, J. Aubert, R. Geller, B. Jacquot and D. van Houtte, Phys.Rev. *A22*, 2403 (1980).
97. S. Bliman, S. Dousson, R. Geller, B. Jacquot and D. van Houtte, J.Physique *42*, 399 (1981).
98. H.J. Kim, P. Hvelplund, F.W. Meyer, R.A. Phaneuf, P.H. Stelton and C. Bottcher, Phys.Rev.Lett. *40*, 1635 (1978).
99. F.M. Meyer, R.A. Phaneuf, H.J. Kim, P. Hvelplund and P.H. Stelton, Phys.Rev. *A19*, 515 (1979).
100. Y. Kaneko, T. Iwai, S. Ohtani, K. Okuno, N. Kobayashi, S. Tsurubuchi, M. Kimura and H. Tawara, J.Phys.B: Atom.Molec.Phys. *14*, 881 (1981).
101. A. Niehaus, *Comments on Atomic and Molecular Physics IX,* 153 (1980).
102. L.M. Kishinevskii and E.S. Parilis, Soviet Phys. JETP *28*, 1020 (1969).
103. H. Winter, Th.M. El-Sherbini, E. Bloemen, F.J. de Heer and A. Salop, Physics Letters *68A*, 211, 1978.
104. P.H. Woerlee, Th.M. El-Sherbini, F.J. de Heer and F.W. Saris, J.Phys.B: Atom.Molec.Phys. *12*, L235 (1979).
105. G.N. Ogurtsov, V.M. Mikoushkin and I.P. Flaks, *Abstracts of ICPEAC XI in Kyoto 1979*, eds. K. Takayanagi and N. Oda, p.650 (North Holland, Amsterdam 1979).
106. A. Niehaus and M.W. Ruf, J.Phys.B: Atom.Molec.Phys. *9*, 1401 (1976).

107. K.H. Berkner, W.G. Graham, R.V. Pyle, A.S. Schlachter and J.W. Stearns, Phys.Rev. *A23*, 2891 (1981).
108. L.D. Gardner, J.E. Bayfield, P.M. Koch, H.J. Kim and P.H. Stelson, Phys.Rev. *A16*, 1415 (1977).
109. H. Knudsen in ref. 2, p.657.
110. H.C. Brinkman and H.A. Kramers, Proc.Acad.Sci.Amsterdam, *33*, 973 (1930).
111. E.J. Shipsey, J.C. Browne and R.E. Olson, J.Phys.B: Atom.Molec. Phys. *14*, 869 (1981).
112. A. Salop and R.E. Olson, Phys.Rev. *A16*, 1811 (1977).
113. J. Vaaben and J.S. Briggs, J.Phys.B: Atom.Molec.Phys. *10*, L521 (1977).
114. C. Bottcher and T. Heil, private communication.
115. H. Ryufuku and T. Watanabe, Phys.Rev. *A19*, 1538 (1979).
116. T.A. Green, E.J. Shipsey and J.C. Browne, Phys.Rev. *A23*, 546 (1981).
117. W. Seim, A. Müller, I. Wirkner-Bott and E. Salzborn, J.Phys.B: Atom.Molec.Phys. *14*, 3475 (1981).
118. I. Wirkner-Bott, W. Seim, A. Müller, P. Kester and E. Salzborn, J.Phys.B: Atom.Mol.Phys. *14*, 3987 (1981).
119. R.E. Olson and A. Salop, Phys.Rev. *A14*, 579 (1976).
120. P. Hvelplund, private communication.
121. L.H. Thomas, Proc.Roy.Soc. *A114*, 561 (1927).
122. L. Vriens in *Case Studies in Atomic Collisions I*, eds. E.W. Mc Daniel and M.R.C. McDowell, p.335 (North Holland, Amsterdam, 1969).
123. R.W. McCullough, W.L. Nutt and H.B. Gilbody, J.Phys.B: Atom. Molec.Phys. *12*, 4159 (1979).
124. R.E. Olson and A. Salop, Phys.Rev. *A16*, 531 (1979).
125. M.B. Shah and H.B. Gilbody, J.Phys.B: Atom.Molec.Phys. *15*, 413 (1982).
126. H.A. Bethe, Ann.Physik *5*, 325 (1930).
127. M. Inokuti, Rev.Mod.Phys. *43*, 297 (1971).
128. A. Schlachter, K.H. Berkner, W.G. Graham, R.V. Pyle, J.W. Stearns and J.A. Tanis, Phys.Rev. *A24*, 1110 (1981).
129. R.E. Olson, K.H. Berkner, W.R. Graham, R.V. Pyle, A.S. Schlachter and J.W. Stearns, Phys.Rev.Lett. *41*, 163 (1978).
130. N. Bohr, K.Dan.Vidensk.Selsk.Mat.Fys.Medd., *18*, no.8 (1948).
131. R.K. Janev and L.P. Presnyakov, J.Phys.B: Atom.Molec.Phys. *13*, 4233 (1980) and Physics Reports *70*, 1 (1981).
132. R.K. Janev, Phys.Letters *83A*, 5 (1981).
133. H. Ryufuku and T. Watanabe, Phys.Rev. *A20*, 1828 (1979).
134. H. Ryufuku and T. Watanabe, JAERI-memo 9454 and H. Ryufuku, Phys.Rev. *A25*, 720, 1982.
135. A.S. Schlachter, K.H. Berkner, W.G. Graham, R.V. Pyle, P.J. Schneider, K.R. Stalder, J.W. Stearns, J.A. Tanis and R.E. Olson, Phys.Rev. *A23*, 2331 (1981).
136. R. Shakehaft, Phys.Rev. *A18*, 1930 (1978).
137. M.B. Shah and H.B. Gilbody, J.Phys.B: Atom.Molec.Phys. *11*, 121 (1978).

138. J.E. Bayfield and G.A. Khayrallah, Phys.Rev. *A12*, 869 (1975).
139. V.V. Afrosimov, G.A. Leiko and M.N. Panov, Soviet Phys.Techn. Phys. *25*, 313 (1980) and *Abstracts of ICPEAC X in Paris*, eds. M. Barat and J. Reinhardt, p.856 (North Holland, Amsterdam, 1977).
140. V.V. Afrosimov, G.A. Leiko, Yu.A. Mamaev and M.N. Panov, Soviet Physics JETP *29*, 648 (1969).
141. W.L. Nutt, R.W. McCullough and H.B. Gilbody, J.Phys.B: Atom. Molec.Phys. *12*, L157 (1979).
142. R. Browning, C.J. Latimer and H.B. Gilbody, J.Phys.B: Atom. Molec.Phys. *2*, 534 (1969).
143. Th.M. El-Sherbini, A. Salop, E. Bloemen and F.J. de Heer, J. Phys.B: Atom.Molec.Phys. *13*, 1433 (1980).
144. P. Hvelplund, H.K. Haugen, H. Knudsen, L. Andersen, H. Damsgaard and F. Fukusawa, Physica Scripta *24*, 40 (1981).
145. G.A. Khayrallah and J.E. Bayfield, Phys.Rev. *A11*, 930 (1975).
146. M.B. Shah and H.B. Gilbody, J.Phys.B: Atom.Molec.Phys. *9*, 1933 (1976).
147. H. Winter, Comments on Atomic and Molecular Physics, in press (1982) and in ref.22.
148. G.A. Murray, J. Stone, M. Mayo and T.J. Morgan, Phys.Rev. *A25*, 1805 (1982).
149. B.H. Bransden and A.M. Ermolaev, Phys.Letters *84A*, 316 (1981).
150. J.L. Barrett and J.J. Leventhal, Appl.Phys.Lett. *36*, 869 (1980) and Phys.Rev. *A23*, 485 (1981).
151. E.J. Shipsey, L.T. Redmon, J.C. Browne and R.E. Olson, Phys.Rev. *A18*, 1961 (1978).
152. A. Müller and E. Salzborn, Phys.Letters *62A*, 391 (1977) and Inst.Phys.Conf.Ser. *38*, 169 (1978).
153. Pan Guang Yan, R. van der Woude, D. Dijkkamp and F.J. de Heer, in ref.22.
154. E. Wigner, Nachr.Gess.Wiss. p.375, Göttingen (1927).
155. E.W.P. Bloemen, Ph.D. Thesis (Leyden, 1980), see also E.W.P. Bloemen, D. Dijkkamp and F.J. de Heer, J.Phys.B: Atom. and Molec.Phys. *15*, 1391, 1982.
156. R.K. Janev, in ref.22 and submitted to Comments on Atomic and Molecular Physics.
157. T.R. Dinterman and J. Delos, Phys.Rev. *A15*, 463 (1977).
158. R.E. Olson, Phys.Rev. *A24*, 1726 (1981).
159. F. Koike, private communication.
160. H. Ryufuku, JAERI-M82-031 report (1982).
161. V.V. Afrosimov, A.A. Basalaev, M.N. Panov and G.A. Leiko, JETP Letters *26*, 536 (1977).

RYDBERG STATES

F. Brouillard

Université Catholique de Louvain
Institut de Physique
B-1348 Louvain-la-Neuve, Belgium

INTRODUCTION

In a fusion plasma at a temperature of some millions of degrees, most of the hydrogen is ionised so that the role of neutral particles might be thought at first to be negligible.

But, firstly, the edges of a plasma are obviously much colder and the processes taking place there could be rather crucial and, secondly, the residual neutral atoms are considerably excited and therefore invested with a large reactivity. This can be a compensation to their relatively small number and make excited neutrals possibly take an important part in a thermonuclear plasma.

Thus, it is of interest to investigate the general properties of highly excited atoms, or RYDBERG atoms, as they are currently named.

In the present context, the study can be essentially restricted to the case of hydrogen, which is by far the most abundant species expected in a fusion plasma. However, Helium will be present to a certain extent and a non negligeable population of Rydberg states can also be formed by electron capture on multiply charged impurities.
Some attention will thus be paid to Rydberg states of atoms or ions with several electrons.
We will consider successively :
the formation of Rydberg atoms,
their radiative decay,
their behavior in electric and electromagnetic fields,
their collisions with atoms and ions.

PRODUCTION OF RYDBERG STATES

Atoms can be formed in Rydberg states through many ways. In fusion plasmas, electron capture by protons or multiply charged ions is likely to be a major source of Rydberg states although direct excitation by electron impact might be significant too.

In laboratory experiments dedicated to the investigation of the properties of Rydberg states, electron impact excitation has been used in the early work (e.g. Cermak and Herman[1], Kupriyanov[2], Hotop and Niehaus[3]) but it is generally not appropriate for precise measurements as the excited population created by electron impact is not well known.

Electron capture into highly excited states, at the other hand, has been extensively used untill recently (e.g. Il'In et al[4], Bayfield and Koch[5], Burniaux et al[6]). In this case, many Rydberg states are still created simultaneously but the population is known.

Production of atoms excited in a single Rydberg state with known principal quantum number n and orbital angular momentum number ℓ has been made possible by the development of powerful dye lasers.

Comment on electron capture into high excited states

The population obtained in that way is characterised by a n^{-3} dependence on the principal quantum number n.
This was already predicted in early theory of charge exchange (Oppenheimer[7]) for the s states. Generalisation to states with non-zero angular momentum was subsequently established (Jackson and Schiff[8], May[9], Chan and Eichler[10]).

The distribution of capture in the different substates has been calculated both quantum-mechanically (Ryufuku and Watanabe[11]) and classically (Salop[12]). It also was predicted by Chan and Eichler[10]).
Experimental data all confirm the n^{-3} law but data are too scarce to assess something regarding the dependence on the angular momentum ℓ and magnetic number m. However, the mutual agreement of the theoretical predictions is good enough to assess that high angular momentum states are preferentially populated up to very high collision energies. This somewhat invalidates the old statement by Brinkman and Kramers[13], that capture takes place in s states only at high energy.

Comment on laser excitation of Rydberg states

With presently available lasers, Rydberg states cannot be

excited directly from the ground state. Two steps, at least, are necessary.

The first step can be a collisional excitation of a metastable state, as in measurements made by Stebbings et al[14] on Xenon, where electrons were used to excite the $6s^3P$ state, from which np and nf Rydberg states could be reached by laser excitation (see figure 1(a)).

The first step can also be laser excitation, as in the investigations performed in Sodium by Littman et al[15]. But in that case, larger power is generally required.
Fortunately, pulsed laser are currently delivering hundreds of kilowatts instantaneous power.

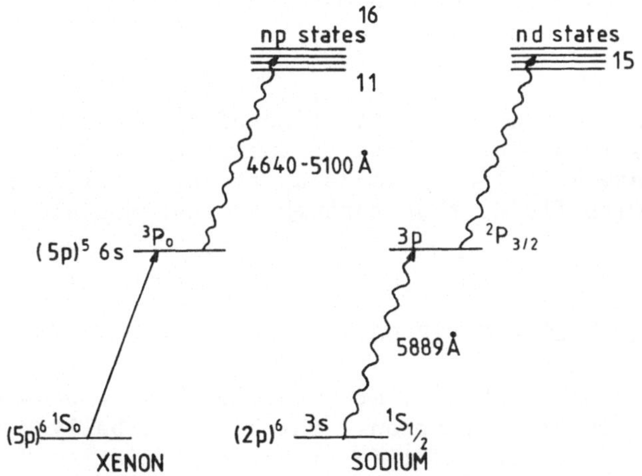

Figure 1.

RADIATIVE LIFE TIME OF RYDBERG STATES

Spontaneous radiative decay of hydrogen is accurately calculable. Oscillator strengths have been tabulated, e.g. by Harriman[16] from which lifetimes can be calculated.
For highly excited states, life times become proportional to n^3. This results from the facts that :
- the square of the dipole moment of any transition ($n \rightarrow n'$) decreases as n^{-3} if n is large and n' relatively small,
- radiative decay to deep lying states is dominant.

Explicit values of the life-times of hydrogen are given in Bethe and Salpeter[17], for n = 1 to 6. For large n, asymptotic values are approximately as follows (in nanoseconds):

ns	np	nd	nf
$2.3n^3$	$0.15n^3$	$0.52n^3$	$1.1n^3$

For atoms with several electrons, the lifetime of highly excited states is somewhat different. For alkali atoms, it can generally by expressed in terms of the effective quantum number n^*, on which a $(n^*)^3$ dependence is expected. For rare gases atoms, the situation is not clear. Hydrogenic behavior has been reported for Rydberg d states of Xenon (Stebbings et al[14]) but a $(n^*)^2$ dependence has been found for the Rydberg d states of Krypton (Delsart et al[18]).

Rydberg states of alkaline earths have been shown to be strongly perturbed by the presence of doubly excited levels (Aymar et al[19]). It should also be kept in mind that life times are considerably affected by even quite weak electric fields (see next section).

In any case the radiative life time of highly excited Rydberg states, ($n > 20$) is so large that, in thermonuclear plasmas, Rydberg states are more likely to decay through collisional processes or interaction with fields than through spontaneous emission of radiation.

RYDBERG STATES IN ELECTRIC FIELDS : QUENCHING AND IONISATION

The eigenstates of an atom are modified by the presence of an electric field. The modifications, which are globally known as Stark effect, lead to altered life times for radiative decay. They also make the atom unstable against ionisation.

The effect is particularly large for highly excited states. Field ionisation has been extensively used to identify and detect

Rydberg states. In fact, it is the major experimental tool for the study of highly excited atoms.

Stark shift and Stark mixing

The Hamiltonian of a one electron atom is an electic field F is

$H = H° + eFz$ where $H°$ is the field free hamiltonian
F the electrid field
e the electronic charge
and z the position in the direction of the field.

It is no longer invariant under rotation excepted around the field direction, so that angular momentum is not a good quantum operator. It at once means that the eigenfunctions of the atom are generally combinations of the field free functions.

One says that the electric field produces a mixing of the field free states. As a result, the dipole matrix elements connecting different states appear as combinations of matrix elements of field free states and life times are therefore also combinations of field free life times.

The amount of mixing is of course dependent on the strength of the perturbation and it is also clear that it becomes increasingly important for higher Rydberg states that have a large spatial extension.

The amount of mixing also depends on the energy separation of the field free states. If the field free states were perfectly degenerated the slightest electric field would produce a complete mixing.
In hydrogenic atoms, all substates of given quantum number n are almost degenerated. They are separated only by fine structure and Lamb shifts factors, but both factors decrease when n increases. Therefore vanishingly small fields are able to produce considerable mixing in hydrogenic Rydberg states. In atoms with several electrons, the absence of degeneracy strongly reduces the mixing.

For hydrogen, the hamiltonian is separable in parabolic coordinates. Eigenstates are characterized by quantum numbers n_1, n_2 and m. Life times of these states have been calculated by Hiskes and Tarter[20].
The energy shift, in a linear approximation is given by

$W = \frac{3}{2} (\frac{n}{Z})(n_1-n_2)a_o eF$ where a_o is the Bohr radius and Z the nuclear charge

or, when W is expressed in (eV) and F in (volt/cm) :

$$W = 7.93 \times 10^{-9}\, n(n_1-n_2)F.$$

An illustration of this formula is given in Figure 2 below for n=5.

n_1	n_2	(n_1-n_2)	m
4	4	0	- 4
4	3	1	- 3
3	4	- 1	
4	2	2	- 2
3	3	0	
2	4	- 2	
4	1	3	- 1
3	2	1	
2	3	- 1	
1	4	- 3	
4	0	4	0
3	1	2	
2	2	0	
1	3	- 2	
0	4	- 4	
3	0	3	+ 1
2	1	1	
1	2	- 1	
0	3	- 3	
2	0	2	+ 2
1	1	0	
0	2	- 2	
1	0	1	+ 3
0	1	- 1	
0	0	0	+ 4

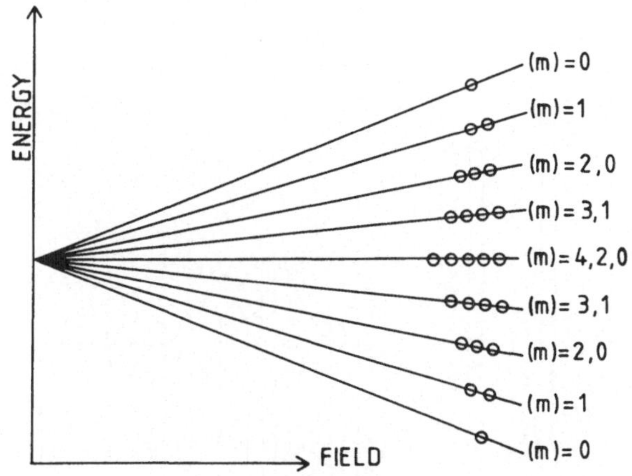

Figure 2.

As a function of the electric field, the levels of a given n form a bundle known as "STARK manifold". Note that the m=o manyfold has the largest divergence. Inclusion of fine structure and Lamb shifts does not modify the overall picture. It just bends the levels at the low field side to correlate them properly to the field free states.

The Stark structure of atoms with several electrons is the subject of several investigations. Precise laser excitation combined with field ionisation (see below) have produce beautiful and didactic experiments. As an exemple of this, Figure 3 reproduces excitation curves measured by Littman et al[15] for Rydberg states of sodium in the vicinity of n=15. In that work, a sodium beam is excited in an electric field by the simultaneous action of two pulsed laser. The second laser is tunable and its frequency determines the energy of the level. After excitation the atoms are ionised by a strong electric pulse. The peaks appearing in the figure correspond to the recorded ionic current.

The figure clearly shows the mixing of the 16s and 16p levels with the n=15 manyfold. Note that only $|m| = 0,1$ states are populated in this experiment.

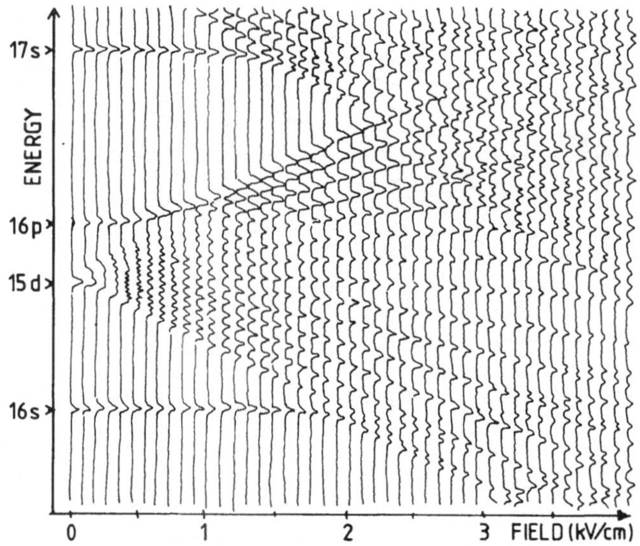

Fig. 3. Stark structure of Na (Littman et al[15]) MIT.

Field ionisation

When an atom is in an electric field, the combined potential of the nuclear charge and the electric field has a local maximum (saddle point).

Quantum mechanically, any state extends to infinity and if the electron is initially in the Coulombic well, it has a non zero probability of tunnelling away. Classically, the situation is even simpler : if the total energy of a state is above the value of the potential V_0 at the saddle point, the state is not bound and the atom becomes ionised (see Fig. 4).

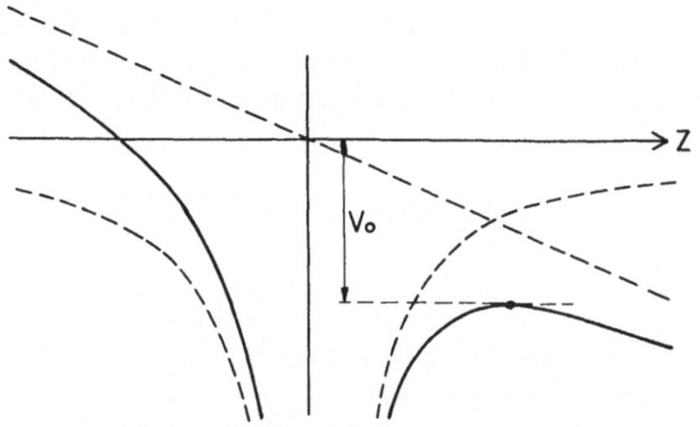

Saddle-point potential

Figure 4.

It is a simple matter to calculate V_o by differentiating :

$$V(z) = -\frac{1}{z} - Fz$$

$$V_o = -2F^{1/2}$$ where V_o and F are in atomic units.

If one neglects the Stark shifts, the energy of the Rydberg states is simply :

$$E = -\frac{1}{2n^2} .$$

Ionisation will then arise for a state (n) if the field is larger than a critical value F^*

$$F^* = \frac{1}{16n^4} \text{ a.u.} = \frac{3.2 \times 10^8}{n^4} \text{ volt/cm.}$$

A refinement of this picture was given by Cooke and Gallagher[21] who consider an effective potential taking the angular momentum of the electron into account. Then :

$$V_o = -2F^{1/2} + |m|F^{3/4} + \frac{3}{16} m^2 F$$ where m is the magnetic quantum number.

The correction is positive so that states with large values of m are more difficult to field ionize.

Quantum tunneling calculations are formulated in terms of an ionisation rate. The field dependence of the ionisation rate is very steep and is illustrated in Figure 5, taken from Bailey et al[22].

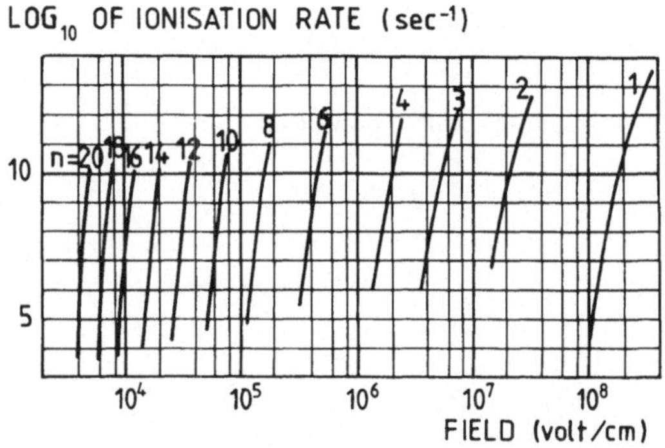

Figure 5.

The ionisation rate can be expressed, at small F, as :

$$I = \frac{4}{F} \exp\,(-2/3F).$$

The critical field which is usually taken as corresponding to an ionisation rate of 10^8/sec, is given by

$$F^* = \frac{6.2 \times 10^8}{n^4} \text{ volt/cm}$$

which is a factor 2 higher than the classical value.
But one must be aware that the classical derivation was oversimplified, as states with an energy appreciably higher than the saddle point energy can survive a long time before they ionise, both in a classical and in a quantum mechanical view.

Now, we must consider the Stark shifts of the levels. In a linear approximation, the states with a given principal quantum n form a manyfold as shown in Figure 6.

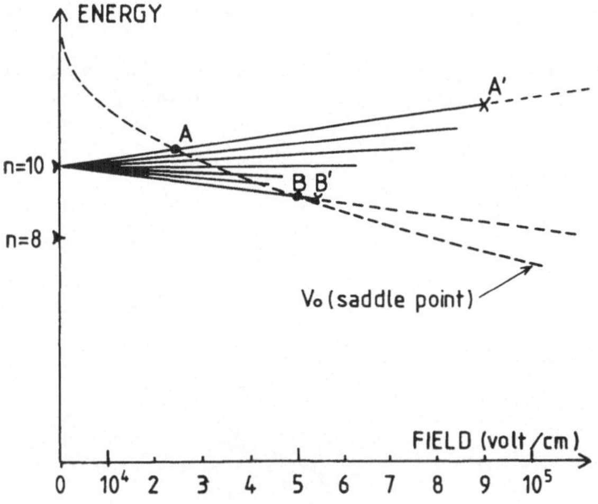

Figure 6.

Following the crude classical picture, a state ionises when its energy reaches the saddle point energy V_o (dashed line in the fig.). The highest component of the manyfold therefore ionises at a lower field (A) than the lowest component (B).
Following quantum calculations, the reverse happens : an ionisation rate of 10^{10}/sec is reached at a lower field for the lower component (B') than for the higher component (A'). This is not surprising, as the electron density of a higher energy state is shifted away from the saddle point region and the state has consequently a lower ionisation rate.
In any case, the critical field is different for different substates and one must be aware of this when field ionisation is used to identify Rydberg states.

An oter important question arises naturally : how does a definite field free Rydberg state evolve when it comes in an electric field ?
First one must solve the problem of the adiabatic correlation of a (j, m_j) state at zero field with a (m_ℓ, m_s) state at high field but there remains another question : is the evolution of the state actually adiabatic ? The answer depends of course on how fast the field is varying.

In fact the slew rates of the fields encountered in experiments are seldom higher than 10^9 volt/cm,sec. and non adiabatic behavior only can take place when pseudo crossings of levels are present.

But Rydberg states precisely offer many possibilities of pseudo-crossings, between the components of adjacent Stark manyfolds (see Figure 7).

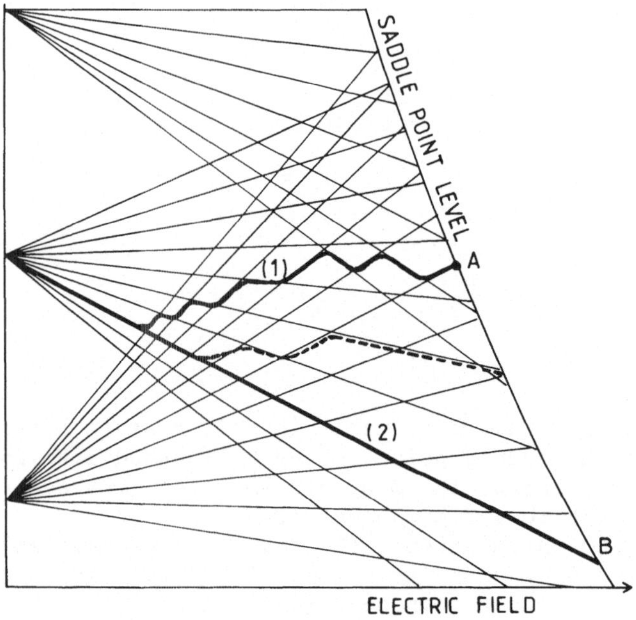

Figure 7.

In hydrogen, however, all the states differ by some true symmetry and the levels therefore really cross wach other with no interact Crossings can therefore be ignored.
By contrast, atoms with several electrons have mostly "avoided

crossings". When the slew rate of the field is small, and the states evolve adiabatically, the path followed to ionisation avoids all the crossings (path 1 in the figure). The saddle point curve is met in A.
For larger slew rates, the probability of passing diabatically through the crossings increases and the ionisation path progressively approaches the fully diabatic path (path 2 in the figure). In that case, the saddle point level is reached in B and ionisation takes place at a higher field.

Diabatic field ionisation of Rydberg states has been reported recently by several authors. For instance, diabatic ionisation has been shown by Jeys et al[23] to be important for the (n=30-36)d states of sodium. Lower states (n=15-20) have been found by Gallagher et al[24] to behave adiabatically. Recently, Neijzen and Dönszelmann[25] also reported diabatic ionisation of Indium.

When diabatic paths are open, field ionisation exhibits several thresholds. Clearly, one must be aware of this when using field ionisation to observe highly excited Rydberg states.

IONISATION IN OSCILLATING FIELDS

A low frequency oscillating field acts on an atom much in the same way as a static field. But the situation becomes different when the frequency becomes comparable with excitation frequencies of the atom.
Then, the quantum nature of the radiation comes in and ionisation as well as excitation are related to the absorption of individual photons.

Absorption of a single photon (photoexcitation and photoionization) is well described by quantum theory but the simultaneous absorption of several photons is not yet well understood. Multiphoton ionisation or excitation is a phenomenon that can only take place in very intense fields, so that the process cannot be properly described by perturbative theories.

Experiments on multiphoton ionisation of ground state or weakly excited atoms require laser power that is not yet available. But the multiphoton ionisation of highly excited Rydberg states have been achieved by means of microwaves with a reasonable intensity (Bayfield and Koch[5], Bayfield et al[26]). Both experiments were done in the frequency range 9-11 GHz with hydrogen atoms in $n > 50$ states. The amplitude of the electric field was 75 volt/cm.

Comparing this field with the nuclear field in the Rydberg n=50 state

$$F = \frac{e}{(a_o)^2 n^4} = 820 \text{ volt/cm},$$

we see that the external field exerces on the electron forces that are comparable with the nuclear attraction and cannot be regarded as a small perturbation. Results obtained for the ionisation rate of state n=48 as a function of the frequency are shown in Figure 8(a). The resonant behavior indicates that multiphoton ionisation dominates the (adiabatic) quantum tunneling.
Fig. 8(b) shows the rate of excitation of the same state n=48 to higher states. It is also structured and has roughly the same magnitude as the ionisation rate.

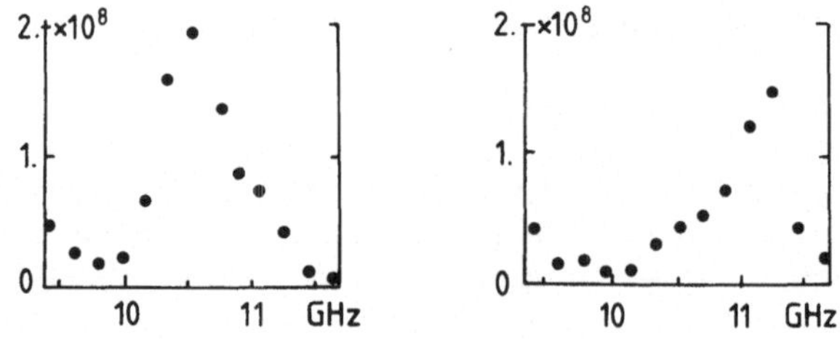

(a) : ionisation rate (b) : excitation rate

Fig. 8. For hydrogen (n=48) as a function of field frequency.

Microwave ionisation of Rydberg atoms has been calculated by Jones et al[27] using classical Monte-Carlo trajectories. In this model, ionisation depends on two parameters only : (ω/ω_a) and (F/F_a) with :

$\omega_a = v_a/r_a$ where r_a is the semi-major axis of the trajectory

$F_a = E_a/r_a$ $\quad v_a$ the root-mean-square velocity of the electron

and $E_a = 1/n^2$ (in a.u.).

In fact ω_a is the angular frequency of the electron and F_a the average force exerted by the nucleus. The result of their calculation is illustrated in Figure 9. Ionisation is seen to move progressively from a frequency-independent process at $F/F_a = 0.2$ to a resonant one at $(F/F_a) = 0.1$.

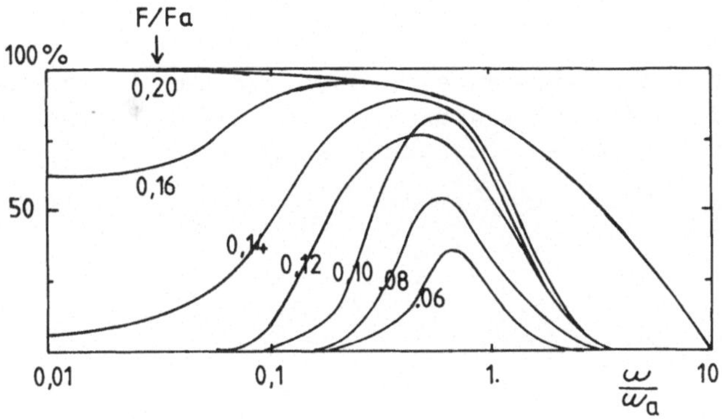

Fig. 9. Percentage ionisation vs ω/ω_a following Jones et al[27].

Finally it must be mentioned that Rydberg states are also quite capable to interact with the black-body radiation surrounding them at room temperature. Absorption and stimulated emission produce a redistribution of the Rydberg states and modify their apparent life-time.

A threefold reduction of the life-time of the 17p and 18p states of Sodium has been reported by Gallagher and Cooke[28]. Calculations of population redistribution has been done by Cooke and Gallagher[29].

THERMAL COLLISIONS OF RYDBERG ATOMS

A large amount of experimental work has been devoted in the recent years to collisions of Rydberg atoms with neutral perturbers at thermal energies. Most of it concerns the destruction (quenching, ℓ-mixing) of Rydberg alkali atoms in collisions with rare gases. Examples among others are the experiments of Gallagher et al[30], Hugon et al[31] and Kellert et al[32].

A first general statement encountered in the literature is that the cross section for collisional destruction of a Rydberg atom should increase with n as n^4, i.e. be proportional to the geometrical section of the atom. This statement is consistent with the following assumptions :

1. The Rydberg electron is contained in a volume which in the average is a sphere of radius $r_n = n^2 a_0$.
2. The only significant interaction is between the neutral perturber and the Rydberg electron.
3. When the impact parameter b is smaller than r_n, the interaction produce a momentum transfer large enough to modify the quantum state.

In fact, the n^4 dependence is observed but for rather small values of n. In Caesium-Caesium collisions for instance (Nakayama et al[33]) the cross section increases as n^4 up to n=14 as shown in Figure 10.

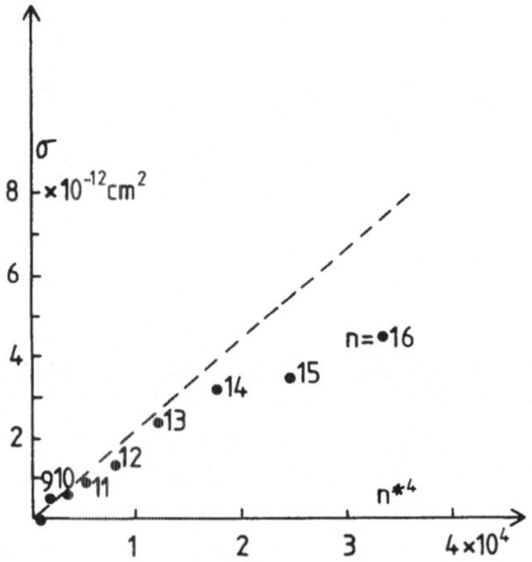

Fig. 10. Depopulation in $C_s(n) + C_s$ collisions, following Nakayama et al[33].

At larger n, the cross sections pass through a maximum and then decreases as illustrated in Figure 11 for the case of sodium (ns) atoms colliding with Argon, investigated by Boulmer et al[34].

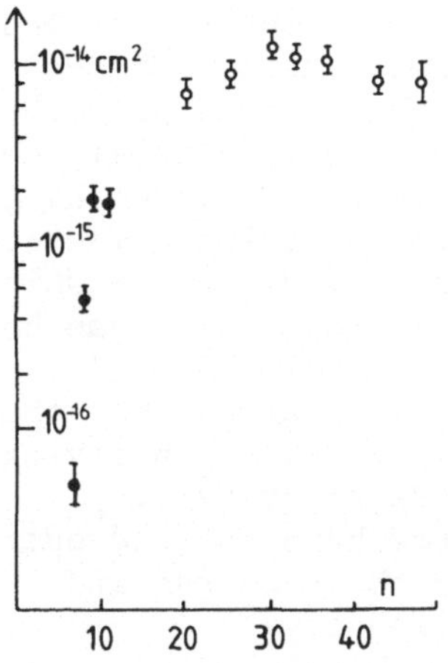

Fig. 11. Depopulation in Na(n) + A collisions, following Boulmer et al[34].

It is easy to understand that the cross section cannot increase as n^4 undefinitely because assumption (3) here above will cease to be valid at large values of n, when the collision time τ will become smaller that the orbiting time τ_n of the Rydberg electron. In fact, the assumption (3) implies that the electron motion sufficiently "fills" the Rydberg sphere so that the passing atom has a chance to approach the electron

$$\tau = r_n/v$$

$$\tau_n = r_n/v_n$$

$$\tau/\tau_n = v_n/v = v_o/nv$$

where v_o and v_n are the orbital velocities in the ground-and the n-state respectively and v is the collision velocity.

Thermal collisions of Rydberg atoms have been studied theoretically both quantum mechanically and semi-classically. Most of them are with neutral perturbers, although extensive studies of ion-Rydberg collisions were made by Percival and his collaborators as reported by Percival and Richards[35]. The theoretical work on collisions with neutrals has been recently reviewed by Matsuzawa[36].
Most of it assumes that the electron behaves as it were free and that its interaction with the neutral perturber is dominant over the ionic core-neutral atom interaction, in all collisional processes. The cross section is therefore formulated by means of the scattering amplitude for electron-atom collision. The specificity of the considered process comes in as a form factor which takes the corresponding momentum into account. That model has been able to explain many experimental data.
However, it has been recently claimed that the picture of a "free electron" is not valid for processes involving large momentum transfers (Li Yu Cheng and van Regemorter[37]).
Furthermore, several authors have pointed out that the contribution of the interaction between the electron and the ionic core is not negligeable (Omont[38] and Flannery[39]).

Thermal collisions of Rydberg atoms can also lead to ionisation. Experimental data on ionisation are very fragmentary. Hotop and Niehaus[3] measured a cross section of the order of 10^{-12} cm^2 for ionisation of "highly excited" atoms of rare gases in collisions with molecular beams (water, ammonia).
Kupryianov[40] investigated the ionisation of excited helium in molecular hydrogen. Kellert et al[41] more recently studied the ionisation of Xenon (nf) states in the range n=25 to 40 in collision with ammonia and found cross sections increasing with n from 10^{-12} cm^2 for n=25 to 2.3 x 10^{-11} cm^2 for n=40 but with a smaller slope at large n. Klucharev et al[42] reported on the ionisation of Rubidium (np) states in collisions with other Rubidium atoms in the range n=8 to 14. In this case the cross section passes through a maximum of some 10^{-13} cm^2 for n=11.
Rubidium (9s) states were found to have a cross section of the same order in the same process by Cheret et al[43].

All these data are more or less well explained by the free electron model, although the disagreement is sometime as large as a factor 5.

When symmetrical systems are considered for ionisation, i.e. in the collision of a Rydberg atom with a ground state atom of the same species, attention must be paid to exchange mechanisms that are likely to be important at low energy. Thus, the channel

$$A(n) + A \rightarrow A + A^+ + e$$

can contribute to ionisation.

This problem has been treated by Janev and Mihajlov[44]. Here, the system is regarded as a quasi-molecule and the reaction between the ionic core of the Rydberg atom and the other atom plays a decisive role. The Rydberg electron interacts with the molecular ion as a whole.
Cross sections were calculated for Hydrogen and Lithium with n ranging from 10 to 25. They have a maximum located between 0.2 and 0.4 eV and are smaller for larger n.
In a subsequent paper (Mihajlov and Janev[45]) a rate constant for the same process with Rubidium atoms was calculated at T=520K for n=6-13. These results could be directly compared with the measurements mentioned above of Klucharev et al[42]. The agreement is quite good. It must be mentioned that the mechanism proposed by Janev and Mihajlov applies also to non symmetrical systems if quasi resonant channels are available.

Finally, the possibility of associative ionisation also exists. But excepted for a few isolated data, experimental information is not existing.

FAST COLLISIONS OF RYDBERG ATOMS

By fast collisions is ment that the relative velocity v of the collision is much higher than the orbital velocity of the electron in the Rydberg atom.

Matsuzawa[36] has shown that the free electron model, with an impulse approximation, is capable of describing fast collisions of Rydberg atoms as well as the slow ones.
Butler and May[46] has already predicted that the cross section for collisional ionisation of Rydberg atoms should coincide with the total scattering cross section of free electrons on the same target atoms and at the same velocity. This point has been confirmed experimentally by Koch[47] who measured the ionisation cross section of Deuterium atoms, excited to Rydberg states with n=35 to 50, in collisions with molecular Nitrogen at C.M. energies ranging from 6 to 11 keV.
Not only are the cross sections close to the electron impact ionisation cross sections but they are n-independent, in accordance with the free electron picture.
At the same time, Koch also established that the cross section for destruction of the Rydberg state is also independent of n and contains a resonant structure as in the case of electron impact ionisation.

Ion-Rydberg atom fast collisions have been treated theoretically by Olson[48] who used the classical trajectory Monte Carlo Method. That work deals with n-changing, ionisation and charge exchange for Rydberg atoms in states n=1,2,5,10 and 20 colliding with ions of charge 1-10.

Collision velocities considered range from one to ten times the orbital velocity of the Rydberg electron, to which all velocities rigourously scale in such theory. The main features of the predictions are :

1. n-changing collisions

They are most frequent to adjacent levels and excitation is globally much more probable than deexcitation. This is illustrated in Figure 12 for ion charge 1 and $v = 2v_n$.

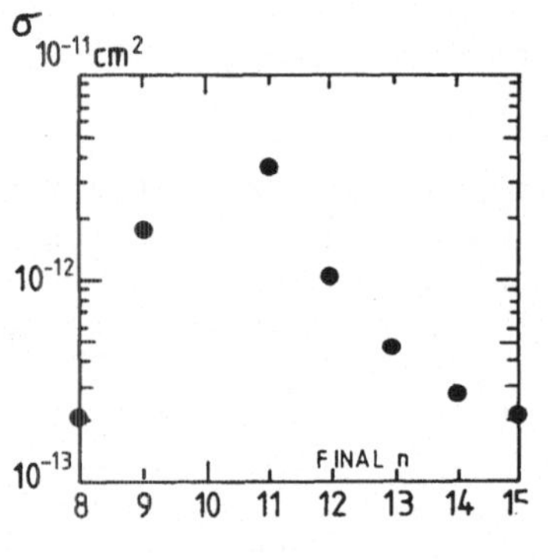

Fig. 12. n-changing cross section initial n = 10, $v/v_n = 2$. Following Olson[48].

2. Ionisation

Ionisation cross sections are proportional to n^4 (for a fixed ratio v/v_n). When (v/v_n) is larger than 5, the cross section is given by the universal formula :

$$\sigma_I = 6\pi\, a_o^2\, n^4 q^2 \left(\frac{v}{v_n}\right)^{-2}.$$

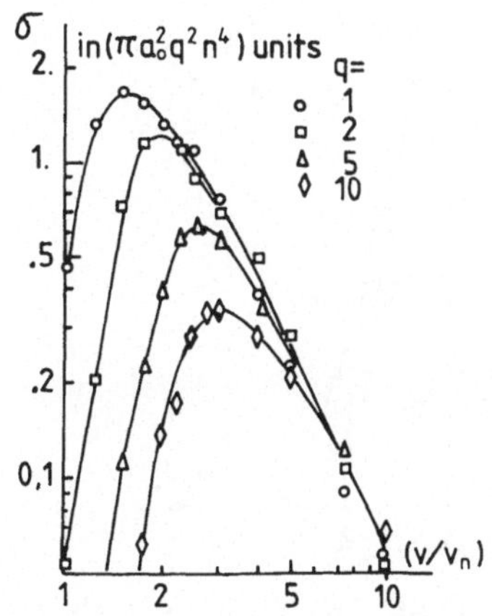

Fig. 13. Ionisation cross section. Following Olson[48].

The maximum value is reached around $v = 2v_n$ as seen in Figure 13.

3. Charge exchange

Charge exchange cross sections are also proportional to n^4. The main feature is that they are velocity-independent at low energy (Figure 14). The low velocity limit is given by the formula :

$$\sigma_{EX} = 5.5\ \pi\ n^4\ q.$$

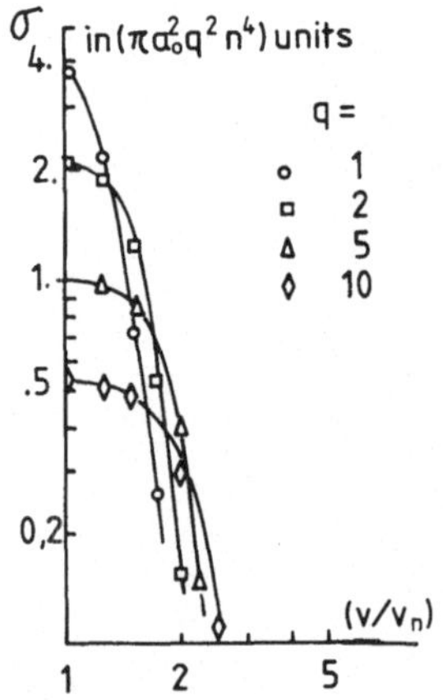

Fig. 14. Charge exchange cross section. Following Olson[48].

This feature is well confirmed by the experimental work of Burniaux et al[6] who measured the charge exchange cross section between He^{++} and H(n) in the range n = 8-25 (Figure 15).

Not only is the energy dependence of the cross section well rendered but the absolute agreement is remarkable. The dotted line in Figure 15 is the value predicted by Olson for n=17 and q=2 and must thus be compared with the experimental data for the group (n = 15-19).

Experimental data also exist for H^+-H(n) collisions (Koch and Bayfield[48]). The agreement with Olson's predictions is less good than in the former case : no clear plateau is reached at the low energy side and the absolute value is about a factor 2 above

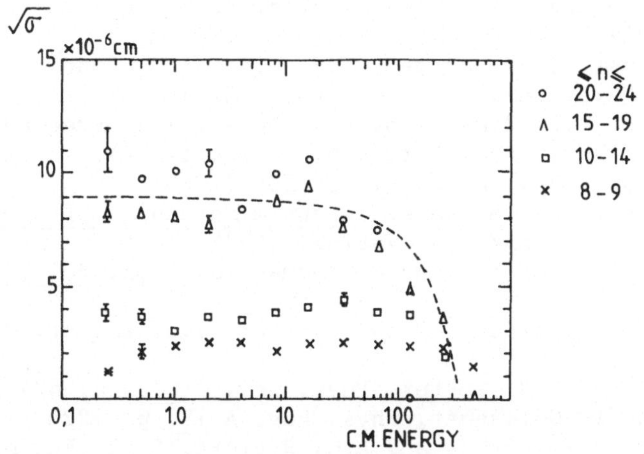

Fig. 15. Charge exchange cross sections for H(n) + He^{++}. Following Burniaux et al[6]. Dash line = Olson's prediction for n = 17, q = 2.

the theoretical one (at $v = v_n$). But it must be mentioned that the measurement included ionisation of H(n) as well as electron capture.

REFERENCES

1. V. Cermak and Z. Herman, Collection Czech. Chem. Comm. 29, p. 953 (1964).
2. S.E. Kupriyanov, Sov. Phys. JETP 21, p. 311 (1965).
3. H. Hotop and A. Niehaus, J. Chem. Phys. 47, p. 2506 (1967).
4. R.N. Il'In, V.A. Oparin, I.T. Serenkov, E.S. Solov'Ev and N.V. Fedorenko, Sov. Phys. JETP 32, p. 59 (1971).
5. J.E. Bayfield and P.M. Koch, Phys. Rev. Lett. 33, p. 258 (1974).
6. M. Burniaux, F. Brouillard, A. Jognaux, T. Govers and S. Szücs, J. Phys. B 10, p. 2421 (1977).
7. J.R. Oppenheimer, Phys. Rev. 31, p. 349 (1928).
8. J.D. Jackson and H. Schiff, Phys. Rev. 89, p. 359 (1953).
9. R.M. May, Phys. Rev. A 136, p. 669 (1964).

10. F.T. Chan and J. Eichler, Phys. Rev. A 20, p. 104 (1979).
11. H. Ryufuku and T. Watanabe, Phys. Rev. A 20, p. 1828 (1979).
12. A. Salop, J. Phys. B 12, p. 919 (1979).
13. H.C. Brinkman and H.A. Kramers, Proc. Acad. Sci. Amsterdam, 33 p. 973 (1930).
14. R.F. Stebbings, C.J. Latimer, W.P. West, F.B. Dunning and T.B. Cook, Phys. Rev. A 12, p. 1453 (1975).
15. M.G. Littman, M.L. Zimmerman, T.W. Ducas, R.R. Freeman and D. Kleppner, Phys. Rev. Lett. 36, p. 788 (1976).
16. J.M. Harriman, Phys. Rev. 101, p. 594 (1959).
17. H.A. Bethe and E.E. Salpeter, Quantum Mechanics of One-and-Two Electrons Atoms, (Springer Verlag, 1957).
18. C. Delsart, J.C. Keller and C. Thomas, J. Phys. B 14, p. 4241 (1981).
19. M. Aymar, R.J. Champeau, C. Delsart and J.C. Keller, J. Phys. B 14, p. 4489 (1981).
20. J.R. Hiskes and C.B. Tarter, Phys. Rev. A 133, p. 424 (1963).
21. W. Cooke and T. Gallagher, Phys. Rev. A 17, p. 1226 (1978).
22. D.S. Bailey, J.R. Hiskes and A.C. Rivière, Nucl. Fusion 5, p. 41 (1965).
23. T.H. Jeys, G.W. Foltz, K.A. Smith, E.J. Beiting, F.G. Kellert, F.B. Dunning and R.F. Stebbings, Phys. Rev. Lett. 44, p. 390 (1980).
24. T.F. Gallagher, L.M. Humphrey, W.E. Cooke, R.M. Hill and S.A. Edelstein, Phys. Rev. A 16, p. 1098 (1977).
25. J.H.M. Neijzen and A. Dönszelmann, J. Phys. B 15, p. L87 (1982).
26. J.E. Bayfield, L.D. Gardner and P.M. Koch, Phys. Rev. Lett. 39, p. 76 (1977).
27. D.A. Jones, J.G. Leopold and I.C. Percival, J. Phys. B 13, p. 31 (1980).
28. T.F. Gallagher and W.E. Cooke, Phys. Rev. Lett. 42, p. 835 (1979).
29. T.F. Gallagher, S.A. Edelstein and R.M. Hill, Phys. Rev. A 15, p. 1945 (1977).
30. W. Cooke and T. Gallagher, Phys. Rev. A 21, p. 588 (1980).
31. M. Hugon, F. Gounand, P.R. Fournier and J. Berlande, J. Phys. B 12, p. 2707 (1979).
32. F.G. Kellert, T.H. Jeys, G.B. McMillian, K.A. Smith, F.B. Dunning and R.F. Stebbings, Phys. Rev. A 23, p. 1127 (1981).
33. S. Nakayama, F.M. Kelly and G.W. Series, J. Phys. B 14, p. 835 (1981).
34. J. Boulmer, J.F. Delpech, J.C. Gauthier and K. Safinya, J. Phys. B 14, p. 4577 (1981).
35. I.C. Percival and D. Richards, Ad. Atom. Molec. Phys. 11, p. 1 (1975).
36. M. Matsuzawa, Electronic and Atomic Collisions, ed. by ODA and Takayanagi (North Holland 1980) p. 493.
37. Li Yu Cheng and H. van Regemorter, J. Phys. B 14, p. 4025 (1981).
38. A. Omont, J. Phys. 38, p. 1343 (1977).
39. M.R. Flannery, J. Phys. B 13, L657 (1980).

40. S.E. Kupriyanov, Sov. Phys. JETP 24, p. 674 (1967).
41. F.G. Kellert, K.A. Smith, R.D. Rundel, F.B. Dunning and R.F. Stebbings, J. Chem. Phys. 72, p. 3179 (1980).
42. A.N. Klucharev, A.V. Lazarenko and V. Vujnovic, J. Phys. B 13, p. 1143 (1980).
43. M. Cheret, A. Spielfiedel, R. Durand and R. Deloche, J. Phys. B 14, p. 3953 (1981).
44. R.K. Janev and A. Mihalov, Phys. Rev. A 21, p. 819 (1980).
45. A.A. Mihajlov and R.K. Janev, J. Phys. B 14, p. 1639 (1981).
46. S.T. Butler and R.A. May, Phys. Rev. A 10, p. 137 (1965).
47. P.M. Koch, Phys. Rev. Lett. 43, p. 432 (1979).
48. R.E. Olson, J. Phys. B 13, p. 483 (1980).

PART III

THE ATOMIC AND MOLECULAR PHYSICS OF CONTROLLED THERMONUCLEAR RESEARCH DEVICES

ATOMIC AND MOLECULAR PROCESSES IN HIGH-TEMPERATURE, LOW-DENSITY MAGNETICALLY CONFINED PLASMAS

H.W. Drawin

Association EURATOM-CEA sur la Fusion
F-92260 Fontenay-aux-Roses, France

1. INTRODUCTION

Atomic and molecular processes appear at different stages and with different degrees of importance in the Physics of high-temperature magnetically confined plasmas.

1. The production of a high-temperature plasma in machines like Tokamaks and Stellerators has to start with a cold molecular hydrogen gas (hydrogen stands for hydrogen and its isotopes deuterium and tritium). Dissociation and ionization of the molecules and atoms are the dominant processes during the phase of plasma production.

2. During the ohmic heating phase proper of a pure hydrogen plasma, the atomic processes are dominated by elastic electron-ion collision (which transfer the energy from the electrons to the ions) and by atom-ion charge transfer collisions which contribute to diffusion and energy losses.

3. During the heating phase gas is constantly or intermittently fed into the plasma in order to maintain the initial density or to increase it. This gas must be dissociated and ionized and contributes to the charge exchange losses.

4. When the plasma has attained its high-temperature regime ($T \geq 10^7$ K) energetic neutral hydrogen atoms created by charge transfer collisions between the hot ions and the neutral atoms hit the walls where they lose their energy. They come back as cold neutrals (so-called recycling) and - while penetrating into the plasma - the ionization and heating process are repeated.

5. Impurity atoms are created mainly by two processes :
a. The plasma hits the limiter inserted in the chamber in order to reduce the size of the plasma column. This leads to sputtering and arcing of the limiter material (mostly stainless steel, but also carbon, tungsten and other elements). Also interaction of the plasma with the divertor plates can liberate metal atoms which eventually diffuse back into the plasma.
b. The high-energy neutral particles hitting the walls will liberate wall material which diffuses into the plasma.

The highly ionized atoms of high-Z elements undergo numerous collision and radiation processes which practically all have a detrimental effect on the stability and on the energy balance of the plasma as a whole.

6. Impurity elements are often artificially introduced for diagnostic purposes. In order to permit a reliable interpretation of the diagnostic measurements quantitative knowledge of the numerous collision and radiation processes is necessary.

7. Finally, the fusion reactions lead to α-particles (helium) which have the beneficial property of heating the plasma. The same particles represent the ash of the fusion - burning process and must be exhausted. Atomic processes will probably play a role in the process of exhausting helium.

The relevance of atomic processes to thermonuclear fusion has been discussed during the past years in several articles and books [1-7]. The present article aims to present the atomic physics problems in thermonuclear fusion research under a unifying aspect starting with the fundamental equations for magnetically confined plasmas.

2. BASIC EQUATIONS FOR MAGNETICALLY CONFINED PLASMAS

Fusion plasmas are submitted to strong magnetic fields of induction $\vec{B} = \mu\vec{H}$ and of electric field strength $\vec{E}$. Electric conduction currents of density $\vec{J}$ build up. The plasma as a highly conducting fluid is globally neutral, i.e. on a macroscopic scale and to a first approximation, electron density n_e and the sum of all ion densities $n^z_{k,i}$ compensate each other (z = ion charge state of particles of chemical element k in internal quantum state $|i\rangle$). The fields obey the Maxwell and the continuity equations for $\vec{B}$ and the displacement vector $\vec{D} = \varepsilon\vec{E}$:

$$\frac{\partial\vec{B}}{\partial t} = -\vec{\nabla}\times\vec{E} \quad \text{(a)} \qquad \frac{\partial\vec{D}}{\partial t} = \vec{\nabla}\times\vec{H} - \vec{J} \quad \text{(b)} \qquad (1)$$

$$\vec{\nabla}\cdot\vec{B} = 0 \quad \text{(a)} \qquad \vec{\nabla}\cdot\vec{D} = \sum_s q_s n_s \cong 0 \quad \text{(b)} \qquad (2)$$

where q_s is the electric charge (either + $e_0 z$ or - e_0) of species s. For an ionized gas : $\varepsilon \cong \varepsilon_0$, $\mu \cong \mu_0$. The magneto-hydrodynamic (MHD) approximation assumes $\partial\vec{D}/\partial t = 0$.

The particle density $n_s(r,t)$ at space point $\vec{r}$ and time t is the velocity integral over the velocity distribution function $f_s(\vec{r},\vec{w}_s,t)$. Let $\vec{w}_s$ be the velocity relative to the laboratory system ; then

$$n_s(\vec{r},t) = \int_{\vec{w}_s} d^3w_s \; f_s(\vec{r},\vec{w}_s,t) \tag{3}$$

with f_s given by the kinetic (Boltzmann) equation

$$\frac{\partial f_s}{\partial t} + \vec{w}_s \cdot \frac{\partial}{\partial \vec{r}} f_s + \frac{\vec{F}_s}{m_s} \cdot \frac{\partial f_s}{\partial \vec{w}_s} = \left[\frac{\partial f_s}{\partial t}\right]_{\substack{\text{collision}\\\text{radiation}}} = \left[\frac{\partial f_s}{\partial t}\right]_{CR} \tag{4}$$

$\vec{F}_s$ is the (self-consistent) force per unit volume acting on a particle of masse m_s :

$$\vec{F}_s = q_s(\vec{E} + <\vec{w}_s> \times \vec{B}) + \text{non electromagnetic forces} \tag{5}$$

where $\vec{E}$ is the sum of externally applied electric fields and of fields created in the plasma by the plasma particles themselves. $<\vec{w}_s>$ is defined by Eq. (7). In Tokamaks and Stellerators, $\vec{B} = \vec{B}_T + \vec{B}_\Theta$, where $\vec{B}_T$ is the toroidal and $\vec{B}_\Theta$ the poloidal magnetic induction. The term on the r.h.s. of Eq. (4) is the collisional-radiative term. It depends on the various collision and radiation processes which contribute to an instantaneous local change of f_s.

2.1 - Rate equations for particle density

Integrating Eq. (4) over $\vec{w}_s$ yields the rate or balance equation for the particle density $n_s(\vec{r},t)$:

$$\frac{\partial n_s}{\partial t} + \vec{\nabla}(n_s <\vec{w}_s>) = \left[\frac{\partial n_s}{\partial t}\right]_{CR} \tag{6}$$

where $<\vec{w}_s>$ is the mean species' velocity. We introduce the mean local velocity $\vec{v}_0$ of the plasma as a whole and the peculiar velocity $\vec{V}_s$ and the mean diffusion velocity $<\vec{V}_s>$ relative to $\vec{v}_0$ by

$$\vec{w}_s = \vec{v}_0 + \vec{V}_s \;,\; <\vec{w}_s> = \vec{v}_0 + <\vec{V}_s> \;,\; \vec{v}_0 = \frac{\sum_s n_s m_s <\vec{w}_s>}{\sum_s n_s m_s} \tag{7}$$

Then the Eq. (6) becomes

$$\frac{\partial n_s}{\partial t} + \vec{\nabla}\cdot(n_s \vec{v}_o) + \vec{\nabla}\cdot(n_s < \vec{V}_s >) = \left[\frac{\partial n_s}{\partial t}\right]_{CR} \tag{8}$$

$< \vec{V}_s >$ and hence, $< \vec{w}_s >$ depend in a complicated manner on the plasma properties and particularly on collision processes.

In the special case of the electron density we have (s = e), i.e.,

$$\frac{\partial n_e}{\partial t} + \vec{\nabla}\cdot(n_e \vec{v}_o) + \vec{\nabla}\cdot(n_e < \vec{V}_e >) = \left[\frac{\partial n_e}{\partial t}\right]_{CR} \tag{9}$$

The r.h.s. of Eqs. (6) (8) (9) describes the instantaneous local rate with which the density n_s (respectively n_e) changes due to collision and radiation processes. The r.h.s. of Eq. (9) writes

$$\left[\frac{\partial n_e}{\partial t}\right]_{CR} = \left[\begin{array}{l}\text{sum of all volume}\\ \text{ionization rates}\end{array}\right] - \left[\begin{array}{l}\text{sum of all volume}\\ \text{recombination rates}\end{array}\right] \tag{10}$$

Summation of all equations (8) yields the following rate equation for the total particle density $n = \sum_s n_s$:

$$\frac{\partial n}{\partial t} + \vec{\nabla}\cdot(n\, \vec{v}_o) + \sum_s \vec{\nabla}\cdot(n_s < \vec{V}_s >) = \left[\frac{\partial n}{\partial t}\right]_{CR} \tag{11}$$

For an isolated system, the r.h.s. can be decomposed as follows

$$\left[\frac{\partial n}{\partial t}\right]_{CR} = \left[\frac{\partial n_e}{\partial t}\right]_{CR} + \left[\frac{\partial n_A}{\partial t}\right]_{CR} + \left[\frac{\partial n_{nucl}}{\partial t}\right]_{CR} \tag{12}$$

The terms on the r.h.s. account respectively for the following processes :

(i) creation and disappearance of electrons due to collision and radiation processes ;

(ii) creation and disappearance of heavy particles (both atoms, ions, and molecules) ;

(iii) creation or capture of particles due to nuclear reactions.

Plasma exhaust and particle injection for additional heating will appear as boundary conditions of the first-order differential equation (11). But it is also possible to add on the r.h.s. of Eq. (12) corresponding source terms which must be calculated consistently with the solutions of Eq. (11).

2.2 - Rate equation for mass density

From Eq. (6) follows for the mass density $\rho_s = m_s n_s$

$$\frac{\partial \rho_s}{\partial t} + \vec{\nabla}\cdot(\rho_s < \vec{w}_s >) = \left[\frac{\partial \rho_s}{\partial t}\right]_{CR} \tag{13}$$

In the absence of nuclear reactions, $[\partial \rho_s/\partial t]_{CR} = m_s [\partial n_s/\partial t]_{CR}$ holds. Summing up all Eqs. (13) yields the rate equation for the total mass density $\rho = \sum_s \rho_s$:

$$\frac{\partial \rho_s}{\partial t} + \vec{\nabla}\cdot(\rho \vec{v}_o) = \left[\frac{\partial \rho}{\partial t}\right]_{CR} = \left[\frac{\partial \rho_{nucl}}{\partial t}\right]_{CR} \tag{14}$$

In the absence of nuclear reactions, $[\partial \rho_{nucl}/\partial t]_{CR} = 0$. The effect of particle injection and exhaust is taken into account by the boundary conditions. But it is also usual to add on the r.h.s. of Eqs. (13) and (14) corresponding source and sink terms which must then consistently be calculated with the solutions at the boundary.

2.3 - Rate equation for momentum density

This rate equation is obtained bu multiplying Eq. (4) by $m_s\vec{w}_s$ and integrating over $\vec{w}_s$. With the mass density $\rho_s = m_s n_s$ follows for species "s" the general rate equation for the momentum density in the laboratory system

$$\frac{\partial}{\partial t}(\rho_s < \vec{w}_s >) + \vec{\nabla}\cdot(\rho_s < \vec{w}_s\vec{w}_s >) - n_s\vec{F}_s = \left[\frac{\partial}{\partial t}(\rho_s < \vec{w}_s >)\right]_{CR} \tag{15}$$

The r.h.s. describes again the rate of change due to collisional - radiative processes. It may be split into two terms: $[\partial/\partial t(\rho_s < \vec{w} >)]_{CR} = < \vec{w}_s > [\partial\rho_s/\partial t]_{CR} + \rho_s [\partial < \vec{w}_s >/\partial t]_{CR}$. Thus, two types of volume processes contribute, those leading to a change of density ρ_s and those changing the velocity vector $< \vec{w}_s >$. Every non equilibrated volume process contributes in principle to the r.h.s. of Eq. (15).

Multiplying Eq. (13) by $\vec{v}_o$ and subtracting the result from Eq. (15) yields the general vector force equation for particles of species "s" :

$$\rho_s \frac{\partial \vec{v}_o}{\partial t} + \rho_s \vec{v}_o\cdot\vec{\nabla}\vec{v}_o + \frac{D}{Dt}(\rho_s < \vec{V}_s > \vec{\nabla}\cdot\vec{v}_o) + \rho_s < \vec{V}_s >\cdot\vec{\nabla}\vec{v}_o + \vec{\nabla}\cdot\vec{P}_s$$
$$- n_s\vec{F}_s = \rho_s\left[\frac{\partial \vec{v}_o}{\partial t}\right]_{CR} + \left[\frac{\partial}{\partial t}(\rho_s < \vec{V}_s >)\right]_{CR} \tag{16}$$

where

$$\frac{D}{Dt} = \frac{\partial}{\partial t} + \vec{v}_o \cdot \vec{\nabla} \tag{17}$$

is the "hydrodynamic derivative" and

$$\vec{\vec{P}}_s = \rho_s < \vec{V}_s \vec{V}_s > \equiv \begin{pmatrix} P_{s,xx} + p_s & P_{s,xy} & P_{s,xz} \\ P_{s,yx} & P_{s,yy} + p_s & P_{s,yz} \\ P_{s,zx} & P_{s,zy} & P_{s,zz} + p_s \end{pmatrix} \tag{18}$$

is the partial pressure tensor. The first two terms of Eq. (16) take into account the force exerted on ρ_s due to the general acceleration of the plasma as a whole, and the first term on the r.h.s. is the collisional-radiative counter part. The term $\vec{\nabla}.\vec{\vec{P}}_s$ accounts for a force due to a gradient of an isotropic partial pressure p_s and due to gradients of the dynamic pressure and of the shear stesses.

In order to obtain the rate equation for the momentum transfer of species "s" in a coordinate frame moving with $\vec{v}_o$, we multiply Eq. (24) for $\vec{v}_o$ by ρ_s/ρ and subtract the result from Eq. (16). We obtain

$$\frac{D}{Dt}(\rho_s < \vec{V}_s >) + \rho_s < \vec{V}_s > \vec{\nabla} \cdot \vec{v}_o + \rho_s < \vec{V}_s > \cdot \vec{\nabla} \vec{v}_o + \vec{\nabla} \cdot \vec{\vec{P}}_s - \frac{\rho_s}{\rho} \vec{\nabla} \cdot \vec{\vec{P}}_s$$
$$- n_s \vec{F}_s + \frac{\rho_s}{\rho} \, {}_s\Sigma\, n_s \vec{F}_s = \left[\frac{\partial}{\partial t} (\rho_s < \vec{V}_s >) \right]_{CR} \tag{19}$$

where $\vec{\vec{P}}$ is the total pressure tensor given by

$$\vec{\vec{P}} = {}_s\Sigma\, \vec{\vec{P}}_s = \begin{pmatrix} P_{xx} + p & P_{xy} & P_{xz} \\ P_{yx} & P_{yy} + p & P_{yz} \\ P_{zx} & P_{zy} & P_{zz} + p \end{pmatrix} \tag{20}$$

Summation over all species "s" gives identical zero as it must be. (Note : ${}_s\Sigma\, \rho_s < \vec{V}_s > = 0$).According to Eq. (5) we have for the partial and total force densities the expressions

$$n_s \vec{F}_s = q_s n_s (\vec{E} + \vec{v}_o \times \vec{B} + < \vec{V}_s > \times \vec{B}) \tag{21a}$$

$${}_s\Sigma\, n_s \vec{F}_s = \rho^c (\vec{E} + \vec{v}_o \times \vec{B}) + \vec{J} \times \vec{B} \tag{21b}$$

where we have admitted an eventually non vanishing local charge density $\rho^c = \sum_s q_s n_s$; further,

$$\sum_s q_s n_s < \vec{V}_s > = \sum_s \vec{J}_s = \vec{J} \tag{22}$$

is the total electrical conduction current density which contributes to ohmic heating (see Eqs. (32) (40)).

The rate equation for the plasma as a whole is obtained by summing Eqs. (15) over all species "s". It follows

$$\frac{\partial}{\partial t}(\rho\vec{v}_o) + \vec{\nabla}\cdot(\rho\vec{v}_o\vec{v}_o) + \vec{\nabla}\cdot\vec{\vec{P}} - \sum_s n_s\vec{F}_s = \left[\frac{\partial}{\partial t}\rho\vec{v}_o\right]_{CR} \tag{23}$$

Taking into account Eq. (14) yields the local force equation for the plasma as a whole :

$$\rho\frac{\partial\vec{v}_o}{\partial t} + \rho\vec{v}_o\cdot\vec{\nabla}\vec{v}_o + \vec{\nabla}\cdot\vec{\vec{P}} - \sum_s n_s\vec{F}_s = \rho\left[\frac{\partial\vec{v}_o}{\partial t}\right]_{CR} \tag{24}$$

This equation can also be obtained directly by summing up all Eqs. (16).

The r.h.s. of Eq. (24) is generally zero, since collisions and radiation processes (assumed to be distributed isotropically) alone cannot modify $\vec{v}_0$. However, if a plasma is submitted to an external particle and/or photon flux (e.g. energetic neutral particle beam for additional heating), the r.h.s. can become different from zero. The rotation of a Tokamak plasma during tangential particle injection is described by the r.h.s. of Eq. (24).

2.4 - Rate equation for energy density

The translational energy density ($\bar{E}^{tr}$), the internal energy density ($\bar{E}^{int}$) stocked in the excitation and ionization levels, and the electromagnetic field energy density contribute to the general power balance. We will only consider $\bar{E}^{tr}$ and $\bar{E}^{int}$.

2.4.1 - Translational and thermal energy density

Multiplying Eq. (4) by $m_s\vec{w}_s\vec{w}_s$ and integrating over $\vec{w}_s$ yields the rate equation for the tensor of twice the translational (or kinetic) energy density of species "s" in the laboratory system :

$$\frac{\partial}{\partial t}(\rho_s < \vec{w}_s\vec{w}_s >) + \vec{\nabla}\cdot(\rho_s < \vec{w}_s\vec{w}_s\vec{w}_s >) - 2n_s < \vec{w}_s\vec{F}_s > = \left[\frac{\partial}{\partial t}(\rho_s < \vec{w}_s\vec{w}_s >\right]_{CR} \tag{25}$$

Owing to Eqs. (7) and (18) this can be expressed as a function of $\vec{v}_o$ and $\overleftrightarrow{P}_s$ and of the partial heat flux tensor $\overleftrightarrow{\vec{Q}}_s$ defined by

$$\overleftrightarrow{\vec{Q}}_s = \rho_s < \vec{V}_s \vec{V}_s \vec{V}_s > \tag{26}$$

It is instructive to discuss briefly the general structure of this equation. Development of Eq. (25) yields

$$\frac{\partial}{\partial t}(\rho_s \vec{v}_o \vec{v}_o) + \vec{\nabla}\cdot(\rho_s \vec{v}_o \vec{v}_o \vec{v}_o) - 2n_s \vec{v}_o \vec{F}_s + \frac{\partial \overleftrightarrow{P}_s}{\partial t} + 2\vec{\nabla}\cdot \overleftrightarrow{Q}_s - 2n_s < \vec{V}_s > \vec{F}_s$$

$$+ \frac{\partial}{\partial t}(\rho_s \vec{v}_o < \vec{V}_s >) + \frac{\partial}{\partial t}(\rho_s < \vec{V}_s > \vec{v}_o) + \vec{\nabla}\cdot(\vec{v}_o \vec{P}_s) + \vec{\nabla}\cdot(\vec{P}_s \vec{v}_o) + \vec{\nabla}\cdot(\rho_s < \vec{V}_s \vec{v}_o \vec{V}_s >$$

$$= \left[\frac{\partial}{\partial t}(\rho_s \vec{v}_o \vec{v}_o)\right]_{CR} + \left[\frac{\partial}{\partial t}(\rho_s \vec{v}_o < \vec{V}_s >)\right]_{CR} + \left[\frac{\partial}{\partial t}(\rho_s < \vec{V}_s \vec{v}_o >)\right]_{CR} + \left[\frac{\partial \overleftrightarrow{P}_s}{\partial t}\right]_{CR} \tag{27}$$

The first two terms are connected with the temporal and spatial variations of the kinetic energy of species "s" due to changes of the plasma as a whole, the first term on the r.h.s. is the collisional-radiative counter part. The fourth and fifth term on the l.h.s. are linked to temporal and spatial variations of the thermal energy (that is kinetic energy relative to a coordinate frame moving with $\vec{v}_o$). The last term on the r.h.s. is its collisional-radiative counter part. $2n_s < \vec{V}_s > \vec{F}_s$ represents the tensor of twice the electric power density directly given to the species "s". All other terms take into account various types of coupling between the species "s" and the plasma as a whole. The Eq. (27) can be simplified using Eq. (24).

We will directly go to the energy tensor of the plasma as a whole. Summing up all Eqs. (27) yields the tensor of twice the kinetic energy density :

$$\frac{\partial}{\partial t}(\rho \vec{v}_o \vec{v}_o) + \vec{\nabla}\cdot(\rho \vec{v}_o \vec{v}_o \vec{v}_o) + \frac{\partial \overleftrightarrow{P}}{\partial t} + 2\vec{\nabla}\cdot\overleftrightarrow{Q} - 2 \sum_s n_s \vec{v}_o \vec{F}_s$$

$$- 2 \sum_s n_s < \vec{V}_s > \vec{F}_s + \vec{\nabla}\cdot(\vec{v}_o \overleftrightarrow{P}) + \vec{\nabla}\cdot(\overleftrightarrow{P} \vec{v}_o) + \sum_s \vec{\nabla}\cdot(\rho_s < \vec{V}_s \vec{v}_o \vec{V}_s >$$

$$= \left[\frac{\partial}{\partial t}(\rho \vec{v}_o \vec{v}_o)\right]_{CR} + \left[\frac{\partial \overleftrightarrow{P}}{\partial t}\right]_{CR} \tag{28}$$

Taking from this equation the trace and dividing by two yields the rate equation for the translational (or kinetic) energy density. The r.h.s. of the resulting equation has the form

$$\left[\frac{\partial}{\partial t}(\frac{1}{2}\rho v_o{}^2)\right]_{CR} + \left[\frac{\partial}{\partial t}\ \frac{3}{2}\ p\right]_{CR} + \sum_{i=1}^{3}\ \frac{1}{2}\left[\frac{\partial}{\partial t}\ p_{ii}\right]_{CR} \tag{29}$$

and shows that collisional-radiative processes intervene in the increase of kinetic energy density of the plasma as a whole (first term) and of the thermal energy density $\bar{E}^{th}$ (second and last term), since

$$\frac{3}{2}\ p \cong \sum_s \frac{3}{2}\ n_s kT_s = \sum_s \bar{E}_s^{th} \equiv \bar{E}^{th} \tag{30}$$

The power injected into a plasma by a beam of energetic neutral particles is absorbed via collisions. These collisions ensure the heating process. However, a part of the injected energy eventually goes into an increase of directional kinetic energy of the plasma as a whole and is then lost for the heating process.

In order to obtain the rate equation for the pressure tensor $\vec{\vec{P}}$ we subtract from Eq. (28) the two tensor equations which are obtained when Eqs. (23) and (24) are multiplied by $\vec{v}_o$ from the left. The result is

$$\frac{\partial\vec{\vec{P}}}{\partial t} + (\vec{\nabla}\cdot\vec{v}_o)\vec{\vec{P}} + (\vec{v}_o\cdot\vec{\nabla})\vec{\vec{P}} + (\vec{\nabla}\cdot\vec{\vec{P}})\ \vec{v}_o + (\vec{\vec{P}}\cdot\vec{\nabla})\ \vec{v}_o - 2\ \vec{v}_o\vec{\nabla}\cdot\vec{\vec{P}}$$

$$+ \sum_s \vec{\nabla}\cdot(\rho_s < \vec{V}_s\vec{v}_o\vec{V}_s >) - 2\sum_s n_s < \vec{V}_s > \vec{F}_s = \left[\frac{\partial\vec{\vec{P}}}{\partial t}\right]_{CR} \tag{31}$$

Taking from this expression the trace and dividing by two yields – without approximations – the following rate equation for the thermal energy density $\bar{E}^{th}$:

$$\frac{\partial}{\partial t}\ \bar{E}^{th} + \vec{v}_o\cdot\vec{\nabla}\ \bar{E}^{th} + (\bar{E}^{th} + p)\ \vec{\nabla}\cdot\vec{v}_o + \vec{\nabla}\cdot\vec{Q}^{th}$$

$$+ \frac{\partial}{\partial t}\ \sum_i \frac{3}{2}p_{ii} + \sum_i v_i\partial_i\ \frac{3}{2}p_{ii} + \sum_{i,j}\ \frac{1}{2}\ (p_{ij}\partial_j v_i + p_{ij}\partial_i v_j)$$

$$= \left[\frac{\partial\bar{E}^{th}}{\partial t}\right]_{CR} + \sum_i\left[\frac{\partial}{\partial t}\ \frac{3}{2}\ p_{ii}\right]_{CR} + \dot{\Omega} \tag{32}$$

where

$$\dot{\Omega} = \sum_s n_s < \vec{V}_s > \cdot\ \vec{F}_s = \text{ohmic power density}\ , \tag{33}$$

$$\vec{Q}^{th} = \sum_s \frac{1}{2}\ \rho_s < \vec{V}_s\cdot\vec{V}_s\vec{V}_s > = \text{thermal heat flux vector}\ , \tag{34}$$

and where v_i, p_{ij}, ∂_i stands respectively for v_{ox}, ..., p_{xy}, ..., $\partial/\partial x$, ...

The collisional-radiative terms of Eq. (32) can in principle not be decoupled from the collisional-radiative term in the rate equation for the internal energy and from the equation of radiative transfer, since thermal energy can be converted in internal energy (i.e. by electronic excitation or ionization) and in photon energy (e.g. bremsstrahlung) via inelastic collisions, and vice versa internal energy can be converted in thermal energy via superelastic collisions and in photon energy via spontaneous radiation and induced collision and radiation processes. It is just the collisional-radiative term in the individual rate equations which ensures the conversion from one energy form into the other.

2.4.2 - Internal energy density

The simplest way to establish the rate equation is to give all internal energy to the heavy particles (molecules, atoms, ions). Thus, electrons possess only translational energy.

Let $E^z_{k,i}$ be the internal energy of a z-times ionized particle of chemical species k in quantum state $| \, i >$. The definition of $E^z_{k,i}$ is given in Fig. 1. For ionized species, $E^z_{k,i}$ contains the sum of the ionization energies of all lower lying stages of ionization. The total internal energy density of all heavy particles is then given by

$$\bar{E}^{int} = \sum_z \sum_k \sum_i n^z_{k,i} \, E^z_{k,i} \equiv \sum_s n_s E^{int}_s \equiv \sum_s \bar{E}^{int}_s \tag{35}$$

Associated with the internal energy is a heat flux vector for the transport of internal energy (reaction heat conductivity) defined by

$$\vec{Q}^{int} = \sum_z \sum_k \sum_i n^z_{k,i} E^z_{k,i} < \vec{V}^z_{k,i} > \equiv \sum_s n_s E^{int}_s < \vec{V}_s > \equiv \sum_s \bar{E}^{int}_s < \vec{V}_s > \tag{36}$$

The rate equation for the internal energy density of species "s" follows from Eq. (6) with n_s replaced by $n_s E^{int}_s$:

$$\frac{\partial n_s E^{int}_s}{\partial t} + \vec{\nabla}\cdot(n_s \, E^{int}_s < \vec{w}_s >) = \left[\frac{\partial n_s E^{int}_s}{\partial t} \right]_{CR} \tag{37}$$

Summing over all (heavy) particles "s" and taking into account the Eqs. (7) (35) and (36) yields for the total internal energy density

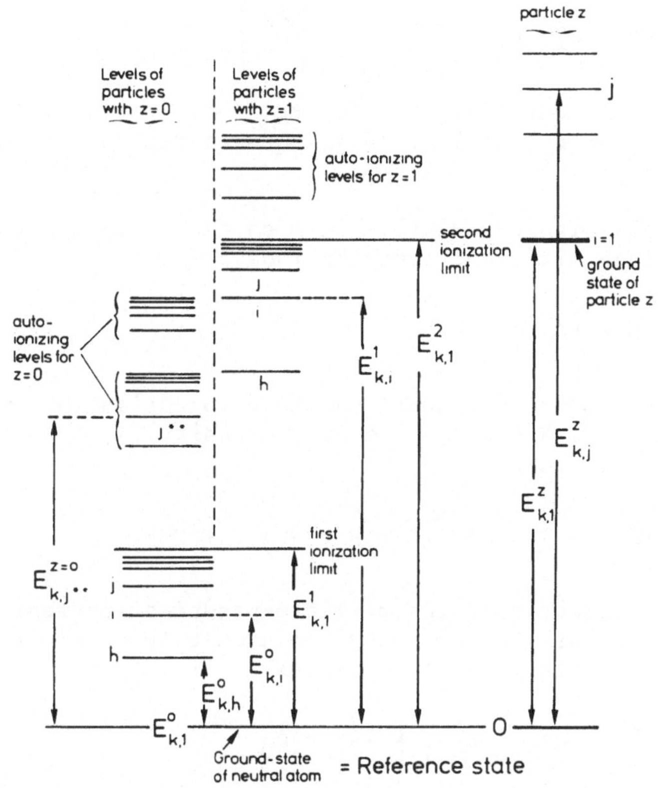

Fig. 1 Definition of the internal energies (excitation and ionization energies) Details are discussed in Ref.[130].

$\bar{E}^{int}$ the rate equation

$$\frac{\partial \bar{E}^{int}}{\partial t} + \vec{v}_o \cdot \vec{\nabla} \bar{E}^{int} + \bar{E}^{int}\, \vec{\nabla} \cdot \vec{v}_o + \vec{\nabla} \cdot \vec{Q}^{int} = \left[\frac{\partial \bar{E}^{int}}{\partial t} \right]_{CR} \quad (38)$$

The r.h.s. ensures the coupling with the thermal energy and with the photon field, since any collisional-radiative change of $\bar{E}^{int}$ changes either the thermal or the radiation energy.

2.4.3 - Thermal plus internal energy density

The Eqs. (32) and (38) give together the rate equation for the total thermal plus internal energy density which we shall denote by $\bar{E}$. With the definitions

$$\bar{E} = \bar{E}^{th} + \bar{E}^{int} \quad (a) \qquad \vec{Q} = \vec{Q}^{th} + \vec{Q}^{int} \quad (b) \qquad (39)$$

the following rate equation is obtained

$$\frac{\partial \bar{E}}{\partial t} + \vec{v}_o \cdot \vec{\nabla}\bar{E} + (\bar{E} + p)\vec{\nabla}\cdot\vec{v}_o + \vec{\nabla}\cdot\vec{Q} = \dot{\Omega} + \left[\frac{\partial \bar{E}}{\partial t}\right]_{CR} \tag{40}$$

where we have omitted all terms from Eq. (32) involving tensor components p_{ii} and p_{ij}. If necessary, these terms can be added to Eq. (40).

The collisional-radiative term on the r.h.s. of Eq. (40) can be split into different contributions :

$$\left[\frac{\partial \bar{E}}{\partial t}\right]_{CR} = \dot{I} + \dot{N} - \dot{R} \tag{41}$$

where $\dot{I}$ = power input by neutral particle injection, $\dot{N}$ = nuclear power heating rate, $\dot{R}$ = radiation power density. It thus follows the important relation

$$\frac{\partial \bar{E}}{\partial t} + \vec{v}_o \cdot \vec{\nabla}\bar{E} + (\bar{E} + p)\vec{\nabla}\cdot\vec{v}_o + \vec{\nabla}\cdot\vec{Q} + \dot{R} = \dot{\Omega} + \dot{I} + \dot{N} \tag{42}$$

At high particle densities the particle-produced isotropic thermal pressure $_s\Sigma n_s kT_s$ must be lowered by an amount Δp to first approximation given by (with ρ_D = Debye radius)

$$\Delta p \approx e_o^2 \left[n_e + \sum_z \sum_k \sum_i z_{k,i}^2 \, n_{k,i}^z\right] \Big/ 24\pi \, \varepsilon_o \rho_D \tag{43a}$$

$$= 5.6\cdot 10^{-31} \left[n_e + \Sigma\, z_{k,i}^2 \, n_{k,i}^z\right]^{3/2} T^{-1/2} \quad \text{Pascal} \tag{43b}$$

(T in Kelvin, n in m^{-3}, 10^5 Pascal ≈ 1 atmosphere).

The Eq. (42) shows that the energy losses due to the escape of radiation are of the same importance as the thermal head losses. $\dot{R}$ contains contributions from free-free (ff), free-bound (fb), di-electronic (di), charge exchange (cx), bound-bound (bb) and cyclotron (c) radiation :

$$\dot{R} = \dot{R}^{ff} + \dot{R}^{fb} + \dot{R}^{di} + \dot{R}^{cx} + \dot{R}^{bb} + \dot{R}^{c} \tag{44}$$

This decomposition is somewhat artificial, since radiation from di-electronic recombination represents pure line (i.e. bound-bound) radiation and charge exchange itself is in the absence of an intense radiation field a radiationless process. Details are given in Section 7.

We have now all equations together for a deeper understanding of the atomic processes and their influence on the plasma properties. Further information on the MHD-equations and the calculation of transport properties for magnetically confined plasmas may be found in Refs.[6,7].

3. ATOMIC AND MOLECULAR PROCESSES DURING THE INITIAL PHASE OF TOKAMAK AND SIMILAR DISCHARGES

In machines of the Tokamak and Stellerator type the plasma production and heating phase must begin with dissociation and ionization of molecular hydrogen at base pressures of the order $1\cdot 10^{-4} \ldots 5\cdot 10^{-4}$ Torr. During this initial phase, the spatio-temporal evolution of the plasma depends on the penetration of the induced toroidal electric field of strength $\vec{E}_T$, on the magnetic field $\vec{B}$ and on the atomic and molecular reactions. In future large Tokamaks, $\vec{E}_T$ is of the order of 1 ... 5 V/m. For comparison : the dissociation energy of H_2° is 4.42 eV, the ionization of atomic hydrogen needs 13.58 eV, the ionization process $H_2^\circ(v = 0) + e^- \rightarrow H_2^+(v = 0) + 2e^-$ requires 15.7 eV. The low electric field strength favours thermalization by Coulomb collisions. In any case, during the first beginning the temperature will be low and deviations from a Maxwellian distribution for the electrons is possible.

The Table 1 summarizes rather schematically the possible reactions for two-body collisions at low and medium energies. Not all reactions are important for the plasma formation process. For diagnostic purposes by emission spectroscopy, however, numerous excitation and further reaction processes must eventually be accounted for depending on the kind of the information desired.

The Fig. 2 gives the cross-sections as a function of relative collision energy for the most important reactions for particles in the ground state. There is still scarce knowledge concerning the production of negative ions under low-pressure laboratory conditions. The maximum values of the cross-sections for reactions N° 6, 12, 17 and 21 are (for the particles in their ground state) :

N° 6 : $H_2^\circ = e^- \rightarrow H^+ + H^- + e^-$, $\sigma_{max} = 3.5\cdot 10^{-20}\,cm^2$, ([18])

N° 12 : $H_2^\circ + e^- \rightarrow H^+ + H^-$, $\sigma_{max} = 4.9\cdot 10^{-18}\,cm^2$, ([22])

N° 17 : $H_3^+ + e^- \rightarrow H_2^+ + H^-$, $\sigma_{max} = 1.6\cdot 10^{-18}\,cm^2$, ([19])

N° 21 : $H^\circ + e^- \rightarrow H^- + h\nu$, $\sigma_{max} = 5.8\cdot 10^{-24}\,cm^2$, ([18])

They can therefore not be made responsible for the high H^- concentrations found experimentally[20-21]. At a neutral gas density of $2\cdot 10^{14}\,cm^{-3}$ and electron temperatures of $kT_e = 0.4 \ldots 8$ eV the H_3^- concentration reached 30 % of the electron density ($n_e \approx 10^{10}\,cm^{-3}$).

Table 1

COLLISION PROCESSES IN A WEAKLY IONIZED PURE HYDROGEN PLASMA

N°	Reaction Process		
1	$H_2^o + e^-$	$\longrightarrow$	$H_2^+ + 2e^-$
2		$\longrightarrow$	$H^o + H^o + e^-$
3		$\longrightarrow$	$H^o + H^+ + 2e^-$
4		$\longrightarrow$	$H^+ + H^+ + 3e^-$
5		$\longrightarrow$	$H^o + H^-$
6		$\longrightarrow$	$H^+ + H^- + e^-$
7	$H_2^o + H_2^o$	$\longrightarrow$	$2H^o + H_2^o$
8	H^o	$\longrightarrow$	$3H^o$
9	$H_2^+ + e^-$	$\longrightarrow$	$H^o + H^o$
10		$\longrightarrow$	$H^o + H^+ + e^-$
11		$\longrightarrow$	$H^+ + H^+ + 2e^-$
12		$\longrightarrow$	$H^+ + H^-$
13	$H_2^+ + H_2^o$	$\longrightarrow$	$H_3^+ + H^o$
14	$H_3^+ + H$	$\longrightarrow$	$H^+ + 2H^o$
15	$H_3^+ + e^-$	$\longrightarrow$	$H_2^o + H^o$
16		$\longrightarrow$	$3H^o$
17		$\longrightarrow$	$H^- + H_2^+$
18	$H_3^+ + H_2^o$	$\longrightarrow$	$H^+ + H_2^o + H_2^o$
19		$\longrightarrow$	$H_2^+ + H + H^o$
20	$H^o + e^-$	$\longrightarrow$	$H^+ + 2e^-$
21		$\longrightarrow$	H^- $(+ h\nu)$
22	$H^- + e^-$	$\longrightarrow$	$H^o + 2e^-$
23		$\longrightarrow$	$H^+ + 3e^-$
24	$H^- + H^+$	$\longrightarrow$	H_2
25 ⋮	numerous excitation processes		

It has been proposed [10,21] that the high H^- production rate proceeds via vibrationally and/or electronically excited levels of the H_2 molecule according to reaction N° 5 specified as follows

$$\text{N° 5a : } H_2(^1X_g^+, v, J) + e^- \rightarrow H_2^- \; (^2\Sigma_u^+, v', J') \rightarrow H\,(1s) + H^-$$

$$\text{N° 5b : } H_2^o(^3\Pi_u, v, J) + e^- \rightarrow H_2^- \; (^2\Pi_u, v', J') \rightarrow H\,(2p) + H^-$$

The reaction 5a is extremely sensitive to changes in vibrational quantum number v and rotational quantum number J. For $v \approx 10$, σ^{max} reaches values of appr. $4.6 \cdot 10^{-16} cm^2$ for both H_2 and D_2, whereas for $v = 0$ the corresponding values are σ_{max} $(H_2) \approx 5 \cdot 10^{-21}$ cm^2, σ_{max} $(D_2) \approx 5 \cdot 10^{-24} cm^2$ ([26]). The reaction 5b yields cross-

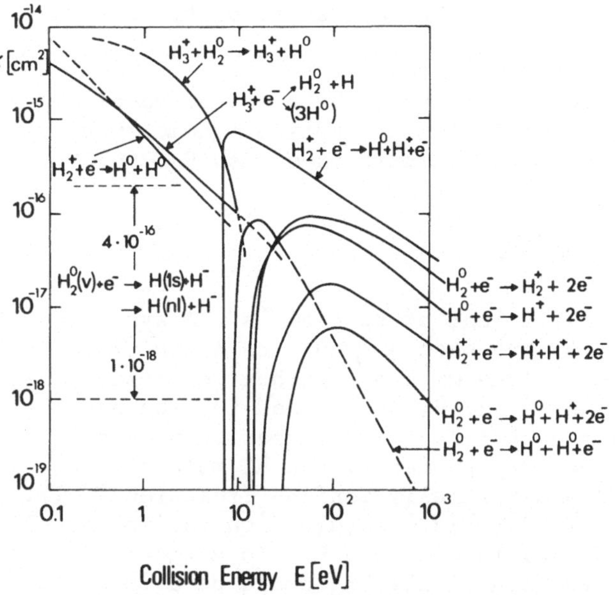

Fig. 2 Cross-sections involving hydrogen molecules, atoms and ions, after different authors.

sections in the range 10^{-18} to $2 \cdot 10^{-17}$ cm^2, see ([20]). This behaviour is schematically indicated in Fig. 2.

The ro-vibrational and electronic processes will influence all together the charge state distribution during the initial state of a discharge in hydrogen. Already the apparently simple dissociation process of a diatomic molecule involves numerous individual reactions and can strongly depend on the excitation of ro-vibrational levels[28]. According to reaction n° 5a, strong ro-vibrational excitation would then determine the H^- concentration. If H^- should be important during the initial phase of plasma production in the next generation of large Tokamaks and Stellerators, the H^- destroying processes must also be considered. The H^- producing processes could be of interest for neutral particle injectors.

For electron temperatures $kT_e < 10$ eV, the H_3^+ ion plays an important role, since H_2^+ which has been created by electron collisions is partially converted to H_3^+. The process is exothermic[27] by 1.17 eV(reaction N° 13). H_3^+ is destroyed via the reactions N° 15 to 19. The reactions N° 15 to 17 involve an intermediate H_3^o molecular state which then dissociates. The corresponding cross-sections have been taken from Refs.[23,24]. (The $H_3^{o\star}$ and $D_3^{o\star}$ molecules have only recently been identified in the discharge of a hollow-cathode [29]). The H_3^+ molecule can also be dissociated (reactions n° 18 and 19), the corresponding cross-sections are nearly energy-independent and are of the order of $1 \cdot 10^{-17} cm^2$ ([25]). Literature to the other cross-section curves in Fig. 2 is given in Ref.[17]. The rate coefficients $< \sigma v >$ calculated for a Maxwellian velocity distribution are shown in Fig. 3.

The particle densities are governed by a set of coupled differential equations of type (6). For cylindrical geometry with r as radial distance from the magnetic axis (see also Eq. (66)) :

$$\frac{\partial n_s}{\partial t} - \frac{1}{r}\frac{\partial}{\partial r}\left(r\, \mathcal{D}_s \frac{\partial n_s}{\partial r}\right) = \left[\frac{\partial n_s}{\partial t}\right]_{CR} \tag{45}$$

where $\mathcal{D}_s$ is the diffusion coefficient of species "s". A system of coupled equations of type (45) has been solved[17] as a function of time for an uniformly filled volume with and without diffusion term, including the six reactions N° 1, 2, 9, 10, 13 and 20. The essential result is that for $kT \geqslant 5$ eV the plasma formation process seems to be primarily a three-step process : the H_2^0 molecule is dissociated (reaction N° 2) and partially ionized (reaction N° 1). Due to the high reactivity of H_2^+, this molecular ion is immediately dissociated (reactions N° 9, 10) into H° and H^+. In a final step, the H° atoms are ionized (reaction N° 20). H_3^+ could be eliminated from the reactions without appreciably changing the solution for the other species. Eliminating the reactions involving H_2^+ leads to much longer ionization times. This is in broad agreement with the ionization processes in a diffusion dominated stationary glow discharge in hydrogen gas[30].

At present, the models suffer from the fact that Maxwellian distribution functions and equal atom, ion an electron temperatures are assumed. A more consistent model should at least include the energy balance equations for the thermal and internal energy of

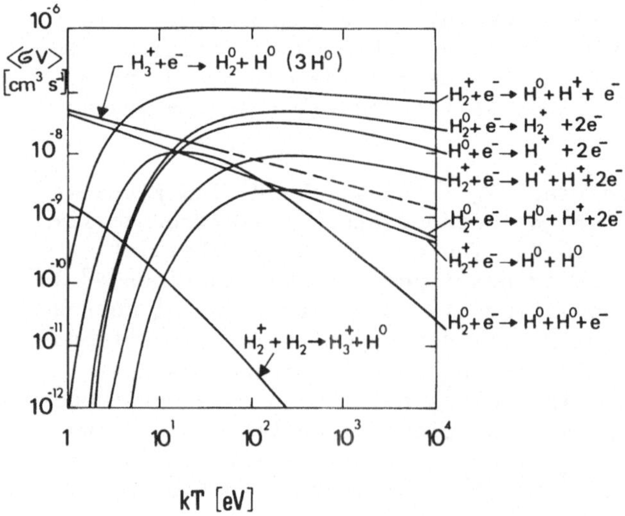

Fig. 3 Rate coefficient for Maxwellian velocity distributions.

each individual species.-Further references are given in [5].

4. SPATIAL DISTRIBUTION OF NEUTRAL HYDROGEN ATOMS IN THE HIGH-TEMPERATURE STATIONARY STATE

4.1 - Estimate from coronal ionization formula

A first estimate of the order of magnitude of the hydrogen atom density over the radius can be obtained from the coronal ionization equation applied to the local stationary values of measured n_e and T_e values. Equation (6) applied to atomic hydrogen in the ground state writes in this special case

$$\frac{\partial n_{H^\circ}}{\partial t} - \vec{\nabla}\cdot(\mathcal{D}_{H^\circ}\vec{\nabla} n_{H^\circ}) = - n_e n_{H^\circ} S_o - n_{H+} n_{H^\circ} P_o + n_e n_{H+} R_+ \qquad (46)$$

with the rate coefficient S_o, R_+ and P_o defined according to the reactions (48). For R_+ one generally assumes (see e.g. Ref.[31]) that all electrons which have been captured in any level of principal quantum number n contribute to R_+, hence $R_+ = {}_n\Sigma R_n = {}_n\Sigma < \sigma_n^R v >$ where $R_n = < \sigma_n^R v >$ is the recombination coefficient for level n. This picture is surely oversimplified, since captured electrons can be re-ejected into the continuum and particles which have been excited from the ground state may be ionized prior to deexcitation. In refined models the kinetics of the excited levels has therefore also been taken into account in the high-temperature regime[32,33,35].

The cross-sections for electronic ionization and recombination are shown in Figs. 4 and 6, the corresponding rate coefficients are given in Figs. 5 and 7. An often applied approximation

$$R_n = 2.1\cdot 10^{-11}\ T_{[K]}^{-1/2}\ n^{-3} \qquad cm^3 s^{-1} \qquad (47)$$

valid for $kT < 10$ eV should be avoided above this temperature, see Fig. 7.

For a stationary ($\partial/\partial t = 0$) and homogeneous ($\vec{\nabla}\cdot = 0$) plasma a collisional-radiative model yields for the ratio n_{H°/n_{H^+} the values shown in Fig. 8. It follows from this that a plasma of temperature of order $kT \approx 1$ keV ($T \approx 10^7$ K) and proton (deuteron, triton) density $n_{H^+} \approx 5\cdot 10^{13} cm^{-3}$ would contain a neutral atom density $n_{H^\circ} \approx 10^6 cm^{-3}$. In present-day Tokamak machines, the actual neutral particle densities lie one to three orders of magnitude above this value due to diffusion which is strongly influenced by charge exchange reactions. The diffusion effect can be taken into account in solving the Boltzmann equation.

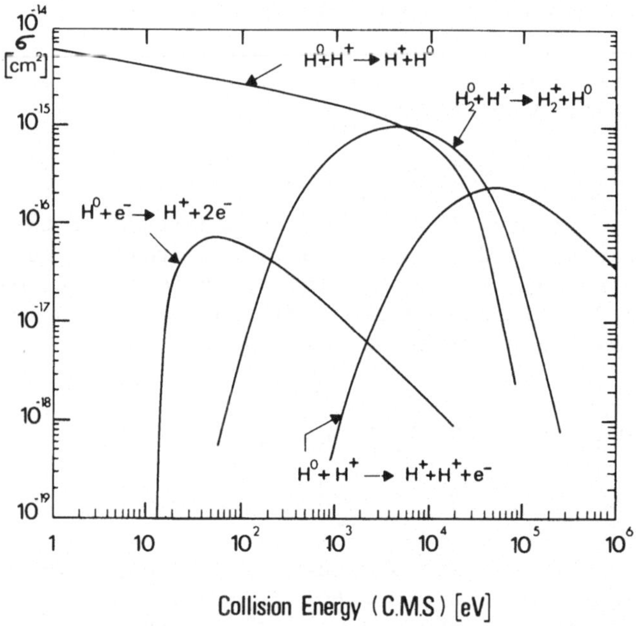

Fig. 4 Cross-sections for ionization of atomic hydrogen, after[34]. C.M.S. means center-of-mass system.

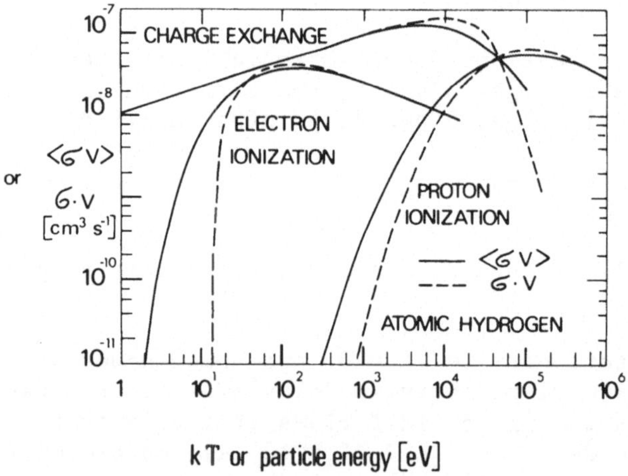

Fig. 5 Rate coefficients for ionization of atomic hydrogen, after[34]. $\sigma \cdot v$ is for a mono-energetic, $\langle \sigma v \rangle$ for a Maxwellian velocity distribution

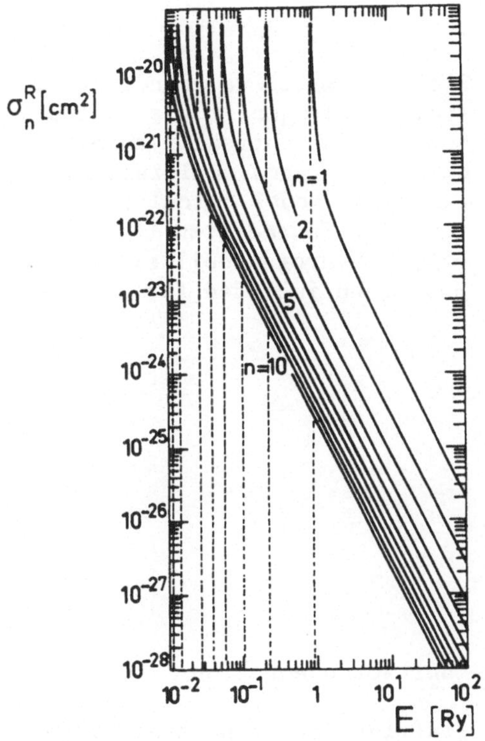

Fig. 6 Cross-sections for radiative recombination of electrons with protons (deuterons, tritons).

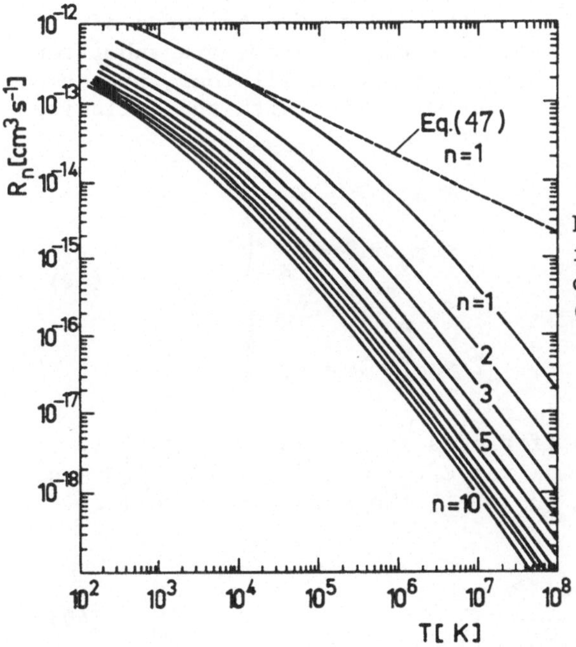

Fig. 7 Rate coefficients for radiative recombination of electrons with protons (deuterons, tritons).

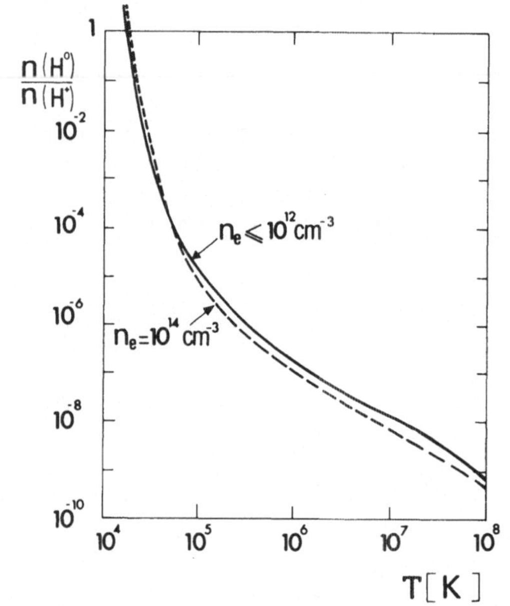

Fig. 8 Relative abundance of atomic hydrogen in a stationary homogeneous optically thin hydrogen plasma. Collisional-radiative model calculations, after[35]. In addition to [35], proton-hydrogen collisions have been included for $T > 10^7$ K.

4.2 - Calculation with the aid of the Boltzmann equation

Finer details of the radial distribution of neutral hydrogen atoms are obtained by solving the Boltzmann equation (4).

In order to get a feeling of the order of magnitudes of the effects involved we will assume that the H° atoms are submitted to electron impact ionization, charge exchange collisions, proton impact ionization and electron-ion recombination with the following cross-sections σ and rate coefficients $< \sigma v >$:

$$\left.\begin{array}{llll} H^{\circ} + e^{-} \rightarrow H^{+} + 2e^{-} & , \quad \sigma_e & , & < \sigma_e v > = S_o \\ H^{\circ} + \underline{H}^{+} \rightarrow H^{+} + \underline{H}^{\circ} & , \quad \sigma_{cx} & , & < \sigma_x \, v > = C_x \\ H^{\circ} + H^{+} \rightarrow H^{+} + H^{+} + e^{-} & , \quad \sigma_p & , & < \sigma_p v > = P_o \\ H^{+} + e^{-} \rightarrow H^{\circ}(n) + h\nu & , \quad \sigma_n^R & , & \sum_n < \sigma_n^R v > = R_+ \end{array}\right\} \qquad (48)$$

The cross-sections and rate coefficients are shown in Figs. 4 to 7. Hydrogen atoms of mean thermal velocity $< w_o > \approx 1 \cdot 10^4 \; T_{[K]}^{1/2} \mathrm{cm \; s^{-1}}$ have in a pure hydrogen plasma ($n_e = n_+$) the effective mean free path

$$\lambda_o = \frac{< w_o >}{n_e \, [S_o + C_x + P_o]} \approx \frac{< w_o >}{n_e C_x} \qquad (49)$$

For $n_e = 10^{14} cm^{-3}$ at $kT \approx 1$ keV, it follows $\lambda_o \approx 3$ cm. The neutral atoms can travel the distance λ_o before being re-ionized. The neutral gas at the boundary can therefore penetrate deep into the hot plasma through a sequence of charge exchange reactions $H^\circ \rightarrow H^+ \rightarrow H^\circ \rightarrow H^+ \rightarrow H^\circ$ ... . The actual local particle density is determined by all reactions (48). - On the other hand, ions which have undergone charge changing reactions can leave the magnetically confined plasma region. This effect is applied for ion temperature measurements.

For the stationary state ($\partial/\partial t = 0$) and no external forces, the Eq. (4) becomes

$$\vec{w}_o \cdot \frac{\partial}{\partial \vec{r}} f_o(\vec{r}, \vec{w}_o) = \left[\frac{\partial}{\partial t} f_o(\vec{r},\vec{w}_o)\right]_{CR} \tag{50}$$

where f_o is the velocity distribution function of neutral hydrogen atoms of velocity $\vec{w}_o$ in direction of the unit vector $\vec{\Omega}$, hence $w_o = \vec{w}_o . \vec{\Omega}$. The operator $\vec{w}_o \cdot \partial/\partial\vec{r}$ acting on f_o can be interpreted as a directional derivative in direction $\vec{\Omega}$.

Introducing the distance s along $-\vec{\Omega}$ from the point defined by $\vec{r}$ (see Fig. 9), the Eq. (50) becomes

$$-\frac{\partial}{\partial s} f_o(\vec{r} - s\vec{\Omega}, \vec{\Omega}, w_o) = \frac{1}{w_o}\left[\frac{\partial}{\partial t} f_o(\vec{r} - s\vec{\Omega}, \vec{\Omega}, w_o)\right]_{CR} \tag{51}$$

For the calculation of the collisional-radiative term we take into account all processes (48), but we neglect scattering ; in other words : it is assumed that the velocity vector of "test-particles" is not modified by the collision processes. When applying the definition Eq. (3), the total reaction rate for collisions of particles (1) interacting with particles (2) is

$$n_1 n_2 < \sigma v > \equiv < f_1 f_2 \sigma v > = \int_{\vec{w}_1} d^3w_1 \int_{\vec{w}_2} d^3w_2 \; f_2(\vec{w}_2)\sigma(|\vec{w}_1 - \vec{w}_2|)\,|\vec{w}_1 - \vec{w}_2| \tag{52}$$

where $v = |\vec{w}_1 - \vec{w}_2|$ is the relative velocity. The expression

$$n_2 < \sigma v > = < f_2 \sigma v > = \int_{\vec{w}_2} d\vec{w}_2 f_2(\vec{w}_2)\sigma(|\vec{w}_1 - \vec{w}_2|)\,|\vec{w}_1 - \vec{w}_2| \tag{53}$$

gives the collisions frequency of a particle (1) of velocity $\vec{w}_1$ interacting with all particles (2).

We consider particularly the propagation of hydrogen atoms with velocity $\vec{w}_o$. Eq. (51) writes

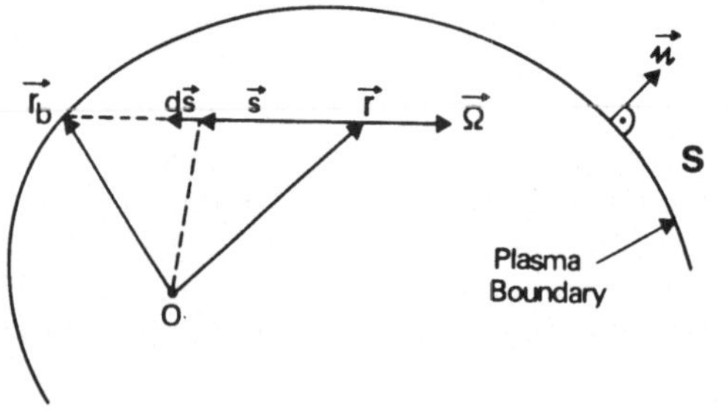

Fig. 9 Definition of the quantities intervening in Eqs. (51) - (63). S denotes the plasma surface.

$$-\frac{\partial}{\partial s} f_o(\vec{r}-s\vec{\Omega},\vec{\Omega},w_o) = - f_o(\vec{r}-s\vec{\Omega},\vec{\Omega},w_o)\left[<f_e\sigma_e v>+<f_p\sigma_p v>+<f_p\sigma_x v>\right]\frac{1}{w_o}$$
$$+ f_p(\vec{r}-s\vec{\Omega},\vec{\Omega},w_o)\left[<f_e\sigma_r v>+<f_o\sigma_x v>\right]\frac{1}{w_o} \quad (54)$$

with $f_e(f_p)$ = electron (proton) velocity distribution function. All functions f are to be taken at $\vec{r} - s\vec{\Omega}$. With the abbreviations

$$Q_o(\vec{r}-s\vec{\Omega},\vec{\Omega},w_o) = [< f_e\sigma_e v > + < f_p\sigma_p v > + < f_p\sigma_x v >] \frac{1}{w_o} \quad (55a)$$

$$S_p(\vec{r}-s\vec{\Omega},\vec{\Omega},w_o) = f_p(\vec{r}-s\vec{\Omega},\vec{\Omega},w_o) [\sum_n < f_e\sigma_n^R v> + < f_o\sigma_x v >] \frac{1}{w_o} \quad (55b)$$

the general solution of Eq. (54) becomes

$$f_o(\vec{r}-s\vec{\Omega},\vec{\Omega},w_o) = f_o(\vec{r}-s_o\vec{\Omega},\vec{\Omega},w_o) \exp\left\{\int_{s_o}^{s} ds'' \, Q_o(\vec{r}-s''\vec{\Omega})\right\}$$
$$+ \int_{s}^{s_o} ds' S_p(\vec{r}-s'\vec{\Omega},\vec{\Omega},w_o) \exp\left\{\int_{s'}^{s} ds'' \, Q_o(\vec{r}-r''\vec{\Omega})\right\} \quad (56)$$

where s_o is some arbitrary distance from the point $\vec{r}$. Putting s = 0 one obtains the value of $f_o(\vec{r},\vec{\Omega},w\ )$ at the point $\vec{r}$ for atoms moving in direction $\vec{\Omega}$ with velocity w_o. The integration constant $f_o(\vec{r}-s_o\vec{\Omega},\vec{\Omega},w_o)$ is determined by the boundary condition. The plasma boundary is given by the position vector $\vec{r}_b$. We choose s_o such that $\vec{r} - s_o\vec{\Omega} = \vec{r}_b$ (hence $s_o = |\vec{r} - \vec{r}_b| \equiv s_b$) and denote the distribution function at the boundary by

$$f_o(\vec{r}_b,\vec{\Omega},w_o) = f_o(\vec{r} - s_o\vec{\Omega},w_o)\Big|_{s_o = |\vec{r} - \vec{r}_b| \equiv s_b} \quad (57)$$

The following final result is obtained

$$f_o(\vec{r},\vec{\Omega},w_o) = f_o(\vec{r}_b,\vec{\Omega},w_o)\ \exp\left\{-\int_0^{s_b} ds''\ Q_o(\vec{r} - s''\vec{\Omega})\right\}$$
$$+ \int_0^{s_b} ds' S_p(\vec{r}-s'\vec{\Omega},\vec{\Omega},w_o) \exp\left\{-\int_0^{s'} ds'' Q_o(\vec{r}-s''\vec{\Omega})\right\} \quad (58)$$

The first term on the r.h.s. is the contribution to $f_o(\vec{r},\vec{\Omega},w_o)$ originating from atoms due to an external source $f_o(r_b,\vec{\Omega},w_o)$ attenuated by the factor $\exp[-\int ds'' Q_o(\vec{r} - s''\vec{\Omega})]$ along the distance s_b, the second term accounts for the production of neutrals from protons inside the plasma at all points $s' < s_b$ and attenuated by the factor $\exp[-\int ds''\ Q_o(\vec{r}-s''\vec{\Omega})]$ along s'.

The solution of $f_o(\vec{r},\vec{\Omega},w_o)$ depends on the plasma-internal source term S_p which is itself a function of f_o, see Eq. (55b). The The closure relation is obtained by putting in Eq. (55b) s = 0. With f_o from Eq. (58) , it follows that

$$S_p(\vec{r},\vec{\Omega},w_o) = \frac{1}{w_o} f_p(\vec{r},\vec{\Omega},w_o) \Big[{}_n\Sigma < f_e \sigma_n^R v >$$
$$+ \int_{w_o'} d^3w_o'\ f_o(\vec{r},\vec{\Omega},w_o')\sigma_x(|\vec{w}_o - \vec{w}_o'|)\,|\vec{w}_o - \vec{w}_o'| \Big] \quad (59)$$

Knowing $f_o(\vec{r},\vec{\Omega},w_o)$ one can calculate the neutral particle density

$$n_{H^o}(\vec{r}) = \int f_o(\vec{r},\vec{\Omega},w_o)d(\vec{w}_o\cdot\vec{\Omega}) = \iiint f_o(\vec{r},\vec{\Omega},w_o)w_o^2\ \sin\vartheta\ \cos\varphi\ d\vartheta\ d\varphi\ dw_o, \quad (60)$$

the neutral flux density

$$\Gamma_o(\vec{r}) = \int w_o\ f_o(\vec{r},\vec{\Omega},w_o)d(\vec{w}_o\cdot\vec{\Omega}) , \quad (61)$$

the neutral current density in a given direction $\vec{\Omega}$

$$\vec{j}_o(\vec{r},\vec{\Omega}) = \vec{\Omega} \int_{w_o} w_o f_o(\vec{r},\vec{\Omega},w_o)dw_o , \quad (62)$$

and the power lost by the plasma due to the escape of neutral atoms through the plasma boundary

$$\dot{P}_o = \int_{\text{surface S}} dS\ \vec{w}\cdot\vec{\Omega} \int_{w_o} \tfrac{1}{2}m_o w_o^2\ w_o f_o(\vec{r}_b,\vec{\Omega},w_o)dw_o\ ,\ \vec{w}\cdot\vec{\Omega} > 0 \quad (63)$$

where $\vec{n}$ is a unit vector normal to the boundary surface S.

Numerical results are shown in Fig. 10 for the ALCATOR Tokamak at a mean plasma density of $5.10^{14}cm^{-3}$ as a function of radial distance from the magnetic axis. (Parameters of ALCATOR[36] : R = 54 cm, a = 12 cm, $B_T \approx$ 6 ... 8.5 Tesla, $kT_e(r = 0) \approx 600 \ldots 1000$ eV, $kT_i(r = 0) \approx 575 \ldots 691$ eV, $I_p \approx 160$ kA). Curve 1 follows when the assumption is made that a coronal-type ionization-recombination balance is established locally. From the measured values of $n_e(r)$ and $T_e(r)$ the radial distribution $n_{H^\circ}(r)$ can immediately be calculated (application of Fig. 8). Curve 2 follows as solution of the Boltzmann equation when all reactions (48) are taken into account. Curve 3 finally is obtained when the recombination process is dropped in the Boltzmann equation. The ALCATOR device is running at relatively high plasma densities, therefore recombination plays a very important role, since the plasma is in the central region "collisionally thick" in relation to charge exchange.

Quite similar computer codes are currently applied in many laboratories for studying radial particle distributions, particle and energy fluxes and energy distribution functions in fusion plasma machines, see e.g.[38,39]. In a consistent treatment the particle, momentum and energy transfer equations for neutral atoms, electrons and ions must be solved simultaneously. This is currently done in solving the velocity integrated Boltzmann equation and its higher velocity moments (i.e. the rate equations for particle density, momentum and energy density described in Section 2 of the present article) see e.g. Refs.[7,40-45].

5. THE DYNAMICS OF IMPURITY SPECIES

The behaviour of impurity species in magnetically confined plasmas depends on several effects which have not yet clearly been identified. The term "dynamics" circumscribes all effects related to time-dependent and time-independent diffusion and to the spatial distribution of the various stages of ionization of the different impurity elements.

The instantaneous local particle density is given by Eq. (6). For a stationary plasma ($\partial/\partial t = 0$) of cylindrical geometry it follows (with $< w_s >_r$ as radial velocity)

$$\frac{1}{r}\,\frac{\partial}{\partial r}(r\, n_s < w_s >_r = \left[\frac{\partial n_s}{\partial t}\right]_{CR} , \tag{64}$$

hence, for the diffusion flux of species "s" relative to the laboratory system :

$$\Gamma_{s,r} = n_s < w_s >_r = \frac{1}{r} \int_0^r r' \left[\frac{\partial n_s}{\partial t}\right]_{CR} dr' \qquad (65)$$

Both the experimental and the theoretical values intervening in the expression $[\partial n_s/\partial t]_{CR}$ are not yet known with sufficient precision in order to determine accurate values of $\Gamma_{s,r}$. Impurity flux evaluation[46] from the measured radial impurity distributions of iron ions in the PLT Tokamak led to the conclusion that an uncertainty of a factor ± 2 in the ionization and recombination rates yields 30 or 1/30 times the neoclassical diffusion coefficient for the region $r \leqslant 15$ cm.

Macroscopically, the diffusion flux can be expressed as a function of the diffusion coefficient $\mathcal{D}_s$ provided the driving forces are known. The latter depend on the spatial distribution of particle densities and temperature and on the species' velocities $<\vec{w}_s>$.

5.1 - Diffusion coefficients

We will approximate the density of the diffusion flux in radial direction by

$$n_s < w_s >_r = \Gamma_{s,r} = - \mathcal{D}_{s,r} \frac{\partial n_s}{\partial r} \qquad (66)$$

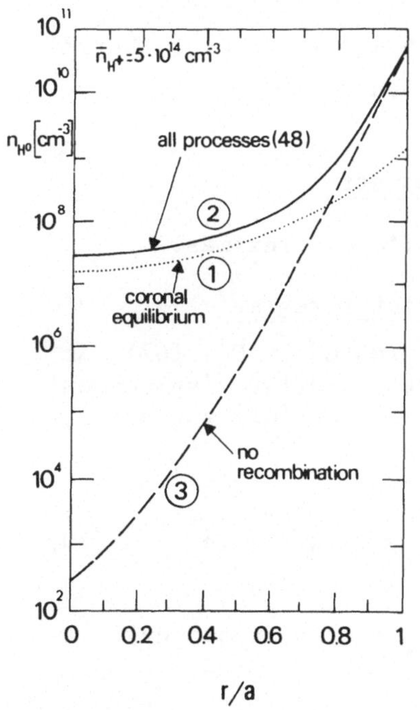

Fig. 10 Radial distribution of the density n_{H° of hydrogen atoms calculated for ALCATOR, after[37]. The mean proton density is $\bar{n}_{H^+} = 5\cdot10^{14}\text{cm}^{-3}$.

The diffusion process can be considered as a random walk across the plasma-containing magnetic field $\vec{B}$, with step length $\Delta\ell$ and the time τ_s between steps, hence

$$\mathcal{D}_{s,r} \approx \frac{\Delta\ell}{\tau_s} = \nu_s \; \Delta\ell \quad , \quad \nu_s = \text{collision frequency} \tag{67}$$

5.1.1 - Classical diffusion across a uniform magnetic field

The step length $\Delta\ell$ is given by the thermal gyro-radius $r_{g,s} = \langle v_s^{th} \rangle / \Omega_s = (2kT_s/m_s)^{1/2}/(q_s B/m_s)$. Every (momentum changing) collision yields a new particle orbit in a distance $\Delta\ell \approx r_{g,s}$ from the one prior to the collision. For the classical diffusion coefficient thus follows the expression

$$\mathcal{D}^{cl}_{s,r} \approx r^2_{g,s} \; \nu_s = \left(\frac{2m_s kT_s}{e_o^2 z_s^2 B^2} \right) \nu_s \tag{68}$$

In the case of Coulomb collisions of an ion "s" with all ions of species s' :

$$\nu_{ss'} = \frac{4\pi \; n_{s'}}{\langle v_s^{th} \rangle^3} \; \frac{e_o^2 z_s^2 \; e_o^2 z_{s'}^2}{(4\pi\varepsilon_o)^2} \cdot \left(\frac{m_s + m_{s'}}{m_s \; m_{s'}} \right)^2 \ell n \; \Lambda \propto (kT_s)^{-3/2} \tag{69}$$

where $\ell n \; \Lambda$ is the Coulomb logarithm[31] and $\langle v_s^{th} \rangle$ the mean thermal velocity of species "s". Further $\nu_s = \sum_{s' \neq s} \nu_{ss'}$.

5.1.2 - Classical diffusion in toroidal geometry

One distinguishes between three different regimes.

(i) The collision dominated or Pfirsch-Schlüter regime

The diffusion coefficient is approximated by Eq. (67). An additional step length arises from the particle drift displacement from a magnetic surface. With the rotational transform angle ι given by (with R = radius of magnetic axis) :

$$\iota(r) = 2\pi \left(\frac{R}{r}\right) \left(\frac{B_\Theta(r)}{B_T}\right) = 2\pi \; q(r) \quad , \; q(r) = \text{safety factor} \tag{70}$$

the additional step length becomes (with Ω_s = gyro-frequency)

$$\Delta \ell = \left(\frac{2\pi}{\iota}\right)^2 \frac{R}{\Omega_s} \quad \nu_s \tag{71}$$

The time τ between two steps is assumed to be the time a particle needs to cross a section of a line of force in which the curvature does not change :

$$\tau \approx \left(\frac{2\pi}{\iota}\right)^2 \left(\frac{R}{< v_s^{th} >}\right)^2 \nu_s \tag{72}$$

This yields the so-called Pfirsch-Schlüter diffusion coefficient

$$\mathcal{D}_{s,r} \simeq \mathcal{D}_{s,r}^{cl} \left[1 + \left(\frac{2\pi}{\iota}\right)^2\right] \quad , \qquad \iota < 1 \tag{73}$$

(ii) The collision-less (or banana) regime

In this regime a class of particles is trapped in the toroidal magnetic mirrors. The diffusion coefficient depends on particle drift displacement and collision frequency. Theory yields[49] for the diffusion coefficient

$$\mathcal{D}_{s,r} = \mathcal{D}_{s,r}^{cl} \; 3.6 \left(\frac{2\pi}{\iota}\right)^2 \varepsilon^{-3/2} \tag{74}$$

where $\varepsilon = a/R$ is the aspect ratio, a the plasma radius. $\varepsilon^{1/2}$ is the fraction of trapped particles.

(iii) The intermediate (or plateau) regime

The diffusion coefficient is nearly independent of the collision frequency and given by[49].

$$\mathcal{D}_{s,r} = 2\pi^{1/2} \frac{r_{g,s}}{a} \varepsilon \left(\frac{2\pi}{\iota}\right) \left(\frac{kT}{e_o z_s B}\right) \tag{75}$$

All three diffusion coefficients Eq. (72) - (74) are so-called "neo-classical diffusion coefficients", but often the term "neo-classical diffusion" is limited to the collision and intermediate diffusion regime.

For further details concerning plasma confinement and diffusion, see e.g. Refs.[6,7,46,50-52].

5.1.3 - Anomalous diffusion

All experiments yield diffusion coefficients in excess to those predicted theoretically. Bohm[53] observed in 1949 in a magnetically confined arc plasma a diffusion rate which could empirically be described by the diffusion coefficient

$$\mathcal{D}_r = \frac{kT_e}{16e_o B} \tag{76}$$

A derivation of the T_e/B dependence is given in Refs.[54,55], for instance.

Impurity measurements on the ASDEX Tokamak have led to the conclusion that the impurity flux density in radial direction is empirically given by[56-57]

$$\Gamma_{s,r} = n_s < w_s >_r = - D'_s \left(\frac{\partial n_s}{\partial r} + \frac{2r}{a^2} n_s \right) \tag{77a}$$

where D'_s is of the order of 4000 cm^2/s independent of the charge state. Equating Eqs. (66) and (77a) yields the following expression for this type of diffusion coefficient :

$$\mathcal{D}_{s,r} = D'_s \left[1 + \frac{2r}{a^2} \frac{n_s}{\partial n_s / \partial r} \right] \tag{77b}$$

An expression similar to Eq. (77a) has been introduced in Ref.[58]. Time-dependent classical diffusion in toroidal configurations is treated in Ref.[59].

5.2 - Diffusion fluxes , a case study

More insight in the particle dynamics is gained by considering the rate equations for particle and momentum density. The instantaneous local particle density is given by Eq. (6). For a stationary hydrogen plasma containing small impurity concentrations ($s \equiv k,i,z$) assumed to be in the ground state ($i = 1$) we have

$$\vec{\nabla}.(n_e < \vec{w}_e >) = \Sigma \text{ (ioniz. rates)} - \Sigma \text{ (recomb. rates)} \tag{78a}$$

$$\vec{\nabla}.(n_p < \vec{w}_p >) = \Sigma \text{ (ioniz. rates)}_{H^o} - \Sigma \text{ (recomb. rates)}_{H^o} \tag{78b}$$

$$\vec{\nabla}.(n_k^z < \vec{w}_k^z >) = \Sigma \text{ (populate rates)}_k^z - \Sigma \text{ (depopul. rates)}_k^z \tag{78c}$$

where the subscript e(p) indicates electrons (protons). Population (depopulation) comprises all process leading to an increase (decrease) of z-times ionized particles of chemical species k. The quasi-neutrality condition imposes $- e_o n_e + e_o n_p + \sum_{z,k} e_o z_k n_k^z = 0$.

In order to solve the rate equations for the particle densities we need not only the rate coefficients for the individual atomic processes but also the velocities $< \vec{w}_s >$. The latter can be obtained from the momentum balance equation (16) in conjunction with Eq. (23). For the stationary state ($\partial/\partial t = 0$) and a fluid inertia-and external momentum source-free plasma it follows* that

$$\vec{\nabla} p_e + e_o n_e (\vec{E} + < \vec{w}_e > x \vec{B}) = - R_{e,p} [< \vec{w}_e > - < \vec{w}_p >] \quad (79a)$$

$$\vec{\nabla} p_p - e_o n_p (\vec{E} + < \vec{w}_p > x \vec{B}) = - R_{p,e} [< \vec{w}_p > - < \vec{w}_e >] \quad (79b)$$

$$\begin{aligned} \vec{\nabla} p_k^z - e_o z_k n_k^z (\vec{E} + < \vec{w}_k^z > x \vec{B}) = & - R_{k,e}^z [< \vec{w}_k^z > - < \vec{w}_e >] \\ & - R_{k,p}^z [< \vec{w}_k^z > - < \vec{w}_p >] \quad (79c) \\ & - \sum_{k',z' \neq k,z} R_{k,k'}^{z,z'} [< \vec{w}_k^z > - < \vec{w}_{k'}^{z'} >] \end{aligned}$$

R_{ab} denotes the friction force constant. When one assumes that only Coulomb collisions contribute to R_{ab} we have for collisions between any species a and b (in SI units)

$$R_{ab} = 2(2\pi)^{1/2} 10^{54} q_a^2 q_b^2 n_a n_b \left(\frac{m_a m_b}{m_a + m_b} \right)^{1/2} T^{-3/2} \ln \Lambda \quad (80)$$

The friction force between impurity species (last terms of Eq. (79c)) can be neglected when the ion charges and impurity concentrations are low. The solutions of the momentum transfer equations yield the mean species' velocities in radial direction of a toroidal configuration in the following general form

$$< w_k^z >_r = - C \frac{1}{B^2 T_s^{3/2}} \left(z_k \frac{\partial p_e}{\partial r} - \frac{n_e}{n_k^z} \frac{\partial p_k^z}{\partial r} \right) \quad (81)$$

where C is a numerical constant. Since p = n kT, both the density and temperature gradients contribute to the diffusion flux. Under actual conditions, an "anomalous flux term" must be added, see above.

---

* The influence of the terms $\rho \vec{v}_0 . \vec{\nabla} \vec{v}_0$ and $\rho [\partial \vec{v}_0 / \partial t]_{CR}$ on the plasma properties is studied in Refs. [60,61], for instance.

5.3 - Radial particle distributions

Knowing the expression for $< \vec{w}_s >$ it is possible to solve the system of rate equations (78c) for the various impurity species $s \equiv k,z$ when the radial electron density and temperature distributions are given. From the solutions $n_k^z(r)$ one calculates the volume emission coefficient ε_L for a particular spectral line emitted by the impurity $s \equiv k,z$. The calculated values can be compared with measurements. Such a comparison is shown in Fig. 11 for a stationary discharge in the ASDEX-Tokamak. For the particle flux Eq. (77a) was applied, with $D' = 4000\ cm^2/s$.

Figure 12 shows as an example the results for a dynamic experiment in ASDEX. At time t = 0, neon gas was injected during 5 ms into a stationary ASDEX-discharge and the line emission was measured as a function of time t. The continuous curves show the measured ε_L- distribution 50 ms after the injection. The broken curves have been obtained from Eqs. (78c) with the flux density given by Eq.(77a). When the second term in Eq.(77a) is dropped* the dotted curve is obtained for Ne X. The second term gives the impurities an additional inward-directed velocity $v'(r) = D'2r/a^2$ and brings the calculations in good agreement with the measurements. In this context it may be mentioned that a parabolic temperature profile $T(r) = T(o)\ (1 - r^2/a^2)$ in connection with Eq. (81) just gives an additional diffusion term proportional to $2r/a^2$.

Quite similar measurements and model calculations are currently made on many Tokamak machines, see e.g. the Refs.[62-68].

6. ATOMIC COLLISION PROCESSES

The temporal and spatial distributions of the particle densities depend in a complicated manner on the ionization, recombination and charge exchange cross-sections. In the frame of the corona model (high temperature, low density, see e.g. Ref.[130,131]) the excitation cross-sections intervene only when the radiation losses are calculated or when spectral line intensities are used for determining absolute and relative impurity densities. In refined models collisional excitation and de-excitation also contribute to the distribution of the densities over the ion charge states and the processes become so involved that it is not further possible to describe a result simply by the algebraic sum of individual reaction rates.

In the following we shall discuss the collision processes to interest for fusion plasmas, with the exception of those already considered in Sections 3 and 4. We will not discuss radiative transition probabilities.

---

*with the diffusion coefficient consistently derived from the hydrogen data.

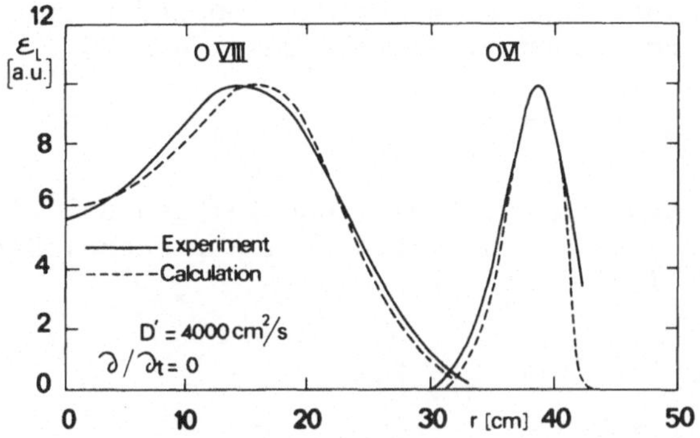

Fig. 11 Radial distribution of line emission coefficient of oxygen ions in a stationary ASDEX discharge, after(57).

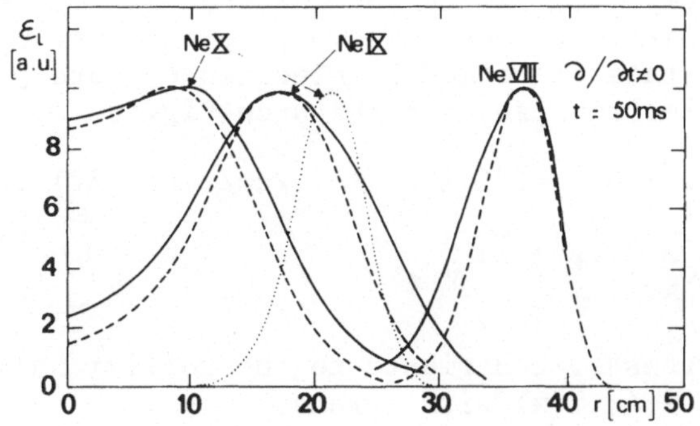

Fig. 12 Radial distribution of line emission coefficient of Neon ions 50 ms after injection of Neon gas into an ASDEX discharges, after(57). Continuous curves : Experiment.

6.1 Electron-impact ionization

We denote by $n_{k,1}^{z}$ the number density of z-times ionized particles of chemical element k in the ground level (i = 1). Electron-impact ionization of a z-times ionized atom initially in the ground level can schematically be described by

$$A^{z+} + e^{-} \xrightarrow{S_{k,1\to j}^{z\to(z+\lambda)}} A^{(z+\lambda)+} \ (j \geqslant 1) + e^{-} + \lambda e^{-} \tag{82}$$

where j > 1 indicates an eventually excited ion $A^{(z+\lambda)+}(j)$. λ = 1 means a single -, λ > 1 a multi-ionization process. Reaction (82) comprises several individual processes :

$$A^{z+}(1) + e^{-} \rightarrow A^{(z+1)+}(j) \quad + 2e^{-} \tag{82a}$$

$$A^{z+}(1) + e^{-} \rightarrow A^{(z+\lambda)}\ (j) \quad + (\lambda + 1)e^{-} \tag{82b}$$

$$A^{z+}(1) + e^{-} \rightarrow A^{z+}(j^{**}) \quad + e^{-} \tag{82c}$$

$$\hookrightarrow A^{(z+1)+}(j) + e^{-} \tag{82d}$$

$$A^{z+}(1) + e^{-} \rightarrow A^{(z+1)+}(j^{**}) \quad + 2e^{-} \tag{82e}$$

$$\hookrightarrow A^{(z+\lambda)+}(j) + (\lambda - 1)e^{-} \tag{82f}$$

$$A^{z+}(1) + e^{-} \rightarrow A^{(z-1)+}(j^{**}) \rightarrow A^{(z+1)+}(j) + 2e^{-} \tag{82g}$$

The reactions (82a,b) are direct knock-on ionizations without populating intermediate states. All other reactions lead to intermediate doubly (triply,...) excited or inner-shell excited states which decay by Auger emission of one or several electrons. In reaction (82g) a doubly excited state is formed by electron capture prior to the ejection of two electrons, see for instance Fig.15.

The ionization coefficient for reaction (82) is

$$S_{k,\ 1\to j}^{z\ \to\ (z\ +\ \lambda)} = \langle \sigma_{k,\ 1\to j}^{z\ \to\ (z\ +\ \lambda)}\ v \rangle \tag{83}$$

The ionization processes contribute to the collisional-radiative term of species s ≡ (k,1,x) with a rate.

$$\left[\frac{\partial n_{k,1}^{z}}{\partial t}\right]_{ioni} = - n_e n_{k,1}^{z} \sum_{\lambda}\sum_{j} S_{k,1\to j}^{z\to(z+\lambda)} + n_e \sum_{\lambda}\sum_{j} n_{k,1}^{(z-\lambda)\to z} S_{k,1\to j}^{(z-\lambda)\to z} \tag{84}$$

Very often reaction (82a) is the most important one. However, the reactions (82c,d,g) give sometimes a larger reaction rate

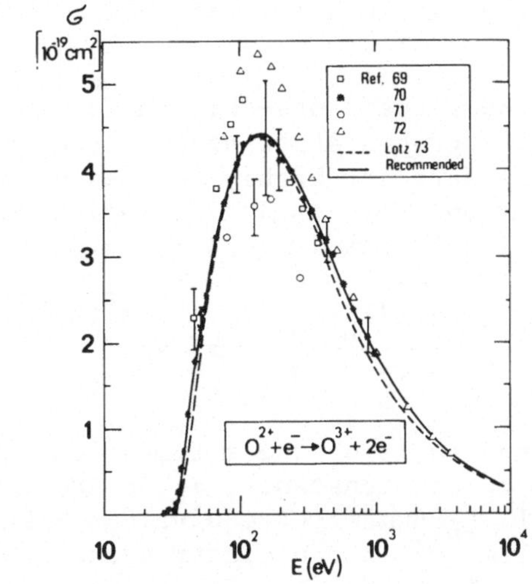

Fig. 13. Ionization cross-section for O^{2+} after different authors. Recommended curve after Ref.[74].

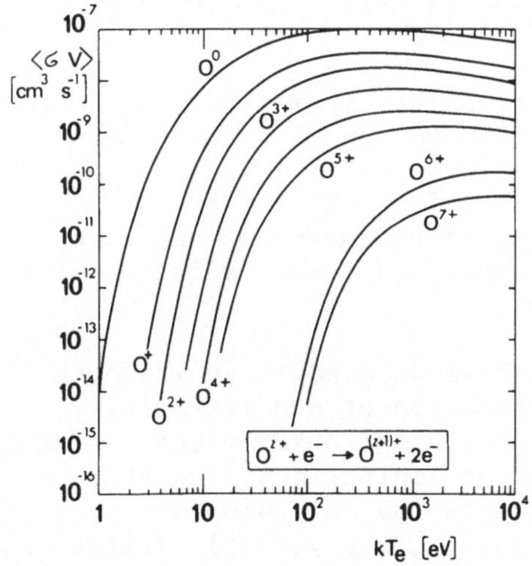

Fig. 14. Ionization rate coefficients for oxygen, for a Maxwellian velocity distribution. Calculated from recommended cross-sections, after Ref.[74].

than reaction (82a), and the reactions (82e,f) can for particular ions dominate the reaction (82b). This happens when inner-shell excitations (especially due to dipole-allowed $\Delta n = 0$ transitions) intervene.

The Fig. 13 shows the cross-section for ionization of the oxygen ion O^{2+} from its ground state after different authors. Since the experimental data for highly ionized atoms are still scarce, the cross-sections for single ionization are often calculated from the (empirical) formula (Lotz-formula)

$$\sigma_{k,1\to1}^{z\to(z+1)} = 4\pi a_o^2 \sum_{L=1}^{N} a_L^z \xi_L^z \frac{\ln(E/I_L^z)}{E\, I_L^z} \left\{ 1 - b_L^z e^{-c_L^z(E/I_L^z - 1)} \right\} \tag{85}$$

where L denotes the sub-shell of the z-times ionized atom to be ionized ; a_L^z, b_L^z and c_L^z are constants, I_L^z is the ionization energy of the L-th sub-shell, ξ_L=number of equivalent electrons in sub-shell L and $E = (1/2)m_e w_e^2$ is the electron energy.

Critical inspection of existing ionization cross-section data (experimental and theoretical) have led to a catalogue of "recommended" cross-sections and rate coefficients ; part I has just been published[74]. The recommended cross-sections have been fitted by the following formula

$$\sigma_{k,1\to1}^{z\to(z+1)} = 4\pi a_o^2 \frac{1}{E\, I_k^z} \left\{ a_1^z \ln(E/I^z) + \sum_{i=1}^{N} b_i^z (1 - \frac{I_k^z}{E})^i \right\} \tag{86}$$

where a_i^z and b_i^z are constants (numerical values are given in Ref.[74]) and I_k^z is the ionization energy which according to Fig. 1 writes $I_k^z = E_{k,1}^{z+1} - E_{k,1}^z$.

The Fig. 14 shows recommended ionization rate coefficients $S_{k,1\to1}^{z\to(z+1)}$ for oxygen atoms and ions, calculated for a Maxwellian velocity distribution.

The Fig. 15 shows an example in which the indirect ionization processes via formation of doubly (triply, ...) intermediate excited states and ejection of Auger electrons are more important than the direct knock-on ionizations. The cross-section suddenly increases every time when the collision energy equals the excitation energy of a doubly (triply, ...) excited state capable to Auger ionize. This occurs the first time at 726 eV when the ground configuration $1s^2\, 2s^2\, 2p^6\, 3s$ is lifted to the first doubly excited states $1s^2\, 2s^2\, 2p^5\, 3s\, 3p\, n's'$ due to excitation with capture. A first electron is ejected according to the reaction

$1s^2\, 2s^2\, 2p^5\, 3s\, 3p\, n's' \to 1s^2\, 2s^2\, 2p^5\, 3s^2 + e^-$

followed by the emission of a second electron :

$$1s^2\ 2s^2\ 2p^5\ 3s^2 \rightarrow 1s^2\ 2s^2\ 2p^6 + e^-$$

Concerning the electron-impact ionization of ions in the ground state, the reader finds an extended list of references in a recent review[76]. It should be mentioned that, at present, practically nothing is known about the cross-sections for multiple ionization of multiply charged ions.

For plasma physical applications one needs also the ionization cross-sections for ions in excited states. Experimental data are practically non existent, only theoretical and semi-empirical ones are available[77-87].

We will consider ionization of an excited z-times ionized atom $A^{z+}(n,\ell)$ leading to a (z + 1)-times ionized atom in the ground state :

$$A^{z+}(n,\ell) + e^- \rightarrow A^{(z+1)+}(1) + 2e^- \qquad (87)$$

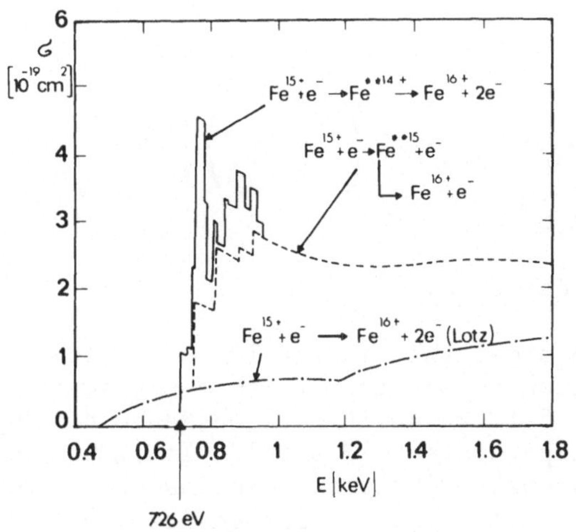

Fig. 15 Ionization cross-section of iron, Fe^{15+}, after Ref.[75].

The cross-section is given by the relation

$$\sigma_{n\ell \to 1}^{z \to (z+1)} = 4\pi a_o^2 \frac{1}{I_{n\ell}^z} G(E, Z_{eff}, n, \ell) \tag{88}$$

where G describes the energy dependence for a particular level. For hydrogenic levels the ionization energy is given by (with $E_1^H = 13.58$ eV) :

$$I_{n\ell}^z = Z_{eff}^2 E_1^H \left[\frac{1}{n^2} + \left(\frac{Z_{eff}^2 e_o^2}{4\pi\varepsilon_o \hbar c n^2} \right)^2 \left(\frac{n}{\ell + 1} - \frac{3}{4} \right) \right] \tag{89}$$

The rate coefficient averaged over all sub-levels ℓ are well approximated by the relation[81]

$$S_{n\to 1}^{z\to(z+1)} = 5.32 \, \pi a_o^2 \left(\frac{E_1^H}{I_n^z} \right)^2 \left(\frac{2kT_e}{\pi m_e} \right)^{1/2} u_n^z \, \psi_1(u_n^z, Z_{eff}) \tag{90}$$

where $u_n^z = I_n^z / kT_e$, and ψ_1 is a dimensionless function. Approximating ψ_1 by[81]:

$$\psi_1 \simeq \left[\frac{e^{-u_n^z}}{1 + u_n^z} \right] \left[\frac{1}{20 + u_n^z} + \ell n \left\{ 1.25 \left(1 + \frac{1}{u_n^z}\right) \right\} \right] \tag{91}$$

we see that with $u_n^z = Z_{eff}^2 E_1^H / n^2 kT_e < 1$ the rate coefficients behave for highly excited levels in as

$$S_{n\to 1}^{z\to(z+1)} \propto \frac{n^2}{Z_{eff}^2} \left[\frac{1}{20} + \ell n \left\{ 1.25 \left(1 + \frac{1}{u_n^z} \right) \right\} \right] \tag{92}$$

6.2 - <u>Ion-impact ionization</u>

The Fig. 5 shows that ionization by proton impact will compete with electron-impact ionization a temperatures $kT \approx 10$ keV. This leads in Fig. 8 to an additional decrease of the ratio $n(H°)/n(H^+)$ for temperatures $T > 3.10^7$K. Ionization of atomic hydrogen by highly charged impurity species can generally be neglected due to the small impurity concentrations, although the cross-section scales with z_s^2.

We consider structureless impurity ions of mass m_s, ion charge $e_0 z_s$ and energy E impinging on a target of ion charge $e_0 z'$ having ξ_i' equivalent electrons in shell i with binding energy I_i'. Further

$$\chi = 1 + \frac{m_e}{m_s}\left(\frac{E}{I_i'} - 1\right) \tag{93}$$

Within a factor of 2 or 3, the ionization cross-section seems to be well described by the semi-empirical relation[88]

$$\sigma_{i\to 1}^{z'\to(z'+1)} = Q^r\, z_s^2 \xi_i' \left(\frac{E_1^H}{I_i'}\right)^2 \left[1 + \frac{m_e}{m_s}\,\frac{1}{1 + (\log_{10}\chi)^2}\right]^{-1} \tag{94}$$

with the reduced cross-section given by

$$Q^r \cong \left[2.284\,\frac{\ell n\,\chi}{\chi} + 2.023\,\frac{\chi - 1}{\chi} - 1.699\,\frac{\chi - 1}{\chi^3}\right] 10^{-16}\text{cm}^2 \tag{95}$$

The cross-section for α-particle-hydrogen atom collisions is four times larger than the one for proton-hydrogen collisions, with the maximum displaced by a factor of ∿ 4 to higher energies. The rate coefficient for α-particle-hydrogen atom collisions can reach values of the one for thermal proton-hydrogen atom collisions for not yet thermalized (i.e. high-energy) α-particles.

6.3 - Radiative electron-ion recombination

We consider the process

$$A^{z+}(1) + e^- \xrightarrow{R_{k,j}^{z\to(z-1)}} A^{(z-1)+}(j) + h\nu \tag{96}$$

The last of the four reactions (48) is a special case of reaction (96). We denote the recombination cross-section by $\sigma_{k,1\to i}^{z\to(z-1)}$ and the rate coefficient by $R_{k,j}^{z\to(z-1)}$. The recombination processes contribute to the collisional-radiative term of species "s" (s ≡ k,1,z) with a rate

$$\left[\frac{\partial n_{k,1}^z}{\partial t}\right]_{\substack{\text{radiat.}\\ \text{recomb}}} = - n_e\, n_{k,1}^z \sum_j R_{k,j}^{z\to(z-1)} = - n_e\, n_{k,1}^z\, R_k^{z\to(z-1)} \tag{97}$$

In the frame of the coronal model, the same expression appears as a positive contribution for the species (k, z − 1).

In most cases, hydrogen -like cross-sections and rate coefficent are applied[31], if necessary modified in order to take into account deviations from hydrogenic energy and to account for partially filled valence shells.

For a number of complex ions, photo-ionization cross-section are available[89-95]. The recombination cross-section can then be calculated from the relation for microreversibility.

6.4 - Di-electronic recombination

Di-electronic recombination has its origin in a direct excitation process with simultaneous capture of the incoming electron, thus leading to a doubly (triply, ...) excited state as shown in reaction (82g). This state can either auto-ionize or it decays spontaneously in emitting a photon (we neglect collisional processes within the system of doubly excited states). Di-electronic recombination is achieved when the spontaneous radiative decay has led to a singly excited state below the first ionization limit of the recombined particle. The emitted photons lead to so-called di-electronic satellite lines. The processes involved can schematically be represented by the following reaction scheme

$$A^{z+}(1) + e^- \underset{\mathcal{A}^{(z-1)\to z}_{k,j^{**}\to 1}}{\overset{\mathcal{C}^{z\to(z-1)}_{k,1\ j^{**}}}{\rightleftarrows}} A^{(z-1)+}(j^{**}) \xrightarrow{\mathcal{D}^{z-1}_{k,j^{**}\to j}} A^{(z-1)}(j>1) + h\nu \quad (98)$$

The doubly excited state j^{**} of ion $A^{(z-1)+}$ is populated with a rate $n_e n^z_{k,1}\,\mathcal{C}^{z\to(z-1)}_{k,1\to j^{**}}$. The total depopulation rate is given by $n^{z-1}_{k,j^{**}}[\mathcal{A}^{(z-1)\to z}_{k,j^{**}\to 1} + \mathcal{D}^{z-1}_{k,j^{**}\to j}]$. The rate at which ions $A^{(z-1)+}(j)$ in the singly excited state j are formed is given by

$$n^{z-1}_{k,j^{**}}\,\mathcal{D}^{z-1}_{k,j^{**}\to j} = n_e n^z_{k,1}\,\mathcal{C}^{z\to(z-1)}_{k,1\to j^{**}}\,\frac{\mathcal{D}^{z-1}_{k,j^{**}\to j}}{\mathcal{A}^{(z-1)\to z}_{1,j^{**}\to 1} + \mathcal{D}^{z-1}_{k,j^{**}\to j}} \quad (99)$$

The expression

$$D^{z\to(z-1)}_{k,j^{**}\to j} = \mathcal{C}^{z\to(z-1)}_{k,1\to j^{**}}\,\frac{\mathcal{D}^{z-1}_{k,j^{**}\to j}}{\mathcal{A}^{(z-1)\to z}_{k,j^{**}\to 1} + \mathcal{D}^{z-1}_{k,j^{**}\to j}} \quad (100)$$

is the di-electronic recombination coefficient for populating a singly excited state j. When this expression is summed over all j^{**} and j one obtains the total di-electronic recombination coefficient $D_k^{z\to(z-1)}$:

$$D_k^{z\to(z-1)} = \sum_{j^{**},j} D^{z\to(z-1)}_{k,j^{**}\to j} \tag{101}$$

Di-electronic recombination contributes to the collisional-radiative term of species s ($s \equiv k,1,z$) with a rate

$$\left[\frac{\partial n^z_{k,1}}{\partial t}\right]_{\substack{\text{di-elec.}\\ \text{recomb.}}} = - n_e\, n^z_{k,1}\, D_k^{z\to(z-1)} \tag{102}$$

In the frame of the coronal model, the same expression appears as positive contribution for the species (k,z-1).

Di-electronic recombination and its impact on the properties of fusion plasmas has recently been reviewed in Refs.[3,5].

The Fig. 16 shows the radiative and di-electronic recombination coefficients for the Ne^{6+} ion after Ref.[97]. The largest contribution to the total di-electronic recombination coefficient D^6_{Ne} originates from the $\Delta n = 0$ transition $n'\ell' \to n\ell$, with $n' = n = 2$, $\ell' = 2p$, $\ell = 2s$. The calculations in Ref.[97] take into account the decay of excited states which themselves can auto-ionize (sometimes termed "secondary ionization") :

$$A^{z+}(1s^2\ 2s) + e^- \rightarrow A^{(z-1)+}(1s^2\ 3p\ n''\ell'') \tag{102a}$$

"second. ionization"

$$A^{z+}(1s^2 3s) + e^- \Longleftarrow \tag{102c}$$

$$A^{(z-1)}(1s^2 3pn''\ell'')\to A^{(z-1)+}(1s^2 3sn''\ell'') + h\nu' \tag{102b}$$

$$\Longrightarrow A^{(z-1)+}(1s^2 2sn''\ell'') + h\nu \tag{102d}$$

Only the reaction (102d) contributes to di-electronic recombination, the reactions (102b,c) lead back to the recombining ion $A^{(z+1)+}$ in an excited state. One obtains a recombination coefficient which is smaller than the one calculated from the Burgess formula (shown as broken curve in Fig. 16).

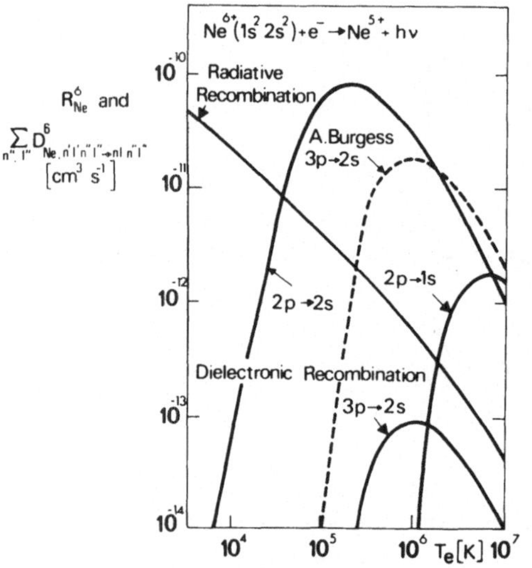

Fig. 16 Radiative and di-electronic recombination coefficient for Ne^{6+}, after Ref.[97].

6.5 - Total radiative and di-electronic recombination

By adding the coefficients for radiative and di-electronic recombination the total recombination coefficient of z-times ionized atoms of chemical element k is obtained :

$$\alpha_k^{z\rightarrow(z-1)} = R_k^{z\rightarrow(z-1)} + D_k^{z\rightarrow(z-1)} \qquad (103)$$

The total recombination rate for the process $z\rightarrow(z-1)$ is given by

$$\left[\frac{\partial n_k^z}{\partial t}\right]_{recomb.} = - n_e \, n_{k,1}^z \, \alpha_k^{z\rightarrow(z-1)} \qquad (104)$$

Useful analytical expressions for the rate coefficients may be found in Refs.[3,31,98,99]. Scaling properties of the di-electronic recombination rates are discussed in Refs.[132,133].

6.6 - Three-body electron-ion recombination

This process plays a dominant role in low-temperature plasmas and in hot plasmas of very high density. The rate coefficient can be calculated by applying the method of detailed balance to collisional ionization and three-body recombination. The latter can generally be neglected in high-temperature low-density magnetically confined plasmas.

6.7 - Charge exchange reactions

We have already seen (Section 4) that charge exchange between proton and hydrogen atoms plays an important role in the formation of the radial density profile of the neutral atoms (see Fig. 10). We consider now charge exchange processes between neutral hydrogen atoms and impurity ions according to the reaction(s)

$$H^\circ(1s) + A^{z+}(1) \underset{C_{cx,j}^{(z-1)+z}}{\overset{C_{cx,j}^{z\to(z-1)}}{\rightleftarrows}} H^+ + A^{(z-1)+}(j) \tag{105}$$

The arrow from left to right gives a net contribution to the recombination of z-times ionized atoms when the recombined system $A^{(z-1)+}(j)$ is not re-ionized, neither by the inverse process (arrow from right to left) nor by other collision and radiation processes, see below.

An order of magnitude of the cross-sections at relative velocities $v_{rel} < v_{1s} = 2.18\cdot 10^8$ cm/s can be obtained by considering the slow collision of a structureless ion of charge e_oz with a hydrogen atom H°(1s), see Fig. 17. The electric field strength at the Coulomb barrier is $e_o/R^2 = ze_o/r^2$, i.e. $r = Rz^{1/2}$. The Coulomb barrier has the potential energy

$$V_c = -\frac{e_o^2}{4\pi\varepsilon_o R} - \frac{e_o^2 z}{4\pi\varepsilon_o r} \tag{106}$$

The hydrogen energy is perturbed by the ion charge and has the approximate value

$$E'_{1s} = \frac{e_o^2}{8\pi\varepsilon_o a_o} - \frac{e_o^2 z}{4\pi\varepsilon_o (r-a_o)} \approx -\frac{e_o^2}{8\pi\varepsilon_o a_o} - \frac{e_o^2 z}{4\pi\varepsilon_o r} \tag{107}$$

The electron can slip over the energy barrier when $|V_c| \geqslant |E_{1s}^{i}|$, i.e. for all values $r \leqslant r_o$ satisfying the condition

$$\frac{e_o^2}{4\pi\varepsilon_o R_o} + \frac{e_o^2 z}{4\pi\varepsilon_o r_o} \geqslant \frac{e_o^2}{8\pi\varepsilon_o a_o} + \frac{e_o^2 z}{4\pi\varepsilon_o r_o} , \qquad (108)$$

hence

$$r_o = R_o\, z^{1/2} = 2a_o\, z^{1/2} \qquad (109)$$

When the electron has passed over the energy barrier it will be captured in the field of the nucleus of charge $e_o z$ and then has the energy $e_o^2 z/8\pi\varepsilon_o a_o j^2$ perturbed by the amount $- e_o^2/4\pi\varepsilon_o r_o$ due to the presence of the proton ; hence, immediately after the capture process into level j the relation

$$\frac{e_o^2}{4\pi\varepsilon_o R_o} + \frac{e_o^2 z}{4\pi\varepsilon_o r_o} = \frac{e_o^2 z^2}{8\pi\varepsilon_o a_o j^2} + \frac{e_o^2}{4\pi\varepsilon r_o} \qquad (110)$$

holds, with $j = n$ as principal quantum number. We define the cross-section by $\sigma_{cx} = \pi(r_o + R_o)^2$ and obtain from Eq. (109) the formula[5]

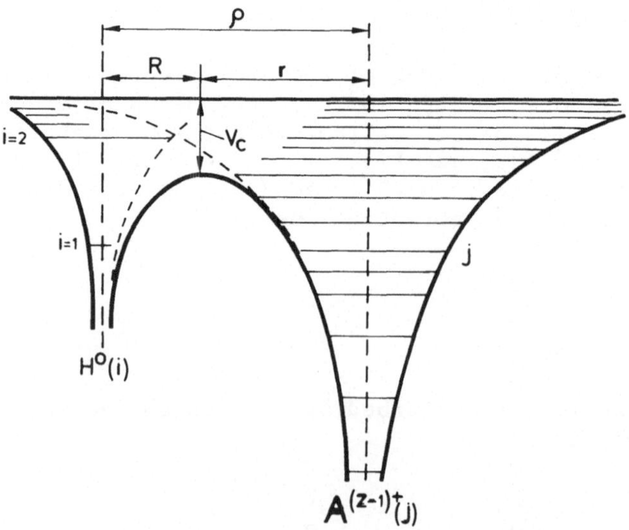

Fig. 17 Definition of the quantities intervening in Eqs. (106) and (107) for adiabatic interaction.

$$\sigma_{cx} = 4\pi a_o^2 \, (1 + z^{1/2})^2 \tag{111}$$

i.e. $\sigma_{cx} \propto z$ for $z \gg 1$. The Eqs. (110) and (111) yield the level j of the captured electron :

$$j = \frac{z}{[\,1 + z^{1/2} - z^{-1/2}\,]^{1/2}} \tag{112}$$

i.e. for $z \gg 1$, $j \propto z^{3/4}$. In practice there is a distribution of levels j and not only one distinct level into which the electron is captured. Therefore Eq.(111) must be considered as a lower bound.

The Fig. 18 shows the total charge exchange cross-sections for collisions of a bare nucleus A^{z+} with hydrogen atoms in the ground state, Fig. 19 gives the distribution over quantum levels $j = n$ and ℓ for four different collision velocities in atomic units.

The general behaviour of the cross-sections with ion charge and relative velocity is rather well described by the Bohr and Lindhard model in detail discussed in Ref.[104] : for collision velocities $v_{rel} \leqslant v_{1s}$ the cross-section is nearly independent of v_{rel}, whereas for $v_{rel} > v_{1s}$ the realtion $\sigma_{cx} \propto v_{rel}^{-7}$ holds. For further details, the reader is referred to the review aticles[105-109]. A catalogue of charge exchange cross-sections involving hydrogen atoms has recently been published[110], theoretical data are given in[134].

Single and double resonant charge exchange in ion-ion collisions have respectively been treated in Ref.[111] and Ref.[112]. The cross-sections are of the order of 10^{-16}cm^2.

6.8 - "Charge exchange recombination" of impurity ions

In order to see whether charge exchange in ion-hydrogen atom collisions really contributes to the recombination of impurity species we have first to evaluate the rate equation for the population density n_{kj}^{z-1}. The excited level j shall be populated by charge exchange, depopulation shall be due to spontaneous radiative decay, electron-impact ionization and inverse charge exchange. For the quasi-homogeneous stationary state follows (with subscript k dropped) :

$$0 = \left[\frac{\partial n_j^{z-1}}{\partial t}\right]_{CR} = n_{H^\circ} n_1^z C_{cx}^{z\to(z-1)} - n_j^{z-1} A_j^{z-1} \left[1 + \frac{n_e S_j^{(z-1)\to z} + n_{H^\circ} C_{cx}^{(z-1)\to z}}{A_j^{z-1}}\right] \tag{113}$$

where A_j^z is the sum of the individual transition probabilities A_{ij}^z.

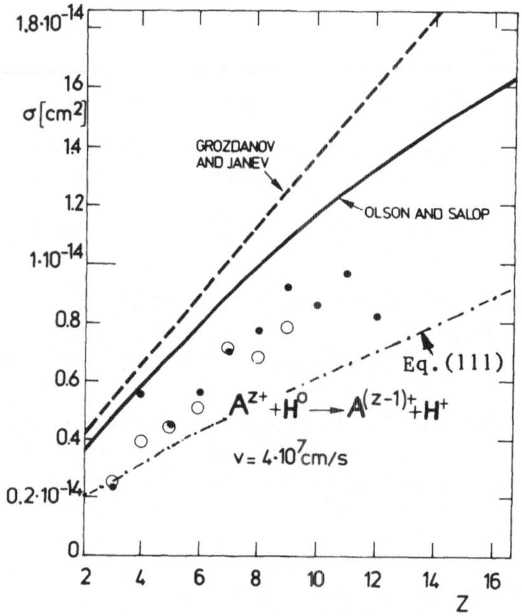

Fig. 18 Charge exchange cross-sections for collisior of bare nucleus A^{z+} with a H° atom. Dots : Experiment[100] ; open circles : Theory[100] ; continuous curve : Theory[102] ; broken curve : Theory[101].

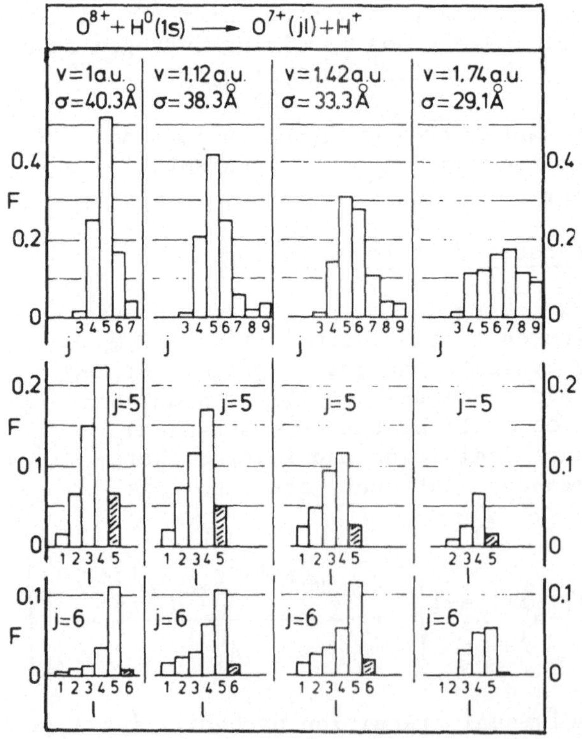

Fig. 19 Charge exchange cross-section for collision of bare oxygen nucleus with a hydrogen atom. Distribution over quantum levels (j = principal and ℓ = orbital quantum number). After Ref.[103]. Hatched region forbidden quantummecanically. F denotes the fractional cross-section.

We see that the captured electron will be re-ejected into the continuum when the second term in the parenthesis is of the order of unity or larger. On the other hand, when the second term is smaller than 0.1 reionization can practically be neglected ; hence the condition for charge exchange recombination becomes :

$$\frac{n_e\, S_j^{(z-1)\to z} + n_{H^\circ}\, C_{cx}^{(z-1)\to z}}{A_j^{z-1}} = \Delta < \frac{1}{10} \tag{114}$$

In order to evaluate this criterion numerically we assume for the recombined ion $A^{(z-1)+}(j)$ a hydrogenic level system with statistical population among sublevels ℓ of given principal quantum number $j = n$ (Under actual conditions this will probably not be the case). The total spontaneous decay rate is then well approximated (with an error of ± 20 %) by

$$A_j^{z-1} = \sum_{i<j} A_{ij}^{z-1} = \frac{4.7\cdot 10^8}{j^3}\left(1 + \frac{10}{j}\right)z^4 \quad s^{-1} \tag{115}$$

The ionization coefficient $S_j^{(z-1)\to z}$ is given by Eq. (90). For the charge exchange coefficient follows with Eq. (111)

$$C_{cx}^{z\to(z-1)} = C_{cx}^{(z-1)\to z} = 4\pi a_0^2 \left(\frac{8kT_+}{\pi\mu}\right)^{1/2} [1 + z^{1/2}]^2 \tag{116}$$

with μ as reduced mass. We substitute in all rate coefficients j by $z^{3/4}$, according to Eq. (112) for $z \gg 1$. We put $\mu \cong m_{H^\circ}$ and find numerically (with T in degree Kelvin, n in cm^{-3}) :

$$\Delta = \frac{n_e\, 2.3\cdot 10^{-5}\, z^{-1/2}\, T_e^{-1/2}\, \psi_1 + n_{H^\circ}\, 5.2\cdot 10^{-12}\, (1+z^{1/2})^2 T_+^{1/2}}{4.7\cdot 10^8\, z^{7/4}\, (1 + 10\, z^{-3/4})} \tag{117}$$

Under high-temperature conditions ($n_{H^\circ}/n_e < 10^{-4}$) the second term in the nominator can always be neglected ; hence, the condition for "charge exchange recombination" of impurity ions colliding with hydrogen atoms becomes

$$\Delta \approx n_e\, 5\cdot 10^{-14}\, \frac{z^{-9/4}\, T_e^{-1/2}\, \psi_1(z, T_e)}{1 + 10\, z^{-3/4}} < \frac{1}{10} \tag{118}$$

For Fe^{26+} at $T_e = 10^8$ K this criterion is fulfilled for $n_e < 5\cdot 10^{18} cm^{-3}$, i.e. all electrons captured in excited states will finally end in the ground state by radiative de-excitation. Thus, radiative, di-electronic and charge exchange recombination contribute to the population of the ground level with a rate

$$\left[\frac{\partial n_1^{z-1}}{\partial t}\right]_{recomb.} = n_1^z \left[n_e\, R^{z\to(z-1)} + n_e\, D^{z\to(z-1)} + n_{H^\circ}\, C_{cx}^{z\to(z-1)}\right] \tag{119}$$

Hence, charge exchange recombination will contribute with less than 10 % to the total recombination rate of ion $A^{z+}(1)$ when the conditior

$$\frac{n_{H^\circ}}{n_e}\;\frac{C_{cx}^{z\to(z-1)}}{R^{z\to(z-1)} + D^{z\to(z-1)}} < \frac{1}{10} \tag{120}$$

is fulfilled. Ions without or with small di-electronic recombination coefficient experience the largest influence by charge exchange.

As an example, for Fe^{26+} at $T_e = T_+ = 10^8$ K the criterion (120) is fulfilled if $n_{H^\circ}/n_e < 2.5\cdot 10^{-7}$. In homogeneous stationary ionization-recombination equilibrium (see Fig. 8) charge exchange will have no influence on the Fe^{26+} concentration. Diffusion and neutral particle injection for heating, however, will lead to high neutral particle density n_{H° which can cause such a strong recombination that the ion charge state distribution is modified. This effect discovered first in astrophysical plasmas[113] has important consequences for fusion plasmas, both for diagnostic purposes and for power loss considerations[114-120]. Under certain conditions and for certain charge states the recombination is entirely controlled by charge exchange with hydrogen atoms[118,119].

6.9 - "Charge exchange ionization" of hydrogen atoms

It can be shown that charge exchange ionization of neutral hydrogen atoms in collisions with impurity ions does generally not modify the neutral hydrogen density under fusion temperature conditions.

6.10 - Electron-impact and proton-impact excitation

In the presence of high-Z impurity ions, electron impact excitation of these ions is a dominant mechanism for producing radiation. Since the high-temperature low-density magnetically confined plasmas are optically thin in the prominent spectral ranges, the produced radiation freely escapes and contributes to the instantaneous energy loss of the plasma. On the other hand, impurity line radiation provides a sensitive diagnostic of plasma temperatures and impurity densities.

In the frame of the corona model which applies to high-temperature low-electron density plasmas, it is assumed that the collisional excitation rate from the ground state (i = 1) is balanced by spontaneous radiative decay. Population of a particular level j is thus due to direct excitation $1 \rightarrow j$ and radiative cascading $m \rightarrow j$ from levels $m > j$. For the rate equation thus follows (with $\partial/\partial t = 0$ and $\vec{\nabla}\cdot = 0$)

$$0 = \left[\frac{\partial n^z_{k,j}}{\partial t}\right]_{exci} = n_e C^z_{k,j1} n^z_{k,1} + \sum_{m>j}^{p} A^z_{k,jm} n^z_{k,m} - \sum_{i<j} A^z_{k,ij} n_{k,j} \qquad (121)$$

where $C^z_{k,j1} = \langle \sigma^z_{k,1\rightarrow j}\, v \rangle$ is the excitation coefficient and $\sigma^z_{k,1\rightarrow j}$ the excitation cross-section of level j from the ground level $i = 1$. Eq.(121) represents a system of coupled equations for the population densities $n^z_{k,j}$, $j = 1, \ldots, p$. The excited state population density is given by

$$n^z_{k,j} = \frac{1}{\sum\limits_{i<j} A^z_{k,ij}} \left\{ n_e C^z_{k,j1}\, n^z_{k,1} + \sum_{m>j}^{p} A^z_{k,jm}\, n^z_{k,m} \right\} \qquad (122)$$

and serves for calculating the volume emission coefficient ε^z_L of a spectral line $j \rightarrow h$ excited by electron-impact :

$$\varepsilon^z_{k,hj} = \frac{1}{4\pi} A^z_{k,hj}\, n^z_{k,j}\, h\nu^z_{k,hj} \qquad (123)$$

Since all densities $n^z_{k,j}$ are proportional to $n^z_{k,1}$, we have

$$\varepsilon^z_{k,hj} \propto A^z_{k,hj}\, n_e\, C^z_{k,j1}\, n^z_{k,1} \qquad (124)$$

i.e. a measurement of a spectral line intensity yields the impurity concentration $n^z_{k,1}$ provided T_e, n_e, $A^z_{k,hj}$ and $C^z_{k,j1}$ are known.

We have seen that additional processes are capable of producing singly excited levels ; these are :

- Radiative recombination (reaction (96)),
- Di-electronic recombination (reaction (82g) and (98)),
- Secondary ionization (reactions (102a and c)),

- Direct ionization with excitation (reactions (82a,b)),
- Inner-shell excitation followed by Auger ionization (reactions (82c-f)),
- Charge exchange with neutral hydrogen atoms (reaction (105)),
- Proton-impact excitation at temperatures $T_+ \geqslant 10^8$ K.

The corresponding rates must be taken into account in Eq.(121) and complicate the numerical solutions. Since not all excitation processes are equally important and since the rate coefficients are not always known with precision is suffices to complete Eq. (121) by the rates for the most prominent reactions. If dominant contributions arise from ions in charge states $z' \neq z$ information about $n^{z'}_{k,1}$ can be obtained. This is, for instance, the case for di-electronic satellite lines.

For a Maxwellian velocity distribution of the electrons, the rate coefficient C_{ji} for excitation of an initial state i to a final state j is[121]

$$C_{ij} = \left(\frac{8kT_e}{\pi m_e}\right)^{1/2} u_{ij}^2 \frac{a_o^2 E_1^H}{gE_{ij}} \int_1^\infty \Omega_{ji}(U) e^{-u_{ij}U} dU \qquad (125)$$

where E_{ij} is the excitation energy, g_i the statistical weight, $U = E/E_{ij}$, $u_{ij} = E_{ij}/kT_e$, E_1^H the ionization energy of hydrogen and

$$\Omega_{ji}(U) = E\,\sigma_{ji}\,g_i \qquad (126)$$

the collision strength for the transition i → j expressed as a function of the excitation cross-section (in units πa_o^2). For the present case, i = 1.

A guide to the literature of electron-impact excitation cross-sections relevant to fusion plasmas may be found in Refs.[3,4,122-124]. Useful formulas for collision strengths and rate coefficients are given in Refs.[99,125-127].

When fine structure transitions are applied for diagnostic studies, proton collisions between fine structure levels must eventually be taken into account[128,129], since the relative sub-level populations are sensitive to both electron and proton collisions.

7. RADIATION LOSSES

The energy losses due to radiation intervene in the energy balance equation (42) as $\dot{R}$ specified in Eq. (44). In

Tokamak, Stellerator and similar machines, reabsorption can be neglected for all types of radiation with the exception of cyclotron radiation which is strongly absorbed in the fundamental frequency. In practice, radiation loss calculations are done in successive steps.

First one has to know the radial distributions of electron density and temperature. These are determined experimentally. When "charge exchange recombination" is important one must also know $n_{H^\circ}(r)$. This quantity is either measured or determined theoretically with the aid of a special computer code (see e.g. Section 4.2). With T(r) known one can calculate all relevant rate coefficients, in general evaluated on the basis of a Maxwellian velocity distribution. In the case of high-energy neutral particles one takes into account the actual velocity distribution of n_{H°.

In a next step one applies the rate equations for the particle densities in their ground state (assuming that all ionization processes accompanied by excitation will finally yield particles in the ground state). The Eq. (6) writes for z-times ionized atoms of chemical element k :

$$\frac{\partial n^z_{k,1}}{\partial t} + \vec{\nabla}\cdot(n^z_{k,1} \langle \vec{w}^z_{k,1} \rangle) =$$

$$+ \sum_{\lambda \geq 1} \sum_j n_e n^{z-\lambda}_{k,1} S^{(z-\lambda)\to z}_{k,1\to 1} + n_e n^{z+1}_{k,1} \alpha^{(z+1)\to 1}_k + n_{H^\circ} n^{z+1}_{k,1} C^{(z+1)\to z}_{k,cx}$$

$$- n^z_{k,1} \left[\sum_{\lambda \geq 1} \sum_j n_e S^{z\to(z+\lambda)}_{k,1} + n_e \alpha^{z\to(z-1)}_k + n_{H^\circ} C^{z\to(z-1)}_{k,cx} \right] \tag{127}$$

This represents a coupled system of rate equations for the $n^z_{k,1}(r)$ (z = 0, 1, ..., Z) relative to each other. The absolute values are fixed either by a measurement of ε^z_L or fixed arbitrarily. The so-called stationary homogeneous coronal ionization-recombination equilibrium solution follows when the ℓ.h.s. is put equal to zero.

With the ground state densities $n^z_{k,1}(r)$ known one can apply the equations which follow hereafter.

An alternate method is to apply directly spectroscopically measured ground state densities without applying Eq. (127).

7.1 - Free-free radiation or bremsstrahlung

The radiated power density resulting from electron-proton

collisions is given by (with kT_e in keV, n in cm^{-3})

$$\dot{R}^{ff,H^+} = 4.8\cdot 10^{-31}\ n_e n_{H^+} (kT_e)^{1/2} \qquad W\ cm^{-3} \tag{128}$$

Species of ion charge $e_o z$ contribute to the radiated power density with (kT_e in keV, n_e and n^z in cm^{-3})

$$\dot{R}^{ff,z} = 4.8\cdot 10^{-31}\ n_e n^z\ z^2 (kT_e)^{1/2} \qquad W\ cm^{-3} \tag{129}$$

The radiation is independent of k.

7.2 - Free-bound radiation

The radiated power density due to the recombination of electrons with z-times ionized atoms of chemical element k is

$$\dot{R}_k^{fb,z} \cong \sum_i R_{k,i}^{z\to(z-1)}\ n_e n_k^z [\alpha kT_e + I_{k,i}^z],\ \alpha \simeq 1 \tag{130}$$

Summation is over all levels i with ionization energy $I_{k,i}^z$.

7.3 - Di-electronic recombination radiation

Di-electronic recombination of electrons with z-times ionized atoms of chemical element k leads to a power density

$$\dot{R}_k^{di,z} = \sum_{j^{**},j} D_{k,j^{**}\to j}^{z\to(z-1)} n_e\ n_k^z \left(E_{k,j^{**}}^{z-1} - E_{k,1}^{z-1}\right) \tag{131}$$

7.4 - Radiation caused by charge exchange

The charge exchange in collisions of hydrogen atoms with z-times ionized atoms of chemical element k leads to (z-1)-times ionized atoms in excited levels j. Under the assumption that all electrons end in the ground level due to radiative cascading, the power density is

$$\dot{R}_k^{cx,z} = \sum_j C_{cx,j}^{z\to(z-1)} \left(E_{k,j}^{z-1} - E_{k,1}^{z-1}\right) n_k^z\ n_{H^\circ} \tag{132}$$

When charge exchange is into sufficiently excited levels only, the excitation energies can be replaced by the ionization energy and we have

$$\dot{R}_k^{cx,z} = C_{cx}^{z\rightarrow(z-1)} n_k^z n_{H^\circ} (E_{k,1}^z - E_{k,1}^{z-1}) = C_{cx}^{z\rightarrow(z-1)} n_k^z n_{H^\circ} I_k^{z-1} \tag{133}$$

7.5 - Bound-bound or line radiation

Multiplication of Eq. (122) by $A_{k,ij}^z(E_{k,j}^z - E_{k,i}^z)$ and summation over all levels (i,j) yields in the frame of the corona model assumption a very simple expression which is independent of the transition probabilities:

$$\dot{R}_k^{bb,z} = \sum_j n_e C_{k,j1}^z n_{k,1} (E_{k,j}^z - E_{k,1}^z) \tag{134}$$

To this expression we have eventually to add the power density originating from the radiative de-excitation of excited levels formed by secondary ionization, direct ionization with excitation and inner-shell excitation followed by Auger ionization (see Section 6.10).

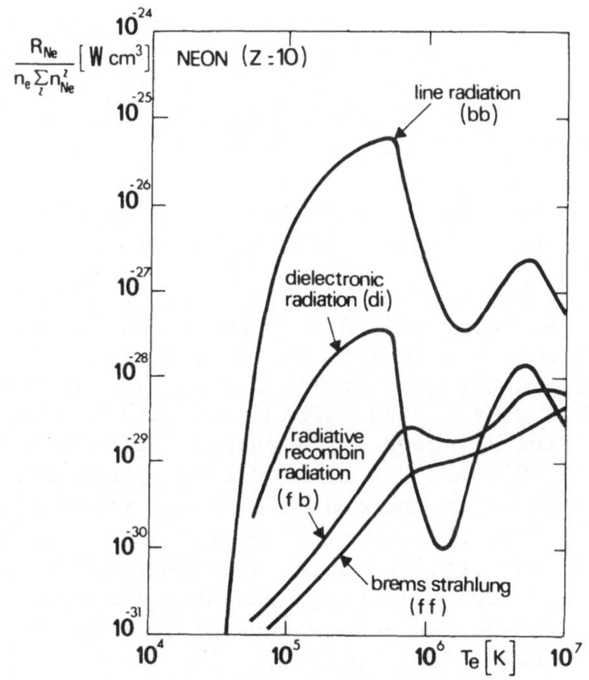

Fig. 20 Radiated power density per electron and per Neon atom, after Ref.[97].

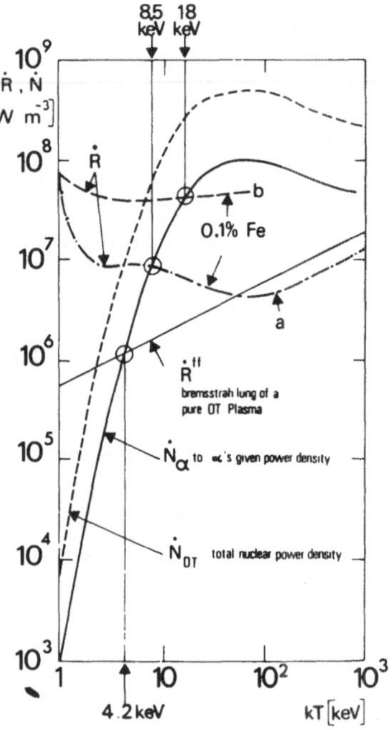

Fig. 21 Radiation power density $\dot{R}$ of electron-proton bremsstrahlung and of 0.1 % iron in comparison with nuclear power density $\dot{N}$. Values for iron from Ref.[119]. Plasma density $n_D = n_T \cong n_e/2 = 5 \cdot 10^{14} cm^{-3}$.

The total radiated power density is obtained by summing the above expressions over all chemical elements k and ion charge states z.

The Fig. 20 shows the radiated power densities per electron and per Neon atom imbedded in form of impurity ions of density n^z_{Ne} in a high-temperature low-density plasma. It has been assumed that the ion charge states obey coronal ionization-recombination equilibrium. The Fig. 21 shows for a pure D-T plasma of density $n_D = n_T = n_e/2 = 5 \cdot 10^{14} cm^{-3}$ the total nuclear power density $\dot{N}_{DT}$, the power density given to the α-particles $\dot{N}_\alpha$, the bremsstrahlung. power density $\dot{R}^{ff}$ without impurities and radiated power density R due to 0.1 % iron impurity. Two cases are distinguished, case a : no charge exchange ; case b : charge exchange included with $n_{H°}/n_e = 10^{-5}$ and an assumed temperature of $T_o = (2/3)T_e$ for the neutral hydrogen atoms, after Ref.[119].

In a future fusion reactor without additional heating the power losses must be compensated by α-particle heating ($\dot{N}_\alpha$). In the presence of 0.1 % iron much higher fusion temperatures are necessary (case a : kT = 8.5 keV, case b : kT = 18 keV) than for a pure D-T plasma (kT = 4.2 keV provided other energy losses are negligible). Important energy losses also originate from (not yet well-understood anomalous) heat conduction, in Eq. (42) globally described by the term $\vec{\nabla}\cdot\vec{Q}$.

This example clearly shows how sensitively the thermonuclear working conditions will depend on the atomic processes when one does not succeed in eliminating high-Z impurities.

REFERENCES

1. M.R.C. McDowell and A.M. Ferendeci (Eds.), Atomic and Molecular Processes in Controlled Thermonuclear Fusion, (Plenum Press, New York, 1980).
2. Equipe TFR, Tokamak Plasma Diagnostics, Nucl. Fusion 18, 647-731 (1978).
3. C. De Michelis and M. Mattioli, Soft X-ray Spectroscopic Diagnostics of Laboratory Plasmas, Nucl. Fusion 21, 677-754 (1981).
4. H.W. Drawin and K. Katsonis (Eds.), Atomic and Molecular Data for Fusion, Physica Scripta 23, 69-214 (1981).
5. H.W. Drawin, Atomic Physics and Thermonuclear Fusion Research, Physica Scripta 24, 622-655 (1981).
6. W.M. Stacey, Jr., Fusion Plasma Analysis (John Wiley, New York, 1981).
7. S.P. Hirshman and D.J. Sigmar, Neoclassical Transport of Impurities in Tokamak Plasmas, Nucl. Fusion 21, 1079-1201 (1981).
8. I.H. Hutchinson and J.D. Strachan, Nucl. Fusion 14, 649 (1974).
9. V.A. Abramov, D.P. Pogutse and E.I. Yurchenko, Sov. J. Plasma Phys. 1, 297 (1974).
10. Equipe T.F.R., Nucl. Fusion 15, 1053 (1975).
11. R. Papoular, Nucl. Fusion 16, 37 (1976).
12. V.A. Abramov, V.V. Vikhrev and D.P. Pogutse, Sov. J. Plasma Phys. 3, 288 (1978).
13. G. Waidmann, Zur Zündphase einer Tokamakentladung, Report Jül-1467 (Kernforschungsanlage Jülich G.m.b.H.,Nov. 1977).
14. D.G. Bulyginskii et al., Sov. J. Plasma Phys. 6, 11 (1980).
15. Yu.A. Ivanov, Yu.A. Lebedev and L.S. Polak, Sov. J. Plasma Phys. 6, 101 (1980).
16. L.P. Kubarev, S.A. Uryupin and L.M. Fisher, Sov. J. Plasma Phys. 6, 106 (1980).
17. D.J. Kaplan and R.D. Bengtson, J. Phys. B : At. Mol. Phys. 14, 1893 (1981).
18. C.F. Barnett, et al., Atomic Data for Controlled Fusion Research, Document ORNL-5207 (Oak Ridge National Laboratory, February 1977).

19. B. Peart, R.A. Forrest and K. Dolder, J. Phys. B : At. Mol. Phys. 12, 3441 (1979).
20. M. Bacal and G.W. Hamilton, Phys. Rev. Letts. 42, 1538 (1979).
21. E. Nicolopoulou, M. Bacal and H.J. Doucet, J. de Physique 38, 1399 (1977).
22. B. Peart and K.T. Dolder, J. Phys. B : At. Mol. Phys. 8, 1570 (1975).
23. M.T. Leu, M.A. Biondi and R. Johnsen, Phys. Rev. A8, 413 (1973).
24. D. Auerbach et al., J. Phys. B : At. Mol. Phys. 10, 3797 (1977).
25. B. Lange, B. Huber and K. Wiesemann, Z. Physik A 281, 21 (1977).
26. J.N. Bardsley and J.M. Wadehra, Phys. Rev. A 20, 1398 (1979).
27. R.E. Christoffersen, S. Hagstrom and F. Presser, J. Chem. Phys. 40, 236 (1964).
28. M. Capitelli and E. Molinari, Kinetics and Dissociation Processes in Plasmas in Low and Intermediate Pressure Range, Article in Topics in Current Chemistry, Vol. 90, edited by S. Veprek and M. Venugopalan (Springer Verlag, Heidelberg 1980).
29. G. Herzberg, J. Chem. Phys. 70, 4806 (1979).
30. B. Dubreuil and A. Cathérinot, J. de Physique 39, 1071 (1978).
31. L. Spitzer, Jr., Physics of Fully Ionized Gases, 2nd ed. Wiley Interscience, New York 1967.
32. L.C. Johnson and E. Hinnov, J. Quant. Spectr. Radiat. Transfer 12, 323 (1972).
33. H.W. Drawin and F. Emard, Physica 85C, 333 (1977).
34. R.L. Freeman and E.M. Jones, Analytic expressions for selected cross-sections and Maxwellian rate coefficients, Report CLM-R137 (Culham Laboratory, Abingdon 1974).
35. H.W. Drawin and F. Emard, Physica 94C, 134 (1978).
36. M. Gaudreau et al., Phys. Rev. Letts. 39, 1266 (1977).
37. Yu.N. Dnestrovskij, S.E. Lysenko and A.I. Kislyakov, Nucl. Fusion 19, 293 (1979).
38. J.G. Gilligan, S.L. Gralnick, W.G. Price, Jr., and T. Kammash, Nucl. Fusion 18, 63 (1978).
39. H. Capes and C. Mercier, Description cinetique des neutres légers dans un plasma de Tokamak, Note interne n° 1220 (Association EURATOM-CEA, Fontenay-aux-Roses, 1981).
40. D.T. Hogan, Multifluid Tokamak Transport Models, Article in Methods in Computational Physics (Eds. B. Alder, S. Fernbach and M. Rotenberg) Vol. 16, 131-164 (Academic Press, New York 1976).
41. J. Killeen, A.A. Mirin and E. Rensink, The Solutions of the kinetic equations for multispecies plasma, Article in Methods in Computational Physics (Eds. B. Alder, S. Fernbach and M. Rotenberg), Vol. 16, 389-432 (Academic Press, New York 1976).
42. D.F. Düchs, D.E. Post and P.H. Rutherford, Nucl. Fusion 17, 565 (1977).
43. P.A. Haldy and J. Ligou , Nucl. Fusion 17, 1225 (1977).

44. C. Mercier, F. Werkoff, J.P. Morera, G. Cissoko and H. Capes, Nucl. Fusion 21, 291 (1981).
45. J.S. Tolliver, E.F. Jeager and C.L. Hedrik, Nucl. Fusion 22, 13 (1982).
46. O. Demokan, N. Demokan and F. Waelbroeck, Application of Linear Programming to Impurity Flux Evaluations, Report Jül-1746 (Kernforschungsanstalt Jülich, 1981).
47. S. Suckewer and E. Hinnov, Phys. Rev. A 20, 578 (1979).
48. E. Hinnov, Spectroscopy of Highly Ionized Atoms in the Interior of Tokamak Plasmas, Article in Ref. 1.
49. A.A. Galeev and R.Z. Sagdeev, Sov. Phys.-JETP 26, 233 (1968).
50. H.A.B. Bodin and B.E. Keen, Experimental Studies of Plasma Confinement in Toroidal Systems, Repts. Progr. Phys. 40, 1415-1565 (1977).
51. R.J. Bickerton, The Containement of Thermonuclear Plasmas, Essays in Physics 6, 113-155 (1977).
52. A. Samain, Les Tokamaks permettront-ils la fusion contrôlée ?, Ann. Physique (Paris) 4, 395-446 (1979).
53. D. Bohm, E.H.S. Burhop, H.S.W. Massey and R.M. Williams, A Study of the Arc Plasma, Article in The Characteristics of Electrical Discharges in Magnetic Fields (Eds. M. Guthrie and R.K. Wakerling) chap. 9, p. 173-333 (Mc Graw Hill, New York, 1949).
54. R.G. Fowler, Phys. Fluids 21, 1972 (1978).
55. C. Deutsch, Phys. Rev. A 17, 909 (1978).
56. W. Engelhardt, K. Behringer, G. Fussmann und das ASDEX-Team, Teilchentransport im Tokamak ASDEX, Fachvortrag, 45. Tagung der Deutschen Physikal. Ges., Hamburg 1981, Verhandlg. DPG (VI) 16, 875-876 (1981). See also Jahresbericht 1981, Institut für Plasmaphysik, MPI, Garching, p. 19 (Garching 1981).
57. K. Behringer, W. Engelhardt, G. Fussmann and the ASDEX-Team, I.A.E.A.-Technical Committee Meeting on Divertors and Impurity Control, July 6-10, Garching 1981), paper I.C6, Conference Proceedings p. 42.
58. H. Coppi and N. Sharky, Nucl. Fusion 21, 1363 (1981).
59. Young-Ping Pao, Phys. Fluids 19, 1177 (1976).
60. H. Troughton, Plasma Physics 20, 943 (1978).
61. W.M. Stacey and D.J. Sigmar, Phys. Fluids 22, 2000 (1979).
62. K. Ando et al., Nucl. Fusion 16, 5 (1976).
63. Equipe TFR, Nucl. Fusion 17, 1297 (1977).
64. TFR Group, Plasma Physics 20, 735 (1978).
65. K. Brau et al., Phys. Rev. A 22, 2769 (1980).
66. R.J. Hawryluk, S. Suckewer and S.P. Hirshman, Nucl Fusion 19, 607 (1979).
67. H.A. Claassen and H. Repp, Nucl. Fusion 21, 589 (1981).
68. E.S. Marmar, J.E. Rice and S.L. Allen, Phys. Rev. Letts. 45, 2025 (1980).
69. M. Hamdan, K. Burkinshaw, and J.B. Hasted, J. Phys. B : At. Mol. Phys. 11, 331 (1978).

70. K.L. Aitken, M.F.A. Harrison and R.D. Rundel, J. Phys. B : At. Mol. Phys. 4, 1189 (1971).
71. A. Müller, E. Salzborn, R. Frodl, R. Becker, H. Klein H. Winter, J. Phys. B : At. Mol. Phys. 13, 1877 (1980).
72. D.L. Moores, J. Phys. B : At. Mol. Phys. 5, 286 (1972).
73. W. Lotz, Z. Phys. 216, 241 (1968).
74. K.L. Bell, H.B. Gilbody, J.G. Hughes, A.E. Kingston and F.J. Smith, Atomic and Molecular Data for Fusion, Part I : Recommended Cross-sections and Rates for Electron Ionization of Light Atoms and Ions, Report CLM-R216 (Culham Laboratory Alingdon, 1982).
75. K.J. LaGattuta and Y. Hahn, Phys. Rev. A 24, 2273 (1981).
76. E. Salzborn, Electron Impact Ionization, Lectures given at the NATO Advanced Study Institute, Baddeck, Nova Scotia, Canada, 13-26 Sept. 1981. To be publ. in NATO ASI Series B, Physics. (Plenum Press, New York).
77. H.W. Drawin, Z. Physik 164, 513 (1961) ; 172, 429 (1963).
78. M.R.H. Rudge and S.B. Schwartz, Proc. Phys. Soc. (Lond.) 88, 563 (1966).
79. K. Omidvar, Phys. Rev. 140, A 26 (1965).
80. J.B. Mann, J. Chem. Phys. 46, 1646 (1967).
81. H.W. Drawin, Collision and Transport Cross-Sections, Report EUR-CEA-FC-383 (Fontenay-aux-Roses, 1966/67).
82. A.E. Kingston, J. Phys. B : At. Mol. Phys. 1, 559 (1968).
83. L.C. Johnson, Astrophys. J. 174, 227 (1972).
84. L.B. Golden and D.H. Sampson, J. Phys. B : At. Mol. Phys. 10, 2229 (1977).
85. L.B. Golden, D.H. Sampson and K. Omidvar, J. Phys. B : At. Mol. Phys. 11, 3235 (1978).
86. D.L. Moores, L.B. Golden and D.H. Sampson, J. Phys. B : At. Mol. Phys. 13, 385 (1980).
87. J.A. Kunc, J. Phys. B : At. Mol. Phys. 13, 587 (1980).
88. O. Bely and P. Faucher, Astron. Astrophys. 18, 487 (1972).
89. W.D. Barfield, Nucl. Fusion 15, 1192 (1975).
90. W.D. Barfield, Astrophys. J. 229, 856 (1979).
91. W.D. Barfield, Int. J. Quant. Chemistry 13, 683 (1979).
92. W.D. Barfield, J. Phys. B : At. Mol. Phys. 13, 931 (1981).
93. R.F. Reilman and S.T. Manson, Astrophys. J. Suppl. Series 40, 815 (1979).
94. L. Armstrong, Jr., Physica Scripta 21, 457 (1980).
95. A. Ron, Y.S. Kim and R.H. Pratt, Phys. Rev. A 24, 1260 (1981).
96. R.J. Gould, Astrophys. J. 219, 250 (1978).
97. V.L. Jacobs, J. Davis, J.E. Rogerson and M. Blaha, Astrophys. J. 230, 627 (1979).
98. J. Dubau and S. Volonté, Di-electronic recombination and its applications in Astronomy, Repts.Prog.Phys. 43, 199-251 (1980).
99. R. Mewe and E.H.B.M. Gronenschild, Astron. Astrophys. Suppl. Series 45, 11 (1981).
100. R.A. Phaneuf, D.H. Crandall and F.W. Meyer, Physica Scripta 23, 188 (1981).

101. T.P. Grozdanov and R.K. Janef, Phys. Rev. A 17, 888 (1978).
102. R.E. Olson and A. Salop, Phys. Rev. A 14, 579 (1976).
103. A. Salop, J. Phys. B : At. Mol. Phys; 12, 919 (1979).
104. H. Knudsen, H.K. Haugen and P. Hvelplund, Phys. Rev. A 23, 597 (1981).
105. B.H. Bransden, Theoretical Models for Charge Exchange, Article in Ref. 1.
106. B.H. Bransden, Article in this volume.
107. F.J. de Heer, Experiments on Electron Capture and Ionization by Ions, Article in Ref. 1.
108. F.J. de Heer, Article in this volume.
109. Dz. Belkiž, R. Gayet and A. Salin, Electron Capture in High-energy Ion-Atom Collisions, Physics Reports 56, 280-369 (1979).
110. Y. Kaneko et al., Cross-Sections for Charge Transfer Collisions Involving Hydrogen Atoms, Report IPPJ - AM-15, Institute of Plasma Physics, Nagoya University (Nagoya, 1980).
111. V.P. Zhdanov, Sov. Phys. Techn. Phys. 21, 117 (1976).
112. R.K. Janev and D.S. Bélic, Phys. Letts. 89 A, 190 (1982).
113. J.W. Chamberlain, Astrophys. J. 121, 390 (1956).
114. R.C. Isler, Phys. Rev. Letts. 38, 1359 (1977).
115. V.V. Afrosimov, Yu.S. Gordeev, A.N. Zinov'ev and A.A. Korotkov, Soviet Phys. JETP Letts. 28, 500 (1978).
116. V.V. Afrosimov, Yu.S. Gordeev, A.N. Zinov'ev and A.A. Korotkov, Sov. J. Plasma Phys. 5, 551 (1979).
117. V.A. Krupin, V.S. Marchenko and S.I. Yakovlenko, JETP Letts. 29, 218 (1979).
118. R.A. Hulse, D.E. Post and D.R. Mikkelsen, J. Phys. B : At. Mol. Phys. 13, 3895 (1980).
119. M.E. Puiatti, C. Breton, C. De Michelis and M. Mattioli, Plasma Physics 23, 1075 (1981).
120. S. Suckewer, E. Hinnov, M. Bitter, R. Hulse and D.E. Post, Phys. Rev. A 22, 725 (1980).
121. R. Mewe, Astron. Astrophys. 20, 215 (1972).
122. I.I. Sobelman, L.A. Vainshtein and E.A. Yukov, Excitation of Atoms and Broadening of Spectral Lines (Springer, Heidelberg, New York, 1981).
123. W.D. Robb, Theoretical Studies of Electron Impact Excitation of Positive Ions, Article in Ref. 1.
124. K. Katsonis, Recent References for Atomic Collision Data of Interest to Fusion, Report IAEA-NDS-AM11 (I.A.E.A., Vienna 1981).
125. R. Mewe and J. Schrijver, Astron. Astrophys. 65, 99 (1978).
126. R. Mewe, J. Schrijver and J. Sylwester, Astron. Astrophys. 87, 55 (1980).
127. R. Mewe, J. Schrijver and J. Sylwester, Astron. Astrophys. Suppl. Series 40, 323 (1980).
128. M.J. Seaton, Monthly Notices Roy. Astron. Soc. 127, 191 (1964).

129. J.G. Doyle, A.E. Kingston and R.H.G. Raid, Astron. Astrophys. 90, 97 (1980).
130. H.W. Drawin, Plasma Impurities and Cooling, Physics Rept. 37, 125-163 (1978).
131. R.W.P. McWhirter, Data Needs, Priorities and Accuracies for Plasma Spectroscopy, Physics Repts. 37, 165-209 (1978).
132. Y. Hahn, Phys. Rev. A 22, 2896 (1980).
133. Y. Hahn, J.N. Gau, R. Luddy, M. Dube and N. Shkolnik, J. Quant. Spectr. Radiat. Transfer 24, 505 (1980).
134. R.K.Janev and B.H.Bransden,Charge Exhange between Highly Charged Ions and Atomic Hydrogen:A Critical Review of Theoretical Data. Report INDC (NDS)-135/GA,IAEA Nuclear Data Section, I.A.E.A. Vienna,July 1982.

ATOMIC PROCESSES IN HIGH-DENSITY PLASMAS

R. M. More

University of California, Lawrence Livermore
National Laboratory, Livermore, California 94550

1. DENSE ATOMIC PLASMAS (INTRODUCTION)

When a solid target is irradiated by an intense pulse of laser light, an inhomogeneous plasma is created (Fig. 1). A low-density ideal plasma ($n_e < 10^{21}$ cm^{-3}) expands toward the laser; this hct blowoff region refracts and absorbs the laser beam. The absorbed energy is conducted into a dense plasma region ($n_e = 10^{21}$ to 10^{23} cm^{-3}) where the heat flow is converted into hydrodynamic motion of the exploding plasma. The large reaction pressure shocks the target surface, compressing it above the initial solid density into a state which may be called high-density matter ($n_e > 10^{23}$ cm^{-3}).

The objective of inertial-confinement fusion research is to achieve energy production in deuterium-tritium fuel compressed by implosion to extreme densities up to 10^2-10^3 times the normal (cryogenic) liquid.[1] These densities are so much greater than those encountered in previous fusion plasmas that they raise many unique questions of atomic physics.[2]

High-density conditions cause dramatic changes in the physical mechanisms which govern most plasma energy-transport and coupling coefficients. It emerges that these coefficients are more sensitive to atomic properties in the dense plasmas.

Conversely, random electric fields in the dense plasma environment significantly perturb atomic bound electrons. When the plasma field exceeds the nuclear field, electrons are liberated from occupied ion-core states by an important process called pressure ionization.[2-5]

Implosion experiments diagnosed by broadening of neon or argon emission lines show plasma electron densities in the range 10^{23} to 10^{24} cm^{-3} (Yaakobi et al.[6]; Hauer, Mitchell et al.[7]; Kilkenny, Lee et al.[8]). The analysis of line widths and intensity ratios is an important specialized branch of the atomic physics of dense plasmas (Griem,[9] Hooper et al.,[10] Hauer,[11] and Key[12]).

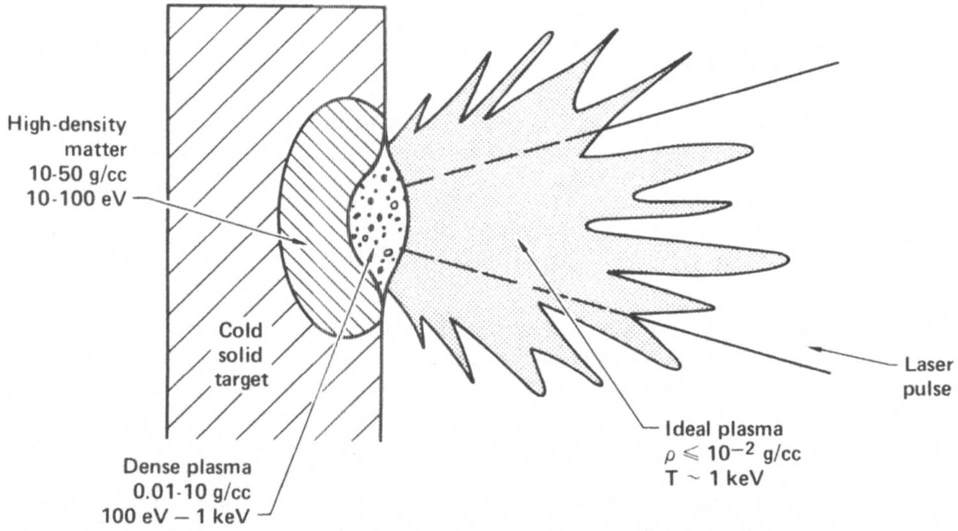

Fig. 1 Schematic representation of laser-irradiated flat target, showing a wide range of density-temperature conditions.

Typical inertial fusion targets are composed of a variety of materials (glass, neon, argon, nickel, plastic, gold) in addition to hydrogen isotopes; this means that we are interested in partially-ionized heavy atoms. For these materials, non-LTE effects are often very substantial. LASNEX simulations of recent gold-disk experiments show that the observed output of 2 keV x-rays is in agreement with non-LTE simulations but falls two orders of magnitude lower than an LTE calculation would predict.[13] A special characteristic of non-LTE in laser-fusion plasmas is a finite or large optical depth.

The high density and significant optical depth of laser plasmas lead to a characteristic problem: it is difficult to make accurate or detailed observations of events occurring in the pellet implosion core. This circumstance compels us to rely heavily upon theory, which is tested primarily by its logical consistency and by overall agreement between experiment and a comprehensive target-implosion simulation.[14]

We can identify three levels of increasing sophistication in the theory of atomic processes in dense plasmas. In the most naive work, it is assumed that atomic rates are proportional to the densities of reacting particles; e.g., electron-impact ionization $\propto n_e$, three-body recombination $\propto n_e^2$, etc. For low-density plasmas, this approach gives well-known and widely-accepted results. At high density, it predicts qualitatively incorrect results, such as complete recombination.

A better strategy is to recompute both eigenstates and cross-sections in a density-dependent model potential, such as the Debye-Hückel or ion-sphere potentials. Calculations at this level include the most important high-density effects: pressure ionization, continuum lowering and free-electron degeneracy. Although the results are more reasonable, this approach leaves many unanswered questions about the fundamental physics.

At the third level of sophistication, attention is directed toward the discrete atomic character of the plasma environment. For each process, a rate is calculated and then averaged over the probable positions of neighboring ions. For example, the rate of electron-ion elastic collisions is expressed as an integral over the ion pair-distribution function (see Eq. 3-2). For static ion Stark-broadening of emission lines, the line profile reduces to a transform of the plasma microfield distribution.[10]

The third approach can also encompass non-static effects associated with a time-varying atomic environment; such effects have proven important in line-broadening theory.[9,10]

At present, not many plasma rate coefficients are calculated at the third level of sophistication, but that is our ultimate objective. In order to build toward this goal, we need information about the structure of partially-stripped ions, their pairwise interactions and the average ion-ion correlations which result from thermal motion of the ions.

2. ION CORES--STRUCTURE, INTERACTIONS, AND CORRELATIONS

The <u>ion core</u> is a nucleus of charge Z together with Z-Q tightly bound electrons. It is surrounded by loosely bound electrons, free electrons and neighboring ions.

For isolated ions, electron eigenvalues and ionization potentials can be calculated by self-consistent field methods. Data is available for a large number of groundstate ions.[15]

For many practical applications, it is adequate to use results from the Thomas-Fermi theory.[16] The TF model of the isolated groundstate ion is a nucleus surrounded by a semiclassical electron gas distributed in a self-consistent potential. The TF electron distribution gives the ion a finite radius R(Z,Q).

One advantage of Thomas-Fermi theory is a scaling law which relates different atomic species so that ion core properties depend only on the fractional charge Q/Z. For example, the TF ion radius R(Z,Q) is scaled by

$$R(Z,Q) = Z^{-1/3}\ R_0(Q/Z) \tag{2-1}$$

This gives the radius of any ion in terms of a universal scaling function $R_0(x)$. Other atomic properties also scale, for example:

total ion binding energy $E_{TOT} = Z^{7/3}E_0(Q/Z)$

ionization potential $I = Z^{4/3}I_0(Q/Z)$

ion dipole polarizability $\alpha = 1/Z\ \alpha(Q/Z)$

average ionization-excitation energy $\bar{I} = ZI_0(Q/Z)$

Appendix A gives more complete definitions and limiting analytic forms for highly stripped ions. Table 1 gives the TF scaling functions in numerical form; it is remarkable that one can summarize the main properties of the ∿4000 ions of the periodic chart on a single page.

How accurate is this Thomas-Fermi data? It is good enough for qualitative estimates, and even some quantitative applications, but not accurate enough to satisfy most theorists. Thomas-Fermi calculations are especially questionable for highly ionized ($Q/Z \rightarrow 1$) and neutral ($Q/Z \rightarrow 0$) limits. We can compare ionization potentials computed with the Thomas-Fermi theory to those obtained from relativistic self-consistent field calculations[15] (Fig. 2): for light atoms there are disagreements of 50 to 100% while for heavy atoms the TF numbers are reasonably accurate.

The average ionization-excitation energy $\bar{I}(Z,Q)$ is used in the Bethe theory of fast-ion stopping[16] to calculate energy transfer to bound electrons; this theory is extended to ion energy-loss in hot plasmas by Nardi, Peleg and Zinnamon.[17] For cold matter, the experimental values of $\bar{I}$ are ∿50% larger than the Thomas-Fermi result $\bar{I}(Z,0) \sim (6.5\ \text{eV})Z$. The same correction factor (∿ 1.5) brings the TF function I(Z,Q) into approximate agreement with quantum calculations for aluminum ions (Q = 1 to 13) performed recently by E. McGuire.[18] The Bethe theory is limited to high projectile ion velocities; the low-velocity range may be treated by extension of the Lindhard-Winther theory[2] using recent calcu-

Table 1. Properties of Thomas-Fermi groundstate ions.[20] The numbers are scaled to Z = 1.

Q/Z	$\bar{I}_0$ (eV)	I_0 (eV)	E_0 (eV)	R_0 (Å)	α_0 (Å^3)
0.1	8.951	0.280	-20.908	5.141	36.006
0.2	11.526	0.882	-20.852	3.267	7.581
0.3	14.374	1.826	-20.720	2.365	2.390
0.4	17.694	3.210	-20.472	1.795	0.861
0.5	21.744	5.206	-20.058	1.383	0.318
0.6	26.953	8.147	-19.400	1.061	0.112
0.7	34.190	12.736	-18.374	0.791	0.0342
0.8	45.664	20.851	-16.739	0.552	7.608×10^{-3}
0.9	69.991	40.336	-13.859	0.321	7.362×10^{-4}

lations of electron-gas stopping due to Skupsky, Arista and Brandt, and Deutsch.[19] Ion stopping is important both for calculations of heavy-ion fusion and for analysis of thermonuclear burn.[20]

For non-degenerate plasmas, the equilibrium (LTE) ionization state $Q(\rho,T)$ can be estimated using an approximate version of the Saha equation:

$$- I(Z,Q) \cong \mu = kT \log \left(\frac{n_e \lambda_e^3}{2}\right) \qquad (2\text{-}2)$$

$I(Z,Q)$ is obtained from Table 1, μ is the chemical potential for non-degenerate electrons, n_e is the free electron density, and

$$\lambda_e = \sqrt{\frac{h^2}{2\pi mkT}} = 6.92 \times 10^{-8} \text{ cm} \sqrt{\frac{1 \text{ eV}}{kT}} \qquad (2\text{-}3)$$

is the electron thermal deBroglie wavelength. Equation (2-2) is only a rough approximation; a more accurate analytic formula for $Q(\rho,T)$ is given in Ref. 2 (Table III-1).

The ion radii in Table 1 can be used to estimate the plasma density at which ions touch their neighbors:

$$\rho = \frac{AM_p}{\frac{4\pi}{3}\,[R(Z,Q)]^3} \tag{2-4}$$

where A = atomic weight, M_p = atomic mass unit. This equation gives a reasonably accurate prediction of the charge state $Q(\rho)$ in the low-temperature degenerate range; i.e., it is a simple model for pressure-induced ionization.

At densities lower than that given by Eq. (2-4), ions come into close contact only during a collision. We are interested in the ion pair-potential U(R) which governs ion-ion collisions.

Ion Interactions

The ion pair-potential can be written generally as

$$U(R) = \frac{Z_1 Z_2 e^2}{R}\,\phi(R) \tag{2-5}$$

where Z_1, Z_2 = nuclear charges and R = internuclear separation. The screening function $\phi(R)$ approaches unity when the ion separation R approaches zero. For large radii, ions interact through their effective charges and $\phi(R) \to Q_1Q_2/Z_1Z_2$.

The pair-potential is defined as the groundstate energy of the electron-ion system at separation R minus the groundstate energy at infinite separation. In Thomas-Fermi theory, the screening function also has a scaling property; for similar ions ($Z_1 = Z_2$, $Q_1 = Q_2$) this reads $\phi = \phi(Q/Z,\ Z^{1/3}R)$.

Several authors have considered pair potentials for collisions of Thomas-Fermi atoms. Variational bounds for U(R) are derived by Firsov (1957, 1958) and an approximate representation of the screening function $\phi(R)$ for collisions of neutrals is given by Lindhard, Nielsen, and Scharff (1963).[22] Lee, Longmire and Rosenbluth[23] calculate U(R) for identical neutral TF atoms at seven radial separations (see Table 2) from numerical solutions of the two-center Thomas-Fermi problem.

In order to calculate the potential between ion cores, we apply a hydrostatic formula[23,24] based on the properties of the electron-gas pressure tensor P_{ij} and the Maxwell stress tensor T_{ij}

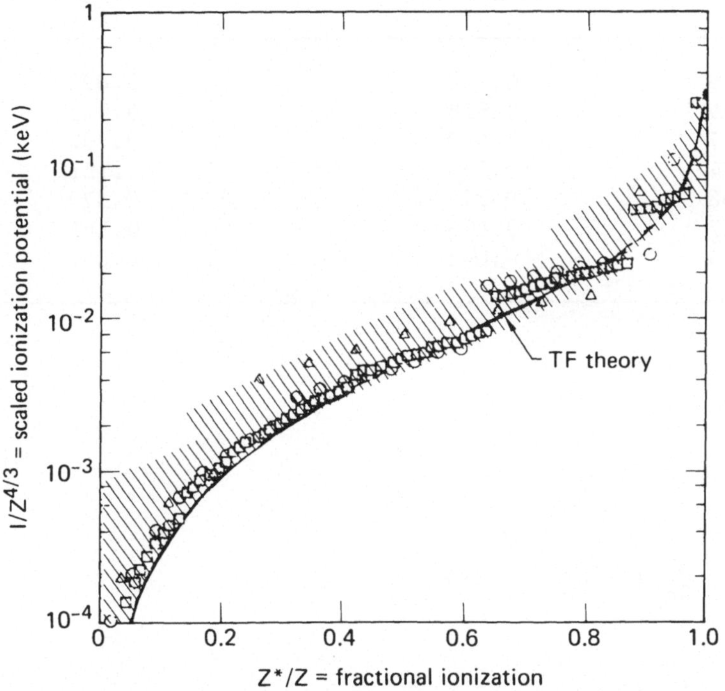

Fig. 2 Scaled ionization potentials for thirty elements fall into the shaded region.[2,15] The solid line is the Thomas-Fermi ionization potential. Triangles, circles and squares respectively represent aluminum, iron and gold ionization potentials.

Table 2. Comparison of Screening Function $\phi(R)$ for Collisions of Neutral TF Atoms (Scaled to $Z_1 = Z_2 = 1$).

Radius R (Å)	$\phi(R)$ Lee, Longmire, Rosenbluth	$\phi(R)$ Superposition Approximation
0.029	0.900	0.892
0.059	0.818	0.812
0.117	0.693	0.690
0.234	0.520	0.521
0.468	0.318	0.327
0.927	0.178	0.154
1.874	0.041	0.051

$$P_{ij} = \frac{2}{5}\frac{\hbar^2}{2m}(3\pi^2 n)^{2/3}\, n(r)\, \delta_{ij} \tag{2-6}$$

$$T_{ij} = -\frac{1}{4\pi}\left(E_i E_j - \frac{1}{2}E^2 \delta_{ij}\right) \tag{2-7}$$

If the electron density n(r) is consistent with the electric field E(r), then

$$\sum_j \frac{\partial P_{ij}}{\partial x_j} = -\, en(r)\, E_i(r) \tag{2-8}$$

This is the condition of hydrostatic equilibrium for the electron fluid. The force on an ion immersed in the electron fluid is given by a surface integral:

$$F_i = -\sum_j \int_s (P_{ij} + T_{ij})\, dS_j \tag{2-9}$$

The integral is independent of the surface S as long as it surrounds (only) the one ion in question. Eq. (2-6) applies to the Thomas-Fermi groundstate; with a modified definition of P_{ij} the other formulas extend to finite temperatures or to improved statistical models which include various quantum corrections.[24]

The electron density n(r) in the colliding ion pair is approximately a superposition of undistorted electron clouds taken from self-consistent calculations for the two separate (isolated) ions:

$$n(\vec{r}) = n_1(|\vec{r} - \vec{R}_1|) + n_2(|\vec{r} - \vec{R}_2|) \qquad (2\text{-}10)$$

The surface integral is taken over the mid-plane separating the (identical) ions (Fig. 3). With this approximation, the potential U(R) is obtained from a radial (one-dimensional) integration:

$$U(R) = \int_{r=R/2}^{\infty} [g_1(r) \left(\frac{\partial V_1}{\partial r}\right)^2 + A\, g_2(r)\, n_1^{5/3}(r)]\, dr$$

where $n_1(r)$, $V_1(r)$ are the electron density and potential in one (isolated) ion,

$$A = \frac{2}{5} \frac{\hbar^2}{2m} (3\pi^2)^{2/3}\, 2^{5/3}$$

$$g_1(r) = 2r \left(r - \frac{R}{2} - \frac{r^3 - (R/2)^3}{3r^2}\right)$$

$$g_2(r) = 4\pi r (r - R/2)$$

We can test the superposition approximation for collisions of neutral atoms by comparing with self-consistent solutions given by Lee, Longmire and Rosenbluth.[23] This comparison (Table 2) shows that superposition is very accurate for strong collisions. For weak collisions or large radii, Eq. (2-10) is less accurate.[25] For ions, however, the weak collisions are governed by the known asymptotic Coulomb potential.

The screening function $\phi(R)$ is shown in Fig. 4 for several ion charges. The calculations assume $Z_1 = Z_2 = 1$; with scaled radius, the data apply to any collision of similar ions. The plots extend over the range of separations $0 < R < 2R(Z,Q)$. In the superposition approximation, $\phi(R)$ is exactly Q_1Q_2/Z_1Z_2 at larger radii.

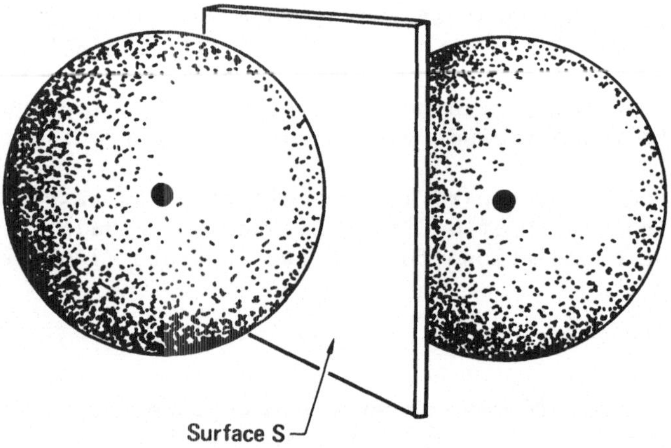

Fig. 3 Geometry for collision of similar ions with integration plane S bisecting the internuclear axis.

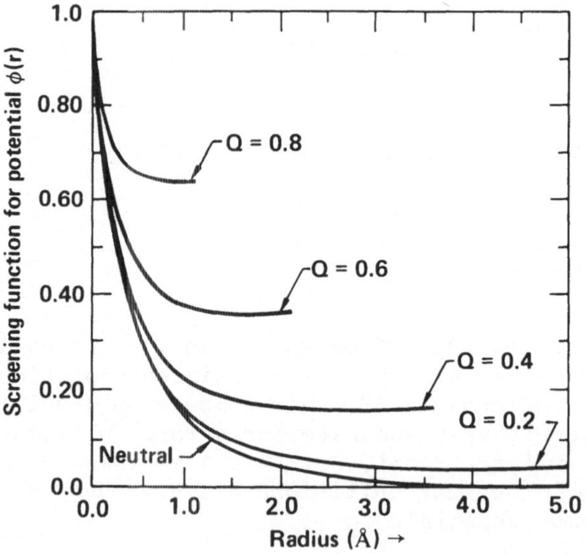

Fig. 4 Numerically calculated screening function $\phi(r)$ for ion-ion interactions as calculated with the superposition approximation. The radius R is scaled to refer to $Z_1 = Z_2 = 1$. Each curve (labelled by the fractional charge of the separate ion) is discontinued at the radius for which the ions first touch.

Figure 4 shows that $\phi(R)$ remains close to Q_1Q_2/Z_1Z_2 even down to about 1/2 the radius at which the ions first touch. Over this range, the ion potential energy is accurately described as a Coulomb potential $U = Q_1Q_2\, e^2/R$.

Ion Correlations

Next, we consider the spatial distribution of neighbors of a given ion core. This is a statistical problem of classical point particles at positions $R_j(t)$. The particles interact through a potential, usually assumed to have the form:

$$U_{TOT} = \frac{1}{2} \sum_{ij} U(|\vec{R}_i - \vec{R}_j|)$$

where $U(|\vec{R}_i - \vec{R}_j|)$ is $\sim Q^2e^2/|\vec{R}_i - \vec{R}_j|$.

A complete analysis of ion dynamics would require examination of many issues, including: 1) the potential energy does not rigorously separate into a sum of additive pairwise interactions; 2) the ion interactions are screened by free electrons,[26] which do not necessarily have the same temperature as the ions[2]; 3) the ion charges fluctuate dynamically due to ionization and recombination[2]; and, 4) the ion core-repulsion deviates from the Coulomb form at small distances, as shown in Fig. 4.

For Coulomb interactions, the thermal equilibrium ion configurational partition function is:

$$Z_c = \int \frac{d^3R_1}{V} \int \frac{d^3R_2}{V} \cdots \int \frac{d^3R_N}{V} \exp\left(-\frac{1}{2}\sum_{ij} \frac{Q^2e^2}{R_{ij}kT}\right) \tag{2-11}$$

and the probability of any specific configuration $\{\vec{R}_j\}$ is

$$\frac{1}{Z_c} \exp\left(-\frac{1}{2}\sum_{ij} Q_{ij}e^2/R_{ij}kT\right)$$

Averages over ion configurations can be evaluated by a Monte Carlo sampling technique introduced by Brush, Sahlin and Teller.[26-28] The thermodynamic averages depend upon a single parameter

$$\Gamma \equiv \frac{Q^2e^2}{R_0kT} \tag{2-12}$$

where

$$R_0 = (3/4\pi n_I)^{1/3} \tag{2-13}$$

is the ion-sphere radius. The ion coupling parameter Γ is the ratio of the Landau length Q^2e^2/kT (= distance of closest approach of two ions having kinetic energy kT) to the actual average separation of neighbor ions $\sim R_0$. If $\Gamma << 1$, the ions move freely in a large volume before colliding with another ion. If $\Gamma > 1$, the ions are being forced together against their Coulomb repulsions; at high Γ the ions are forced into a tightly correlated arrangement which minimizes their electrical energy.

The ion pair-correlation function g(r) is defined such that $n_I g(r)$ is the average density of ions a distance r away from one given ion. With this definition, $g(0) = 0$ and $g(r) \to 1$ as $r \to \infty$. In terms of g(r), the static structure factor S(k) is

$$S(k) \equiv 1 + n_I \int g(r)\, e^{-i\vec{k}\cdot\vec{r}}\, d^3r \qquad (2\text{-}14)$$

g(r) or S(k) give a summary description of the average local distribution of neighbors. These functions are spherically symmetric averages and do not retain any information about fluctuating anisotropies in the plasma environment.

A more delicate measure of the local environment is given by the probability distribution $P(\vec{E})$ for the electric microfield produced by nearby ions and electrons.[10] This field distribution is used in calculations of line-broadening produced by the Stark effect, a useful plasma spectroscopic diagnostic.[6-12]

At low densities, the ion correlation is described by Debye-Hückel theory. The ion Debye length D_I is

$$D_I = \left(\frac{kT}{4\pi n_I e^2}\right)^{1/2} = (3\Gamma)^{-1/2} R_0 \qquad (2\text{-}15)$$

and the number of ions in a Debye sphere

$$\mathcal{N} = \frac{4\pi}{3} n_I D_I^3 = (3\Gamma)^{-3/2} \qquad (2\text{-}16)$$

The Debye-Hückel theory is a good approximation when $\Gamma << 1$, so that $D_I >> R_0$ and $\mathcal{N} >> 1$. In this limit, the ion pair-correlation function g(R) and structure factor S(k) are:

$$g(r) = 1 - \frac{Q^2e^2}{rkT}\, e^{-r/D_I} \qquad (2\text{-}17)$$

$$S(k) = k^2 D_I^2/(1 + k^2 D_I^2) \tag{2-18}$$

In the Debye-Hückel approximation the average potential around a test ion is $V(r) = (Qe/r) \exp(-r/D_I)$. (The potential is modified by additional electron screening.)

The high density limit ($\Gamma \gg 1$) is often described by an ion-sphere picture.[2,3,27-31] In the simplest version of this model, one assumes

$$g(r) = \begin{cases} 0 & r < R_0 \\ 1 & r \geq R_0 \end{cases}$$

that is, no ion penetrates closer than a distance R_0. Assuming the free electrons are spatially uniform, this leads to a model poten- tial

$$V(r) = \frac{Qe}{r} + \frac{1}{2}\left(\frac{Qe}{R_0^3}\right) r^2 - \frac{3}{2}\frac{Qe}{R_0} \tag{2-19}$$

Properly used, the ion-sphere model can give reasonable results for dense plasmas;[28,30] however, the assumed pair-correlation function is not much like the correct high-Γ form (see Fig. 5).

For dense plasmas, $g(r)$ and $S(k)$ are calculated numerically by Monte Carlo or molecular-dynamics methods. A typical result is shown in Fig. 5. (This case corresponds to $\Gamma = 100$.) The salient features are: few ions at small separations, a ring of nearest neighbors at distance $\sim 1.7\ R_0$, and the less sharply defined rings of more distant neighbors.

Molecular Dynamics (MD) simulations integrate classical equations of motion for ion coordinates and give additional information concerning the time-dependence of ion correlations or the dynamic structure $S(k,\omega)$. Recent analytic and numerical results are reviewed by Baus and Hansen.[28] Hansen and McDonald[32] simulate the complete electron-ion plasma for strong coupling ($\Gamma = 2$) by molecular dynamics. The tendency of classical point electrons to recombine in the attractive $1/r$ potential is inhibited by use of a pseudopotential.[33]

Another approach for the static correlation functions is the method of integral equations. Recent applications to the Coulomb system are described by Baus and Hansen,[28] and DeWitt and Rosenfield.[34]

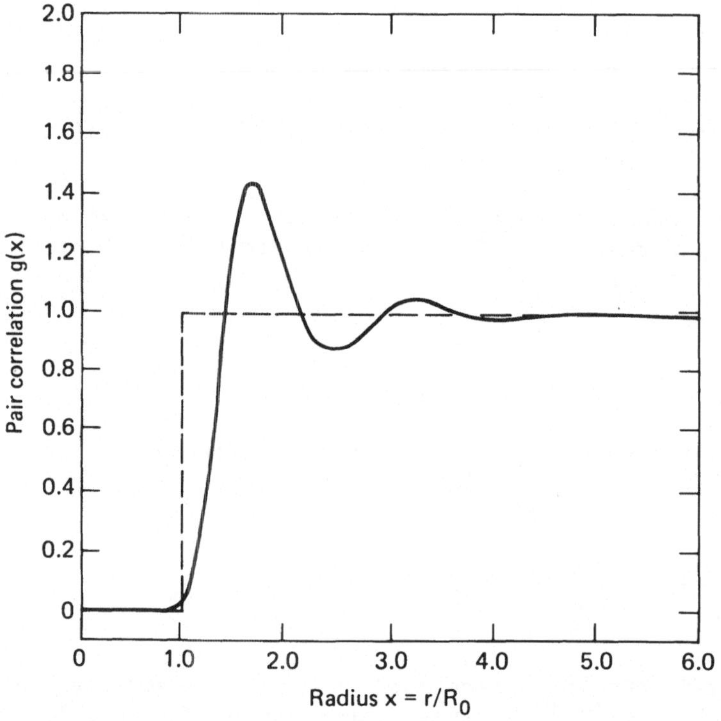

Fig. 5 Ion pair distribution function g(r) for the one component plasma (Γ = 30) (data supplied by Dr. H. DeWitt). Dashed line indicates the ion-sphere approximation.

A wide range of density-temperature conditions occur in laser-produced plasmas (Fig. 1). In low-density plasma directly heated by the laser, Γ is usually < 1. DT fuel is also normally in a weak-coupling state ($\Gamma < 1$) because of the small nuclear charge. The ablation plasma, however, often reaches high densities corresponding to $\Gamma = 1$ to 50.

3. FREE ELECTRON COLLISION PHENOMENA

We next consider scattering of free electrons by the ion cores. In conventional plasmas this is simply Rutherford scattering produced by the Coulomb potential Qe/r. However, in dense plasmas the scattering is more sensitive to ion-core structure and to the spatial arrangement of nearby ions.

A simple geometric estimate shows why core structure is more important in dense plasmas: the ion cores occupy a larger fraction of the total volume, as measured by the ratio $(R(Z,Q)/R_0)^3$. A more detailed analysis[2] shows that the fraction of classical electron orbits which intersect the ion core, i.e., reach the region $r < R(Z,Q)$, is

$$f \cong \frac{1}{3} \frac{\Gamma^2 kT}{Q^2 I} \sim \frac{1}{3} \frac{\Gamma}{Q} \frac{R(Z,Q)}{R_0}$$

In a low-density plasma, $f \ll 1$, and most collisions involve only the asymptotic Coulomb potential. In a dense plasma, f may approach unity, and calculation of the electron mean free path requires detailed information about ion core positions and potentials.

Consider an electron interacting with an assembly of many ions. To the accuracy of the superposition approximation (see Eq. (2-10) and Table 2), the total electrostatic potential is a superposition of contributions from the individual ions:

$$V(\vec{r}) = \sum_j v(\vec{r} - \vec{R}_j) \tag{3-1}$$

The ion cores are assumed to be located at fixed positions $\vec{R}_j$ during the scattering. The individual ion potential $v(\vec{r} - \vec{R}_j)$ is determined by some model for ion-core structure such as the Thomas-Fermi theory.

When a free electron scatters from the assembly of ions, the scattered waves interfere. In the Born approximation, the resulting cross-section per ion, in differential form, is

$$\frac{d\sigma}{d\Omega} = \left(\frac{d\sigma}{d\Omega}\right)_{\substack{\text{isolated}\\ \text{ion}}} S(q)$$

where

$\hbar k = mv$ = initial (electron) momentum

$\hbar q = 2mv \sin\theta/2$ = momentum transferred

$v(q)$ = Fourier transform of one-ion potential

$S(q)$ = plasma structure factor, defined in Eq. (2-14)

The momentum transport cross-section, $\sigma_{tr} = \int (1 - \cos\theta)\, d\sigma/d\Omega\, d\Omega$, is then

$$\sigma_{tr} = \frac{\pi}{k^4} \left(\frac{em}{2\pi\hbar^2}\right)^2 \int_0^{2k} q^3 \,|v(q)|^2\, S(q)\, dq \tag{3-2}$$

In condensed-matter physics, Eq. (3-2) is often called the Ziman formula. The interpretation is that small-angle scattering is limited by interference of scattered waves originating on neighbor ions. The structure-factor S(q) operates as a form-factor describing the spatial distribution of neighbor ions.

Equation (3-2) can be evaluated analytically in the Debye-Hückel limit. For fully stripped ions, the potential v(q) is simply:

$$v(q) = 4\pi Qe/q^2$$

The structure factor S(q) is given in Eq. (2-18). With these approximations, the electron mean-free-path is:

$$\frac{1}{\ell} = n_I \sigma_{tr} = \frac{4\pi Q^2 e^4 n_I}{m^2 v^4} \left[\frac{1}{2} \ln \left(1 + \frac{4m^2 v^2 D_I^2}{\hbar^2}\right)\right] \tag{3-3}$$

This is the usual result of multiple-scattering theory for the Lorentz gas. The logarithmic factor is called the Coulomb logarithm.

It is reassuring that the Ziman formula reduces to the usual cross-section for low-density fully-ionized plasmas, but inertial fusion applications will require data for dense and partially-ionized material. At high densities, the structure factor S(q) is the Fourier transform of a strongly-peaked pair-distribution g(r) [see Fig. 5]. Information about the ion core enters Eq. (3-2) through the potential V(q), which deviates from the pure-Coulomb form for partially-stripped ions. Together, these two corrections dramatically change the mean free path at high-density conditions.

Eq. (3-2) can be evaluated by combining numerical structure-factors S(q) from the one-component plasma together with electron-ion potentials v(r) obtained from a self-consistent field method. This recipe is not completely satisfactory because the Born approximation is inaccurate when applied to low-energy electrons. One can improve Eq. (3-2) by replacing the matrix-element V(q) of the potential by a T-matrix element T(k,q) calculated by partial-wave scattering theory.[35]

A comprehensive study of the transport coefficients for degenerate low-Z dense plasmas (appropriate to white-dwarf interiors) is given by Minoo, Deutsch and Hansen.[36] These authors survey the earlier astrophysical literature and compare several approximations. More recent calculations aimed at the range $\Gamma \lesssim 1$ are given by Boercker, Rogers and DeWitt. This work,[37] and the interesting work of Hansen, et al. concentrate on exploring a dynamical screening theory including frequency-dependent structure factors $S(k,\omega)$ and electron-ion correlations.

A practical formula for high-density thermal conductivity is given by Brysk, et al.[38] Brysk's approach was recently extended provide a comprehensive package of transport coefficients (including thermoelectric and thermomagnetic effects) which are approximately valid for arbitrary plasma conditions and composition.[39]

So far, the discussion considers only elastic scattering of electrons by ions. We are also interested in density effects on inelastic cross-sections. Preliminary investigations indicate that plasma screening changes the cross-sections for inelastic processes involving small energy-transfers.[40]

Conduction Inhibition

Electron thermal conduction is very important to the establishment of an ablation flow profile in laser-target interaction (see Fig. 1). However, the temperature gradient at high laser intensities is as large as

$$\frac{\partial T}{\partial x} \sim \frac{1 \text{ keV}}{10 \, \mu} \sim 10^{10} \, {}^\circ\text{K/cm}$$

and in this circumstance it is no surprise that the usual linear transport theory is inadequate. The expansion in powers of $\partial T/\partial x$ has failed.

Theoretical expectations based on kinetic theory suggest that the heat current density would saturate at a maximum

$$q_{max} \simeq f \, n_e \, kT \sqrt{\frac{kT}{m}}$$

where $f \simeq 0.6$ follows from a Knudsen gas model. Experiments do not support this expectation; a large number of spherical and planar laser irradiation experiments are empirically replicated by simulations in which the heat current is strongly inhibited to values corresponding to $f \sim 0.05$. The heat flow inhibition is probably the most important unexplained phenomenon observed in laser-target interactions.[41]

Recently, it has been realized that the majority of the heat current is carried by energetic electrons ($E \geq 10$ kT) in the tail of the thermal distribution. In an ablation plasma, collisional ionization removes these electrons; it may remove them so rapidly that they cannot be replaced by electron-electron collisions. In this way, atomic ionization may contribute significantly to conduction inhibition along with other mechanisms which distort the electron distribution function.[42]

4. ELECTRON STATES I

For a more detailed analysis of dense-plasma processes, it is necessary to calculate atomic properties such as electron energy spectra, line-strengths and electron-impact cross-sections. This data is needed for an arbitrary atom in a plasma of arbitrary density and temperature. It is natural to attack this problem by extending the usual self-consistent field method. Using this approach, Liberman[43] considers a single nucleus centered in a spherical cavity in a positive charge background. The electron energies ε_s and wave-functions $\psi_s(r)$ are found by solving the one-electron Schroedinger equation:

$$-\frac{\hbar^2}{2m}\nabla^2\psi_s + [V(r) + V_{xc}(r)]\,\psi_s = \varepsilon_s\psi_s(r) \tag{4-1}$$

It is usual to include an exchange or local-density exchange-correlation potential V_{xc}. (Liberman solves the Dirac equation and uses the zero-temperature Kohn-Sham exchange potential.)

For thermal equilibrium (LTE) plasmas, each electron state is independently occupied according to Fermi-Dirac statistics,

$$f_s = [1 + \exp\,(\varepsilon_s - \mu)/kT]^{-1} \tag{4-2}$$

From this assumption, it follows that the average electron density is

$$n(r) = \sum_s f_s\,|\psi_s(r)|^2 \tag{4-3}$$

This electron density is required to be consistent with the original potential $V(r)$.

At first sight this may appear to be a reasonably rigorous approach, but the reader will notice that Eq. (4-2) implies that electron states are occupied by fractional numbers of electrons. The calculation yields one-electron eigenvalues representing the spectrum of the so-called <u>average atom</u>. Electron eigenvalues are averaged over ionization and excitation states of the central ion, as well as the entire local plasma environment. The average eigenvalues $\varepsilon_s = \varepsilon_s(\rho,T)$ depend on density and temperature primarily because of shifts in the dominant ionization stage $Q(\rho,T)$.

Assuming we adopt the average-atom viewpoint, many interesting questions of detail arise. Liberman explores two alternative formulations of the cavity model which differ in the way atomic properties are distinguished from those of the background. Choices must be made for the manner in which neutrality is enforced, for the pressure formula, and for boundary conditions imposed on the wavefunctions.

Alternative versions of the sperical-cell model are described by Perrot,[44] Berggren and Froman,[45], Rozsnyai[46] and others.[40,47]

Bound electrons have energy less than the asymptotic potential, which may be taken to be zero; outside the cavity, bound-state wave-functions decrease exponentially with radius. Because of the spherical symmetry of the potential, the eigenvalue $E_{n,\ell,j}$ has degeneracy $2j + 1$.

At positive energies, there are continuum states composed of an incident plane wave scattered into various angular-momentum channels. Resonance scattering is produced by the centrifugal barrier surrounding the inner electrostatic well. Scattering resonances contribute peaks to the continuum density of states; in a nonrelativistic theory,

$$g(\varepsilon) = c_1\sqrt{\varepsilon} + \frac{2}{\pi} \sum_{\ell} (2\ell+1) \frac{d\delta_\ell}{d\varepsilon}$$

and a narrow resonance gives a sharply-defined peak. Early cell-model calculations neglected the resonance contribution and consequently predicted unphysical discontinuities during pressure ionization.

It is useful to develop a simple representation of the one-electron energy-levels emerging from self-consistent field calculations. For non-relativistic spectra, a WKB approximation gives:

$$E_{n,\ell} = E_n^0 - \frac{Q_n^2 e^2}{2a_0 (n - \Delta_{n\ell})^2} \qquad (4\text{-}4)$$

$$r_n = a_0 n^2 / Q_n \qquad (4\text{-}5)$$

r_n is the classical (Bohr) orbit radius, determined as the harmonic average over a WKB wave-function, i.e., $1/r_n = \langle n,\ell | 1/r | n,\ell \rangle$. Defining a screening function $Z(r)$ from the atomic potential by $V(r) = Z(r)e/r$, we have

$$Q_n = \left[Z(r) - r \frac{dZ}{dr} \right]_{r = r_n} \quad (4\text{-}6)$$

$$E_n^0 = -e^2 \left[\frac{dZ}{dr} \right]_{r_n} \quad (4\text{-}7)$$

where Q_n is the charge corresponding to the electric field at radius r_n and $\Delta_{n,\ell}$ is a quantum defect. For high quantum numbers (Rydberg states), $\Delta_{n,\ell}$ is approximately independent of n and is associated with the inner turning point of the classical orbit. For inner-shell electrons, $\Delta_{n,\ell}$ has contributions from both inner and outer turning points. The screened charge Q_n and quantum defect $\Delta_{n,\ell}$ depend on plasma density and temperature.

More elaborate screening models are discussed by Rogers[48] and Huebner[49]. Huebner shows that photoelectric cross-sections are governed by the inner-screening charge Q_n (rather than the effective charge $Z(r_n)$ defined from the potential).

A simpler screening model finds extensive application in fusion research (Mayer,[50] Post et al.,[51] Lokke and Grasberger[52]). The screened hydrogenic model is obtained[5] from Eq. (4-4) if one neglects the quantum defects:

$$E_{n\ell} \to E_n = E_n^0 - \frac{Q_n^2 e^2}{2a_0 n^2} \quad (4\text{-}8)$$

$$r_n = a_0 n^2 / Q_n \quad (4\text{-}9)$$

The screened charges Q_n are calculated as linear combinations of the shell populations (now simply functions of the principle quantum number n):

$$Q_n = Z - \sum_m \sigma(n,m) P_m \quad (4\text{-}10)$$

The screening coefficients $\sigma(n,m)$ represent the overlap of wave-functions of nth and mth shells. Mayer[50] gives screening coefficients based on perturbation theory for hydrogenic atoms; an alternative set[5] gives more accurate ionization potentials for nearly-neutral atoms.

Eq. (4-8) describes a Bohr atom with effective charges. The electrostatic potential at radius r_n consists of the potential outside a core of charge $Q_n e$ plus a sum of potentials inside shells having charge $-eP_m$ located at radii $r_m > r_n$:

$$V(r_n) = \frac{Q_n e}{r_n} - \sum_{m>n} \frac{P_m e}{r_m} \sigma(m,n) \tag{4-11}$$

The electron orbits are quantized by the Bohr formulas:

$$mv_n^2/r_n = Q_n e^2/r_n^2 \tag{4-12}$$

$$mr_n v_n = n\hbar \tag{4-13}$$

these assumptions give energies

$$E_n = \frac{1}{2} mv_n^2 - eV(r_n) = -\frac{Q_n^2 e^2}{2a_0 n^2} + E_n^0 \tag{4-14}$$

in agreement with Eq. (4-8). The electron-electron energy is evaluated using the harmonic-mean property of r_n; the result is a simple expression for the total ion energy $E_{TOT} = \sum_n E_n P_n - U_{ee}$:

$$E_{TOT} = \sum_n \left(-\frac{Q_n^2 e^2}{2a_0 n^2}\right) P_n \tag{4-15}$$

The eigenvalue E_n is the derivative of E_{TOT} with respect to the population P_n; i.e., if Eq. (4-15) is regarded as an effective or model Hamiltonian, it satisfies Koopman's theorem with the eigenvalue E_n of Eq. (4-14). This property is required for thermodynamic consistency.

For thermal equilibrium applications of the average-atom model, the integer population P_n is replaced by an average population calculated from Fermi statistics,

$$\overline{P}_n = \frac{D_n}{1 + \exp(E_n - \mu)/kT} \tag{4-16}$$

where $D_n = 2n^2$ is the shell degeneracy. The energy E_n appearing in (4-16) is obtained by consistently solving Eqs. (4-10, -11, -14) with the average populations $\overline{P}_n$. This is essentially the finite-temperature Hartree approximation for the model Hamiltonian (4-15).

Average-atom eigenvalues depend strongly on temperature: as temperature rises, bound electrons are thermally ionized and the screened charges Q_n increase. As a result, the energy-levels increase with temperature, so that spectral lines shift toward higher energies. For the quantum self-consistent field theories (Eqs. 4-1 to 4-3), the ℓ-splitting of levels decreases as the hydrogenic limit is approached (for high-Z atoms, there is relativistic spin-orbit splitting even in the hydrogenic limit).

The density-dependence of average-atom eigenvalues is much weaker. Numerical results from self-consistent field calculations show that the density-dependence is principally caused by recombination which reduces the effective charges Q_n.

Smaller contributions are produced by two other density effects. First, at higher density there is a higher degree of thermal excitation, and this excitation alters the self-consistent potential, in turn producing a shift of energy levels. Second, a small shift is associated with inner screening of bound electrons by free electrons, often called the plasma polarization shift. Both effects are too small to be easily resolved in average-atom self-consistent field calculations.

Pressure Ionization and Density Effects on Outer Electrons

Electron states with large quantum numbers are obviously very sensitive to events in the surrounding plasma.

These states are important in three ways: first, the atomic partition function would diverge if states with high quantum-number were included; second, the highly-excited bound states are important for line absorption and dielectronic recombination; third, at high density, the plasma begins to perturb occupied core levels important to the thermodynamic energy of the plasma.

Many authors calculate energy-levels, collision cross-sections, and oscillator-strengths for a Debye-screened atomic potential. However, there are objections to this model even at low densities ($\Gamma << 1$) where the Debye-Hückel theory is valid.[5] The basis of these objections is that, for large radii ($r > R_0$), the Debye potential is the small average of a large fluctuating field. A

*It may clarify this situation to point out that when $\Gamma >> 1$, the Debye theory correctly gives the average potential and also the average of any linear function of V(r), such as the plasma coupling energy. The Debye potential is not satisfactory when used to describ quantities (such as boundstate eigenvalues (which depend on the potential in a nonlinear way.

bound electron having large orbit radius $r_n > R_0$ has low velocity $v_n \sim Q_n e^2/n\hbar$ and responds much more strongly to the fluctuations than to the small average potential.* For example, the mean free path of a Rydberg electron is less than the orbit circumference whenever as $r_n > R_0$.

Quantum cell calculations based on Eqs. (4-1, 2, 3), show that electron eigenstates are forced into the continuum at densities where the ion-sphere radius R_0 becomes less than the orbit radius $r_n = a_0 n^2/Q_n$. This simple geometric criterion is algebraically equivalent to an energetic criterion which compares the binding energy $-Q_n^2 e^2/2a_0 n^2$ to the potential $-Qe^2/R_0$ by which the neighbor ions lower the continuum edge.[2]

A practical model for pressure ionization can readily be constructed for the average-atom model.[4] In this picture, boundstates become continuum resonances as soon as they move to positive energy. The resonance states are no longer completely orthogonal to continuum wave-functions, especially for broad resonances.[53] In practice, this transition is modelled by reducing the shell degeneracy D_n from $2n^2$ to zero as the atom is compressed past the critical radius $R_0 = r_n$.

The remainder of the state density $(2n^2 - D_n)$ appears in the lowered continuum. Because fractional populations are essential to this method (D_n is a smooth function of density), it applies to the average-atom model but would not make sense for the original Saha theory.

This pressure-ionization model suffices for description of light atoms but is not very satisfactory for transition metals or heavier atoms. A more detailed electronic energy spectrum, including subshell splitting, is required to calculate thermodynamic properties of high-Z plasmas.

5. ELECTRON STATES II

The average-atom model (AA) is heavily used in practical calculations for both magnetic and inertial fusion, whether in the quantum cell-model formulation of Eqs. (4-1, 4-3) or in the algebraic screened hydrogenic model of Eqs. (4-8) to (4-15).

For low-Z or nearly-ionized atoms, detailed atomic energy-levels are known, and most scientists prefer to work in this representation (often called detailed configuration accounting or DCA method). It is useful to compare the AA and DCA pictures in order to identify essential limitations of the AA description.

Few-Electron Ions

We first consider a low density, low-Z plasma which contains ions in hydrogen-like and helium-like ionization stages.

In hydrogen-like ions, 1s → 2p transitions have an energy

$$h\nu = (Z^2 - \frac{1}{4} Z^2)\ e^2/2a_0 \tag{5-1a}$$

The helium-like ions have a corresponding transition $1s^2$ → 1s2p at an energy near

$$h\nu = [(Z - 0.66)^2 - (Z - 1)^2/4]\ e^2/2a_0 \tag{5-1b}$$

The average-atom model gives instead a single 1s → 2p absorption line at an average energy determined by the relative number of hydrogenic and helium-like ions in the plasma. If the energy of this transition is calculated from the Mayer screening model (using the initial populations),

$$h\nu_{AA} = [(Z - 2\sigma_{11}P_1)^2 - \frac{1}{4}(Z - \sigma_{21}P_1)^2]\ e^2/2a_0 \tag{5-1c}$$

This is equivalent to an average of Eqs. (5-1a, b) except for (1) possible minor differences in the screening constants, and (2) a difference between $\langle P_1\rangle^2$ and $\langle P_1^2\rangle$, i.e., nonlinearity in the average.

The cross section for line absorption with excitation of a 1s electron to a 2p state is

$$\sigma_{abs} = P_{1s}(D_{2p} - P_{2p})\frac{\pi e^2}{mc} f_{1s \to 2p} \tag{5-2}$$

where $f_{1s\to 2p}$ = average oscillator strength, and D_{2p} = 6 = final state degeneracy. Equation (5-2) applies in a detailed configuration description, where the populations are integers, and must be averaged over ions. The average-atom model calculates the absorption cross-section from the same equation, with $\langle P_{1s}(D_{2p} - P_{2p})\rangle$ replaced by $\langle P_{1s}\rangle (D_{2p} - \langle P_{2p}\rangle)$; this is understood to give the total cross-section for transitions 1s → 2p for any ion.

Although the average photon energy (5-1c) and average absorption cross-section are reasonable, it is dangerous to replace two absorption lines by one stronger line in a calculation of line-transport or line-formation.

In certain cases, the averaged version of Eq. (5-2) is very misleading. For example, consider a plasma consisting of H-like, He-like and Li-like groundstate ions. The existence of hydrogen-like ions implies $\langle P_{1s} \rangle < 2$ and the presence of Li-like ions implies $\langle P_{2s} \rangle > 0$. If each ion is in its groundstate, the actual emission rate is zero.

In the average-atom description of this plasma, $\langle P_{2s} \rangle \neq 0$ and $\langle P_{1s} \rangle < 2$ and so the average atom emits radiation (with the aid of Stark splittings which break the usual selection rules).[31] In this case, the AA prediction is totally incorrect, a failure which may be traced to the assumption $\langle P_{2s}(D_{1s} - P_{1s}) \rangle = \langle P_{2s} \rangle (D_{1s} - \langle P_{1s} \rangle)$.

In dense plasmas, there are excited H-like, He-like ions present together with the Li-like ions. In this case the average-atom rates are closer to correct.

From the limited information given by the average populations, it is not possible to work backward to the individual ion densities, unless an additional assumption (such as LTE) is added.

Another interesting problem arises for ions with metastable levels such as the 1s2p triplet state of helium. The DCA description is able to distinguish between 1s2p singlet and triplet populations, but the average-atom model has no degree of freedom to describe this difference. Rate equations for the average atom predict that an isolated excited atom will decay to its groundstate without any vestige of metastability.

In a dense plasma collisions may break up metastable states and electric fields may be large enough to enable forbidden transitions (such as 2s → 1s). Nevertheless we must be very cautious in using the results of average-atom calculations.

Complex Atomic Spectra

When the spectrum becomes complex with many thermally excited electrons, the number of configurations rises dramatically. Under these conditions, it is impractical to accurately calculate all the energy levels and oscillator-strengths for transitions between them. In this case the average-atom description is attractive because of its computational simplicity. The question is whether it affords an accurate description of the atomic statistical problem. This question has been addressed in the context of LTE plasmas by Mayer,[50] Green,[54] Grimaldi and Grimaldi-Lecourt[55] and More.[2]

Let us consider a comparatively heavy atom (niobium) in thermal equilibrium at 200 eV. We will compare AA and DCA calculations for the atomic spectrum generated by Eqs. (4-8 to 14).[2]

In the DCA, atomic states are characterized by the populations of various shells: a group of states is labelled by integer populations P_n where $0 < P_n < 2n^2$. The ion charge is $Z^* = Z - \sum_n P_n$ and the effective charges (defined by Eq. 4-10) are $Q_n = Z - \sum_m \sigma(n,m) P_m$. The ion energy is determined by the approximate Hamiltonian of Eq. (4-15):

$$E_{TOT} = - \sum_n \left(\frac{Q_n^2 e^2}{2a_0 n^2}\right) P_n$$

The group of states so described is highly degenerate. The number of quantum states is

$$C = \prod_n \frac{D_n!}{P_n!(D_n - P_n)!}$$

These states are generated by permutations of the electrons over allowed single-particle states consistent with the stipulated shell populations $\{P_n\}$.

The probability of the set of populations $\{P_n\}$ is, for a thermal equilibrium plasma,

$$P(\{P_n\}) = \underline{N}\, C \exp\left(-\frac{E_{TOT} + F_0}{kT}\right)$$

where C = combinatorial factor, F_0 = free energy of Z^* free electrons in the average atomic volume V, and N = normalization factor (reciprocal of the partition function). With this probability distribution, we calculate average populations $\langle P_n \rangle$ and other averages. A typical calculation requires summation over 10^5 sets of populations $\{P_n\}$:

Some conclusions which emerge from this study are:

1) The averages of P_n are within 10% of values calculated using Fermi functions as in Eq. (4-16).

2) The number of thermally excited configurations rises very rapidly with plasma density so that the average of C is $\underset{\sim}{\sim} 10^{10}$ for Nb at 200 eV, 1 g/cc.

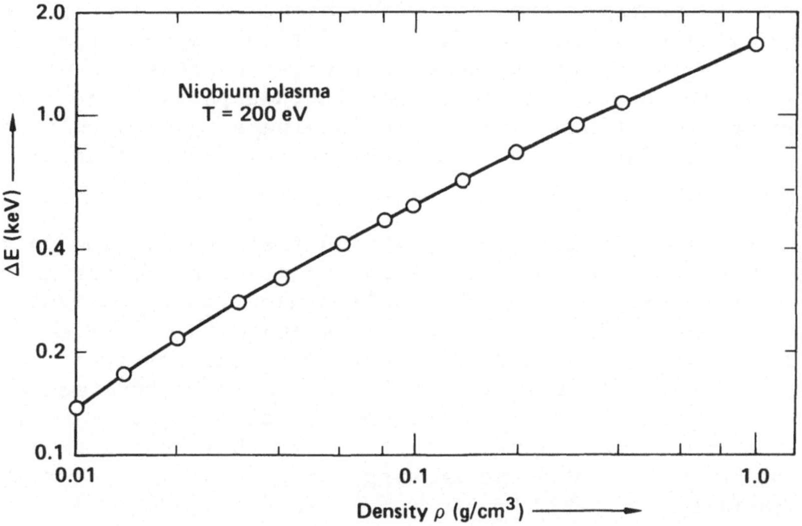

Fig. 6 Average atomic excitation energy (relative to groundstate ions) for an LTE niobium plasma as calculated by the detailed configuration method.

3) The internal excitation energy rises rapidly with density for atoms having many bound electrons (Figure 6).

The summation over all sets of populations includes states with arbitrarily inverted populations (e.g., vacancies in k-shell) and thus includes many autoionizing states. It is found that the average excitation energy is large enough, for Nb at 200 eV and 1 g/cc, to allow many (3-5) autoionizations.

Elaborate analytic investigations of the corrections to average-atom populations are reported by Green[54] and Grimaldi and Grimaldi-Lecourt.[55]

6. NONEQUILIBRIUM PLASMA STATES

Calculations of non-equilibrium atomic processes are often performed using a non-LTE version of the screened hydrogenic average-atom model described in section 4. The equilibrium equation for average shell populations, Eq. (4-16), is replaced by a rate equation including transitions caused by collisional and radiative mechanisms. Details of this model are described in the lecture of G. Zimmerman,[56] and in various other reports.[1,2,51,52]

The non-LTE average-atom model is subject to the list of critical comments given in section 5. In addition, we note that (1) hydrogenic rate coefficients and oscillator-strengths are not very accurate; (2) the usual model does not include possible density effects on cross-sections; and (3) it is always assumed that the majority of free electrons have an equilibrium (Maxwellian or Fermi-Dirac) distribution.[42]

Laser-produced plasmas depart from thermal equilibrium in many ways, and there is a natural temptation to develop increasingly comprehensive kinetic models, which allow numerical treatment of an arbitrary nonequilibrium states. This approach may be the best method to improve the fundamental accuracy of our physical modelling, but there is also a role for approximate treatments which bring out the physical character of realistic but simplified modes of departure from thermal equilibrium. Results of this second type have the advantage of suggesting intuitive ideas that we can work with.

Non-LTE Electron Populations

Because of the relevance to astrophysical and low-density laboratory plasmas, there is a great deal of literature analyzing the steady coronal or collisional-radiative plasma.[9,12,31,51,52,57] In this model, an atom is coupled to Maxwellian free electrons at a fixed temperature T_e and allowed to radiate freely (i.e., coupled to a radiation field at $T_R = 0$). Salzmann[58] formulates the CR model in an interesting way which emphasizes the density-dependence of bound-electron populations in optically-thin matter.

High-density plasmas of ICF interest are often not optically thin, and a coronal or collisional-radiative model is too extreme a simplification. A greater degree of realism is provided by a model in which the radiation field has a black-body distribution at a temperature somewhat lower than T_e. We therefore consider an atom coupled collisionally to free electrons having a high temperature T_e and coupled radiatively to a black-body distribution at a lower temperature T_R.

To greatly simplify the formulas, consider a two-level atom which has eigenstates 1> and 2>. P_1 and P_2 denote the populations of each state ($P_1 + P_2 = 1$). [The final result is more general than implied by this simple model.] The populations evolve due to collisionally induced transitions:

$$\left(\frac{dP_1}{dt}\right)_e = - aP_1 + a\, e^{E/kT_e} P_2 \qquad (6-1)$$

and radiatively induced transitions

$$\left(\frac{dP_1}{dt}\right)_R = - bP_1N_\nu + bP_2(N_\nu + 1) \qquad (6\text{-}2)$$

where $E = h\nu$ = energy difference between the levels, and N_ν = number of photons per mode = $\left(e^{h\nu/kT_R} - 1\right)^{-1}$. The notation can be simplified by defining

$$\tilde{a} = bN_\nu \qquad \tilde{a}\, e^{E/kT_R} = b(N_\nu + 1)$$

so that

$$\left(\frac{dP_1}{dt}\right)_R = - \tilde{a}\, P_1 + \tilde{a}\, e^{E/kT_R}\, P_2 \qquad (6\text{-}3)$$

For the equations given, only photons of frequency $h\nu$ are coupled to the atom; in a more realistic treatment the atom is coupled to photons of arbitrary frequencies via photoelectric processes.

The steady-state populations are determined by requiring the time-derivative P_1 to vanish. This is a linear equation for P_1 whose solution is:

$$P_1 = \frac{a\, e^{E/kT_e} + \tilde{a}\, e^{E/kT_R}}{a(1 + e^{E/kT_e}) + \tilde{a}(1 + e^{E/kT_R})} \qquad (6\text{-}4)$$

Entropy is produced in three ways in the system considered. If the atomic populations change, then the atomic entropy production is

$$\dot{S}_{atom} = - k\, \dot{P}_1 \log (P_1/P_2) \qquad (6\text{-}5a)$$

Entropy produced in the heat baths is given by

$$\dot{S}_e = \frac{h\nu}{T_e} \left(\frac{dP_1}{dt}\right)_e \qquad (6\text{-}5b)$$

$$\dot{S}_R = \frac{h\nu}{T_R} \left(\frac{dP_1}{dt}\right)_R \qquad (6\text{-}5c)$$

If the time-derivatives are eliminated via Eqs. (6-1, 6-3), then the rate of entropy production is expressed as a function of the atomic population P_1.

We can vary the population P_1 to determine the state of minimum rate of entropy production. In general, the resulting state is not identical to the steady-state determined in Eq. (6-4). However, in the special case $|T_e - T_e| << T_e$ where the system is close to equilibrium, it turns out that the state of minimum entropy production is identical, to first order in $\Delta T = |T_R - T_e|$ to the steady-state given by Eq. (6-4).

Therefore, principle of minimum entropy production characterizes the steady nonequilibrium state to first order in the departure from complete thermal equilibrium.[59,60]

In qualitative terms, one can say that the atomic system occupies the state which minimizes the steady-state rate of conversion of electron energy into radiation. This statement is not an exact paraphrase of the result established because the conversion efficiency is actually the rate of energy transfer, not quite identical to the rate of entropy production. In real laser-interaction plasmas, there are additional physical complexities which change the situation, including effects of spatial gradients, transient or dynamic effects, and departures from the limit $\Delta T << T$.

Within the linear regime ($\Delta T << T$), we have found that the minimum entropy production principle also characterizes the steady state of more complicated models including the complete non-LTE average atom model described above.

Electron-Ion Nonequilibrium

Several plasma processes induce an electron-ion temperature difference:

1) Laser radiation heats the free electrons,

2) Free electrons cool by emission of bremsstrahlung and recombination radiation, and

3) Ions are heated by strong shock-waves or by adiabatic compression in an implosion.

Electron-ion heat transfer is strongly impeded by the inequality of electron and ion masses. Estimates of collisional heat exchange show that the time to equilibrate electron and ion temperatures often exceeds laser deposition or plasma flow time-scales.

If the plasma density is low, it is reasonable to assume a two-fluid model of interpenetrating ideal gases having separate temperatures (T_e, T_i) and additive properties. In dense plasmas, the Coulomb interactions are strong enough to significantly modify the ideal-gas equation of state and a better description is needed.

We consider a nonequilibrium steady plasma characterized by unequal electron and ion temperatures. The screened ion pair-potential in such a two-temperature plasma may be taken as

$$U(R) = \frac{Q_1 Q_2 e^2}{R} \exp\left(-\frac{R}{D_e}\right) \qquad (6\text{-}6)$$

where $D_e = D_e(T_e)$ is the Debye-length associated with electron screening. Any sudden energy-deposition into the electrons raises their temperature and then is communicated to the ions on two separate time-scales:

1) Nearly instantaneously (time $\sim$ 1/electron plasma frequency) as an alteration of the screening length D_e, the ion pair-potential U(R) and the subsequent thermal motion of the ions. The Coulomb interactions transfer energy in the form of electrical work between the two species.

2) More slowly, through collisions that transfer random thermal energy (heat) from electrons to ions.

A straightforward statistical treatment[2] shows that pressure, energy, and other variables are functions of electron temperature T_e, ion temperature T_I, and volume V. Changes of the energy E = $E(T_e, T_I, V)$ obey

$$dE = T_e dS_e + T_I dS_I - p dV \qquad (6\text{-}7)$$

where S_e and S_I are separate entropies associated with the phase space occupied by electron and ion distributions. The pressure and energy do not separate into additive contributions except at low densities.

A generalized condition of thermodynamic consistency is found to be

$$\left(\frac{\partial E}{\partial V}\right)_{T_e,T_I} = T_e\left(\frac{\partial p}{\partial T_e}\right)_{T_I,V} + T_I\left(\frac{\partial p}{\partial T_I}\right)_{T_e,V} - p \qquad (6\text{-}8)$$

This equation is a constraint on any approximate model for the thermodynamic properties of nonequilibrium plasmas. The partial derivative $\partial p/\partial T_e$ is defined as a derivative for constant volume V and ion temperature T_I; for the case of thermal equilibrium ($T_e = T_I$), Eq. (6-8) reduces to the usual thermodynamic consistency relation.

Practical specific heat coefficients can be defined by

$$dT_e = A_{ee}dQ_e + A_{eI}dQ_I \tag{6-9}$$

$$dT_I = A_{Ie}dQ_e + A_{II}dQ_I \tag{6-10}$$

$dQ_e = T_e dS_e$ = heat added to electrons, $dQ_I = T_I dS_I$ = heat added to ions. These equations apply to an infinitesimal transformation occurring at constant volume V. Through use of a Maxwell relation generated from Eq. (6-7), we can show

$$\frac{A_{eI}}{T_e} = \frac{A_{Ie}}{T_I} \tag{6-11}$$

If an amount of heat dQ_e is suddenly added to the electrons (e.g., by a short laser pulse), then Eqs. (6-9, 10) assert that both electron and ion temperatures will rise. The change in ion temperature is produced indirectly by the sudden reduction in the electron screening which strengthens the potential U(R) and immediately is reflected in the ion kinetic energy.

Equations (6-7) through (6-11) are derived[2] by assuming a canonical ensemble distribution function for the position and velocity distribution of the ions based on the potential of Eq. (6-6). The Born-Oppenheimer approximation is essential to the derivation.

The two-temperature effects are largest at dense-plasma conditions, where the electron-ion interaction energy is a significant perturbation of the ideal-gas equation of state. Analytic formulas for the electron-ion energy are available only in the low-density limit.

7. DENSE ATOMIC PLASMAS (SUMMARY)

The dense plasma is normally defined as a strongly coupled electron-ion system in which the ion correlation length (Debye length) is so short that the screening is complete within a few atomic spacings. The Coulomb system is essentially classical and forms a background or environment for the atomic physics. The dense atomic plasma has the additional ingredient of partial ionization.

For partially ionized plasmas, the ionization state Q is a self-adjusting quantity. For example, when the electron-ion interaction, measured by Qe^2/R_0, becomes large enough to be comparable to kT, free electrons recombine and the interaction is reduced. As shown by Eq. (2-2) the ionization adjusts itself so that I(Z,Q) is a multiple of kT--typically, $\sim$10 kT in a low-density LTE plasma, but as little as $\sim$2 kT in a dense plasma. In low-density plasmas, the large ratio I(Z,Q)/kT implies a strong energetic decoupling of bound electrons from the free particles of the plasma.

Let us summarize changes which occur when a nondegenerate equilibrium (LTE) plasma is compressed at constant temperature.

As the density rises, the average ion charge state Q decreases. Because of this, the ionization potential of the outermost bound electron falls and the ratio I/kT decreases. The excitation energies are submultiples of the ionization potential, and so the typical excited state is closer to the groundstate in a dense plasma. Because of the reduced energy required for excitation, the average number of excited electrons increases.

In low-density plasmas, electron-ion scattering is controlled by the simple Coulomb potential Qe^2/R for most collisions. Contrawise, in dense plasma, the non-Coulomb core potential is important because the cores fill a large fraction of the plasma volume.

In dense plasma, excited states near the continuum are more often occupied. These states interact strongly with neighbor ions, and can be understood only through more elaborate models than those used for more deeply bound electrons. Collisional ionization and recombination rates greatly increase in dense plasmas (approximately proportional to density squared), so that ionization and bound populations fluctuate on shorter time-scales.

Thus, the atom in a dense plasma has more active degrees of freedom and plays a more vigorous dynamical role than in low-density plasmas. The conclusion emerging from this analysis is very clear: dense plasmas containing partially-stripped ions are very interesting from the atomic point of view.

APPENDIX A.
THE ISOLATED GROUNDSTATE TF ION

The model describes an ion of nuclear charge Z and ion charge Q. The Thomas-Fermi equations are:

$$\nabla^2 V = 4\pi e n(r) \tag{A1}$$

$$n(r) = \frac{1}{3\pi^2}\left(\frac{2m}{\hbar^2}\right)^{3/2} (\mu + eV(r))^{3/2} \tag{A2}$$

where V(r) = electrostatic potential, e = electron charge, n(r) = electron number density, μ = electron chemical potential. Eq. (A1) expresses electrostatic self-consistency, and Eq. (A2) gives the zero-temperature (i.e., groundstate) electron density of a semiclassical electron gas. The isolated ion is spherically symmetric. The boundary conditions for Eqs. (A1, A2) are:

$$V(r) \rightarrow \frac{Ze}{r} \qquad r \rightarrow 0 \tag{A3}$$

$$V(r) = \frac{Qe}{r} \qquad r > R_i \tag{A4}$$

where R_i = radius at which n(r) first equals zero. From Eq. (A2),

$$\mu = - Qe^2/R_i \tag{A5}$$

we also have

$$Z - Q = \int n(r)\, d^3r \tag{A6}$$

The contributions to the ion's energy are:

1) Electron-gas kinetic energy:

$$K = \int \left[\frac{3}{5}\,\frac{\hbar^2}{2m}\,(3\pi^2 n(r))^{2/3}\right] n(r)\, d^3r \tag{A7}$$

2) Electrostatic interaction of electrons with nucleus:

$$U_{en} = - \int \frac{Ze^2}{r}\, n(r)\, d^3r \tag{A8}$$

3) Electron-electron interaction:

$$U_{ee} = \frac{e^2}{2} \int \frac{n(r)\, n(r')}{|r - r'|} d^3r\, d^3r' \quad \text{(A9)}$$

Eq. (A9) can easily be reduced to a one-dimensional (radial) integral.

The total ion energy is $E_{TOT} = K + U_{en} + U_{ee}$. There is a variational theorem by which $E_{TOT} = E[n(r)]$ is minimal for the correct electron density n(r), subject to the constraint of Eq. (A6). From Eq. (A2) it follows that:

$$\mu(Z - Q) = \frac{5}{3} K + U_{en} + 2U_{ee} \quad \text{(A10)}$$

The virial theorem yields the relation

$$2K + U_{en} + U_{ee} = 0 \quad \text{(A11)}$$

The virial theorem is a consequence of Eq. (A1, A2) and not an additional assumption. Equations (A10, A11) give

$$K = - E_{TOT} \quad \text{(A12)}$$

$$U_{en} = \frac{7}{3} E_{TOT} - (Z - Q)\mu \quad \text{(A13)}$$

$$U_{ee} = - \frac{1}{3} E_{TOT} + (Z - Q)\mu \quad \text{(A14)}$$

and so a table of Q, E_{TOT} and $I = -\mu$ together give all the atomic energies. The ionization potential $I = -\mu = d\, E_{TOT}/dQ$. This data is given in Table 1.

For the neutral free atom, $Q = \mu = 0$ and a numerical calculation yields

$$E_{TOT} = - 0.768745\, Z^{7/3} \frac{e^2}{a_0} = - 20.92 \text{ eV } Z^{7/3} \quad \text{(A15)}$$

The limit of highly charged ions is treated analytically by recognizing that when $N_B = Z-Q$ is << Z, then the nuclear potential dominates and

$$n(r) \cong \frac{1}{3\pi^2} \left(\frac{2Z}{a_0}\right)^{3/2} \left(\frac{R_i - r}{rR_i}\right)^{3/2} \quad \text{(A16)}$$

According to Eq. (A6) the volume integral of n(r) must be Z-Q, which yields

$$R_i = \frac{a_0}{Z^{1/3}} (18)^{1/3} \left(1 - \frac{Q}{Z}\right)^{2/3} \quad \text{(A17)}$$

and hence

$$I = (18)^{-1/3} \frac{e^2}{a_0} \frac{Z^{4/3}}{(1 - Q/Z)^{2/3}} = 10.38 \text{ eV} \frac{Z^{4/3}}{(1 - Q/Z)^{2/3}} \quad \text{(A18)}$$

$$E_{TOT} = \frac{3}{(18)^{1/3}} \frac{e^2}{a_0} Z^{7/3} (1 - Q/Z)^{1/3} \quad \text{(A19)}$$

These limiting forms illustrate the scaling property discussed in the main text.

The Bloch constant or mean ionization-excitation potential, denoted $\bar{I}$, is conventionally defined as:

$$\log \bar{I} = \frac{1}{Z-Q} \int n(r) \log [\hbar\omega_p(r)] \, d^3r \quad \text{(A20)}$$

where

$$\omega_p(r) = \sqrt{4\pi e^2 \, n(r)/m} \quad \text{(A21)}$$

is the local electron-gas plasma frequency. Equation (A20) for $\bar{I}$ follows from the high-velocity limit of the inhomogeneous electron-gas stopping power according to a formula proposed by Lindhard and Winther. Comparison of calculated values of $\bar{I}$ with experiments shows a substantial discrepancy which may tentatively be attributed to the (unknown) gradient corrections. The Thomas-Fermi values of $\bar{I}$ are too small by a factor of $\sim$ 2. A limiting form of $\bar{I}(Z,Q)$ for highly stripped ions is found with the use of Eq. (A16):

$$\bar{I} = \left(e^{1/2}\sqrt{\frac{8}{9\pi}}\right)\frac{e^2}{a_0}\frac{Z}{(1-\frac{Q}{Z})^{1/2}} = 23.86\ \text{eV}\ \frac{Z}{1-Q/Z} \qquad \text{(A22)}$$

(The symbol e within parentheses is 2.718...).

The dipole polarizability of the Thomas-Fermi ion is calculated by assuming a small perturbation of n(r), V(r) having the form

$$eV_1(\vec{r}) = \frac{1}{r} f(r) \cos\theta \qquad \text{(A23)}$$

From Eqs. (A1, A2) we deduce

$$\frac{d^2 f}{dr^2} = \left[\frac{2}{r^2} + K_0^2(r)\right] f(r) \qquad \text{(A24)}$$

where $K_0(r)$ is determined by the unperturbed density $n_0(r)$:

$$K_0^2(r) = \frac{4}{\pi a_0}\left(3\pi^2 n_0(r)\right)^{1/3} \qquad \text{(A25)}$$

For $r \to 0$, the solution is required to be regular. For $r > R_i$, the solution has the form

$$f(r) = Ar^2 + \frac{B}{r} \qquad \text{(A26)}$$

The first term is the potential of a uniform (homogeneous) applied field $E(\infty)$ and the second is the potential of the induced dipole $P = \alpha E(\infty)$ with

$$\alpha = -\frac{B}{A} \qquad \text{(A27)}$$

Numerical data for α are given in Table 1. The analytic result for highly charged ions is

$$\alpha = \frac{63}{16}\frac{a_0^3}{Z}\left(1-\frac{Q}{Z}\right)^3 \qquad Q \to Z \qquad \text{(A28)}$$

An interesting theorem concerns the penetration of the applied field $E(\infty)$ to the region near the nucleus. The local dipolar field near the nucleus $E(o)$ is determined by examining the coefficient of r^2 in the series expansion of f(r) near r = 0. The result is

$$\beta = \frac{E(o)}{E(\infty)} = \frac{Q}{Z} \tag{A29}$$

Extensive discussions of Thomas-Fermi theory are given by March and Gombas.[16] In recent years, several investigators have examined quantum corrections to the Thomas-Fermi theory which improve the accuracy at the expense of the TF scaling property. Accounts of this work are given by Kalitkin,[61] Kirzhnitz,[62] More[2] and Schwinger.[63]

REFERENCES

1. Nuckolls, J., Wood, L., Thiessen, A. and Zimmerman, G., Nature 239, 139 (1972); Lawrence Livermore National Laboratory Annual Reports UCRL-50021; see also the series Laser Interactions and Related Plasma Phenomena, vols. 1-6, edited by H. Hora and H. Schwartz, Plenum Press.

2. R. More, UCRL-84991, "Atomic Physics in Inertial-Confinement Fusion", March 1981, to appear in Applied Atomic Collision Physics, Vol. II, ed. by H. S. Massey, 1983.

3. Proceedings of Workshop Conference on Lowering of the Ionization Potential, JILA Report 79, University of Colorado, Boulder, Colorado, 1965.

4. G. Zimmerman and R. More, J.Q.S.R.T. 23, 517 (1980).

5. R. More, J.Q.S.R.T. 27, 345 (1982).

6. B. Yaakobi, D. Steel, E. Thorsos, A. Hauer and B. Perry, Phys. Rev. Letters 39, 1526 (1977); Yaakobi et al., Phys. Rev. A19, 1247 (1979); Yaakobi et al., Phys. Rev. Letters 44, 1072 (1980).

7. A. Hauer, K. Mitchell, D. van Hulsteyn, T. Tan, E. Linnebur, M. Mueller, P. Kepple and H. Griem, Phys. Rev. Letters 45, 1495 (1980); Mitchell et al., Phys. Rev. Letters 42, 232 (1979).

8. J. Kilkenny, R. Lee, M. Key and J. Lunney, Phys. Rev. A22, 2746 (1980); Kilkenny et al. in Spectral Line Shapes (p. 367), ed. by B. Wende, deGruyter and Co., Berlin (1981).

9. H. R. Griem, M. Blaha, and P. Kepple, Phys. Rev. 19A, 2421 (1971); H. Griem, in Laser Interaction and Related Plasma Phenomena, vol. 5, Plenum Publishing Co., (1981); H. Griem, Phys. Rev. A20, 606 (1979); H. Griem, Plasma Spectroscopy, McGraw-Hill, New York (1964); H. Griem, Spectral Line Broadening by Plasmas, Academic Press, New York (1974).

10. C. H. Hooper, Phys. Rev. 149, 77 (1966); Phys. Rev. 165, 215 (1968); Woltz, Iglesias and Hooper, J.Q.S.R.T. 27, 233 (1982).

11. A. Hauer (preprint LA-UR-80-1660).

12. M. Key and R. Hutcheon, Advances in Atomic and Molecular Physics, vol. 16, Academic Press (1981).

13. M. Rosen, et al, Physics of Fluids 22, 2020 (1979).

14. G. Zimmerman, presentation at this conference; G. Zimmerman and W. Kruer, Comments on Plasma Physics 2, 85 (1975); see also articles in Ref. 1.

15. T. A. Carlson, C. W. Nestor, Jr., N. Wasserman, and J. D. McDowell, Atomic Data 2, 63 (1970); J. S. Scofield, Phys. Rev. 179, 9 (1969); J. Scofield, unpublished data tables.

16. H. Bethe and R. Jackiw, Intermediate Quantum Mechanics, 2nd Ed., W. A. Benjamin, Inc., New York, (1968); N. H. March, Self-Consistent Fields in Atoms, Pergamon Press, Oxford (1975); P. Gombas, Die Statistiche Theorie Des Atoms und Ihre Andwendungen, Springer-Verlag, Vienna (1949).

17. E. Nardi, E. Peleg, and Z. Zinnamon, Physics of Fluids 21, 574 (1978).

18. E. McGuire et al., Phys. Rev. A26, 1318 (1982).

19. S. Skupsky, Phys. Rev. A16, 727 (1977); N. Arista and W. Brandt, Phys. Rev. A23, 1898 (1981); C. Deutsch, G. Maynard and H. Minoo, to appear in Laser Interaction and Related Plasma Phenomena, vol. 6, ed. by H. Hora, Plenum Press (ca. 1983).

20. R. More, Y.-T. Lee and D. S. Bailey, unpublished preprint UCRL-87147.

21. H. D. Betz, Methods of Experimental Physics 17, 73 (1980); W. E. Lamb, Phys. Rev. 58, 696 (1940).

22. O. Firsov, Soviet Physics JETP 5, 1192 (1957) and JETP 6, 534 (1958); J. Lindhard, V. Nielsen and M. Scharff, Det Kong Danske Vidensk Selsk 36, No. 10 (1968).

23. C. Lee, C. Longmire and M. Rosenbluth, unpublished report LAMS-5694, Los Alamos National Laboratory (1974).

24. R. More, Phys. Rev. A19, 1234 (1979).

25. J. P. Biersack and J. F. Ziegler, Nuclear Instruments and Methods 194, 93 (1982) give a recent review of interatomic potentials for strong collisions of neutral atoms.

26. J.-P. Hansen, Journal de Physique 36, L-133 (1975); S. Galam and J.-P. Hansen, Phys. Rev. A14, 816 (1976); H. DeWitt, Strongly Coupled Plasmas, ed. by G. Kalman, Plenum Publishing Co., New York (1978).

27. S. Brush, H. Sahlin, and E. Teller, J. Chem. Phys. 45, 2102 (1966).

28. A comprehensive review of the properties of the Coulomb fluid is given by M. Baus and J.-P. Hansen, Physics Reports 59, 1 (1980).

29. J. Stewart and K. Pyatt, Astrophys. J. 144, 1203 (1966).

30. P. Vieillefoisse, J. Physique 42, 723 (1981).

31. J. Weisheit, to be published in Applied Atomic Collision Physics, vol. II, ed. by H. S. Massey, Academic Press (1983).

32. J.-P. Hansen and I. R. McDonald, Phys. Rev. Lett. 41, 1379 (1978).

33. C. Deutsch, Physics Lett. A60, 317 (1977).

34. H. DeWitt and Y. Rosenfeld, Physics Lett. A75, 79 (1979); Y. Rosenfeld, Journal de Physique, 41, C2-77 (1980); Y. Rosenfeld and A. Baram, J. Chem. Phys. 75, 427 (1981).

35. Both Born and partial-wave calculations were performed for dense neon plasmas by Pauline Hsu Lee, unpublished doctoral thesis, University of Pittsburgh (1977).

36. H. Minoo, C. Deutsch and J.-P. Hansen, Phys. Rev. A14, 840 (1976).

37. D. Boercker, F. J. Rogers and H. DeWitt, Phys. Rev. A25, 1623 (1982); J.-P. Hansen and J. McDonald, Phys. Rev. Lett. 41, 1379 (1978); M. Baus, J.-P. Hansen, L. Sjogren, Phys. Lett. 82A, 180 (1981).

38. H. Brysk, P. Campbell and P. Hammerling, Plasma Physics 17, 473 (1975).

39. Y.-T. Lee and R. More, unpublished.

40. J. Davis and M. Blaha, J.Q.S.R.T. 27, 307 (1982); N. F. Lane, J. Weisheit and B. Whitten, unpublished; see also reference 31.

41. A review of the experimental situation and some proposed explanations is given by W. Kruer, Comments on Plasma Physics 5, 69 (1979).

42. A Fokker-Planck calculation of transport by non-Maxwellian free electrons has recently been performed by J. Albritton at LLNL. The calculations indicate that distortion of the tail of the Maxwellian tends to inhibit electron thermal conduction.

43. D. A. Liberman, J.Q.S.R.T. 27, 335 (1982); D. A. Liberman, Phys. Rev. B20, 4981 (1979).

44. F. Perrot, Phys. Rev. A26, 1035 (1982).

45. K.-F. Berggren and A. Froman, Arkiv fur Physik 39, 355 (1969).

46. B. Rozsnyai, Phys. Rev. 145, 1137 (1972).

47. W. Zink, Astrophys. J. 162, 145 (1970); T. Carlson, D. Mayers, and D. Stibbs, Mon. Not. Roy. Ast. Soc. 140, 483 (1968).

48. F. J. Rogers, Phys. Rev. A23, 1008 (1981).

49. W. Huebner, J.Q.S.R.T. 10, 949 (1970); W. Huebner, M. Argo, and L. Ohlsen, J.Q.S.R.T. 19, 93 (1978).

50. H. Mayer, unpublished Los Alamos Report LA-647.

51. D. Post, R. V. Jensen, C. B. Tarter, W. Grasberger, and W. A. Lokke, Atomic Data and Nuclear Tables 20, 397 (1977).

52. W. A. Lokke and W. H. Grasberger, preprint UCRL-52276, Lawrence Livermore National Laboratory (1977).

53. R. More, Phys. Rev. A4, 1782 (1971); R. More and E. Gerjuoy, Phys. Rev. A7, 1288 (1973).

54. J. Green, J.Q.S.R.T. 4, 639 (1964).

55. F. Grimaldi and A. Grimaldi-LeCount, J.Q.S.R.T. 27, 373 (1982).

56. G. Zimmerman, unpublished lecture at NATO Workshop, Palermo, 1982.

57. R. McWhirter in Plasma Diagnostics, ed. by Huddlestone, Academic Press, New York (1965). R. Landshoff and J. Perez, Phys. Rev. A13, 1619 (1976).

58. D. Salzmann, Phys. Rev. A20, 1704 (1979); Phys. Rev. A20, 1713 (1979); Phys. Rev. A21, 1761 (1980).

59. The result given here is based, in part, on an unpublished study of nonequilibrium thermodynamics by F. Keffer and R. More.

60. The conclusion reached here contradicts an analysis given by C. Kittel, Elemental Statistical Physics, p. 165-168, J. Wiley, New York, 1958. A careful comparison will show that the textbook treatment omits the spontaneous emission factor included in our Eq. (6-2). This omission is effectively equivalent to selecting a very high radiation temperature (so that $N_\nu >> 1$), but in this case one cannot achieve $\Delta T << T_e$. Thus it is not surprising that the (incorrect) rate equations given by Kittel are incompatible with the principle of minimum entropy production. Footnote 59 applies also to these remarks.

61. N. N. Kalitkin and L. Kuźmina, Sov. Phys. Sol. State 7, 287 (1972); F. Perrot, Phys. Rev. A20, 586 (1979); R. More, Phys. Rev. A19, 1234 (1979).

62. D. A. Kirzhnitz, Yu. Lozovik and A. Shpatakavskaya, Sov. Phys. Uspekhi 18, 649 (1976).

63. J. Schwinger, Phys. Rev. A22, 1827 (1980).

THE PLASMA BOUNDARY REGION AND THE ROLE OF ATOMIC AND MOLECULAR PROCESSES

M.F.A. Harrison

UKAEA/Euratom Fusion Association
Culham Laboratory, Abingdon, Oxon OX14 3DB, England

1. INTRODUCTION

Confinement of plasma particles and energy by a magnetic field is degraded in a region close to the wall of the containment vessel due to localised processes which give rise to enhanced losses of charged particles and energy. Losses of thermal energy occur when the magnetic field lines intercept the wall of the vessel and also because plasma ion bombardment of the wall releases radiating impurity elements due to processes such as sputtering. Neutralisation of plasma ions* in collisions with the wall is the principal source of neutral atoms within the vessel and subsequent charge exchange of these atoms with ions in the adjacent plasma provides a mechanism by which energetic neutral particles can cross the magnetic field and thereby recycle between the plasma and the wall. This recycling of energetic atoms is a source of energy loss from the plasma and also an additional mechanism for the release of impurity elements. Thus the concentration of neutralised plasma ions and also of impurity atoms and ions is peaked close to the wall where inelastic atomic collisions between electrons and these atomic species result in localised cooling of the plasma electrons. The principal cooling processes are excitation followed by spontaneous emission of photons which are absorbed by the wall. In

*Plasma ions in fusion research are predominantly protons whose neutralised products are thus either "hydrogen" atoms or molecules. The plasma in a DT fusion reactor will be formed from deuterium and tritium but these species will be referred to here as "hydrogen" unless a distinction must be made between the isotopes.

addition, ionisation yields cold secondary electrons and thereby reduces the average temperature of the plasma.

The boundary region is somewhat indeterminate but for the basis of the present discussion it will be assumed that it can be defined as that part of the plasma where plasma transport is predominantly in the direction of the magnetic field. This implies that the magnetic flux tubes in the boundary must intercept the walls of the vessel and so these magnetic field lines are termed "open". The interface between the boundary region and the confined plasma therefore occurs at the "separatrix". This is the magnetic surface that bounds the region wherein the field lines are "closed" so that plasma transport towards the wall must take place by motion across the magnetic field. For the sake of conciseness the subsequent discussion will be restricted to toroidal confinement devices and, because of the breadth of existing studies, particular emphasis will be placed upon the tokamak configuration.

Plasma behaviour in the boundary region plays a major role in determining the rate of build-up and the ultimate steady state concentration of impurity ions within the confined plasma. Control of this impurity concentration will be a particularly important factor in near term experiments such as TFTR and JET where the plasma heating period will extend for several seconds. Recycling of neutral particles within the boundary influences strongly the ability of external vacuum pumps to exhaust neutral gas from the edge of the plasma and, since the plasma in longer term devices such as INTOR[1] must ignite and burn for 100 to 200 s, it will be necessary not only to control impurities but also to fuel with neutral DT and to exhaust the helium gas produced from the $^{3}D + ^{3}T \rightarrow ^{4}He + n$ fusion reaction. These requirements have, during recent years, provided a stimulus for developing self-consistent concepts of the boundary plasma in which the interactive mechanisms of plasma transport are coupled by atomic processes to plasma-surface interactions. By these means it has been possible to evolve self-consistent models of the boundary region which can be used as a design base for next generation devices and also as a means for interpreting present-day experiments.

The objective of this paper is to describe in outline the basic processes of plasma transport and plasma surface interactions and to discuss the atomic processes of particular relevance to the boundary region. Incorporation of these basic processes into self-consistent models of the boundary region will be briefly described and some examples will be given of the boundary conditions predicted. The self-consistent amalgamation of these disparate subjects into a general description of the plasma is a relatively new concept; recent surveys can be found in articles by Post[2] and Harrison[3] and an appreciation of the present status can be gained from the proceedings of specialised conferences such as Ref.4.

2. MAGNETIC TOPOLOGY OF THE PLASMA BOUNDARY

An elemental tube of magnetic flux in the boundary region of a toroidal plasma can be envisaged in the manner indicated in Figure 1(a) where a ribbon-like flux tube is shown wrapped over the separatrix surface. The magnetic field consists of two components; the toroidal field B_{tor} and the poloidal field B_{pol}. In the case of the tokamak, B_{tor} is generated using external windings but B_{pol} is generated by a current that is induced to flow in a toroidal direction within the plasma. The flux tube twists around the torus and the length of a closed tube in the direction z parallel to

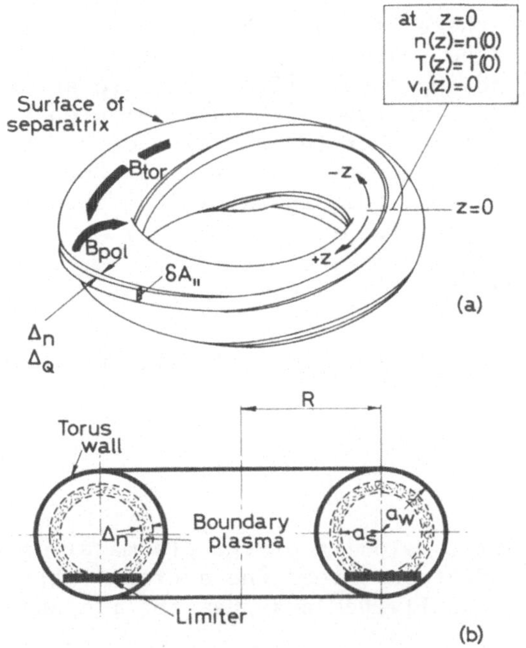

Fig.1 Topology of a conceptual boundary plasma.
(a) shows an elemental tube of magnetic flux. The tube has an area $\delta A_{||}$ normal to the field and its radial extent is either Δ_n or Δ_Q depending upon whether the flow of particles or energy is being considered; the total length of the tube is $2\pi Rq(a_s)$.
(b) shows the boundary plasma associated with a toroidally symmetric limiter plate located at the bottom of the torus. The radial position a_s defines the position of the separatrix.

the resultant magnetic field can be expressed as

$$L_{||} = 2\pi Rq(r) \tag{2.1}$$

where

$$q(r) = \frac{r}{R}\,\frac{B_{tor}}{B_{pol}}$$

is the safety factor of the plasma of major radius R and minor radius r. It is customary to protect the torus wall by a limiter so that the separatrix radius a_s, shown in Figure 1(b), is inboard of the wall radius a_w. At the separatrix, the field lines are open and, since the limiter shown in Figure 1(b) is symmetrical in the toroidal direction, the length of these open lines is

$$L_{||}(a_s) \approx 2\pi Rq(a_s).$$

It is assumed that there is negligible flow of plasma current in the boundary region outboard of the separatrix, so that those plasma electrons and ions that diffuse outward across the separatrix will enter this flux tube with an equal probability of drifting towards the limiter in either the positive or negative direction of z. Thus the average length of boundary flux tube traversed by plasma which flows across the separatrix and then flows to the limiter surface can be taken as

$$\tfrac{1}{2}\, L_{||}(a_s) = \pi Rq(a_s). \tag{2.2}$$

The residence time for plasma particles within the boundary can then be expressed as

$$\tau_{||} = \frac{\pi Rq}{v_{||}} \tag{2.3}$$

where $v_{||}$ is the drift velocity of the plasma along the flux tube. It will be seen in Section 3 that the average drift velocity in the z direction will generally be less than the ion sound speed,

$$C_s = \left(\frac{ZkT_e + kT_i}{m_i}\right), \tag{2.4}$$

*Throughout this paper the symbol T is used whenever kT is expressed in eV and in general Gaussian units are used for other parameters. The symbols e, k, etc, have their conventional meaning. The electron and ion masses are m_e and m_i and their temperatures are T_e and T_i. For a DT mixture m_i is taken as 2.5 m_i(proton). The ion charge state is denoted by Z.

where T_e and T_i are respectively the electron and ion temperatures.* The particle residence time $\tau_{||}$ is typically $\sim 10^{-3}$ s and it is appreciably shorter than the characteristic time, $\tau_{\perp}$, for plasma particle transport across the magnetic field; $\tau_{\perp}$ ranges from $\sim 10^{-2}$ s in present-day experiments to predicted values of ~ 1 s in large devices such as INTOR. Even so, some cross field transport must take place during the time taken for the boundary plasma to drift along the open field lines and the scale length of the radial gradient of plasma density in the boundary can, in its simplest form, be expressed as

$$\Delta_n = (D_{\perp}\tau_{||})^{\frac{1}{2}} \tag{2.5}$$

where $D_{\perp}$ is the diffusion coefficient for cross field diffusion. Typically Δ_n is a few cm.

Particles in the boundary plasma will thus flow in a region that extends a distance $\sim \Delta_n$ outboard of the separatrix and the volume of this boundary region is about $4\pi^2 Ra_s\Delta_n$. This volume is filled with flux tubes whose length is $2\pi Rq(a_s)$ and so it is convenient to envisage that plasma flows along a channel which comprises a group of elemental flux tubes each of which has a radial thickness Δ_n and elemental area $\delta[A_{||}]_n$ normal to the direction z. The total area of the channel along which plasma particles flow to the limiter is thus given by

$$[A_{||}]_n \approx \frac{4\pi^2 Ra_s\Delta_n}{2\pi Rq(a_s)} \approx \frac{2\pi a_s\Delta_n}{q(a_s)} \tag{2.6}$$

and, if magnetic flux is conserved, $[A_{||}]_n$ is uniform throughout the length of the flow channel.

3. TRANSPORT TO THE LIMITER SURFACE OF PLASMA PARTICLES AND ENERGY IN THE DIRECTION PARALLEL TO THE MAGNETIC FIELD

The principal characteristics of transport along the magnetic field are conveniently discussed in terms of the linear representation of the flow channel which is illustrated schematically in Figure 2. The face of the channel which lies in contact with the separatrix is assumed to be fed uniformly by the outward flow of particles and of energy that are transported into the boundary plasma by transport across the magnetic field. Within the boundary, plasma flow is predominantly in the positive and negative directions of z and plasma sheaths will be established at the ends of the channel where the plasma is in contact with the limiter surface.

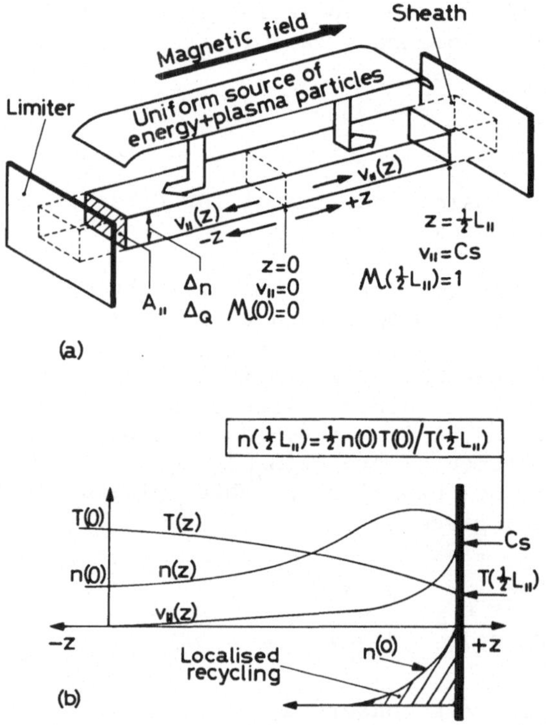

Fig.2 Linear representation of the flow channel for collisional plasma transport along the magnetic field in the boundary region. (a) shows a simple one-dimensional concept of flow parallel to z which is terminated by plasma sheaths at the surface of a limiter. (b) shows the dependence upon z of the plasma density, n, temperature, T, and flow velocity $v_{||}$. The conditions sketched refer to powerful recycling in the locality of the limiter; $n^{(o)}$ being the density of neutral particles formed by plasma bombardment of the limiter surface. Sources of charged particles and energy in other regions are assumed to be uniformly distributed in the z direction.

Properties of the plasma sheath

Plasma contained in a vessel with electrically conducting walls will, in the absence of a magnetic field, assume a positive potential with respect to the walls whenever $T_e \gtrsim T_i$. The potential ensures that the ready flow of charge carried by mobile electrons to the wall is reduced (due to the repulsive electrostatic force) so that it equals that carried by ions. This condition is likely to pertain to flow parallel to the magnetic field in the boundary plasma. The nett loss of charge is then zero and an electrostatic field is established in the vicinity of the wall, the extent of this field being related to the screening properties of the plasma. The scale length of the potential gradient is characterised by the Debye screening length, λ_D, which can be expressed as

$$\lambda_D = \left(\frac{kT_e}{4\pi n_e e^2}\right)^{\frac{1}{2}} = 7.43 \times 10^2 \left(\frac{T_e}{n_e}\right)^{\frac{1}{2}} \quad [\mathrm{cm}] \qquad (3.1)$$

where n_e is the number density of plasma electrons. The sheath potential U can be determined from Poisson's equation in which are substituted the densities of electrons and ions in the region close to the wall (see for example Chen[5]). A solution,

$$U = \frac{kT_e}{2e} \ln\left(\frac{m_i}{2\pi m_e}\right), \qquad (3.2)$$

is representative of singly charged ions that are incident upon a plane surface and uninfluenced by a magnetic field. For a "hydrogen plasma" $U \approx 3T_e/e$.

There is also a longer range, ion accelerating region called the "pre-sheath" over which a potential difference $U' \gtrsim \frac{1}{2}\, kT_e/e$ is established in order to impart a directed drift velocity to ions in the adjacent plasma; this ensures that the sheath region is fed with ions at such a rate that a balance of negative and positive charge density can be maintained in the electrostatic field close to the surface. The balance in charge density requires that the plasma drift velocity at the sheath edge is at least equal to the ion sound speed.

If ions have a Maxwellian distribution of velocities, then those which enter the sheath carry an amount of energy equal to $2kT_i$ because high energy particles preponderate over those with the average energy ($1.5\ kT_i$). Positively charged ions are accelerated by the sheath potential and, if their charge state is Z, strike the surface with an energy E_i which is given by

$$E_i = 2kT_i + Ze(U + U'). \qquad (3.3)$$

The repulsion of electrons by the electrostatic field reduces the density of electrons in the sheath region but each electron carries about $2kT_e$ of energy to the surface. Thus the total energy carried to the surface by each plasma ion and its associated electrons can be expressed as

$$E_{ie} = 2kT_i + Ze(U + U') + \chi_i + Z\phi_w + Z2kT_e$$

where χ_i is the ionisation threshold energy and ϕ_w is the work function of the surface. In practice the ions cannot deliver all of their energy to the surface because many of them are back-scattered as energetic atoms. If the energy reflection coefficient is R_E (see Section 5) then the energy $\mathcal{E}_{ie}$ lost to the surface by each ion and its associated electrons is

$$\mathcal{E}_{ie} = [2kT_i + Ze(U + U')] (1 - R_E) + \chi_i + Z\phi_w + Z2kT_e. \quad (3.4)$$

Release of impurity atoms is strongly dependent upon the energy of ions incident upon the boundary surface (see Section 5) and in this context it is advantageous to minimise the sheath potential. The potential is reduced if the plasma electrons are cooled due to inelastic atomic collisions with released impurity atoms and with recycling "hydrogen" but another mechanism is the release of secondary electrons from the surface. These electrons are accelerated by the sheath potential and enter the plasma where they can thermalise and then return to the surface. The nett loss of charge from the plasma must remain zero and so the electron current flowing through the sheath in the direction from the plasma to the surface must be greater than the accompanying flow of ion current by an amount equal to the secondary emission current. The sheath potential will therefore decrease in order to maintain this inequality (see Hobbs and Wesson[6]). However, it must be stressed that this effect is highly sensitive to the topology of the magnetic field; for example, if the field lines graze the surface then many secondary electrons are likely to be suppressed because their gyro-radius is smaller than the sheath thickness.

Plasma density ($n = n_e = n_i$) in the boundary of a typical tokamak lies in the range 10^{13} to 10^{14} cm^{-3} and the electron temperature in the range 10 to 200 eV. The Debye length is $\sim 10^{-3}$ cm and hence the sheath region is very much smaller than the length of the plasma flow channel in the boundary because $L_{||}(a_s)$ lies in the range 0.5 to 4.0×10^3 cm.

Particle transport in a collisional boundary plasma

Properties of plasma transport within the flow channel which feeds the sheath region are sensitive to the exchange of energy and momentum in collisions between the plasma electrons and ions. Such exchange takes place predominantly due to multiple, small-angle Coulomb scattering collisions whose effective mean free path (for electron-electron collisions) can be expressed as

$$\lambda_{ee} = \frac{4 \times 10^{13}\, T_e^{\,2}}{n Z^2 \ell n \Lambda} \qquad [\mathrm{cm}] \tag{3.5}$$

where $\ell n\Lambda$ is the Coulomb logarithm ($\sim$ 10). The magnitude of λ_{ee} is typically a few 10^2 cm. The electron-ion collision mean free path is of comparable magnitude and in many instances plasma flow along the magnetic field from the separatrix to the edge of the plasma sheath at the limiter can be regarded as collisional. The principal aspects of such flow are shown firstly by the fluid equation for continuity,

$$\frac{d}{dz}\left(n v_{||} A_{||}\right) = S A_{||}\,, \tag{3.6}$$

where S is a volume source term for charged particles (ie, number of particles/unit volume/unit time), and secondly by the associated momentum equation,

$$m\, n\, v_{||} \frac{dv_{||}}{dz} = -\frac{dP}{dz} - m\, v_{||}\, S. \tag{3.7}$$

Here, $m = (m_i + m_e)$, $P = (nT_e + nT_i)$ is the plasma pressure and it is assumed that the particle source does not add momentum to the flow. Manipulation (see Morgan and Harbour[7]) yields

$$\frac{1}{v_{||}} \frac{dv_{||}}{dz} = -\left(\frac{1}{1 - M^2}\right) \frac{1}{A_{||}} \frac{dA_{||}}{dz}$$

$$+ \frac{(1 + M^2)}{(1 - M^2)} \left[\frac{S}{n v_{||}} + \frac{1}{(T_e + T_i)} \frac{d}{dz}(T_e + T_i)\right] \tag{3.8}$$

Here M is the Mach number of the flow and all parameters are functions of z.

The flow must start from rest at the point of symmetry which implies that $v_{||} = 0$ and $dv_{||}/dz = 0$ when $z = 0$ so that $M(o) = 0$. The flow is therefore subsonic and accelerates up to $M(\frac{1}{2}L_{||}) = 1$ at the sheath edge. In order to illustrate trends in behaviour it is useful to simplify Eq. (3.8) by the assumptions that $M^2 \ll 1$,

$dA_{||}/A_{||} = 0$ and $d(T_e + T_i)/dz = 0$. This leads to the expression

$$\frac{dv_{||}}{dz} = \frac{S}{n}, \tag{3.9}$$

which shows that the velocity gradient $dv_{||}/dz$ is directly dependent upon the local value of the particle source $S(z)$ whenever n is invariant. If such invariance of density is valid, then it is apparent that $v_{||}$ increases linearly with z where the flow channel in the boundary plasma is subjected to a uniform source of charged particles, ie $S(z)$ = constant where the channel is fed by plasma that flows across the separatrix. However, in the region close to the limiter, the channel is fed locally by ionisation of neutral particles that arise due to plasma bombardment of the limiter surface and in general $S(\frac{1}{2}L_{||}) > S(z)$. Thus there is a corresponding increase in the gradient of $v_{||}$ close to the limiter and the plasma flow accelerates until the drift velocity equals C_s at the sheath edge. Typical conditions predicted for a boundary where $S(\frac{1}{2}L_{||}) > S(z)$ = (constant) are sketched in Figure 2(b).

Transport of plasma energy to the limiter

Transport of energy along a collisional flow channel to the sheath edge is dominated by the powerful effect of electron thermal conduction parallel to the magnetic field. Collisionality due to electron-electron collisions is assured because most electrons are reflected by the boundary sheaths and ultimately leave the plasma only as a consequence of electron-electron collisions. The heat flux conducted along the flow channel can be expressed as

$$-\kappa_o T^{5/2} \frac{dT}{dz} = \frac{\alpha Q}{[A_{||}]_Q} \tag{3.10}$$

where $T = T_e = T_i$, Q is the total flow of energy in the channel assumed to be equal to the flow Q_1 across the separatrix, $[A_{||}]_Q$ is the effective area of the energy transport channel and $\kappa_o T^{5/2}$ is the thermal conductivity. The coefficient κ_o is given by Spitzer[8] as

$$\kappa_o = \frac{3.15 \times 10^2}{Z_{eff}\ \ell n\Lambda} \qquad [W(eV)^{-7/2}\ cm^{-1}] \tag{3.11}$$

where Z_{eff} is the effective charge state of the plasma. The parameter α is the source distribution function of the input energy and, for a uniformly distributed flow of energy into the toroidal boundary, $\alpha = z[\pi R q(r)]^{-1}$. It is apparent that the energy flux in collisional flow is strongly dependent upon T and it is unlikely that T will appreciably exceed a few 10^2 eV even in a reactor where $\sim 10^2$ MW must be transported by conduction to the limiter. Equation (3.11) also indicates that dT/dz is small where T $\sim$ 100 eV and only becomes substantial close to the limiter where inelastic collisions in the recycling plasma cause localised cooling of the plasma electrons and concomitant reduction in $\kappa_0 T^{5/2}$.

This behaviour which is also sketched in Figure 2(b) provides justification for the preceding simplification that n(z) tends to invariant throughout most of the length of the flow channel. Indication of the variation in density can be obtained from manipulation of Eq.(3.8) for conditions where $T_i = T_e$ and both S and $A_{||}$ are independent of z; this yields

$$\frac{n(o)}{n(z)} = \frac{[1 + M(z)^2]T(z)}{[1 + M(o)^2]T(o)}$$

and, since $M(\frac{1}{2}L_{||}) = 1$, so

$$n(\tfrac{1}{2}L_{||}) = \tfrac{1}{2}\, n(o)\; T(o)/T(\tfrac{1}{2}L_{||}). \qquad (3.12)$$

The ready ability of the electrons to transport power throughout the boundary plasma implies that $[A_{||}]_Q$ can be significantly smaller than $[A_{||}]_n$ and it follows that, in collisional transport parallel to the magnetic field, the scale length for radial transport of energy, Δ_Q, will generally be smaller than Δ_n.

Convective transport

Equation (3.5) shows that λ_{ee} becomes greater than $\frac{1}{2}L_{||}$ where either n is small or T is large and in these conditions the free-streaming flow of particles and energy along the channel will not be impeded by Coulomb collisions. Even so, electrons remain the major transporters of energy due to their high mobility in the channel between the boundary sheaths. The energy flow transported by free-streaming electrons can be expressed as

$$Q_e \approx \frac{n_e}{4} A_{||}\, \Omega\, kT_e \left(\frac{8kT_e}{\pi m_e}\right)^{\frac{1}{2}}, \qquad (3.13)$$

where Ω is the numerical constant ($\lesssim 2$ and dependent upon the distribution of electron velocities). Collisionless transport can become significant in the plasma that recycles close to the limiter surface because in this region the transport path tends to be less than λ_{ee}.

The sheath region is always collisionless because $\lambda_{ee} >> \lambda_D$ and so Q_s, the energy flow through the sheath which links the channel to the limiter surface, must be by convective transport. The flow is expressed as

$$Q_s = \Gamma_s E_{ie} = n(\tfrac{1}{2}L_{||})A_{||} C_s E_{ie} , \tag{3.14}$$

where Γ_s is the plasma particle flow to the limiter and E_{ie}, which is described by Eq.(3.4), is the energy deposited upon the surface by each ion and its associated electrons.

Modelling of the boundary plasma

Behaviour of the boundary plasma has been predicted by numerical codes, for example Post[2] and Morgan and Harbour[7], but the physical processes involved can perhaps be identified more readily in analytical approaches, for example Harrison et al.[9] In this matter approach the plasma conditions at the separatrix n(o), $A_{||}$, and the total input energy flow $2Q_{\perp}$ over the two channels of total length $L_{||}$, are taken as input parameters that define a particular tokamak. Equations (3.14), (3.12) and (3.10) are then employed in an iterative manner to predict $n(\tfrac{1}{2}L_{||})$ and $T(\tfrac{1}{2}L_{||})$ at the sheath. The energy lost from the plasma by atomic processes, $Q_{(atomic)}$, must be assessed in a self-consistent manner so that the energy deposited by charged particle transport to the surface of the limiter can be determined from the relationship

$$Q_s = Q_{\perp} - Q_{(atomic)} \tag{3.15}$$

To determine $Q_{(atomic)}$ it is necessary to take account of plasma surface interactions; in particular the sputtering of atoms of limiter material and also the return of plasma ions in the form of both energetic atoms and cold atoms (or molecules).

4. DIVERTORS

The wall of the toroidal vessel should be effectively shielded from direct impact of plasma particles whenever the limiter geometry, see Figure 1(b), is such that $(a_w - a_s) \gtrsim \Delta_n$. However, the flow of boundary plasma is concentrated upon the limiter surface which is liable to release impurity ions into the adjacent region of confined plasma so that the limiter acts as a source of impurities within the torus. This source can be reduced by the use of a "divertor" which is a device that deflects or "diverts" the plasma flow channel into a separate chamber appended to the containment vessel.

The concept of a divertor action is most simply appreciated in the case of a poloidal divertor. The poloidal field in a tokamak is produced by a current that is induced to flow within the plasma

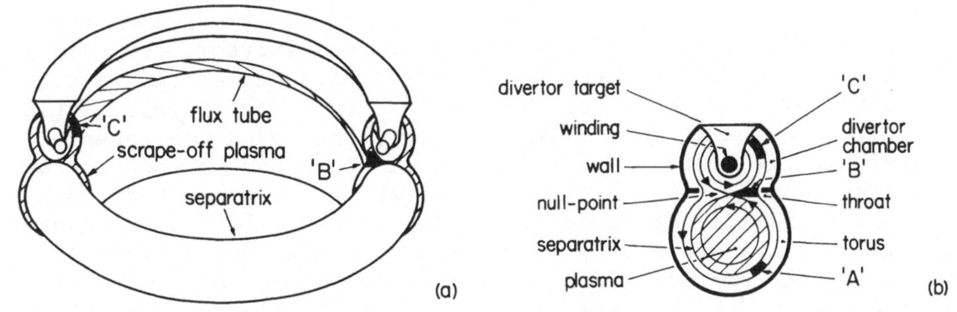

Fig.3 Schematic illustration of a tokamak with a single-null poloi dal divertor.

The perspective view (a) illustrates the toroidal symmetry and the rotation of the magnetic flux tubes around the separatrix. The view in the poloidal plane (b) shows the configuration of the poloidal magnetic field and the boundary surfaces of the containment vessel together with the location of the divertor winding relative to the plasma.

and this field can be opposed (at least over a smsall range of poloidal angle) when a complementary current is passed in the same direction through a nearby external conductor. The poloidal divertor relies upon such local annulment and its principles are illustrated in Figure 3 where a single divertor winding is shown lying above but parallel to the magnetic axis of the plasma. Location of "the null-point" (ie, the region where $B_{pol} = 0$ in the poloidal plane) defines the separatrix and it is dependent upon the spacing between the winding and the plasma and also upon the relative magnitude of the currents. The walls of the vessel are necked-in around the null-point so that communication between the divertor chamber and the containment vessel is restricted; the system is symmetrical in the toroidal plane and additional divertor windings can be provided to increase the number of null-points. The open magnetic field lines outboard of the separatrix are twisted, in the manner shown in Figure 1(a), so that plasma at position 'A' is rotated to 'B' and then to 'C' as it moves in the toroidal direction around the torus. In effect, plasma that diffuses across the separatrix into region 'A' is scraped out of the torus and deposited upon the divertor target; indeed, the toroidal boundary plasma is termed the "scrape-off" region and the diverted field serves the role of a "magnetic limiter". The effective length of the flow channel in the scrape-off plasma is about $\pi Rq(a_s)$ and the channel area $A_{||}$ is given by Eq.(2.6).

Other forms of divertor are based upon different magnetic configurations and convenient compilations of information can be found in Harbour[10] and in Ref.4.

Modelling techniques previously discussed for the plasma transport to the limiter are also applicable to the divertor, the principal difference being that neutral particles produced at the target of the divertor tend to be locally ionised and hence recycle within the divertor chamber. Thus $S(\frac{1}{2}L_{||}) >> S(z)$ and, as a consequence, $v_{||}(z)$ within the torus is likely to be smaller than in the case of the limiter. In general, $\Delta_n \sim (a_w - a_s)$ and it is predicted that divertor action will not be particularly efficient in its transport of charged particles directly from the separatrix to the divertor chamber, ie, the particle "unload efficiency" is rather low. However, the energy "unload efficiency" tends to be high because electron thermal conduction is not effected by the low drift velocity in the scrape-off region. Release of impurities by processes such as sputtering is strongly dependent upon the energy flux incident upon the boundary surfaces so that impurity atom release is concentrated at the divertor target. These impurities tend to be retained within the divertor chamber due to localised ionisation followed by entrainment of the ions in the plasma flow to the target.

5. PLASMA SURFACE INTERACTIONS

Conditions in the boundary plasma are influenced by plasma-surface interactions. The most significant are interactions associated with the return of atoms and molecules of "hydrogen" to the plasma and processes associated with the release of atomic impurities, for example, sputtering of the boundary material and desorption of impurity species. Some processes are initiated by the impact of single atoms or ions whereas others are of a macroscopic nature, eg, evaporation, unipolar arcing and flaking. A further distinction can be made between processes which involve momentum transfer (eg, ion impact) and those processes associated with irradiation of the boundary surfaces by photons and energetic electrons. A comprehensive review of the processes and data pertinent to tokamak experiments has been produced by McCracken and Stott[11] but, in the context of the present discussion, the scope of this broad subject area is reduced to considerations of the return of incident plasma particles and to the physical sputtering of boundary material by energetic atoms and ions.

Return of incident particles

Ions and atoms of moderate incident energy ($\sim 10^2$ eV) can pass through the surface of a solid and then be scattered by inelastic collisions within the lattice of the material. Some of the incident particles backscatter to the surface, from which they emergy with reduced kinetic energy, whereas the remainder slow down to thermal energies and are thus trapped within the lattice. Lindhard et al.[12] proposed that both the range and energy loss of the incident particles could be characterised by a reduced energy, ϵ, given by

$$\epsilon = \frac{M_2}{(M_1 + M_2)} \frac{a}{A_1 A_2 e^2} E \tag{5.1}$$

where M_1,A_1 and M_2,A_2 are the mass and atomic numbers of the incident particle and target atom respectively and E is the incident energy. Lindhard postulated that the parameter a should be set equal to the Thomas-Fermi screening length, ie,

$$a = 0.4685 \left(A_1^{2/3} + A_2^{2/3}\right)^{-\frac{1}{2}},$$

so that

$$\epsilon = 32.55 \frac{M_2}{(M_1 + M_2)} \frac{1}{A_1 A_2 \left(A_1^{2/3} + A_2^{2/3}\right)^{\frac{1}{2}}} E(\text{keV}). \tag{5.2}$$

The probability for backscattering or "reflection" of an incident particle of energy E is described by means of a particle reflection coefficient R_N and, if E is expressed in units of ϵ, then R_N for various combinations of projectile and target has been shown to lie approximately on a universal curve. This is a function of ϵ and it

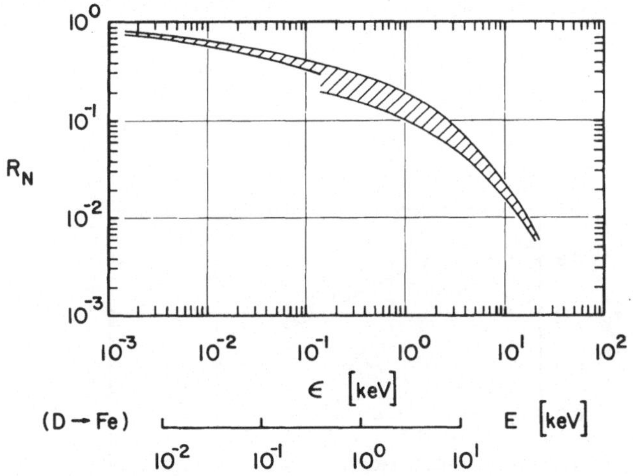

Fig.4 The reflection coefficient R_N for light atoms and ions incident normally upon surfaces plotted as a function of reduced energy ϵ.

The shaded region illustrates the spread in data presented by Eckstein and Verbeek[13] and ϵ (in keV) is determined from Eq.(5.2). The scale of incident energy E for D → Fe is also shown.

has the form shown in Figure 4. The shaded region indicates the spread in data for light ions (biased somewhat in favour of measured values) that are taken from a compilation by Eckstein and Verbeek[13]. These data refer to particles in the energy range below 20 keV which are incident normal to the surface, but the data allow for the fact that emerging particles have an angular distribution which is close to cosine. It is apparent that ions with low incident energy are more readily backscattered than trapped and that backscattering is greatest for targets with large atomic number. The reflection coefficient is predicted to increase with deviation from normal incidence; at grazing incidence $R_N \to 1$. The reflection coefficient for incident atoms should be similar to that for ions because the charge state has little significance once the particle enters the influence of the lattice system.

The emerging particles are predominantly neutral due to electron capture within the solid; appreciably less than 1% of incident protons emerge as charged particles in the energy range of interest, ie, E < 1 keV.

The energy distribution of the backscattered particles is described by a coefficient R_E which is defined as the fraction of the incident energy that is reflected, ie, emerges from the surface. The coefficient R_E, expressed as a function of ϵ, also lies approximately upon a universal curve. The average energy of the backscattered atoms is R_E/R_N and this ratio is shown in Figure 5 using data taken from McCracken and Stott[11]. The fraction of energy carried

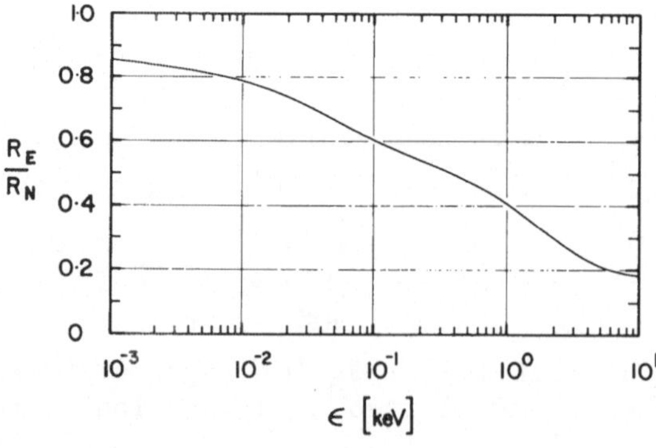

Fig.5 The average energy of reflected particles (R_E/R_N) plotted as a function of reduced energy ϵ.
Data are taken from McCracken and Stott[11].

away by backscattered particles becomes greater as the incident energy is reduced because slow particles penetrate but weakly and so can lose little of their energy by collisions with the lattice.

Particles are trapped within the lattice system when they lose most of their initial kinetic energy but the range of penetration is small ($\sim 10^{-6}$ cm) and so the peak number density of particles within the penetration depth rapidly becomes comparable to that of the lattice atoms and the system quickly reaches equilibrium. The gradient of particle concentration is much steeper towards the surface than into the bulk material and trapped particles diffuse more readily towards the surface. In equilibrium, each incident particle is either backscattered or releases a particle that has been trapped in the solid; there is no nett retention of particles by the solid and the ratio of outgoing fluxes of backscattered to de-trapped particles is given by $R_N/(1 - R_N)$.

The de-trapping mechanisms within the lattice structure appear to be a combination of thermal release and release induced by energetic particles. The energy of the de-trapped particles is likely to approximate to the temperature of the wall and consideration of the surface binding energy favours the release of "hydrogen" in the form of molecules; this concept is frequently accepted in modelling of boundary plasma conditions.

Details of backscattering and de-trapping differ between incident ions or atoms. For normally incident particles of energy E, both species give rise to a fraction $R_N(E)$ which is backscattered with an average energy $E(R_E/R_N)_E$ and a fraction $(1 - R_N)_E$ of low energy, de-trapped particles. However, atoms tend to impact with randomly distributed incident directions and with a velocity distribution corresponding to Maxwellian. Since ions are accelerated through the sheath their energy is more closely mono-energetic and their velocity tends somewhat towards normal incidence.

Physical sputtering

Physical sputtering occurs due to the ejection of a surface atom which in some manner receives an impulse from an incident particle. Sputtering by plasma light ions has recently been reviewed by Roth[14] and suffice to state here that the sputter yield has been shown to consist of two components: one due to direct transfer of momentum from the ion to an atom at the surface and the second due to backscattered projectiles that transfer momentum whilst they are exiting through the surface. The first mechanism is strongly evident at grazing incidence and gives rise to the ejection of atoms into an anisotropic cone in the forward direction. The second mechanism, which is more evident at normal incidence and higher energy, tends to display a cosine type distribution of ejected atoms. The sputter yield (atom per ion) is greatest at angles close to grazing incidence where its value can be as high as 20 times that at normal incidence.

There is an identifiable threshold energy below which insufficient energy is transferred to the lattice atoms for sputtering to occur. This threshold, E_{th}, can be related to the sublimation energy of the solid, E_w, in the manner

$$E_{th} \approx \frac{E_w}{\zeta(1 - \zeta)} \tag{5.3}$$

where

$$\zeta = \frac{4M_1M_2}{(M_1 + M_2)^2} .$$

A recent compilation of data for sputtering by light ions at low energy (Roth et al.[15]) lists values of E_{th}, for example: H→C = 9.9 eV; H→Fe = 64 eV; H→Mo = 164 eV; and H→W = 400 eV. Assessment of data for low energy H^+, D^+, $^3He^+$ and $^4He^+$ impact upon a wide range of targets yields a universal curve for the sputtering yield, Y, at normal angle of incidence,

$$Y = 6.4 \times 10^{-3}\, M_2 \zeta^{5/3} \left(\frac{E}{E_{th}}\right)^{1/4} \left(1 - \frac{E_{th}}{E}\right)^{7/2} \text{ atoms/ion.} \tag{5.4}$$

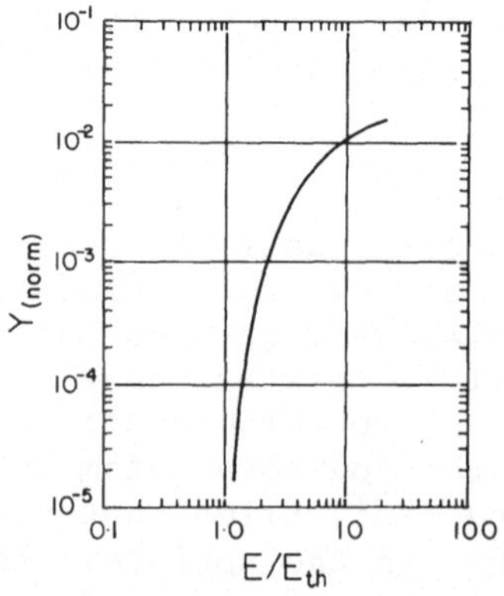

Fig.6 The normalised sputter yield $Y_{(norm)} = Y/(M_2\zeta^{5/3})$ plotted as a function of the energy parameter E/E_{th}.

Data are taken from Roth et al.[15] and the parameters are discussed in the text.

The expression describes the sputter yield to within a factor of two when $M_1/M_2 < 0.4$ and $1 < (E/E_{th}) < 20$; the dependence of the normalised yield $Y_{(norm)} = Y/(M_2\zeta^{5/3})$ upon E/E_{th} is shown in Figure 6.

The magnitude of Y is illustrated in Figure 7 for various ions incident upon a nickel target. Curves through the data points show the fit of the energy dependence of Eq.(5.4). It is apparent that the sputter yields for "hydrogen" ions are not substantial in the energy range appropriate to the boundary plasma but the corresponding yield for self-sputter of Ni by Ni can exceed unity. Effects of self-sputtering can be significant because the initially sputtered atom becomes ionised by collisions with the plasma electrons and returns to the surface upon which it impacts with the energy gained due to acceleration through the plasma sheath, see Eq.(3.3). During this recycling period it is likely that the ion will suffer several ionising collisions with electrons which raise its charge state to a moderately high level (see Section 6); self-sputtering therefore becomes significant even at modest electron temperatures, eg, T_e ranging from 20 to 60 eV. This recycling of sputtered material introduces radiating impurity ions into the boundary regions which in

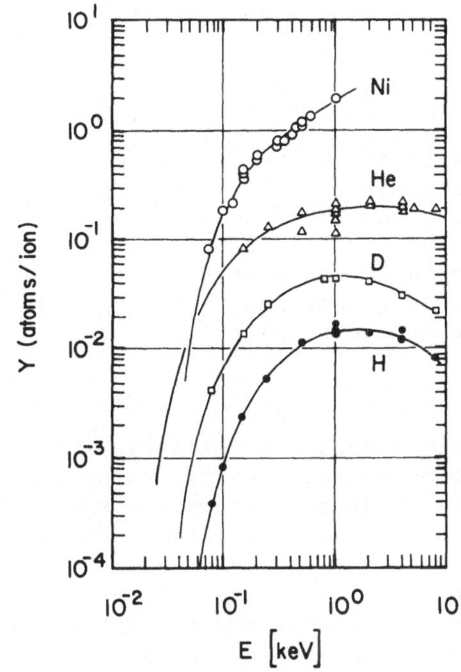

Fig.7 The sputtering yield Y plotted as a function of energy E for ions at normal incidence upon nickel.
Data are taken from Roth et al.[15]

turn cause localised cooling of the plasma electrons and a concomitant reduction of the sheath potential and hence the sputter yield. The system may be self-regulating in the sense that the sputtering rate becomes stabilised when power carried to the localised region of impurities by electrons in the plasma is balanced by power radiated from the impurities. The consequences of this action are considered in Section 7.

6. ATOMIC PROCESSES IN THE BOUNDARY PLASMA

Significant effects of atomic reactions in the boundary region relate predominantly to free-bound collisions and the most prolific bound species is "hydrogen" (either atomic or molecular). There are also smaller concentrations of impurity atoms and ions. Low mass impurities such as oxygen and carbon arise from de-trapping of surface gases or from sputtering of carbon surfaces, medium mass elements such as iron are produced by erosion of the containment vessel which is frequently constructed from stainless steel whereas tungsten (or other high mass refractory elements) may be present due to erosion of limiters or divertor targets. In a DT burning reactor there will also be helium. These atoms enter the plasma with an initial velocity that is governed by surface interactions and their subsequent collision rate with charged plasma particles can be expressed as

$$K = n\, n_o \langle \sigma \bar{v} \rangle \; . \tag{6.1}$$

Here n and n_o are respectively the density of the plasma and the neutral species, $\bar{v}$ is their relative velocity, σ is the cross section for the process and $\langle \sigma \bar{v} \rangle$ is averaged over the Maxwellian velocity distribution of the collisions. In general, collisions with electrons are the most frequent because of the comparatively high electron velocity but some heavy particle collisions, such as symmetric, resonant charge exchange

$$H + H^+ \rightleftharpoons H^+ + H, \qquad \text{(i)}$$

can also be significant because of the large magnitude of the cross section at low collision energy. Charge exchange between impurity ions and high velocity "hydrogen" atoms can result in appreciable deposition of power in the boundary during neutral beam heating.

The plasma can be regarded as transparent to most radiation emitted by free-bound transitions and so photon induced reactions can be neglected. Inelastic atomic collisions involving electrons remove thermal energy from the electron component of the plasma by exciting radiative transitions in "hydrogen" atoms and in atoms and unstripped ions of impurities. Electron energy is also dissipated in ionising collisions. Electron-ion recombination in the boundary is unlikely, firstly because the electron density is insufficient to support three body recombination and secondly because the characteristic time, τ_α, for two body radiative recombination,

$$\tau_\alpha = [n_e\ \alpha(T_e)]^{-1}, \qquad (6.2)$$

is too great for recombination to occur at a significant rate. The two body recombination rate coefficient $\alpha(T_e)$ of unstripped ions is relatively large due to the contributions from dielectronic recombination for which $\tau_\alpha > 10^{-2}$ s. This is several orders less than τ_α for electron-"proton" two body radiative recombination but, since the unstripped impurity ions tend to be concentrated close to the boundary surfaces, the ion residence time, τ_{imp}, must be appreciably less than the residence time of the bulk of the boundary plasma for which $\tau_{||} \sim 10^{-3}$ s (see Section 2). Thus, τ_{imp} is appreciably less than τ_α and two body recombination can be neglected.

Dissipation of electron energy due to collisions of electrons with "hydrogen" atoms

The most significant atomic processes associated with electron collisions with "hydrogen" atoms are:

a. Excitation of an atom from level (p) to an upper level (q),

$$H(p) + e \rightarrow H(q) + e, \qquad \text{(ii)}$$

which may be followed by spontaneous radiative decay,

$$H(q) \rightarrow H(p) + h\nu\ , \qquad \text{(iii)}$$

and b. Ionisation of an atom in level (p),

$$H(p) + e \rightarrow H^+ + e + e. \qquad \text{(iv)}$$

An important associated reaction is collisional de-excitation which is the reverse of (ii) and this is a super-elastic process which returns energy to the colliding electron.

The implications of these electron impact processes are discussed in detail in papers by Bates et al.[16], Bates and Kingston,[17] McWhirter and Hearn[18] and Hutcheon and McWhirter[19]. It is postulated that electron collisions cause ionisation either directly, by transferring a bound electron to the continuum, or indirectly through a sequence of level transitions which terminates in the continuum. The cross sections for such level transitions are large, proportional to n^4 where n is the principal quantum number, and so

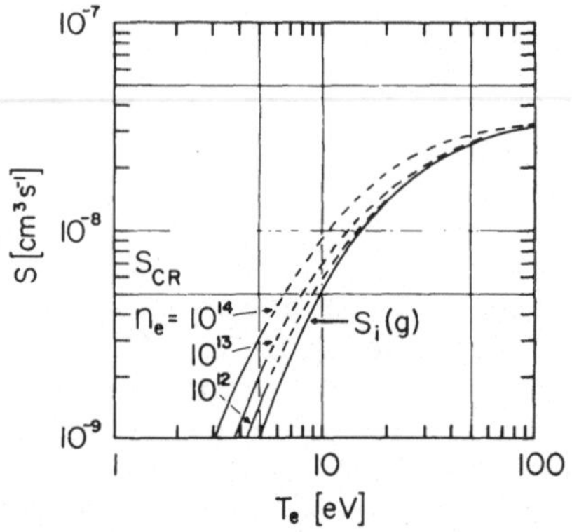

Fig.8 The rate coefficient for electron impact ionisation of hydrogen atoms.
The collisional-radiative ionisation coefficients S_{CR} are taken from Bates et al.[16] and shown for electron densities of 10^{12}, 10^{13} and 10^{14} cm^{-3}. $S_i(g)$ is the ground state ionisation coefficient corresponding to conditions of low electron density.

the collision time,

$$\tau_n = (n_e \langle\sigma_n v_e\rangle)^{-1}, \tag{6.3}$$

becomes shorter than the radiative lifetimes of all but the lowest lying excited states; the lifetime of high n states is proportional to n^3. The probability of spontaneous emission is thereby reduced in favour of (a) ionisation by a chain of non-radiative upward transitions and (b) repopulation of the ground state by a complementary chain of non-radiative downward transitions. Within the boundary plasma τ_n is appreciably less than the particle residence time so that the probability of ionisation is enhanced by non-radiative transitions and the ionisation rate can be expressed as

$$K_i = n_e\, n(g) S_{CR} . \tag{6.4}$$

Here n(g) is the density of ground state species and S_{CR} is a composite coefficient, called the "collisional radiative ionisation coefficient," which allows for the processes outlined above. S_{CR} is compared with the ground state coefficient $S_i(g)$ in Figure 8.

The preceding discussion leads to the concept that electrons colliding with a partially ionised "hydrogen" plasma lose energy by ionisation and by radiation from a few low lying radiative levels and the average amount of energy, ξ_i, expended in producing one proton-electron pair must include contributions from both processes.

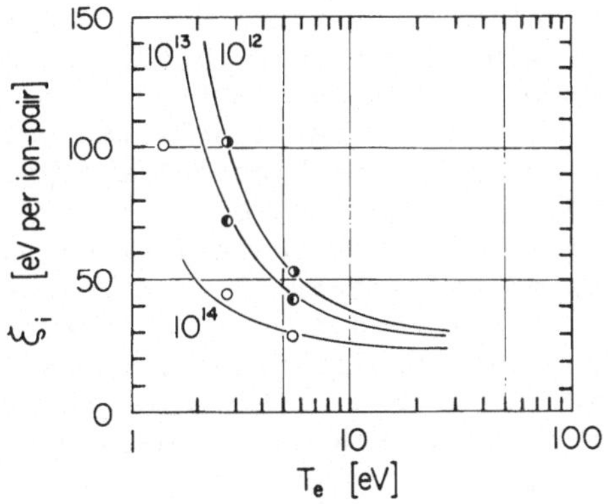

Fig.9 The average electron energy ξ_i dissipated in producing one proton-electron pair in atomic hydrogen plotted as a function of electron temperature T_e.

Solid lines show the data of McWhirter and Hearn[18] for hydrogenic ions Data for H atoms (taken from Bates et al.[16] and Bates and Kingston[17]) are shown by circles.

McWhirter and Hearn[18] have evaluated ξ_i for hydrogen-like ions in conditions where recombination is insignificant; the appropriate expression is

$$\xi_i = \frac{(\chi_i S_{CR} + P_1 n_e^{-1})}{S_{CR}}, \qquad (6.5)$$

where $P_1(T_e,n_e)$ is a coefficient that allows for radiative power loss from ground state ions. The data appropriate to $Z = 1$ (ie, equivalent to "hydrogen") are shown as a function of T_e (for $n_e = 10^{12}$, 10^{13} and 10^{14} cm^{-3}) by the solid lines in Figure 9. The parameters scale for fully stripped ions with $Z > 1$ according to

$$[T_e]_z \equiv Z^2 T_e; \; [\xi_i]_z \equiv Z^2[\xi_i]_H \text{ and } [n_e]_z \equiv n_e Z^{-7} \qquad (6.6)$$

and, because of the strong sensitivity of $[n_e]_z$ upon Z, it is unlikely that non-radiative processes will influence the ionisation rates of fully stripped impurity ions.

This analysis is directed towards electron hydrogenic-ion collisions and so employs Coulomb-Born cross sections which overestimate excitation and ionisation in electron-atom collisions except at high T_e. To show the significance of this effect data for H atoms (S_{CR} from Bates et al.[16] and P_1 from Bates and Kingston[17]) have been substituted into Eq.(6.5) and the results plotted as circles in Figure 9.

It is also worth noting that each ionising event involved in the recycling of "hydrogen" atoms corresponds to an amount of energy $(\xi_i - \chi_i)$ extracted from the plasma electrons and subsequently radiated to the wall, whereas χ_i is transferred from the electrons and stored as potential (not kinetic) energy in the plasma ions. The initial energy of the ions corresponds to that of their parent atoms whereas most of the secondary electrons are emitted with low energy (a few eV being typical). Subsequent Coulomb collisions with the bulk plasma tend to bring the energies of the ions and secondary electrons into equilibrium with the plasma temperature. The time, t_T, required for a test particle of temperature T' to reach the temperature T of the plasma has been given by Spitzer[8] as

$$t_T = \frac{7.34 \times 10^6 \; M'M}{n \; Z'^2 \; Z^2 \; \ln\Lambda} \left(\frac{T'}{M'} + \frac{T}{M}\right)^{3/2} \qquad [s] \tag{6.7}$$

and so the temperature attained by the recycling ions and electrons is governed by the ratio of their residence times to the appropriate value of t_T.

The influence of charge exchange and the recycling of 'H' atoms

The rate coefficients for charge exchange and electron impact ionisation are such that, in a homogeneous plasma, the frequency of charge exchange collisions is greater than that of ionising collisions. During each charge exchange collision a plasma proton captures an electron and the resultant atom moves freely across the magnetic field in a direction which is almost randomly distributed around the collision site. In effect, the collision scatters the plasma "proton" so that it moves (as an atom with energy corresponding to the ion temperature) either outward towards the wall or inward towards the region of hotter plasma. A sequence of such charge exchange collisions may take place before the transported atom is ionised and at each collision the resultant atom has the local temperature of the plasma "protons." Molecules of "hydrogen" do not play a direct role in this cycle because their cross-section for charge exchange is relatively small; nevertheless, they are readily dissociated in collisions with electrons and their product atoms subsequently undergo charge exchange. These atomic processes govern the distribution of energy amongst those "hydrogen" atoms that return to the wall.

Transport of atoms can be assessed using Monte Carlo techniques which take account of the changes in both direction and energy that result from collisions of atoms with both plasma ions and boundary surfaces; see, for example, Heifetz et al.[20] Alternative approaches, eg, Harrison et al.[9] are based upon analytical methods that describe a random walk but the significance of atomic processes can be illustrated by using a simplified version,

Harrison[3], that assesses the probability that an atom released from the wall will recycle back to the wall as a "daughter" atom.

Consider two conditions, an initial one (a), where an atom has a velocity v_0 characteristic of its release from boundary surface and a subsequent one (b), where the velocity has been changed by collision with a plasma proton so that the "daughter" atom corresponds to a randomly directly particle with thermal velocity v_{th}. The plasma is assumed to be homogeneous and of infinite extent and motion is considered only in the direction x normal to the surface. The flux Γ^a of condition (a) atoms at a distance x is given by

$$\Gamma^a(x) = \Gamma^a(o) \exp^{-\frac{x}{\Delta_a}}$$

where

$$\frac{1}{\Delta} = \frac{1}{\lambda^a_{cx}} + \frac{1}{\lambda^a_i} \tag{6.8}$$

and λ^a_{cx} and λ^a_i are respectively the mean free path for charge exchange and electron ionisation. In an element of extent dx at x, there is a rate,

$$K^a(x) \simeq \Gamma^a(x)\frac{dx}{\lambda^a_{cx}} ,$$

of charge exchange collisions which produce condition (b) daughter atoms that move with equal probability either more deeply into the plasma (where they are assumed to be trapped by ionisation) or else backwards towards the surface. Motion in condition (b) can be treated as a diffusive problem in which the diffusion step length corresponds to the charge exchange mean free path λ^b_{cx}. Diffusion need be considered only in one direction and so the scale length, Δ_b, of diffusive transport can be expressed as

$$\Delta_b \approx \left(\frac{N_b}{3}\right) \lambda^b_{cx} . \tag{6.9}$$

Here N_b is the average number of charge exchange collisions prior to ionisation and it is equal to the ratio of collision times, namely,

$$N_b \approx \frac{\tau^b_i}{\tau^b_{cx}} \approx \frac{\lambda^b_i}{\lambda^b_{cx}} \approx \frac{1}{G_b}$$

where G_b is the ratio of the electron ionisation to the charge exchange rate coefficients, ie, $G_b = S^b_{CR} / S^b_{cx}$. The atoms returning from dx at x are in condition (b) and their flux at the wall, $\Gamma^b(x)$, is given by

$$\Gamma^b(x) \approx \tfrac{1}{2} K^a(x) \exp^{-\frac{x}{\Delta_b}} .$$

Manipulation and integration yields the probability Γ^b/Γ^a that a daughter atom returns to the wall, namely,

$$\frac{\Gamma^b}{\Gamma^a} \approx \tfrac{1}{2} \frac{1}{\lambda^a_{cx}} \cdot \frac{\Delta_a \Delta_b}{(\Delta_a + \Delta_b)} , \tag{6.10}$$

which can be expressed as

$$\frac{\Gamma^b}{\Gamma^a} \approx \tfrac{1}{2}\left[1 + G_a + (3G_b)^{\frac{1}{2}} \frac{\lambda^a_{cx}}{\lambda^b_{cx}}\right] , \tag{6.11}$$

where G_a is the ratio of rate coefficients for condition (a) atoms.

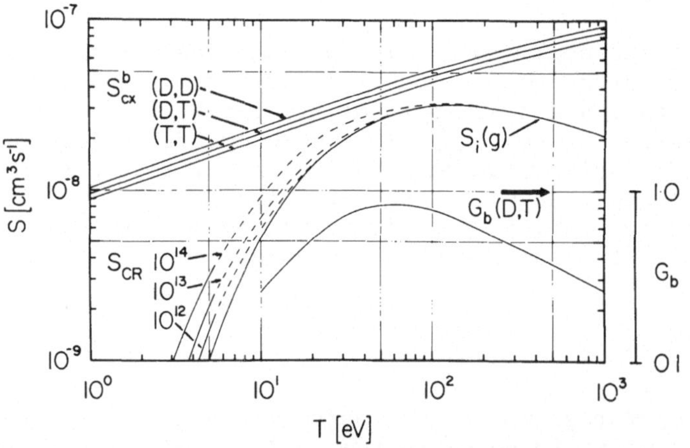

Fig.10 Rate coefficients for charge exchange and electron impact ionisation plotted as a function of plasma temperature T.

The analytical form of the charge exchange cross section is taken from Riviere[21] and the influence of isotope mass is shown by S^b_{cx} plotted for DD, TT and DT. The ratio $G_b = S^b_{CR}/S^b_{cx}$ is shown for D→T collisions.

The charge exchange mean free paths are given by

$$\lambda^{a}_{cx} = \frac{v_o}{n_i \langle \sigma_{cx} \bar{v}_{op} \rangle} \quad \text{and} \quad \lambda^{b}_{cx} = \frac{v_{th}}{n_i \langle \sigma_{cx} \bar{v}_b \rangle} ,$$

where $\bar{v}_{op}$ is the relative velocity of a collision involving an atom with velocity v_o and a plasma "proton" with thermal velocity v_{th}. For condition (b) collisions, $\bar{v}_b$ is given by

$$\bar{v}_b = \left[\frac{8kT_i}{\pi} \frac{(m_o + m_i)}{m_o m_i} \right]^{\frac{1}{2}}$$

where m_o and m_i are respectively the masses of the atom and plasma ion.

Sensitivity of H atom recycling to atomic processes can be identified from Figure 10 where S^b_{cx}, S_{CR} and G_b are shown as functions of plasma temperature $T = T_e = T_i$. The ratio G for both conditions (a) and (b) is rather insensitive to T over the range 20 to 200 eV which typifies many boundary regimes. The preceding simple assessment indicates that a typical value of the probability of Γ^b/Γ^a appropriate to fast backscattered atoms (ie, $v \to v_{th}$) is about 0.17 but this probability increases as the atom release velocity is reduced to a limiting value of about 0.3 when $v_o \to 0$.

Effects of molecular "hydrogen"

Inelastic collisions of electrons with neutral molecules of "hydrogen" extract thermal energy from the electron component of the plasma in a manner comparable to electron "hydrogen" atom collisions. There is a lack of data pertinent to the radiative power losses associated with the production of an (electron + H_2^+) ion-pair and so the difference between H_2 and H is not usually distinguished when determining radiative power losses. However, it is important to assess the degree of molecular dissociation in order to account for the component of "H" atom recycling that arises as a consequence of the presence of "hydrogen" molecules.

The most significant collisions between electrons and neutral "hydrogen" molecules are:

dissociation, $H_2 + e \to H + H + e$ $[S^o_d]$

ionisation, $H_2 + e \to H_2^+ + e + e$ $[S^o_i]$

and dissociative ionisation, $H_2 + e \to H^+ + H + e + e$ $[S^o_{di}]$.

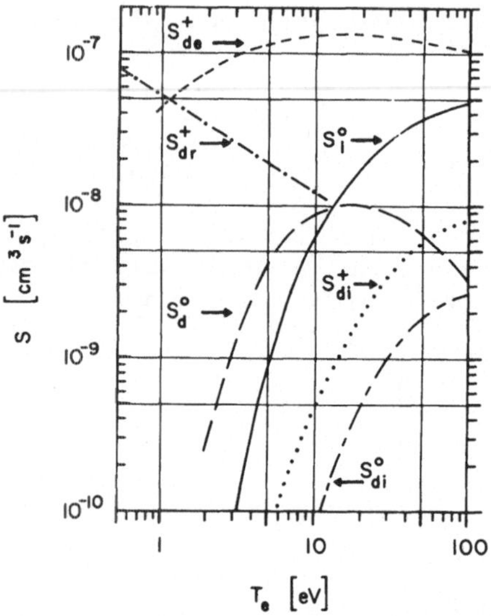

Fig.11 Rate coefficients for electron collisions in H_2 and H_2^+ plotted as a function of electron temperature T_e.

Rate coefficients for neutral molecules are taken from Jones[22] and those for ions, S^+, are discussed by Harrison.[3] The subscripts identify the reactions discussed in the text.

Experimental data for these reactions are available and the rate coefficients, S^o_d, etc, calculated by Jones[22] are plotted in Figure 11. It is evident that dissociation into two H atoms predominates at low electron temperature whereas formation of H_2^+ is the most significant reaction where $T_e \gtrsim 10$ eV. These data refer to molecules in their ground vibrational state and they can be employed in plasma modelling with the reservation that some of the detrapped molecules may be vibrationally excited.

The most significant collisions involving H_2^+ are:

dissociative recombination,

$$H_2^+ + e \rightarrow H + H \qquad [S^+_{dr}]$$

dissociative excitation,

$$H_2^+ + e \rightarrow (H_2^+)^* + e \rightarrow H^+ + H + e + e \qquad [S^+_{de}]$$

and dissociative ionisation,

$$H_2^+ + e \rightarrow H^+ + H^+ + e + e \qquad [S^+_{di}].$$

Experimental data for H_2^+ have been reviewed by Dolder and Peart[23] and the status of the corresponding rate coefficients has been discussed by Harrison.[3] The appropriate rate coefficients for the molecular ion, S^+_{dr}, etc, are plotted in Figure 11 and refer to ions which have a distribution of vibrationally excited levels that is closely determined by the appropriate Frank-Condon factors (see for example, Dunn[24]). Dissociative excitation of H_2^+ into H^+ and H is the most powerful process when $T_e \gtrsim 5$ eV and so any H_2^+ formed from neutral H_2 molecules will almost immediately dissociate. Dissociative recombination into H atoms is the dominant process for the destruction of H_2^+ at low T_e but relatively few H_2^+ ions can be formed at such low temperatures.

It is reasonable to express the rate coefficient $S^o(H^+)$ for the formation of protons from H_2 neutral molecules as

$$S_2^o(H^+) \approx S_i^o\left(\frac{S^+_{de} + 2S^+_{di}}{S^+_{de} + S^+_{di}}\right) + S^o_{di} \tag{6.12}$$

and that for the formation of H atoms as

$$S_2^o(H) \approx 2S^o_d + S^o_{di} + S^o_i\left(\frac{S^+_{de}}{S^+_{de} + S^+_{di}}\right) . \tag{6.13}$$

Thermal energy neutral "hydrogen" molecules that are formed by detrapping of "hydrogen" from boundary surfaces will penetrate but weakly into the plasma because of their low velocities. The charged dissociation products are trapped by the magnetic field and become entrained in the plasma flow that drifts to the surface. The neutral "H" atom products can penetrate more deeply into the plasma.

The effects of electron collisions with impurity elements

Electrons dissipate energy by exciting radiative transitions in atoms and unstripped ions of the impurity components of the plasma. Moreover, the charge state reached by the impurity atoms before they return to the boundary surface has a powerful effect upon the rate of surface erosion (see Section 5). Impurity ions within the boundary plasma will not be in local-thermal-equilibrium with plasma electrons because the ion residence time, $\tau_{||}^+$, is shorter than the electron-ion recombination time τ_α and the coronal equilibrium conditions for charge state populations will not apply, ie,

$$\frac{n_Z}{n_{(Z-1)}} \neq \frac{S_{i,(Z-1)}(T_e)}{\alpha_Z(T_e)} . \tag{6.14}$$

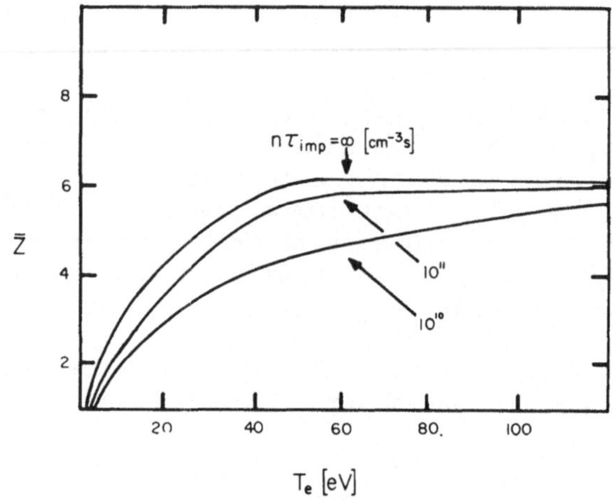

Fig.12 The average charge state $\bar{Z}$ of oxygen ions plotted as a function of electron temperature T_e.

Data for $\bar{Z} = \Sigma n_Z Z/\Sigma n_Z$ are taken from Abramov[26] and are shown for several values of the product $n\tau_{imp}$ where the ion residence time τ_{imp} is independent of Z. Conditions when $n\tau_{imp}\to\infty$ correspond to coronal equilibrium.

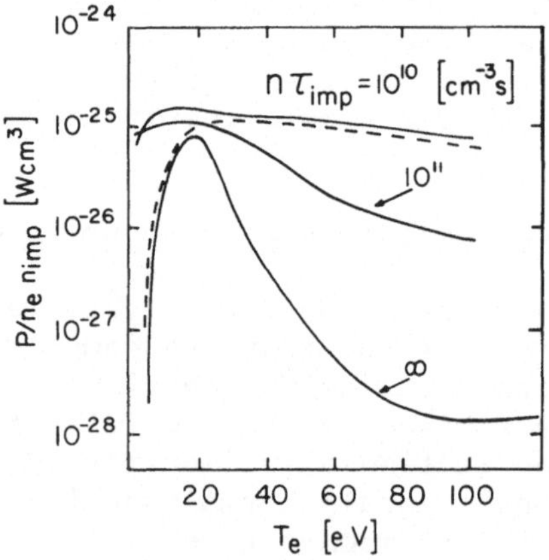

Fig.13 The radiated power function $P/n_e n_{imp}$ for oxygen plotted versus the electron temperature T_e.

Data of Abramov[26] are shown by the solid lines, those of Shimada et al.[25] by the dashed line.

To account for this situation Shimada et al.[25] and Abramov[26] introduce an ion loss term into the rate equations for charge balance, eg,

$$n_e n_{(Z-1)} S_{i,(Z-1)} - n_e n_Z S_Z + n_e n_{(Z+1)} \alpha_{(Z+1)} - n_e n_Z \alpha_Z - n_Z \tau_{imp}^{-1} = 0. \qquad (6.15)$$

The temperature dependence of the average charge state of oxygen determined by Abramov[26] for $n\tau_{imp}$ values of 10^{10} and 10^{11} cm^{-3} s is shown in Figure 12; the average charge state is lower than predicted by the coronal equilibrium model. The radiated power function, ie, $P(n_e n_{imp})^{-1}$ where $n_{imp} = \Sigma n_Z$, is shown in Figure 13 and it is apparent that these non-coronal models predict substantially greater power losses due to radiation when $T_e > 30$ eV.

An alternative approach, Harrison,[3] is based upon the determination of the characteristic loss of energy associated with each stage of ionisation experienced by the impurity particle during its residence time, τ_{imp}, within the plasma. Electron-ion recombination is neglected and it is assumed that ionisation proceeds in a stepwise manner in which ΔZ is unity. The total amount of energy associated with the ionisation cycle of each released impurity atom can then be expressed as

$$\xi_{imp} = P_{01}(\xi_o^r + \chi_o) + P_{01} P_{12}(\xi_1^r + \chi_1) + P_{01}P_{12}P_{23}(\xi_2^r + \chi_2) + \ldots \quad [eV/atom] \qquad (6.16)$$

where ξ_Z^r is the radiative loss associated with each ionisation step and the upper limit of charge state is determined by the residence time or by the fully stripped condition of the ion. The parameters P_{01}, P_{12}, etc, denote the probability that the impurity atom passes through that particular stage of ionisation during its time within the plasma and it is implicit that $P_{01} = 1$. If the residence time, τ_{imp}, is known, then the probabilities can be determined from

$$P_{Z\to(Z+1)} = 1 - \exp\left(-\frac{\tau_{imp}}{\tau_i}\right)_Z, \qquad (6.17)$$

where

$$\tau_{i,Z} = (n_e S_{i,Z})^{-1}$$

is the characteristic time for ionisation.

The energy, ξ_Z, extracted from the plasma electrons during each

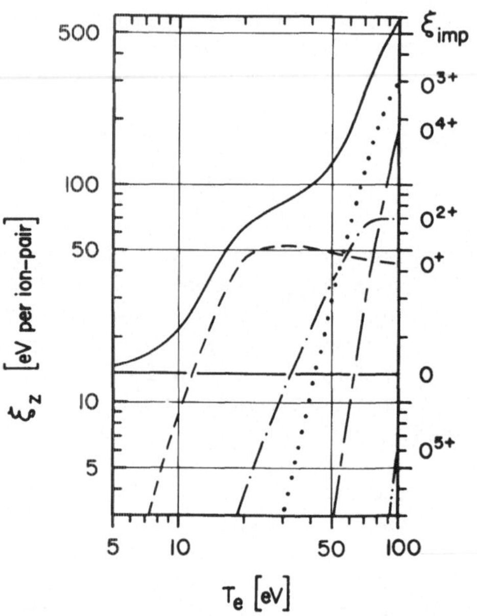

Fig.14 Electron energy dissipated by recycling oxygen ions.
Data are taken from Harrison[3] and refer to oxygen atoms whose initial energy is very much less than T_e. Atomic data are taken from Summers and McWhirter.[27]

stage of ionisation can be determined from

$$\xi_Z = \chi_Z + \left(\frac{P_{LZ}(n_e,T_e)}{S_{C\alpha}(n_e,T_e)}\right)_Z \tag{6.18}$$

where P_{LZ} is a "line radiated power loss coefficient" and $S_{C\alpha}$ is the "collisional dielectronic ionisation coefficient" which is analogous to S_{CR} used previously for ionisation of hydrogenic ions.

The residence time τ_{imp} is dependent upon the entrainment of impurity ions within the flowing boundary plasma and also upon the topology of the magnetic field and boundary surfaces (a brief description can be found in Harrison et al.[9]) but, to indicate the power loss, the data for ξ_Z presented in Figure 14 has been determined using the assumption that the impurity atoms are emited with a temperature $T' \ll T$ and that $[\tau_{imp}]_Z \approx [t_T]_Z$ where $[t_T]_Z$ is the thermal equilibration time given by Eq.(6.7). Contributions from each charge state up to Z = 5 are shown and it is apparent that the effects of charge states greater than Z = 3 are not likely to be significant in conditions when $T_e \lesssim 50$ eV.

The preceding data take no account of the fate of secondary electrons produced by ionisation. These will be heated by electron-

electron collisions and the cooling effect upon the plasma can be substantial if the ions become highly charged. For example, an ionisation cycle that extends to Z = 4 in a plasma where T_e = 100 eV is likely to introduce an additional energy loss of 600 eV per emitted atom.

7. INTERACTION BETWEEN THE BOUNDARY AND THE CONFINED PLASMA

The preceding discussions have treated the boundary plasma in isolation from the confined plasma inboard of the separatrix. In reality these regions are closely coupled, the boundary acting as a sink for energy transported outward across the separatrix and as a neutralising region for the outward flow of plasma ions. However, "hydrogen" atoms may recycle between the two regions and the energetic daughter atoms from the hot central plasma can cause serious erosion of the torus wall. There is also an inward flow of ions from the boundary due to cross field diffusion of the boundary plasma as it drifts to the limiter or divertor. This latter aspect is particularly significant in the case of impurity ions which may diffuse inboard of the separatrix and radiate energy from the central plasma. The acceptable concentration of impurity ions within the central plasma is dependent upon the balance between the power input to the plasma, eg, the power deposited by α-particle heating within a burning DT reactor, and the power lost to the walls by atomic radiation. As a guideline, the concentration of low mass elements, eg, beryllium, carbon, etc, may be about 10^{-2} but high mass elements, eg, tungsten, should not exceed 10^{-4}.

Inward transport of impurities due to plasma-surface processes can be reduced by using a divertor although the effect is likely to be less powerful for impurities released by atom recycling to the torus wall because, subsequent to ionisation within the boundary, these particles take some time to drift into the protective environment of the divertor chamber. A limiter does not provide the advantageous segregation of the plasma bombarded surface from the separatrix and it is therefore more likely to cause impurity contamination. However, it may be possible to reduce this effect by operating the boundary plasma in conditions of high density and low temperature. At low temperature, bombardment by both ions and atoms is unlikely to cause substantial sputtering and, if the density is high, the torus walls are shielded from the direct impact of those energetic daughter atoms which move outward from the central plasma. To attain these conditions it is necessary to cool the edge plasma (ie, the boundary and possibly the inboard region adjacent to the separatrix) and this cooling can be performed by radiating impurity ions. However, it is essential that the radiative losses occur predominantly in the edge plasma and that excessive accumulation of impurity ions does not occur in the very hot core of the central plasma. If the limiter surface is formed from a medium or high mass metal, then the electron temperature of the boundary plasma may be

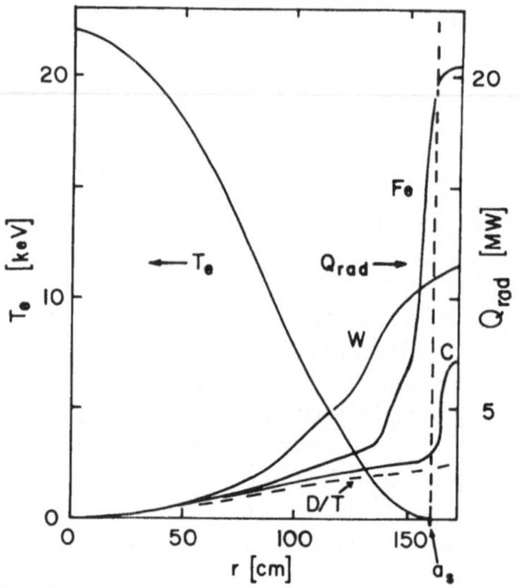

Fig.15 Radial characteristics of the radiated power loss Q_{rad} and the electron temperature T_e predicted for a limiter bounded plasma in JET.

Data for limiters made from either carbon, iron or tungsten are taken from Neuhauser et al.[30]; the heating power is 28 MW.

self-regulated by the effects of self-sputtering discussed in Section 5.

Radiatively cooled plasma edge

The preceding concept of a limiter in a radiatively cooled plasma edge has been proposed as a suitable operational mode for the JET experiment, see Gibson,[28] and the scheme has been named the "cool plasma mantle." To predict plasma performance it is necessary to link the radial properties of plasma transport across the magnetic field to the physical processes discussed here in respect to transport along the field to the boundary surfaces. To date this has not been achieved using a fully self-consistent description of the boundary region but simplified versions of the boundary have been linked to one-dimensional radial transport codes, eg, Watkins et al.[29] Another example from Neuhauser et al.[30] is illustrated in Figure 15 for conditions where the limiters are constructed of either carbon, iron or tungsten. The noteworthy features of this prediction are that significant radiation losses from carbon occur only within the boundary plasma, iron radiates only within the edge plasma whereas a substantial fraction of tungsten radiation emanates from the hotter plasma core. It is particularly interesting to note that the powerful atomic radiating ability of tungsten is compensated by the lower sputter yield so that the concentration of tungsten is low and the total radiation losses are less than those from iron impurity ions.

8. CONCLUSION

It is apparent that the boundary behaviour has substantial impact upon the performance of a magnetic confinement fusion device. Aspects such as the adoption of a divertor as a limiter have bearing upon the capital cost of a reactor and these considerations are at present critical issues in the forward planning of fusion research. An appreciation of the problems calls for a self-consistent interrelation of plasma transport, surface interactions and atomic processes. These must also be linked to the constructional and maintenance requirements of the containment vessel; such aspects as thermal and electromagnetic stresses must be taken into account and consideration given to the exposure of the system to neutron bombardment.

Another important aspect is the need to exhaust neutral helium atoms from a DT burning reactor. Studies of this requirement are based upon prediction of the neutral particle fluxes to the wall of containment vessel coupled to assessment of the ability of the vacuum pumps to extract the helium component; examples of such studies are Heifetz[20] and Harrison et al.[9].

Consideration of the boundary plasma as an integral subject for study is a relatively new concept but it has now become an area of considerable activity and it is one which offers stimulating problems for the future.

REFERENCES

1. International Tokamak Reactor: Phase One, IAEA Vienna (1982).
2. D. E. Post, D. Heifetz and M. Petravic, in: "Proc. 5th Conf. on Plasma Surface Interactions in Controlled Fusion Devices", Gatlinburg, May 1982 (to appear in J.Nucl.Mater.)
3. M. F. A. Harrison (1982) in: "Applied Atomic Collision Physics" ed. H. S. W. Massey, B. Bederson and E. W. McDaniel (Academic Press) Vol.2.
4. IAEA Technical Committee Meeting on Divertors and Impurity Control (ed. M. Keilhacker and U. Daybelge, Max-Planck-Institut für Plasmaphysik, Garching, 1981).
5. F. F. Chen (1974) "Introduction to Plasma Physics", Plenum Press, New York.
6. G. Hobbs and J. Wesson (1967), Plasma Physics 9, 85.
7. J. G. Morgan and P. J. Harbour (1980) in: "Fusion Technology" (Proc. 11th Symp., Oxford) Vol.2, Pergamon, p.1187.
8. L. Spitzer (1962), "Physics of Fully Ionised Gases", John Wiley and Sons Inc., New York.

9. M. F. A. Harrison, P. J. Harbour and E. S. Hotston (1981b), in: "European Contributions to the Conceptual Design of the INTOR Phase 1 Workshop", Euratom, EURFUBRU/XII-132/82/EDV2. Brussels, 1982) 231. (To appear in Nuclear Technology/Fusion.)
10. P. J. Harbour (1981) in: "Plasma Physics for Thermonuclear Fusion Reactors" (ed. G. Casini) Harwood Academic Publishers, Paris (for the Commission of the European Communities) 255.
11. G. M. McCracken and P. E. Stott (1979), Nucl.Fusion 19, 889.
12. L. Lindhard, M. Sharff and H. E. Schiøtt (1963), Mat.Fys.Medd. 33, 39.
13. W. Eckstein and H. Verbeek (1979), Max-Planck-Institut für Plasmaphysik, Report IPP9/32.
14. J. Roth (1980), "Proceedings of the Symposium on Sputtering" Vienna 1980.
15. J. Roth, J. Bohdansky and W. Ottenberger (1979), Max-Planck-Institut für Plasmaphysik, Report IPP9/26.
16. D. R. Bates, A. E. Kingston and R. W. P. McWhirter (1962), Proc.Roy.Soc. A267, 297.
17. D. R. Bates and A. E. Kingston (1963), Planet and Space Science 11, 1.
18. R. W. P. McWhirter and A. G. Hearn (1963), Proc.Phys.Soc. 82, 641.
19. R. J. Hutcheon and R. W. P. McWhirter (1973), J.Phys.B. 6, 2668.
20. D. Heifetz, D. Post, M. Ulrickson and J. Schmidt, in: "Proc. 5th Conf. on Plasma Surface Interactions in Controlled Fusion Devices" Gatlinburg, May 1982 (to appear in J.Nucl.Mater.)
21. A. C. Riviere (1971), Nucl.Fusion 11, 363.
22. E. M. Jones (1977), Culham Laboratory Report, CLM-R175.
23. K. T. Dolder and B. Peart (1976), Reports on Progress in Physics 39, 697.
24. G. H. Dunn (1966), J.Chem.Phys. 44, 2592.
25. M. Shimada, M. Nagami, K. Ioki, S. Izumi, M. Maeno, H. Yokomizo, K. Shinya, H. Yoshida, N. H. Brooks, C. L. Hsieh, A. Kitsunezaki and N. Fujisawa (1981), in: "Japanese contributions to the 3rd meeting of the INTOR Phase IIA workshop" IAEA Vienna (Euratom, EURFUBRU/XII-2/81/EDV71, Brussels, 1982).
26. V. A. Abramov (1982), in: "USSR contributions to the 5th meeting of the INTOR Phase IIA workshop" IAEA Vienna (Euratom, EURFUBRU/XII-132/82/EDV23, Brussels, 1982).
27. H. P. Summers and R. W. P. McWhirter (1979), J.Phys.B. 14, 2287.
28. A. Gibson (1978), J.Nucl.Mater. 73/77, 92.
29. M. L. Watkins, A. E. P. M. Abels van Maanen and P. M. Stubberfield (1982), in: "Plasma Physics and Controlled Nuclear Fusion Research (Proc. 9th Int. Conf. Baltimore, 1982) Paper D-2-1.
30. J. Neuhauser, K. Lackner and R. Wunderlich (1982) in: "European contributions to the 5th meeting of the INTOR Phase IIA workshop", IAEA Vienna (Euratom, EURFUBRU/XII-132/82/EDV20, Brussels, 1982).

NEUTRAL PARTICLE BEAM PRODUCTION AND INJECTION*

D. Post and R. Pyle*

Princeton Plasma Physics Laboratory,
Laboratory, University of California

INTRODUCTION

Intense neutral beams are used to heat, fuel, adjust electric potentials, and diagnose fusion plasmas. They may be used to sustain currents in plasmas. We shall comment on some of the ways that atomic physics enters into the design, diagnosis, and application of neutral beam systems. It will be apparent that the treatment is selective and superficial, but we hope to mention most areas of interest, and indicate that there is a continuing need for new ideas and new techniques.

This paper is divided into two sections: The first is a discussion of the interactions of neutral beams with confined plasmas, the second is concerned with the production and diagnosis of the neutral beams. In general we are dealing with atoms, molecules, and ions of the isotopes of hydrogen, but some heavier elements (for example, oxygen) will be mentioned. The emphasis will be on single-particle collisions; selected atomic processes on surfaces will be included.

The two chief plasma physics requirements that a fusion reactor must meet are adequate confinement and a temperature of

* This work was supported by in part by the Director, Office of Energy Research, Office of Fusion Energy, Development and Technology Division, of the U.S. Department of Energy under Contract No. DE-AC03-76-SF00098, and in part by PPPL DOE Contract No. DE-AC02-76-CH03073.

about 10 keV or greater. The requirements for confinement ($n\tau_E > 10^{14}$ sec/cm^3, where n is the plasma density and τ_E is the confinement time) occurs because the energy loss rate of the plasma cannot exceed the self heating rate by alpha particles produced by DT fusion in a fusion reactor.[1] The second requirement of a temperature of 10 keV occurs because the D-T (deuterium-tritium) fusion reaction rate is small for temperatures below 10 keV.[2] A D-D (deuterium-deuterium) reactor would require temperatures of 30 keV or more. The issues of interest to fusion research are thus not only of plasma confinement but plasma heating. A typical reactor would be heated in steady state by the slowing down of the 3.5 MeV alpha particles produced in the fusion reaction. However, the plasma will have to be heated to the 10 keV, or so, at which the plasma heating is significant and, of course, present experiments need to be heated. Neutral injection has been the most successful heating method to date in magnetic fusion research.

Tokamaks require a toroidal current for confinement and equilibrium. This current is usually supplied by a change of flux in a coil. The need for the flux swing in the transformer coils sets a time limit on the tokamak pulse of about 10^3 - 10^4 seconds for a reactor. Then the tokamak plasma must be terminated and the flux reversed in the current driving coil. The termination and restarting of the plasma burn introduces thermal cycling stresses in the structure of the reactor, increasing the design requirements. There is thus great interest in a "steady state" tokamak in which the current is driven by a means other than by transformer action by a flux swing in a coil. Neutral beams can inject momentum into the plasma, and have been proposed for current drive in tokamaks.[3]

Experiments and reactor designs based on the mirror concept[4,5] require a large high energy population of ions with relatively low velocities parallel to the magnetic field. The high energy of the ions is necessary to maintain a large electrostatic field for electron confinement. In most current mirror experiments and reactor designs, this population of ions with $v_\parallel / v_\perp$ small is provided by neutral injection.

A typical neutral injection system involves an ion source, an acceleration system, a neutralization system, and a beamline to connect to the torus (Fig. 1). The energetic neutral atoms cross the magnetic field that confines the plasma, and are captured in the plasma by electron and ion impact ionization and charge exchange. The captured fast ions slow down, gradually heating the plasma by elastic coulomb collisions with the plasma ions and electrons. If energetic deuterons are injected into a plasma containing tritium ions, some of the deuterons can undergo fusion reactions with the tritons as they slow down.

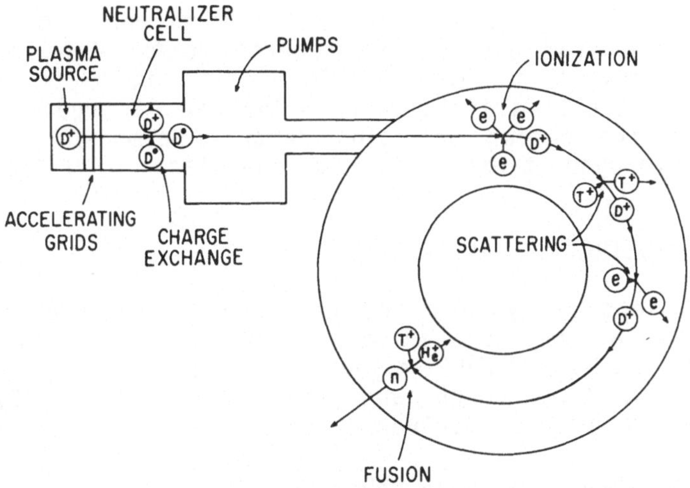

Fig. 1. Neutral injection by the acceleration of H^+ and neutralization by charge exchange, and the subsequent stopping of the beam in the plasma by coulomb scattering and nuclear reactions. (taken from Ref. 6)

The fast ions from the injected beam often have large orbits which must be confined. In addition, lack of axisymmetry due to the discrete toroidal field coils in the tokamak or the inherent lack of symmetry in the magnetic field topology of a mirror or stellerator can lead to unconfined orbits.[6] This loss mechanism for beam ions is not expected to be severe for large, high current tokamaks, but these losses are an area of active research for stellerators.[7]

The injection of neutral hydrogen atoms can enhance the impurity radiation in plasma experiments through such reactions as[8]

$$C^{+6} + H^{0} > C^{+5}$$
$$C^{+5} + e^{-} > (C^{+5})^{*} > C^{+5} + h\nu$$

The hydrogen atoms enhance the recombination rate of the carbon (or other impurity ions). The recombined ions can then be excited by electron impact with background electrons, often leading to an increase in the impurity radiation losses in the fusion experiment.

Present neutral injection systems are based on positive ions. The low neutralization efficiency of such ions at energies above ~ 80 keV/amu limits the usefulness of positive ion systems.[9] Future large tokomaks and mirror experiments would profit from the use of high energy beams with energies above 100 keV/amu. Such beams would be based on negative ions which can be neutralized efficiently. The use of negative ions also opens up the option of using neutral beams of elements heavier than hydrogen at extremely high energies (10-20 MeV).

NEUTRAL BEAM PENETRATION AND FAST ION ORBITS

Neutral injection systems are designed to produce a well collimated beam of neutral hydrogen atoms. The neutral atoms freely cross the confining magnetic field where they are captured by charge exchange and ionization[10] (Fig. 2) the attenuation along the beam is then given by

$$I = I_0 \exp(-n_e \sigma_{eff} x) = I_0 \exp(-x/\lambda).$$

This attenuation is averaged over each flux surface to produce a source of hot ions for each flux surface.[11,12] The local beam deposition rate, normalized by the volume averaged beam deposition rate, is defined as H(r). H(r) > 1 implies greater than average local heating (Fig. 3).

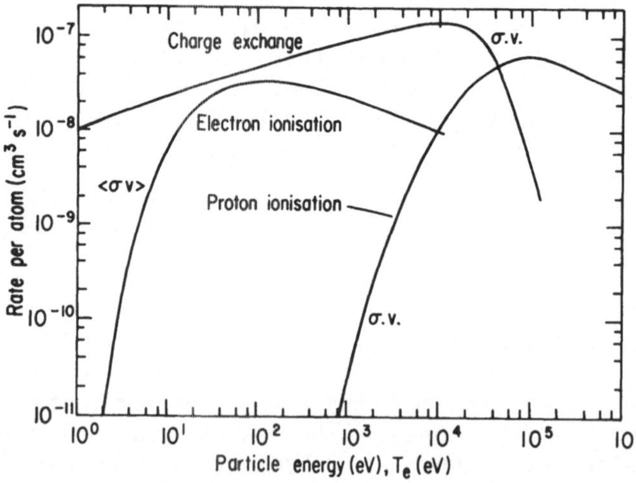

Fig. 2. Reaction rates for atomic hydrogen for electron impact ionization, proton impact ionization, and charge exchange with protons as a function of lab hydrogen energy. The energy scale for the electron ionization rate is the electron temperature.

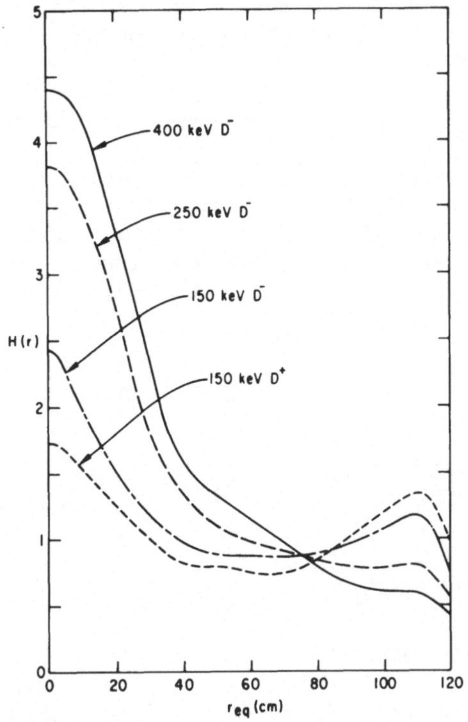

Fig. 3. Neutral beam deposition profiles for 150, 250, and 400 keV D^- based beams and for a 150 keV D^+ based beam on small version of the proposed FED tokamak.[13]

The presence of impurities can affect the ionization of the injected beam. The two processes of interest are charge exchange and ion-impact ionization of H_0 with multiply-ionized impurity ions[14]

$$A^{+q} + H^0 \rightarrow A^{+q} + H^+ + e^-$$
$$\rightarrow A^{+q-1} + H^+.$$

An enormous amount of theoretical and experimental work has gone into determining these cross sections.[15] It was expected that the dominant process might be ion impact ionization which would scale as Z^2. Thus the cross section for beam penetration would scale as $Z_{eff} = \dfrac{n_H + Z^2 n_Z}{n_e}$ where n_H is the hydrogen density, n_Z is the impurity density and n_e is the electron density. Typical plasmas often have $Z_{eff} \sim 2 - 5$. A mean free path that scaled as $\lambda = \lambda_H/Z_{eff}$ would preclude heating of most plasma experiments by neutral beams since the beam ions would all be deposited on the plasma edge.

Fortunately, it turned out that the dominant process at the energies of interest is charge exchange, which scales more like Z than Z^2. Thus we have

$$\frac{1}{\lambda} = n_e\,\sigma_{eff} = n_H\,\sigma_{ion} + n_H\,\sigma_{cx} + \frac{n_e < \sigma v >_e}{v_b} + \sum_Z n_i(\sigma_{Zion} + \sigma_{Zcx})$$

where σ_{ion} is the hydrogen ion impact ionization cross section, σ_{cx} is the hydrogen charge exchange cross section, $<\sigma v>_e$ is the electron impact ionization cross section, and the last term represents the impurity ionization and charge exchange cross sections. Noting that $\sigma_{Zcx} \sim Z\sigma_H$, we have approximately, in the range where $\sigma_{cx} << \sigma_{ion}$,

$$n_e\,\sigma_{eff} = n_H\,\sigma_{ion} + \frac{n_e <\sigma v>_e}{v_b} + n_Z\,Z\,\sigma_{ion}$$

$$= (n_H + Z\,n)\ \sigma_{ion} + \frac{n_e <\sigma>_e}{v_b}$$

$$= n_e\,(\sigma_{ion} + \frac{<\sigma v>_e}{v_b})\ .$$

Thus by ignoring the impurities and treating the plasma as a pure hydrogen plasma, we would have obtained almost the right answer (Fig. 4). Nonetheless, it was crucial that the question be resolved as the fusion community was spending ~ \$ 2 x 10^9 on TFTR, JT-60 and JET, and these experiments would not work well with beam injection if the penetration cross section scaled as Z_{eff}. At high injection energies of 300-1000 keV and higher, the dominant process is ion impact ionization which does scale as Z^2, so this process will be important for high energy beams.

An approximate formula for the trapping length in a large tokamak has been worked out by the INTOR group[16] for a deuterium beam:

$$\lambda = 2.8 \times 10^{13}\ \frac{E_b\ (\text{KeV})}{n_e\ (\text{cm}^{-3})}\ \ \text{cm.}$$

For perpendicular injection, $\lambda \sim a/2$, where a is the plasma minor radius, ensures reasonable beam penetration. This implies that the injection energy should scale as

$$E_b(\text{KeV}) = 180 \times 10^{-16}\ na,$$

yielding an energy of ~ 300 KeV for INTOR parameters.[17] The good penetration requirement can be ameliorated by initially injecting into a low density plasma and gradually building up the plasma density as the alpha heating increases, and by the shift outward of the flux surfaces as the plasma heats up, which

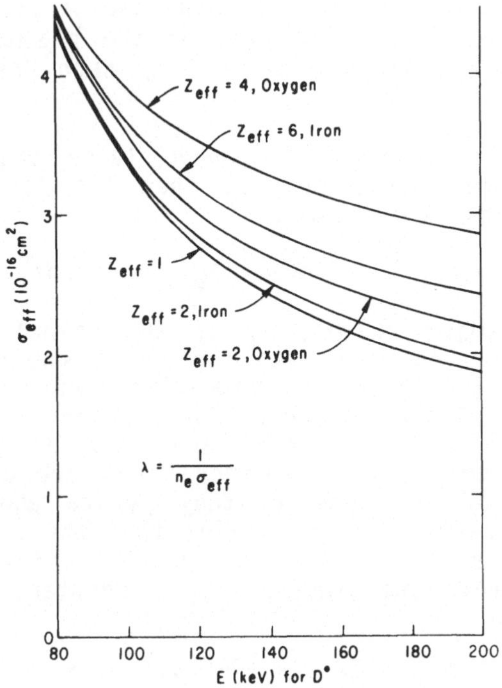

Fig. 4. Effective trapping cross section versus neutral hydrogen beam energy, defined as:

$$\sigma_{eff} = \frac{Z-Z_{eff}}{Z-1}(\sigma^H_{cx} + \sigma^H_{p\text{-}ion}) + \frac{Z_{eff}-1}{Z(Z-1)}(\sigma^{imp}_{cx} + \sigma^{imp}_{ion}) + \frac{\langle\sigma v\rangle_{e\text{-}ion}}{v_{beam}}$$

for a deuterium neutral beam traversing a deuterium plasma with an impurity of charge Z. σ^H_{cx} is the hydrogen charge exchange cross section, σ^H_{p}-ion is the proton impact ionization cross section, σ^H_{cx} is the $_H$ impurity-hydrogen charge exchange cross section and σ_{ion} is the ion impact ionization cross section.[10,13]

reduces the column density of plasma to the plasma center.[18]

However, a high beam energy is still desirable as it makes heating the plasma center less problematical.

As can be seen from above, the mean free path is roughly proportional to the beam energy. The presence of hydrogen atoms with one-half and one-third of the acceleration potential in the beams, formed from H_2^+ and H_3^+ in the ion source, greatly reduces the heating effectiveness of the beam.

Mirror plasmas are usually sufficiently small that penetration is not a problem. In fact, in a number of reactor designs, only 10-20% of the beam is ionized by the plasma. The upper limit on the beam energy is often set in such designs by this requirement.

Fast ions in a tokamak often have large orbits. If the tokamak is completely axisymmetric, then the toroidal angular momentum of the ions is conserved:[19]

$$P_\phi = mv_\phi R + (ZeR/c) A_\phi = \text{constant},$$

where R is the major radius, v_ϕ is the velocity around the torus, m is the mass of the beam ion, Z is the charge of the beam ion, and A_ϕ is the vector potential around the torus due primarily to the plasma current. The conservation of P_ϕ implies that the orbits are periodic and thus, if the beam particles don't hit the wall or limiter, they are confined. The angular momentum of the ions as they gyrate about the field lines is also conserved. This implies that the magnetic moment of the particles is conserved as the magnetic field strength is varied along a field line since $M v_\perp \rho_L$ = constant together with the gyroradius $\rho_L = \frac{v_\perp}{eB/mc}$ implies that $\frac{mv_\perp^2}{2B} = \mu$, a constant,

where μ is the magnetic moment of the particle and $v_\perp$ is the component of the velocity of the particle perpendicular to the field line.

The conservation of μ implies for particles of a given $v_\perp$ and total velocity v, there is an upper limit to the magnetic field strength for their orbits. Writing,

$\varepsilon = \frac{1}{2} m(v_\parallel^2 + v_\perp^2)$ and substituting μ for $v_\perp^2$, we see that

$$v_\parallel = [2(\varepsilon - \mu B)/m]^{1/2},$$

and that for a constant ε and μ, there is maximum B at which $v_\parallel = 0$ and above which v is imaginary. Thus these particles are reflected from regions of high field strength toward regions of lower field strength along a field line. This is the basis for confinement in the mirror concept[4].

Since the magnetic field obeys Ampere's law, $\nabla \times B = 4\pi/c\ j$, a toroidal magnetic field falls off as 1/R, where R is the distance from the center line of the torus. (BR = constant). We can approximate $R = R_0 (1 + r \cos \theta)$, where R_0 is the major radius of the magnetic axis of the torus, r is the minor radius, and θ is the angle r makes with toroidal plane (Fig. 5). Conservation of magnetic moment implies

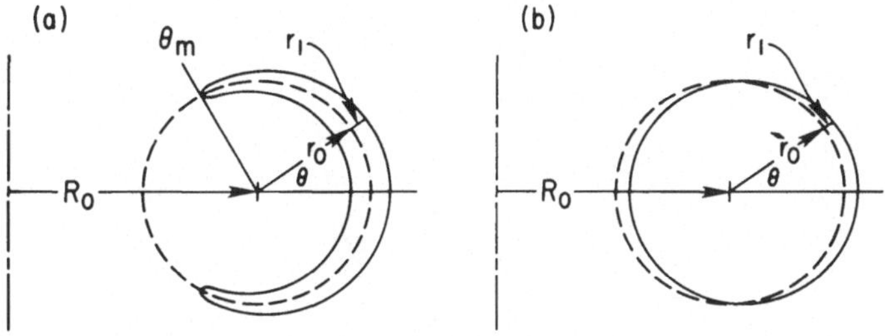

Fig. 5 (a). Particles with $\left(\frac{V}{V_\perp}\right)^2 < 2r/R$ at θ=0 are trapped and their guiding centers make banana shaped trajectories about the flux surfaces.

(b). Particles with $\left(\frac{V_\parallel}{V_\perp}\right)^2 > 2\,\frac{r}{R}$ at θ=0 are not trapped, "passing," and have orbits that are shifted depending on the sign of V . Taken from Ref. 6.

$$\frac{mv_\perp^2}{2} = \mu B_T = \mu\, B_0 \frac{R_0}{R} = \mu B_0\, (1 - \frac{r}{R_0} \cos \theta).$$

As θ increases toward $\pm\,\pi$ (Fig. 5) R decreases and B increases, so that beam ions with

$$\frac{v_\parallel^2}{v_\perp^2} \leq 2r/R_0 \text{ at } \theta = 0$$

are trapped (as in a mirror) on the large R side of the plasma (Fig. 5), similar to the particles in the Van Allen belt.[4] Tangentially injected beam ions ($v_\perp << v_\parallel$) will execute orbits which are circles with centers displaced by a distance Δ from the magnetic axis (Fig. 6).[6]

The consequences of these shifts is that all of the coinjected (injected parallel to the plasma current) beam ions deposited outside the region of width 2Δ are confined. This is usually the case. In contrast, all of the counter-injected beam ions deposited in the outside region of width 2Δ are lost. The lost fast ions often sputter limiter and wall material with the result that the observed impurity levels in discharges with counter injection are often higher than with co-injection.

In order to achieve better penetration with a given energy beam, the beam is often oriented perpendicular to the torus to minimize the distance to the magnetic axis. Then the fast ions

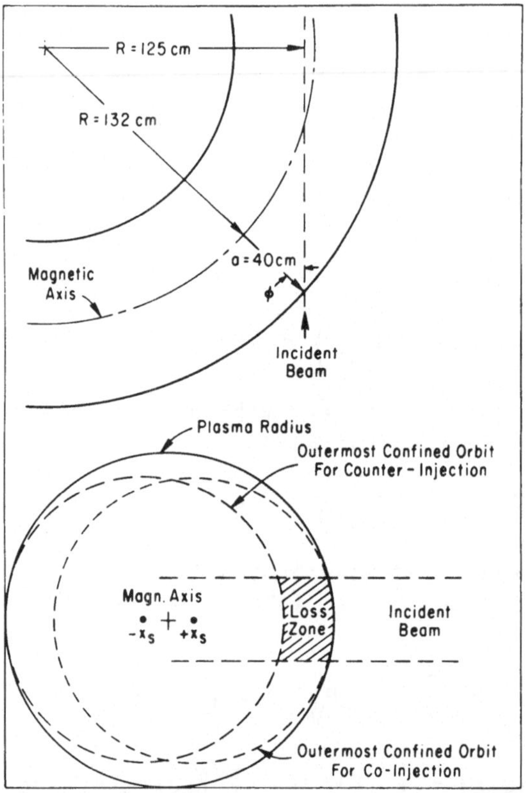

Fig. 6. Schematic of neutral injection on PLT. All of the counter injected ions that are captured in the shaded region have orbits that intersect the wall of limiter and are lost. Taken from Ref. 6.

have $v_\perp >> v_\parallel$ and have trapped (Fig. 5a) orbits. Unless the plasma current is large many of the ions captured near the plasma edge can scatter (due to collisions with plasma ions) into orbits that intersect the limiter and are lost.

One problem with perpendicular injection arises from the fact that the toroidal field in tokamaks is produced with discrete coils. The toroidal field "bulges out" slightly between coils with the result that the field strength varies along a field line and the field is not axisymmetric. The loss of axisymmetry implies that the canonical angular momentum around the torus is not conserved and the particle orbits are not necessarily periodic. Ions may be locally trapped between two coils in local minima in the field or scatter off the irregularities in the field and eventually drift out of the plasma (Fig. 7). Perpendicular injection produces initially trapped ions which can be strongly affected by the ripple.

Injection into inherently non-axisymmetric systems such as stellarators, mirrors, and bumpy toruses suffers from many of the same problems as rippled tokamaks. Nonetheless, neutral injection is being used successfully on stellarators and mirrors.

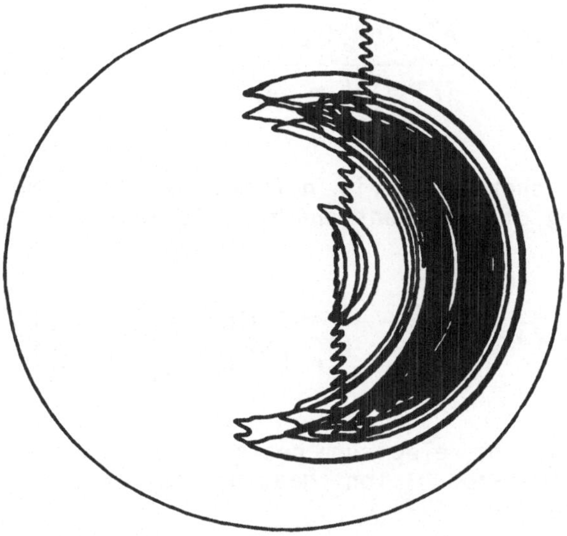

Fig. 7. A trapped ion injected near perpendicularly into a 30 cm plasma with a 3% field ripple is easily trapped and detrapped in local magnetic mirrors by small angle scattering due to collisions with the ions and escapes by vertical drift. (Taken from Ref. 6, originally calculated by R. Goldston).

FAST ION SLOWING DOWN AND PLASMA HEATING

The fast ions from neutral injection slow down by elastic coulomb collisions with the background ions and electrons. The rate at which a typical fast ion of energy E_f, charge Z_f and atomic number A_f slows down in a plasma composed of ions of atomic number A_i and charge Z_i, and electron temperature T_e, is given by[20]

$$\frac{dE_f}{dt} = - \frac{\alpha}{(E_f)^{1/2}} - \beta E_f = - \beta \left(E_f + \frac{E_c^{3/2}}{E_f^{1/2}}\right)$$

where the alpha term represents slowing down due to collisions with the ions and the beta term represents slowing down due to collisions with the electrons. α and β are defined as:

$$\alpha = 4\pi e^4 \ln \Lambda_i \left(\frac{A_f}{2m_p}\right)^{1/2} Z_f^2 \sum_i \frac{Z_i^2 n_i}{A_i}$$

and

$$\beta = \frac{8}{3} (2\pi m_e)^{1/2} \frac{e^4 \ln\Lambda_e n_e Z_f^2}{(KT_e)^{3/2} m_p A_f} .$$

The critical energy E_c, is defined as the beam energy at which beam heating of electrons and ions is equal:

$$E_c = (\alpha/\beta)^{2/3} = 14.8 A_f kT_e \left[\Sigma \frac{(\ln \Lambda_i) Z^2{}_i n_i}{(\ln \Lambda_e) A_i n_e}\right]^{2/3} .$$

When $E_f > E_c$ the electron heating is greater, and when $E_f < E_c$, the instaneous ion heating is greater. The two coulomb logorithms are

$$\ln \Lambda_e = 23.9 + \ln (Te (eV)/ (n_e (cm^{-3}))^{1/2})$$

and

$$\ln \Lambda_i = -5.2 + \ln \left(\left(\frac{T_e (eV}{n_e (cm^{-3})}\right)^{1/2} \frac{v_f^2 \text{ cm/sec } A_f A_i}{Z_f Z_i (A_f + A_i)}\right) .$$

For typical tokamak reactor conditions with $n = 10^{14}$ cm^{-3}, $T_e = 10^4$ eV, and 120 KeV D^0 injection, $\ln \Lambda_i = 22.7$, and $\ln \Lambda_e = 17$.

The thermalization time is defined as the time it takes an ion to slow down from the initial energy E_0 to 0.

$$\tau_{Th} = - \int_{E_0}^{0} \left(\frac{dE_f}{dt}\right) dt = \frac{\tau_s}{3} \ln \left(1 + \left(\frac{E_0}{E_c}\right)^{3/2}\right)$$

where τ_s is a slowing down time for the electrons defined as

$$\tau_s = \frac{2}{\beta} = .371 \frac{A_f}{Z_f^2} \left(\frac{T_e}{10\ \text{KeV}}\right)^{3/2} \left(\frac{10^{14}}{n_e}\right) \left(\frac{17}{\ln \Lambda_e}\right) \text{sec}$$

and E_0 is the injection energy. Typical thermalization times in present day experiments are .015 - .030 seconds, primarily due to the low T_e [Table 1].

Table 1. Beam and injection parameters of the PLT AND TFTR experiments (Z_{eff} = 2).

	PLT	TFTR	TFTR	TFTR
Beams	40 KeV H	120 KeV D	120 KeV D	3.5 MeV α particles
n	5 x 10^{13}cm	5 x 10^{13}cm	10^{14} cm	10^{14}cm
T	2 KeV	5 KeV	10 KeV	10 KeV
τ_s	.066 sec	.52 sec	.74 sec	.37 sec
E_c	42 KeV	210 KeV	220 KeV	470 KeV
Th	.015 sec	.062 sec	.084 sec	.377 sec

For the next generation of tokamak fusion experiments such as TFTR, the thermalization times are roughly .06 seconds, while 3.5 MeV alpha particles in TFTR will have thermalization times of roughly .4 seconds. The alpha particles will primarily heat the electrons ($E_\alpha \gg E_{crit}$) whereas the fast beam ions will primarily heat the plasma ions.

Neutral beams are used for heating on PLT, PDX, ISX, D-III, TFR, ASDEX, DITE, JFT-2, JIPPT-II, and T-11.[6] On PLT the ion temperature was heated to approximately 7 KeV with 2.5 MW of D^0 neutral beams injected into a hydrogen plasma[2] (Fig. 8). TFTR, JET and JT-60 will all use neutral beams.

TFTR[6] is designed to take advantage of the fact that the DT fusion cross section peaks between 100 KeV and 200 KeV. Thus, 120 KeV deuterium beam ions have a significant probability of reacting with the tritium ions while they are slowing down. Most of the contribution to the fusion reaction rate comes from the high energy tail (near 100 KeV) of the Maxwell-Boltzman distribution, and with 120 KeV D^0 neutral injection, all of the beam ions are injected at near the optimum velocity for fusion. Taking advantage of the reacting beams, scientific break-even (defined as the condition where the fusion power produced in the plasma is comparable to the heating power can be achieved with $n\tau_E$ ~ 10^{13} sec/cm^3 instead of the more severe requirement that $n\tau_E$ ~ 10^{14} sec/cm^3 when the fusion power must be produced by thermonuclear reactions[22].

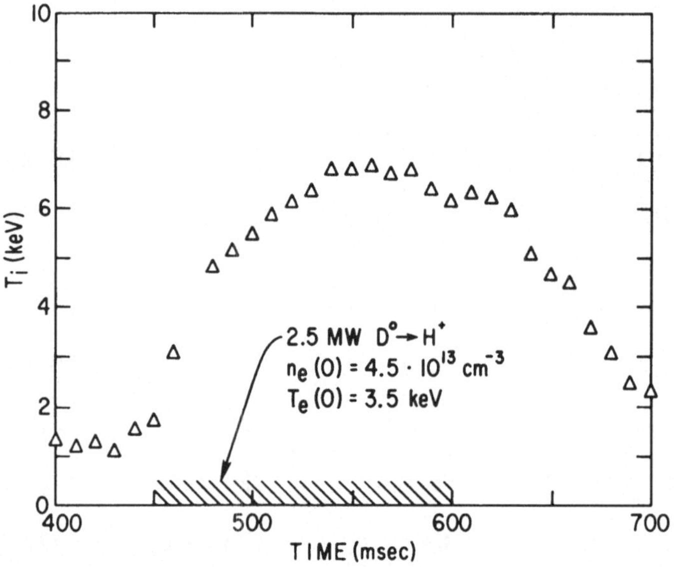

Fig. 8. Charge-exchange ion temperature in PLT for the injection of 2.5 MW D^0 beams into an H^+ plasma, measured by a mass selective (H^+) fast neutral detector. (Taken from Refs. 6,21).

Neutral beams are a crucial part of producing confined particles in mirror experiments[4]. As we have seen, particles with small enough $v_\parallel/v$ at the center low field region are confined. From conservation of energy and magnetic moment one can show that particles with pitch angles $\xi_c = \frac{v_\parallel}{v} < \sqrt{1 - B_c/B_m}$ are confined, where B_c is the field at the low field central point along the field line at which ξ_c is determined and B_m is maximum field along the field line. Particles with larger ξ_c are unconfined since $v_\parallel$ never goes to zero. Thus a mirror confined plasma consists of particles with $v_\parallel/v < \xi_c$. Collisions which change the pitch angle can change the particle from confined to unconfined. Thus the particles are confined for a collision time. The ion and electron collision times for 90° changes in the pitch angle are given below.

$$\tau_{ii} = \frac{2 \times 10^{11} \sqrt{M}}{n \, \ell n\Lambda \quad Z_1^2 \, Z_2^2} E_i^{3/2} \text{ sec}$$

and

$$\tau_{ee} = \frac{10^{10}}{n \, \ell n\Lambda} T_e^{3/2} \text{ sec}$$

where $Z = q/e$ and M is the ion mass in atomic units; E_i is the ion energy in keV; T_e is the electron temperature in keV. For $T_e \lesssim E_i$, as is the usual case for beam heated mirrors, $\tau_{ee} << \tau_{ii}$, and the electrons quickly become unconfined. The electrons thus would leak out of the mirror much more quickly than the ions except that the positive charge of the ions holds the electrons back. Thus a positive potential, ϕ, forms which holds the electrons back (confining them electrostatically) so that the ion and electron loss rates are equal. The electrons have to be confined for many electron collision times which requires the electrostatic potential $e\phi >> T_e$. However $e\phi$ must be less than the ion energy E_i if the ions are to be confined so one must have $T_e << e\phi < E_i$. Since the ions are hotter than the electrons, the ions slow down and heat the electrons by collisions. The fast ions thus must be heated continuously, usually with neutral beams. Taking all of these effects into account[4], we find that for mirror machines the $n\tau$ (product of density and energy confinement time) is given by[4]

$$n\tau = 2.6 \times 10^{10} E_0^{3/2} \log_{10} (B_m/B_c) \text{ sec/cm}^3$$

where E_0 is the injected neutral beam energy in keV. Thus high energy neutral beams are useful for good confinement in mirrors. The ions could also be heated by radiofrequency methods and preliminary experimental work on this has begun[4,5].

CHARGE EXCHANGE RECOMBINATION

In Section II, we saw that the reaction

$$A^{+q} + H^0 > A^{+q-1} + H^+$$

where A^{+q} is an impurity of charge q, had cross sections of ~ $Z^{1.2} \times 1.4 \times 10^{-16}$ cm^2. While this cross section was not large enough to greatly alter the neutral beam penetration situation, the cross section for this charge exchange could still be 10^{-15} to 10^{-14} cm^2. This charge exchange is an additional recombination mechanism for the impurity ions which can alter the charge state distribution of the impurities.

Since the impurities would be more recombined, and have more electrons to excite than would be the case without charge exchange recombination, the total impurity radiation could be increased by the injection of neutral beams into a plasma containing some impurities.[8]

A convenient way of parameterizing the effect is in terms of the ratio of the neutral density to the electron density. In coronal equilibrium, the relative abundance of adjacent ionization states is determined by the ratio of their total ionization and recombination rates[8]

$$\frac{n_{q-1}}{n_q} = \frac{R_q}{I_{q-1}} = \frac{\alpha_q^{Rad} + \sigma_q^{Die} + (n_0/n_e)\,\alpha_q^{cx}}{K_{q-1}}$$

where α_q^{Rad} is the radiative recombination rate coefficient, α_q^{Die} is the dielectronic recombination rate coefficient, K_q is the electron ionization rate, and

$$\alpha_q^{cx} = \sigma_{cx}^{q}\, v_{beam} \; .$$

Thus the usual coronal equilibrium charge state distribution and radiation losses are modified (Fig.9). Very little of the radiation enhancement comes from the charge exchange event and the subsequent radiative decay. Most of the radiation comes from subsequent electron excitation and radiative decay of the recombined ions.

This radiation has been observed on several tokamaks.[23,24] For the most practical situations, the increase in the impurity radiation caused by charge exchange recombination during neutral beam injection is only a small fraction of the beam heating power.

However, in DITE the increase in the radiative losses was as large as the beam heating power, resulting in no heating with the beam.[25] Fortunately, DITE had somewhat atypical conditions compared to most other beam injected tokamaks. The beam energy was low, decreasing v_0 and therefore increasing n_0, the neutral density. DITE also had relatively high impurity levels, especially C, O and Ti. The radiation in DITE was also observed to be toroidal assymetric. The radiation rate near the beam was as much as a factor of 3 or more higher than the radiation 180°around the torus from the beam. Detailed modeling[25] showed that this was consistent with charge exchange recombination in the beam line of sight and subsequent ionization as the impurity ions diffused along the field lines away from the beam.

Although charge exchange recombination enhanced radiation is not a major part of the power balance in most beam heated tokamaks, it can be important if the conditions are right (low n_e, high n_0, high impurity concentrations). The process also affects the spectroscopically observed charge state distribution, and must be included when using the observed charge state distributions to determine the effects of beam injection on impurity transport.

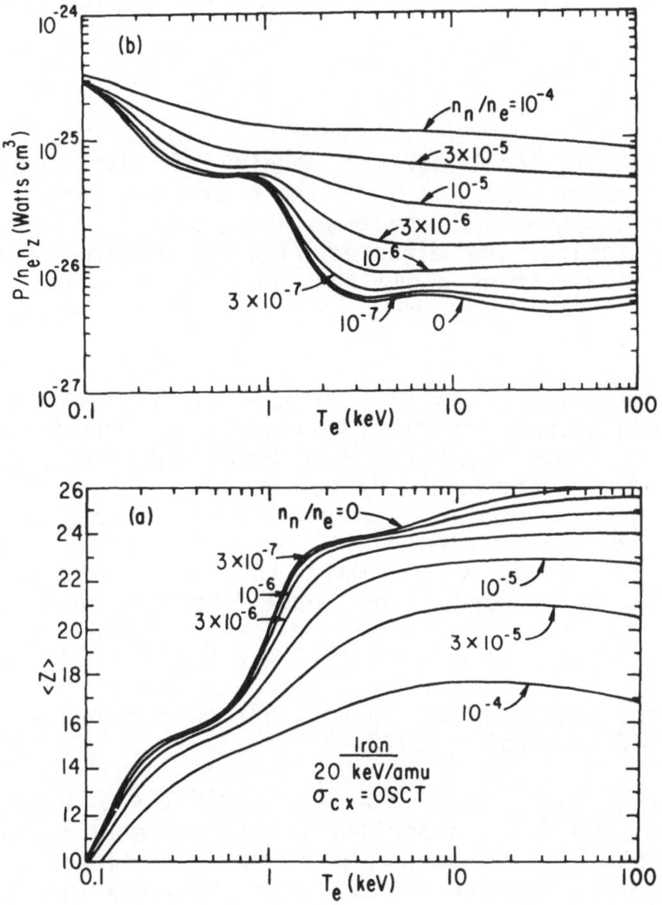

Fig. 9. Neutral beam modified coronal equilibria for iron. The beam energy is 20 keV/AMU. The curves are parametrized by the neutral fraction n_0/n_e. Shown are the average charge state <Z>, and (b) the overall radiation rate coefficient $P/n_e n_z$, both as functions of the electron temperature. (Taken from Ref. 8).

CURRENT DRIVE AND HIGH ENERGY BEAMS

It is becoming a common perception that if tokamaks are to be seriously considered for fusion reactor designs, they must be steady state. The thermal cycling stresses produced when the tokamak must shut down and start up again cause enormous engineering problems. Thus a method for driving the current in a tokamak, or any other concept which relies on induced internal currents such as spheromaks, reversed field pinches and reversed field mirrors, would enhance the reactor prospects of that fusion containment approach.

Tangentially injected neutral beams form a circulating ion current [3]

$$j_{circ} = \frac{n_b < v_{\parallel} > e\, Z_B}{2\pi R_0},$$

where j_{circ} is the local current density of fast ions, n_b is the density of beam ions, $<v_{\parallel}>$ is their average parallel velocity, Z_B is the charge of the beam ions, and R_0 is the major radius of the torus. The directed flow of the fast ions will try to drag the electrons along through coulomb collisions. In the absence of collisions, this electron current would cancel the ion current.

However, the electrons will slow down due to collisions with the background plasma. For the classical case, the friction of the beam ions on the electrons will equal the friction of the electrons on the background plasma for $Z_B = Z_{eff}$. The inclusion of trapped electrons (which cannot circulate toroidally as easily as the passing electrons), leads to a reduction in the electron current which would imply that a current would be driven even if $Z_B = Z_{eff}$. The total beam driven current is then

$$j_b = j_{circ}\,(1 - Z_b/Z_{eff}\,(1 - G\,(Z_{eff}, E)))$$

where G is the neoclassical correction which depends on Z_{eff} and $E = r/R_0$. G is usually ~ 1/2.[26]

This current has been observed on DITE.[27] Experimental verification is difficult since the experiment must be done for a skin time (time for the magnetic field to resistively diffuse through the plasma) and the typical skin times are long compared to the pulse length of most machines. Clear signs that neutral beams have injected momentum into the plasma have been observed. Rotation speeds of up to 1.2×10^7 cm/sec have been observed on PLT with co-injection[28] (Fig. 10).

The optimum conditions for neutral beam current drive involve a complicated set of tradeoffs. However recent studies[26] indicate that the optimum beam energy for current drive in a reactor is ~ 1 MeV for D^0, with about 100 MW of injected power required.

Given the interest in good penetration and therefore high energy for heating and high energy for current drive, considerable effort is being spent on the development of negative ion based neutral beams that would have essentially the same plasma physics as positive ion based neutral beams.

Recently studies have been done of the advantages and feasibility of using either very high energy neutral beams of atoms with Z > 1 based on negative ions or on very high energy single

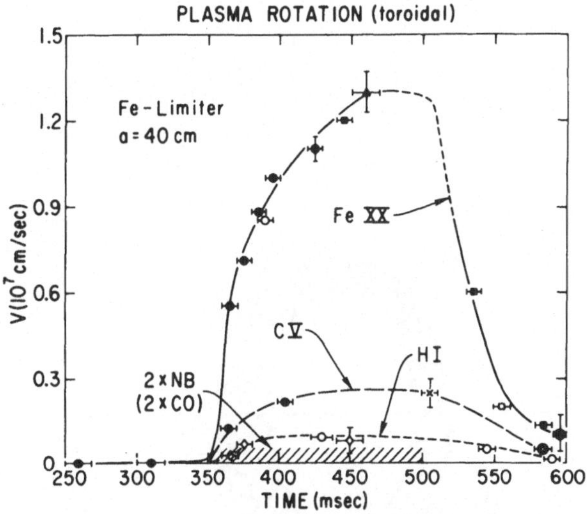

Fig. 10. The toroidal rotation velocity in PLT as a function of time for 1.5 MW of coinjected neutral beams (with an iron limiter). The different impurity ions and hydrogen light refer to different radial positions. (Taken from Ref. 28.)

ionized beams of such ions as B^+, Na^+, etc. for plasma heating.[29,30] The major advantage of these beams is that they would have a very high energy (~ 1 MeV/amu) and therefore would have very small currents for a given power compared to 100 KeV to 500 KeV neutral hydrogen beams (basically $1/Z^2$ less current and Z^2 higher energy for a given beam power).

The light atom neutral beam would penetrate into the plasma as a high energy neutral. Then the atom would be singly ionized, and begin to drift in the magnetic field. As the singly ionized ion was successively ionized to progressively higher charge states, the gyroradius would shrink ($\rho \propto 1/Z$), and the orbit would change. The excursions from the flux surface would also become smaller (Fig. 11). The combination of a penetration as a high energy neutral and as a drifting ion with a shrinking gyroradius can lead to very centrally peaked heating profiles.

The accurate computation of these orbits requires a knowledge of the ion impact ionization cross sections for the injected atoms at energies of ~ 1 MeV/amu. The computations (Fig. 11) were done using a semi-classical prescription[31] (Fig. 12). More accurate cross sections would be useful if they become available.

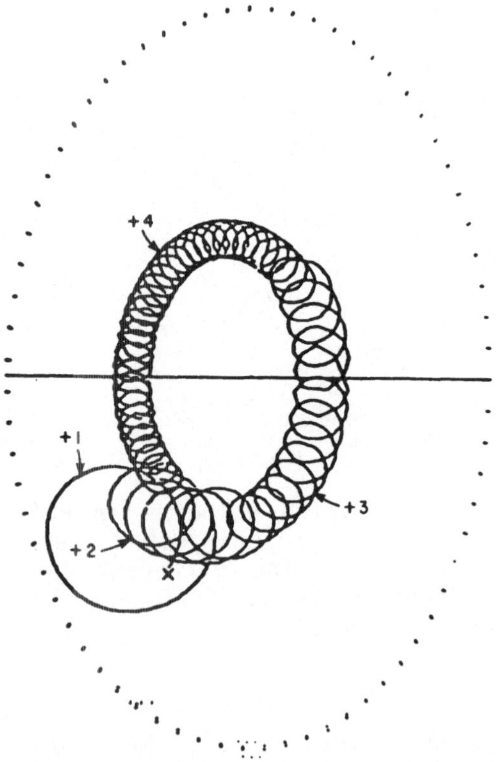

Fig. 11. A trajectory of a 32 MeV oxygen neutral atom injected into a larger plasma. The ion is progressively ionized as it drifts toward the plasma center. (Taken from Ref. 29).

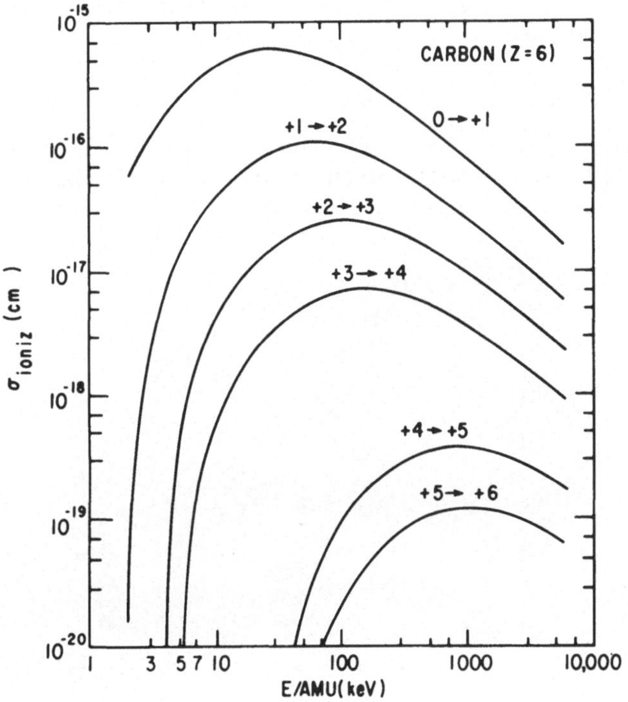

Fig. 12. Proton impact ionization cross sections for carbon as a function of the relative energy. (Taken from Ref. 29, based on Ref. 31).

NEUTRAL BEAM SYSTEMS

GENERAL PARAMETERS

The large fusion experiments that will operate during the 1980's require tens of megawatts of neutral hydrogen or deuterium beams with particle energies of about 40-80 keV/nucleon, and pulse lengths up to 30 seconds. These beams are obtained by charge exchange in low-pressure gases. Preliminary designs for future experiments and reactors use injection energies of more than 100 keV/nucleon; it is assumed that these beams will be obtained by collisional- or photo-detachment of electrons from intense, energetic beams of negative ions.

For a number of technical reasons neutral beam systems are built and operated as a number of separate modules. Typical parameters for a positive-ion module are: accelerated current ~ 40 - 100 A, active area (accelerator area) ~ 200 - 400 cm^2, accelerator transparency ~ 50%, current density at the plasma ~

0.2 - 0.5 A cm^{-2}. Details of two typical modules are given in references 32 and 33.

Positive-Ion Systems

An artists conception of the Princeton Plasma Physics Laboratory TFTR experiment with four neutral beam injection lines (three shown at the right) is given in Fig. 13.

Fig. 13. Artist's conception of the Princeton Plasma Physics Laboratory TFTR experiment.

Each of the neutral injection lines has three 120 kV, 65 A(power supply drain) plasma source/accelerator modules. Four beamlines (12 source modules) will inject about 20 MW (total) of 120 KeV D^0 atoms into the confined plasma.[34] The cost of such a system is several dollars per watt of neutral beam.

A system of this type, based on the production, acceleration, and neutralization of positive ions is shown schematically in Fig. 14. A moderately dense plasma ($n_i = n_e > 10^{12}$ cm^{-3}) is produced by a d.c. or r.f. discharge in a chamber containing hydrogen or deuterium at low pressure, typically 1-10 mtorr. Neither the energy distributions of the electrons, ion, atoms and molecules, nor the composition of this partially-ionized gas are known in detail. The bulk of the electrons have a temperature of a few electron volts, and there is a population with energies up to about 100 eV. The diassociation of the hydrogen in the discharge chamber is inferred from a model, and is very uncertain, perhaps 50%. The ion temperature is inferred from the angular distribution of an accelerated beam, and is also subject to large uncertainties, but is less than one electron volt.

In addition to the atomic and molecular hydrogen ions, the plasma also contains impurities, especially oxygen, that may be accelerated and neutralized, and/or may affect the composition of the hydrogen plasma ($H^+:H_2^+:H_3^+$). Present plasma sources typically give ion beams which are about 1% oxygen, in the form of a water ion. This is unacceptably large for many planned experiments. There is little information about other contaminants.

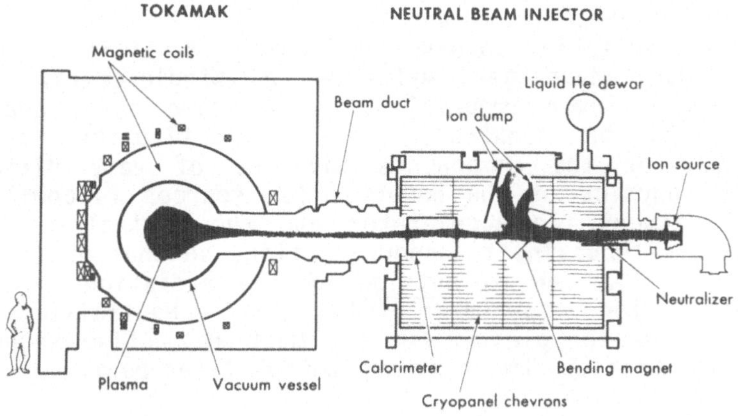

NEUTRAL BEAM INJECTOR AND THE TOKAMAK (TFTR)

Fig. 14. Schematic of a positive-ion-based neutral beam line, from the viewpoint of the neutral beam developer.

Ions and electrons in the plasma eventually reach the walls of the discharge chamber. Attached to one of the walls is an accelerator structure containing three or four electrodes, each containing many slots or circular holes. The plasma generator and accelerator combine into a single module often called an "ion source". Ions and neutral gas from the plasma generator pass through the accelerator into the neutralizer. The neutralizer is a region one or two meters long, which contains hydrogen or deuterium gas at an average pressure of a few millitorr, i.e., the accelerated ions enter a region containing about 10^{16} molecules cm^{-2}, in which neutralization by electron capture can occur. Molecular ions also produce neutrals by collisional dissociation.

The remaining part of the neutral beam system, by far the largest and most expensive part, consists of components to separate the neutral and residual ion beams, cryogenic vacuum pumps, diagnostics, power supplies, and computers for data acquisition and control. A description of a complete system is given in Ref. 35. The required gas pumping-speed in a long-pulse system, such as is shown in Fig. 14 is larger than 30 m^2 of cryopump at a few degrees Kelvin.

Negative-Ion Systems - Neutral beam systems based on negative hydrogen ions will be used for acceleration and neutralization at energies above approximately 100 keV/nucleon, because of the higher neutralization efficiencies that are possible with negative ions.

Negative-ion-based beams are potentially attractive at lower energies, if suitable powers and power densities can be achieved. The advantages will be a single-energy beam, (hopefully) with fewer impurities. The same basic beamline components will be required, as for positive-ion systems. However, many of the components will be of very different design. The physics of the negative ion sources is completely different. Between the accelerator and the neutralizer there may be a strong-focussing transport section, so that most of the neutral-beam system can be outside of the radiation shielding around the fusion experiment (Fig. 15). Neutralizers may contain either gas or plasma, but the most attractive technique at present is photodetachment by a powerful laser beam.

Negative-ion-based neutral beam systems are in the early stages of development. The first applications on fusion experiments are expected to be in the mid-1990's. Present development activities on negative-ion sources is about evenly split between two production mechanisms: electron capture in a metal vapor[36,37], or production on low-work-function surfaces.[38,39] A third production technique, dissociative attachment in a discharge[40], is in the research phase and shows considerable promise.

Efficiencies - There are several types of efficiency to consider, e.g. the neutralization efficiency, and the system efficiency. From an application viewpoint, it is the system cost and the system efficiency that are of most interest. However, for the present discussion, the topic of interest is the neutralization efficiency. Calculated curves of this neutralization efficiency, defined as neutral-beam power divided by accelerated-beam power, are shown as a function of energy for several ions and neutralizers in Fig. 16. The potential advantage of negative ions, and especially a negative-ion beam

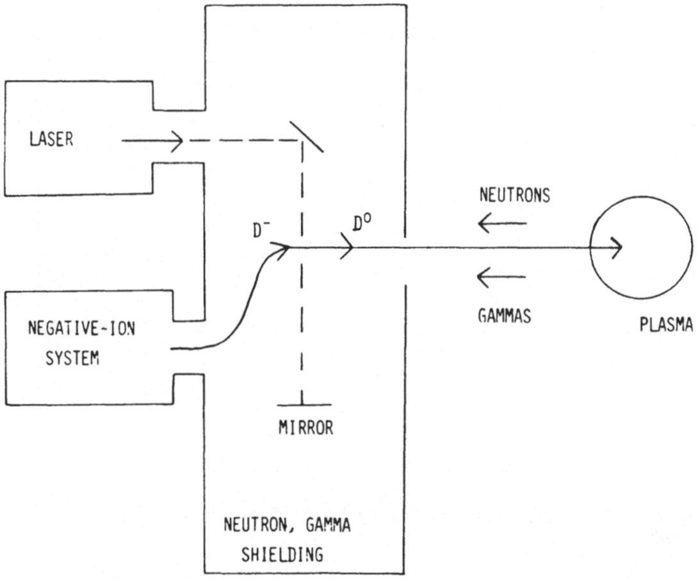

Fig. 15. Schematic of a negative-ion-based neutral beam system with negative-ion accelerator and neutralizer (laser components out of the line of sight for neutrons and gammas.

with a photodetachment neutralizer, is clear.

With one exception, no useful role has been found for molecular ion (D_2^+, D_3^+) beams, in fact, it is almost always desireable to minimize the D_2^+ and D_3^+ components from ion sources. This is because D_2^+ ions produce neutrals with half of the accelerated energy and D_3^+ ions give one-third-energy neutrals. These low-energy neutrals rarely penetrate far enough into the target plasma to do anything useful.

POSITIVE-ION BASED SYSTEMS

In this section some details of atomic physics processes in positive-ion systems will be summarized. An excellent review of this topic is given in Ref. 41; this paper contains many references which will not be repeated here. The variety of physical processes, especially in the plasma source, is large, and knowledge of the physical conditions (the electron-energy distribution, for example) is usually poor. Moreover, in most ion sources the wall interactions play a decisive role, and the physical condition of these surfaces is unknown, and perhaps unknowable. Finally, many of the phenomena are controlled as

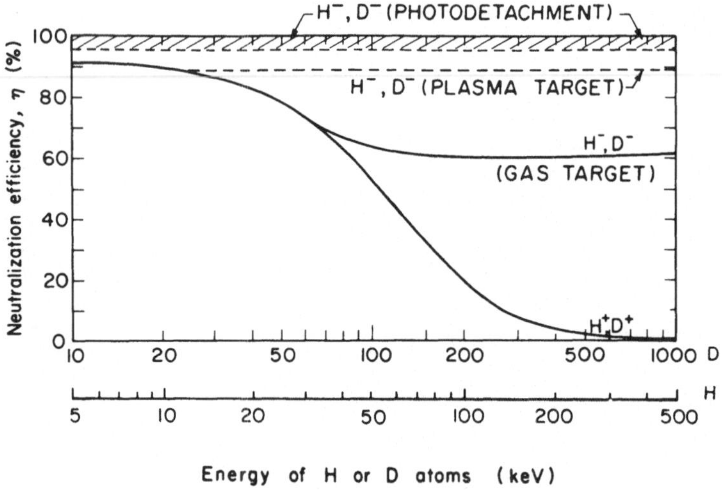

Fig. 16. Neutralization efficiencies for several ion beams and neutralizers.

much by plasma effects as by atomic physics. Still, progress requires at least a qualitative understanding of the physics involved.

ION SOURCE

Volume Atomic Processes - A list of what is believed to be the more important reactions taking place in the volume of a hydrogen plasma is given in Table 2. Cross sections for some of these are given in Figures 17a and b. Some points of special interest are the thresholds for production of H^+, H_2^+, and H_3^+ ions, because in general we want to maximize H^+ production and minimize H_2^+ and H_3^+ production, and sometimes want to maximize the H_2^+.

Impurities also are ionized by the electrons. These impurities come from the cathodes and walls of the discharge chamber, the largest component being oxygen. From an analysis of the accelerated beams, we infer that oxygen impurity ions, primarily hydride-ions, constitute several percent of the plasma under typical conditions.

Surface Atomic Processes - Atomic processes at the discharge chamber walls play an important role, and, in principle, could play a dominant role in the determination of the composition of

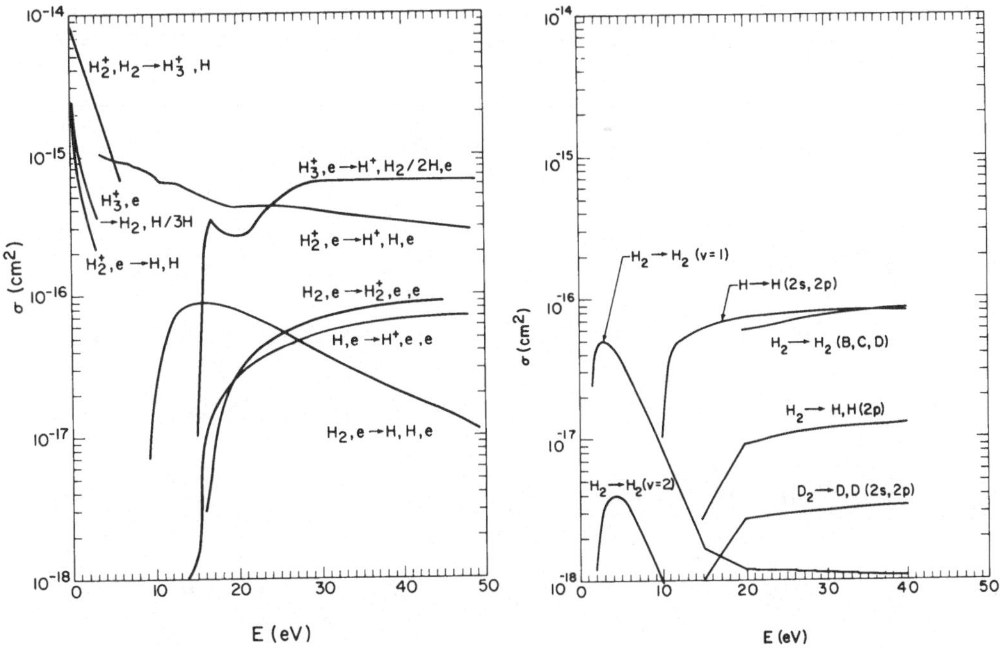

Fig. 17. Cross sections of reactions occurring in the volume of a hydrogen plasma.

the plasma. Hydrogen molecules introduced into the chamber are partially dissociated by the discharge. For the neutral hydrogen densities existing in these sources, about 10^{14} cm^{-3}, most of the atoms and molecules go to the walls without gas collisions, and then return to the volume in their original or different states. If recombination did not occur at the walls, then the neutral hydrogen atom component of the gas could be increased, and the molecular-ion fraction of the beam decreased. So far, we have not found a practical way to do this. Heating the walls to ~ 2300C is a possible, but difficult engineering approach. Oxygen found in neutral beams is assumed to come from chemical reactions between atomic hydrogen and oxides.

Source Model - We need a model of the plasma in an ion source in order to predict how to make improvements. For example, for some applications it is important that the atomic (H^+, D^+) ion percentage of the accelerated beam be 90% or greater. The properties of the discharge are strongly dependent on plasma and atomic physics in the volume and at the walls. In one model[42] cross sections for ten atomic and molecular volume processes (Fig. 17), a recombination coefficient for the walls

Table 2. Hydrogen reactions occurring in the volume of an ion-source plasma. Three types of excitation reactions are included.

$H + e \rightarrow H^+ + 2e$	$H_3^+ + e \rightarrow H^+ + H_2 + e$
	$H^+ + 2H + e$
$H_2 + e \rightarrow 2H + e$	$H_3^+ + e \rightarrow H_2 + H, 3H$
$H_2 + e \rightarrow H_2^+ + 2e$	$H_3^+ + e \rightarrow H_2 + H, + e$
$H_2^+ + e \rightarrow 2H$	$H + e \rightarrow H^*(2s,2p) + e$
$H_2^+ + e \rightarrow H^+ + H + e$	
$H_2^+ + H_2 \rightarrow H_3^+ + H$	$H_2 + e \rightarrow H_2^*(B,C,D) + e$
	$H_2 + e \rightarrow H_2^*(v=1,2,3) + e$

and a self-consistent calculation of the electron energy distribution yield D^+, D_2^+, and D_3^+ current densities at the walls. The reaction rates for the processes used are given in Fig. 18 a,b. With the model of Reference 42 it is possible to predict the ratios of H^+, H_2^+, and H_3^+ over a wide range of arc power if the wall recombination coefficient is assumed to be $\gamma = 0.2 \pm 0.1$.

Magnetic Filter - The concept of a "magnetic filter"[43] is shown in Fig. 19. High energy electrons (50-100 eV) enter the plasma from a sheath at the filaments, and would produce H^+, H_2^+ and H_3^+ ions throughout the volume if no barrier existed. The inclusion of a weak transverse magnetic field (~ 100G) prevents most of the high-energy electrons from reaching the accelerator, because they must diffuse across the "magnetic filter" and experience energy-degrading collisions in the process. The electrons near the accelerator have too low an energy to produce many molecular ions, but help destroy the molecular ions that diffuse from the filament side. In this way the H^+ fraction at the accelerator can be raised by as much as ten percentage points. There is, of course, a price: The arc

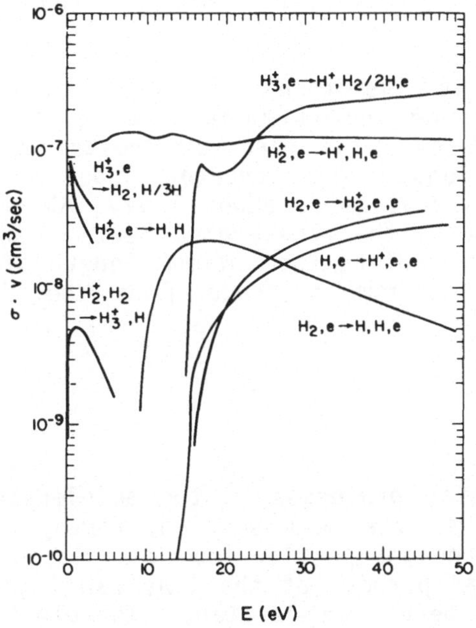

Fig. 18. Hydrogen reaction rates used in a positive-ion source model.

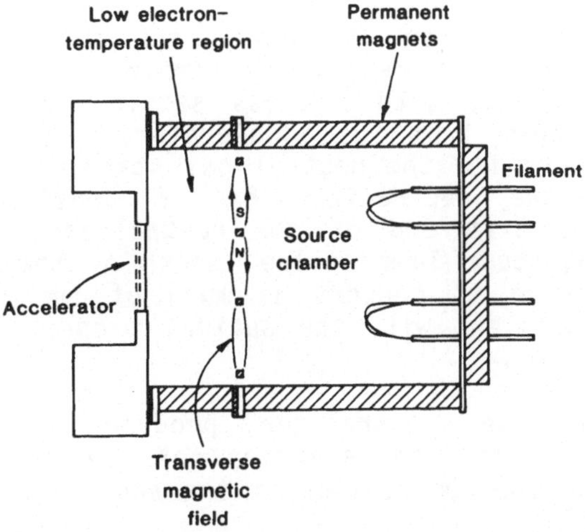

Fig. 19. "Magnetic-filter" plasma source, designed to enhance the H^+, D^+ fraction of ions later accelerated and neutralized.

power must be raised to maintain the same ion density at the accelerator.

Experimentally it was found in the first try that the electron temperature on the filament side is ~ 10 eV and the accelerator side ~6 eV; six electron volts is sufficient to produce molecular ions. The computer model shows that the optimum electron temperature is about 3 eV. At the optimum the H^+ fraction of ions in the accelerated beam is about 90%. The computer model also predicts the optimum length of the filtered plasma. i.e., the optimum distance from the filter to the accelerator.

ACCELERATOR

The atomic physics processes in the accelerator are of the same kind as given in the next section, namely ionization and charge exchange in the gas passing from the plasma source to the neutralizer. Several percent of the ions can interact with the neutral gas while being accelerated. Ion-electron collision products can be troublesome, e..g., electrons are accelerated in the backward direction and may produce x-rays or melt the ion source.

NEUTRALIZER

The neutralizer section of nearly all positive-ion-based neutral beam systems is a relatively simple mechanical device one or two meters long, with a cross section slightly larger than that of the beam from the accelerator. At the position where the beam enters it, the neutral gas streaming through the accelerator from the neutralizer has a density of about 10^{14} cm^{-3}. At the exit end of the neutralizer, the density is an order of magnitude lower. The length of the neutralizer is set by the desire to convert as much of the ion beam to neutrals as is consistent with the optical properties and cost of additional beamline.

Collisions in the neutralizer produce a plasma ($n \sim 10^9$-10^{10}cm^{-3}) with electron energies of several eV. The effects, if any, on the composition of the neutralizing gas have not been determined.

Cross Sections - In the neutralizer there is a competition between the conversion of atomic positive ions to neutrals, governed by the cross section $\sigma 10$ (Fig. 20a) and the

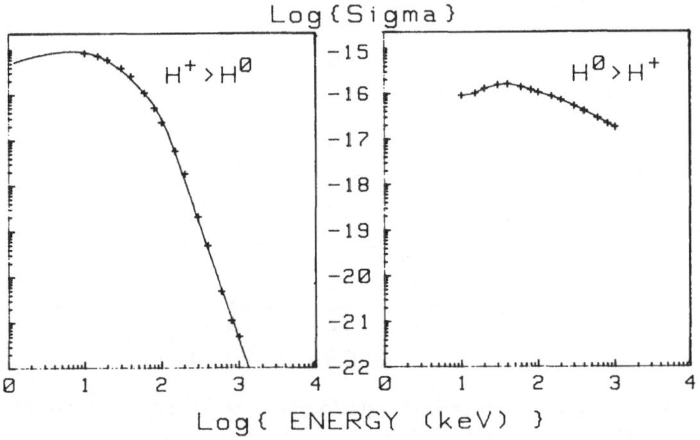

Fig. 20. Cross sections for conversion of H^+ to H^0, σ_{10}, in H_2(a) and conversion of H^0 to H^+, σ_{01}, in H_2(b).

destruction of neutrals, σ_{10}, (Fig. 20b). Usually we can ignore the cross sections for production and destruction of negative ions when calcuating the neutral yield (this is not true when one is interested in the emerging ion beams; at lower energies (~ 10 keV) approximately one percent of the accelerated positive ion beam can be converted to negative ions). Molecular ions are dissociated or neutralized before equilibrium of the high-energy component is achieved. Because actual neutralizers are not of equilibrium thicknesses, some molecular ions survive and, in fact, are useful for system diagnostics[44]. However, we do not include the cross sections here; these data can be found in compilations such as the series Atomic Data for Fusion[45].

<u>Neutral Fractions</u> - From a differential equation including electron capture and electron loss, we find that the neutral fraction of a beam origionally consisting of H^+ ions, after traversing a neutralizer with a "thickness" of π molecules/cm^2, is

$$F^0 = \frac{\sigma_{10}}{\sigma_{10} + \sigma_{01}} \exp\left[1 - e^{-(\sigma_{10}+\sigma_{01})\pi}\right] ,$$

where σ_{10} and σ_{01} are electron capture and loss cross sections, and π is the neutralizer thickness in molecules cm^{-2}. In the limit of an infinitely thick target

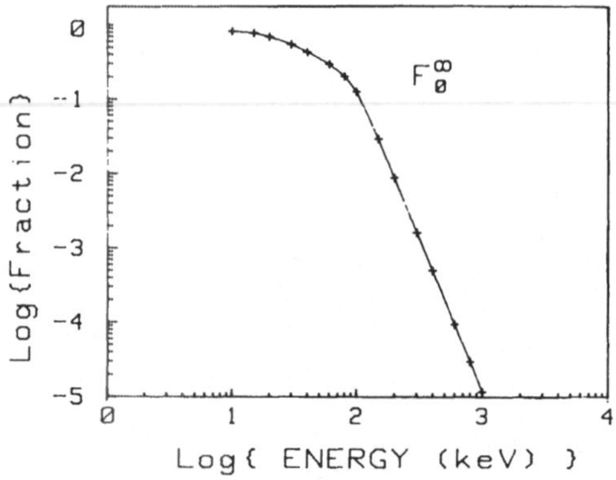

Fig. 21. The neutralization efficiency for a thick target $\sigma_{10}/(\sigma_{10} + \sigma_{01})$.

$$F^{0}_{\infty} = \frac{\sigma_{10}}{\sigma_{10} + \sigma_{01}} .$$

F^{0}, obtained from the data of Fig. 20, is shown in Fig. 21. Note that the efficiency falls off approximately as E^{-4} at high energies. The approach to F^{0}_{∞} as the neutralizer thickness is increased is shown in Fig. 22 for 80 keV/AMU ions. As a practical matter, increasing the neutralization efficiency usually means increasing the gas throughput, and hence the pumping required. As a compromise, system designers may use a neutralizer giving about 90% of F^{0}_{∞}.

Gases other than hydrogen may be used as neutralizers. Some years ago several large mirror experiments built up a confined plasma in an ultra-high vacuum by "Lorentz-ionization" of neutral beams. Beams of highly-excited atoms (magnesium vapor is a good neutralizer for producing them) were ionized by the action of the equivalent electric field $\overline{\varepsilon} = \overline{v} \times \overline{B}$ in the trapping region. In general use, however, F^{0}_{∞} is pretty much independent of the gas used, and hydrogen is the common choice.

Region between Neutralizer and Plasma - To accomodate the ion-deflection magnet, ion-beam stops, calorimeters and duct through the coils of the confinement experiment, the beam line following the neutralizer may be six or more meters long. To prevent appreciable re-ionization of the neutral beam, the average background gas (mostly hydrogen) density must be such that $\sigma nL \ll 1$ for all neutral particles (Fig. 20b). This can be achieved by large cryopumps and careful analysis and control of

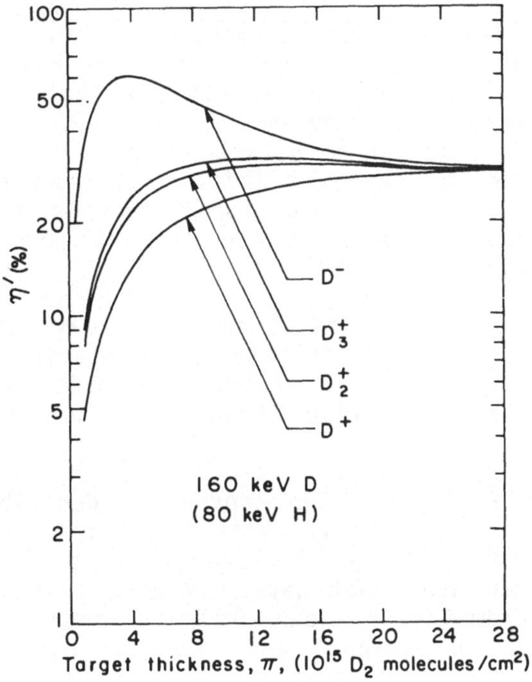

Fig. 22. The approach to equilibrium with increasing target thickness.

where stray beams strike surfaces and cause gas to be evolved.

This region presents many challenges to the engineer and to the physicist developing neutral beam diagnostics.

NEGATIVE ION BASED SYSTEMS

There is no operational high-current neutral beam system based on negative ions, and may not be for 5-15 years because the main incentive for this expensive development is a future fusion experiment not yet approved, and perhaps not yet conceived. However, the perceived need and required lead time are so great that the research and development programs around the world are growing rapidly. There are many areas requiring work by atomic and plasma physicists.

Negative-Ion Generator

Three ways to form large sources of negative hydrogen ions are currently being considered; they are, in the order in which they are being tried:

Conversion of Low-Energy Positive Ions to Negative Ions by Charge Exchange - The research described in Refs. 36 and 37 is based on the production of a high-current low-energy positive-ion beam which passes through an alkali or alkaline-earth-metal vapor, and emerges as a mixture of H^+, H^0, and H^-. The negative-ion yields that can be obtained from a thick target are quite large at low ion energies (Fig. 23),[46] as much as 50%. From a practical viewpoint, additional factors must be considered, eg. obtainable positive-ion current densities, scattering, space charge effects, etc.

The complication of the addition of a heavy-metal charge-exchange cell, and the possibility of contamination of the fusion experiment by the metal vapor, have made this negative-ion approach less attractive than the surface-production technique.

Production on Surfaces - The approaches described in Refs. 38 and 39 make use of the fairly large emission of negative ions from surfaces bombarded by low-energy ions and neutrals, especially those surfaces that have low work functons and high atomic-masses. Examples of pure alkali metal negative-ion secondary emission coefficients are given in Fig. 24.[47,48] A

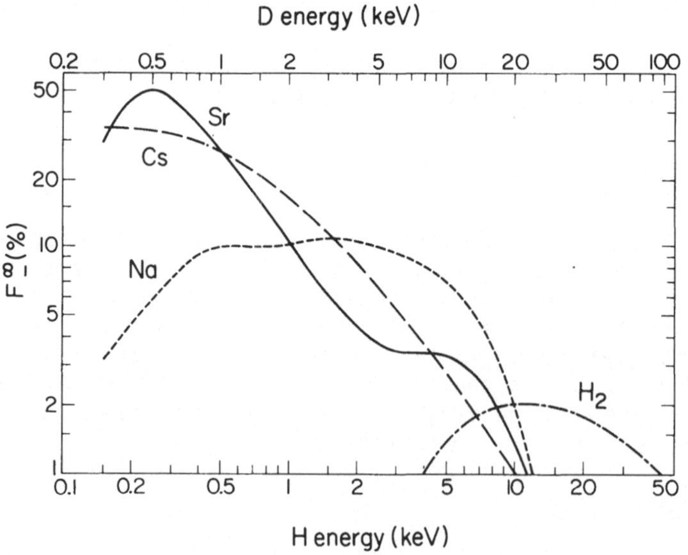

Fig. 23. The percentage conversion of H^+ to H^- in thick alkaline-earth targets.

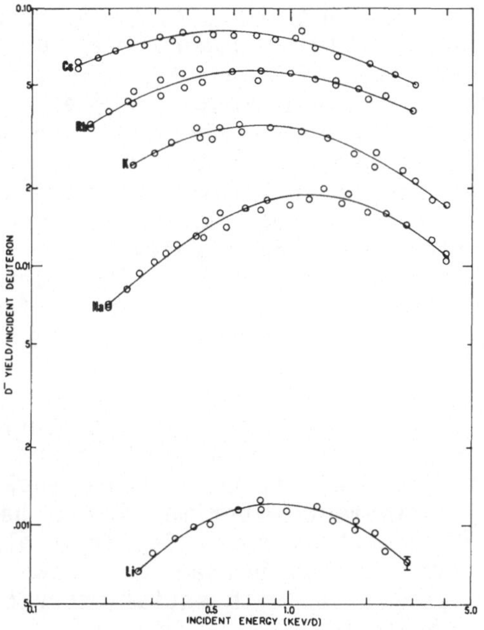

Fig. 24. Negative ion secondary emission coefficients for pure alkali metals.

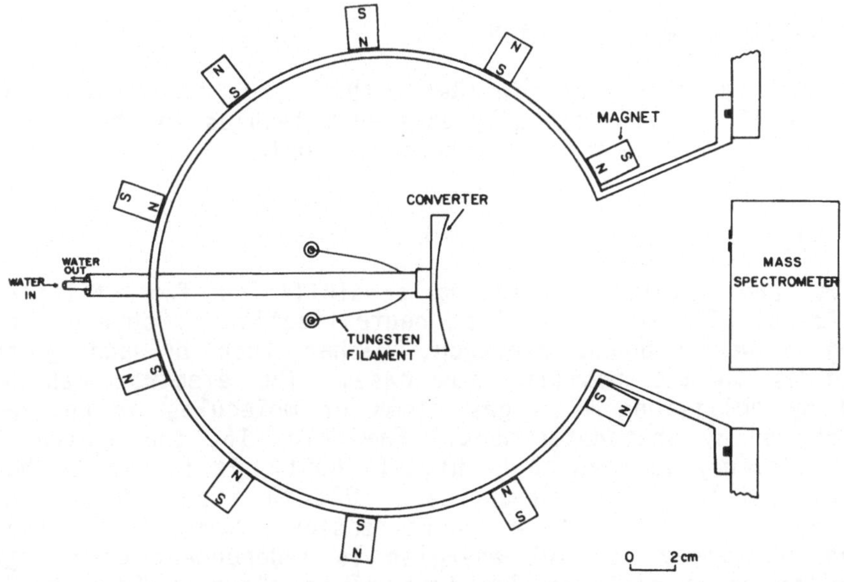

Fig. 25. Schematic of a surface-production negative ion source. The Mo "converter" is in a plasma containing hydrogen and cesium ions, and is biased positive with respect to the plasma.

model has been developed that explains these results, and those for partial coverages yielding lower work functions (for example, a partial monolayer of Cs on Mo). We also know that the H^- yield for thermal bombardment energies is rising with energy, but very low. Unfortunately, data are missing for the critical region between about 3 eV and 100 eV. From experiments with an actual ion source containing hydrogen and cesium[39] in which the energy spectra of negative ions are measured carefully, it appears that the negative ions are produced by scattering of < 10 eV neutrals, or by a different process, e.g., collisional desorption of H^- from the surface. These H^- production roles in sources consequently are still research topics.

Volume Production Charge-Exchange and Surface - The H^- production techniques require the use of cesium on other metal vapors and plasmas. We would rather not use such materials near accelerators and confinement experiments. It has been shown[40] that quite large H^- densities can exist in a plasma. (Whether currents of interest to fusion can be drawn from a volume production source, and if so, with satisfactory electron and gas control, remains to be seen.)

The proposed formation mechanism is dissociative attachment of low-energy electrons to vibrationally excited H_2 molecules:

$$e + H_2(V > 6) \rightarrow H_2^- \rightarrow H^- + H$$

This topic is discussed in Ref. 49. The production and destruction of the vibrationally-excited molecules in the plasma and at the walls is an important research topic.

Neutralizer

There are several intriguing possibilities for efficient neutralizers. The reason is that neutralization is obtained by removing a weakly-bound electron, rather than by adding an electron as in the positive ion case. The electron can be removed by collisions with gas atoms or molecules or charged particles, or by photodetachment. From Fig. 16, the neutralization efficiency is seen to be high (> 60%). It is even higher in an ionized gas, and can approach 100% in a photodetachment neutralizer. In all of these neutralization schemes the neutralization efficiency at high energies is independent of energy because the cross sections for loss of an electron from an H^- ion and from an H^0 atom have the same energy dependence.

Figure 22 shows that there is a value of the target thickness for which the neutralization efficiency is a maximum. If we neglect cross sections other than the two-electron loss cross sections for the H^- ion and H^0 atom, σ_{-10} and σ_{01}, respectively, then

$$\varepsilon(\pi) = \frac{\sigma_{-10}}{\sigma_{01} - \sigma_{-10}} \left[e^{\frac{1}{2}\pi(\sigma_{01}-\sigma_{-10})} - e^{\frac{1}{2}\pi(\sigma_{01} - \sigma_{-10})} \right] e^{-\frac{1}{2}\pi(\sigma_{01}+\sigma_{-10})}$$

This expression has a maximum when

$$\pi = \frac{\ln(\sigma_{-10}/\sigma_{01})}{\sigma_{01} - \sigma_{-10}}$$

A more accurate result is obtained by including collisions in which two electrons are lost, with the cross section σ_{-11}. These can be included in the formula for η by replacing σ_{-10} by $\sigma_{-10} + \sigma_{-11}$ everywhere except in the numerator pf the first term. At low energies a negative ion production term must be included, but this process is negligible at energies of interest to fusion.

The practical use of lasers for photodetachment is discussed in Ref. 50. The ionization cross section peaks at a photon energy of about 1.5 eV, which is not far from the wavelength of the powerful steady-state iodine chemical lasers.

In addition to offering the possibility of approximately 100% neutralization, the photodetachment approach offers other advantages. The absence of left-over ion beams to dispose of makes the neutral beam system much simpler and cheaper. Moreover, the photoionization process is selective, so that impurities may be eliminated from the neutral beam.

DIAGNOSTICS

Good diagnostics of physical conditions in various parts of the neutral beam line is important in the research and development phase, and in the operations phase. Because power densities in present neutral beams are tens of kilowatts per square centimeter, and pulses are > 1 S, it is difficult or impossible to sample the neutral beams for composition or impurities with solid analyzers, especially when the beams are being injected into the plasma targets. Techniques are in use, or need to be developed, which utilize natural emission from the beam, or non-perturbing optical or particle probes.

The most important neutral beam properties are the currents and angular distributions of the full-, half-, and one-third-energy hydrogen or deuterium atoms that reach the plasma, and the currents of neutral impurities. No good techniques for measuring impurity currents have been developed, and the very useful doppler-shift optical technique measurements (described next) are made in the neutralizer rather than at the exit end of the beam line. We expect laser techniques to provide powerful diagnostic aids in coming years.

Doppler-Shifted Light--The energy and spatial distributions of a neutral beam can be measured by observing the doppler-shifted Hα and Hβ light emitted by the beam as it transverses the neutralizers.[51,52] Accelerated H^+, H_2^+, H_3^+ ions interact with the gas in the neutralizer and produce atoms with energies E, 1/2E, and 1/3E, where E is the final energy of the accelerated ion beam. By observing the radiation emitted by the beam at an angle to the beam direction, three separate Doppler-shifted spectra can be recorded.

To obtain neutral currents, it is necessary to know the following cross sections (others turn out to be unimportant for usual conditions).

1. $H^+ + H_2 > H^*$ (3S, 3P, and 3D levels)
2. $H^0 + H_2 > H^*$ (")
3. $H_2^+ + H_2 > H^*$ (")
4. $H_2^0 + H_2 > H^*$ (")
5. $H_3^+ + H_2 > H^*$ (")
6. $H_2^+ + e > H^*$ (3s) + H(1S) or H^+

The light intensity of the ith component (i = 1, 2, 3, corresponding to full, half, and third energy) is proportional to N(n), (the population density of the nth level of excited hydrogen atoms with velocity v_i). Using an index j to refer to species considered in this model

$$N(n)_i = \frac{1}{v_i} \sum_j n_0(z - x)\, \sigma_j(n)\, J_j(x) \exp\left(- \frac{x}{v_i T_n}\right) d_x,$$

where v_i is the velocity of the ith component, z is the distance between the observation point and the exit grid of the source, n_0 is the density of the gas in the neutralizer, $\sigma_j(n)$ is the optical excitation cross section for the jth species in the mixed beam $J_j(x)$ is the current density of the corresponding species, T_n is the lifetime of the nth level of hydrogen, and j is summed over those species contributing to the ith component of the beam.

The data are processed by a computer to yield (1) the neutral currents (from the areas under the curves), (2) the angular distribution for each energy fraction (from the widths of the peaks), and (3) the mean directions of the neutral beams (from the centroids of the peaks).

A signal corresponding to a hydrogen atom energy of about E/18 is believed to be from the breakup of accelerated water ions, and hence gives a measure of the oxygen impurity in the neutral beam.

Special Topics

Ideas for beams different from those discussed so far ie, the conventional neutral H or D beams, have arisen recently. One of them, the use of intense tritium beams, does not change the atomic physics appreciably, but will present new engineering problems. Some other suggestions are

1. "Light" atoms, e.g., Li through Ne; the application of such beams has been discussed in Section II of this article.
2. Polarized neutral beams.

Polarized Neutral Beams--One of the most intriguing ideas is to inject polarized neutral beams into a polarized plasma. Both the nuclear reaction rates and the angular distributions of the reaction products could be favorably modified.[53] Several ways for obtaining nuclear polarization in conventional polarized ion sources are given in Ref.[54] Whether these, or other techniques, can be applied to intense neutral beam systems remains to be seen.

ACKNOWLEDGMENTS

The authors are grateful for discussions with and assistance from Drs. D. Mikkelsen, C. Singer, L. Grisham, R. Goldston, and K. Berkner, F. Burrell, W. S. Cooper, K. Ehlers, K. N. Leung, and Mr. J. W. Stearns.

This work was supported by in part by the Director, Office of Energy Research, Office of Fusion Energy, Development and Technology Division, of the U.S. Department of Energy under Contract No. DE-AC03-76-SF00098, and in part by PPPL DOE Contract No. DE-AC02-76-CH03073.

REFERENCES

1. J. D. Lawson, Proc. Phys. Soc., London, Sect. B 70, 6 (1957).
2. S. Greene, 1967, UCRL-702522, Lawrence Livermore Laboratory, Livermore, California.
3. T. Ohkawa, Nucl. Fus. 10, 85 (1970).
4. T. K. Fowler, "Mirror Theory", Fusion, Vol. 1, ed. E. Teller, Academic Press (1981).
5. R. F. Post, "Experimental Base of Mirror Confinement Physics, Fusion, Vol. 1, ed. E. Teller, Academic Press (1981).
6. H. Furth, "The Tokamak," Fusion, Vol. 1, ed. E. Teller, Academic Press (1981).
7. J. Shohet, "Stellerators," Fusion, Vol. 1, ed. E. Teller, Academic Press (1981).
8. R. Hulse, D. Post, D. Mikkelsen, J. Phys. B 13, 895 (1980) V. Krupin,. V. Marchemko, and S. Kakovlenko, JETP Lett. 29, 3895 (1979).
9. W. Kunkel, "Neutral Beam Injection," Fusion, Vol. 1, ed. E. Teller, Academic Press (1981).
10. R. Freeman and E. Jones, CLM R-137, Culham Laboratory, Abingdon, England (1974).
11. J. Rome, J. Callen, and J. Clarke, Nucl. Fus. 14, 141 (1974).12.G. Lister, D. Post, and R. Goldston, Symp. Plasma Heat. in Toroidal Devices, 3rd (ed. E. Sindoni) p. 303, E'ditvice, Compositori, 1976.
13. L. D. Stewart, et al. "Proc. of 2nd International Symposium on the Production and Neutralization of Negative Hydrogen Ions and Beams," T. Sluyters, ed. BNL-51304, 1980.
14. R. Olson, K. Berkner, W. Graham, R. Pyle, A. Schlachter, and J. Stearns, Phys. Rev. Lett. 41, 163 (1978).
15. H.B. Gilbody, Physica Scripta 23, 143 (1980).
16. INTOR Group, Rep. Int. Tokamak Reactor Workshop, (IAEA, Vienna) 1980.
17. M. Menon, Proc. of IEEE 69, 1012 (1981).
18. J.A. Holmes, J. Rome, W. Houlberg, Y-K. Peng, and S. Lynch, Nucl. Fus. 20, 59 (1980).
19. T. Stix, Plasma Physics 14, 367 (1972).
20. D. Sivukhin, in "Reviews of Plasma Physics," (M.A. Leontovich, ed.) Vol. 4, Pergammon, Oxford (1966).
21. H. Eubank, et al., Phys. Rev. Lett. 43, 270 (1979).
22. J. Dawson, H. Furth, and F. Tenney, Phys. Rev. Letter 26, 1157 (1971).
23. R.C. Isler, C. Crume, Phys. Rev. Lett. 41, 1296 (1978).
24. S. Suchever, E. Hinnov, M. Bitter, R. Hulse, D. Post, Phys. Rev. A (to be published).
25. W. Clark, J. Cordey, M. Cox, S. Fielding, R. Gill, R. Hulse, P. Johnson, J. Paul, W. Peacock, B. Powell, M. Stamp, and D. Start, Nucl. Fusion 22, 333.

26. D. Mikkelsen and C. Singer, "Optimization of Steady-State Beam-Driven Tokamak Reactors," (to be published).
27. W. Clark, J. Cordey, M. Cox, R. Gill. J. Hugill, J. Paul, and D. Start, Phys. Rev. Lett. 45, 1101 (1980).
28. S. Suchewer, et al., Phys. Rev. Lett. 43, 207 (1979).
29. L. Grisham, D. Post, D. Mikkelsen, and H. Eubank, Nuclear Technology/Fusion 2, 199 (1982).
30. J. Dawson and K. MacKenzie, Proceedings of the 2nd Joint Grenoble-Varenna International Symposium, EUR-742-EN, p. 953 (1980).
31. M. Gryzinski, Phys. Rev. Lett. 138, A336 (1965).
32. K. H. Berkner et.al., Proc. 8th Symp. Engin. Prob. Fusion Research, San Francisco, IEEE Pub. Mo. 79 CH 1441-5NPS (1979) p. 214.
33. W. Gardner, et.al., RSI 53, 424 (1982).
34. P. Reardon, "Status Report on TFTR", Proc. 3rd Topical Meeting on the Technology of Nuclear Fusion, Santa Fe (1978) p. 621.
35. K. H. Berkner, et.al. Proc. 9th Symp. Engin. Prob. Fusion Research, Chicago IEEE Pub. No. 81CH1715-2NPS (1981) p. 763.
36. E. B. Hooper, P. Poulsen, P. Pincosy, JAP 52 7027 (1981); E. B. Hooper, P. Poulsen, O. A. Anderson, Nucl. Tech. Fusion 2, 362 (1982).
37. M. Delannay, et al. "A New Type of Neutral Injector Based on the Production of D^- by Double Elextron Capture," Heating in Toroidal Plasmas, Proceeding of the 2nd International Symp., Como, Italy, Sept. 1980, p. 831.
38. Yu I. Belchenko and V. G. Dudnikov, Proc. of the 15th Intl. Conf. on Phenomena in Ionized GAses, Minsk (1981) p. 1504.39. K. N. Leung and K. W. Ehlers, RSI 53, 803 (1982).
40. G. W. Hamilton and "Proceedings of the Second Intl. Symp. onProduction and Neutralization of Negative Hydrogen Ions and Beams, Brookhaven Natl. Lab. (1980) p. 90.
41. P. Raimbault and J. P. Girard, "Atomic and Molecular Physics Problems in Fast Neutral Injection," Physica Scripta 23, 108 (1982).
42. C. F. Burrell, C. F. Chan and W. S. Cooper (LBL), private communication.
43. K. W. Ehlers and K. N. Leung "Effect of a Magnetic Filter on Hydrogen Ion Species in a Multicusp Ion Source. LBL Report (LBL-12255) (1981).
44. J. W. Stearns, private communication.
45. D. H. Crandall and C. F. Barnett (ORNL) and W. L. Wiese (NBS)"Atomic Data for Fusion," Controlled Fusion Atomic Data Center, ORNL.
46. R. McFarland, A. S. Schlachter, J.W. Stearns, R. Olson, "D^- Production by Charge Transfer in Thick Alkaline-Earth Targets," to be published in Physical Review A.
47. P. J. Schneider, et al., Phys. Rev. B 23, 941 (1981).

48. J. R. Hiskes, P. J. Schneider, Phys. Rev. B 23, 949 (1981).
49. J. R. Hiskes, et al., "Hydrogen Vibrational Population Distributions and Negative Ion Concentrations in a Medium Density Hydrogen Discharge, "Lawrence Liverore National Laboratory Report UCRL-86873 (1981).
50. J. H. Fink, "Evaluating Laser-Neutralized Negative Ions as a Source of Neutral Beams for Magnetic Fusion Reactors," Lawrence Livermore National Laboratory Report UCRL-87301 (1982).
51. C. F. Burrell, et al., "Doppler Shift Spectroscopy of Powerful Neutral Beams, RSI 51, 1451 (1980).
52. G. A. Cottrell, A. R. Martin, and C. Padget "Optical Diagnosis of a High-Current Beam," Heating in Toroidal Plasmas (1980) p. 945.
53. R. M. Kulsrud, et al. "Fusion Reactor Plasmas with Polarized Nuclei," submitted to Phys. Rev. Lett.
54 A. D. Krish and A. T. M. Lin, Editors, Polarized Proton Ion Sources, American Institute of Physics, New York, (1982).

SPECTROSCOPIC PLASMA DIAGNOSTICS

H.W. Drawin

Association EURATOM-CEA sur la Fusion
F-92260 Fontenay-aux-Roses, France

1. INTRODUCTION

The ultimate aim of the various diagnostic methods applied in fusion research is to obtain information, as detailed as possible, on the local thermodynamic state and of the transport properties of the plasma investigated. Very different diagnostic methods are applied, very often making use of rather different physical effects. There exist authoritive articles and books in which specific or various methods are treated, see e.g. Refs.[1-13]. In this article some specific diagnostic problems and results will be discussed which are based on the spectroscopy of photons emitted by atoms (ions). First, however, we make some general remarks concerning plasma spectroscopic methods.

2. PASSIVE AND ACTIVE DIAGNOSTIC METHODS

Atomic spectroscopy represents an important tool in the field of plasma diagnostics. As in particle diagnostics, we distinguish between passive and active methods. Passive methods rely on the photon fluxes escaping freely from the plasma without perturbing the latter. In active methods one measures and evaluates the photon fluxes due to localized or global plasma perturbations imposed from the outside. Active methods are, for instance, the determination of the electron temperature T_e by scattering of laser photons (Thomson scattering) or the determination of completely stripped impurity species by spectral line emission due to charge exchange with injected neutral hydrogen (deuterium) atoms. This latter example may also show that it is sometimes difficult to distinguish clearly between particle and photon diagnostics, since the emitted photons can also be used for analyzing indirectly the injected atomic beam.

3. PHYSICAL MODELS FOR THE INTERPRETATION OF MEASURED DATA

In order to obtain information on the plasma properties, physical models are applied which link the measured quantities to the plasma parameters. Two extreme models are the hypothesis of "complete local thermodynamic equilibrium" (C.L.T.E.) and the hypothesis of "homogeneous stationary coronal ionization - recombination equilibrium". Under actual laboratory conditions, neither of these models applies in full generality and, thus, additional assumptions must be introduced in the models which all rely on the various atomic processes and transport properties. For a review, see e.g. Refs.[1,4,9,14-18]. An often applied model is the so-called "collisional-radiative model".

4. DETERMINATION OF LOCAL PLASMA EMISSION PROPERTIES

Very often, the observed photon fluxes emanate from extended plasma regions (for instance, when observing along a chord through the plasma). In such a case, special deconvolution methods must be applied (e.g. "ABEL-inversion") in order to get the photon flux originating from a well-defined local region. There exist numerous publications about this problem, see e.g. Refs.[1,2,19-23]. Error propagation is treated in Refs.[29,30], in Ref.[30] the solution of an ABEL-type integral equation in the presence of noise is calculated, and the reliability of local determinations of plasma emission intensity is discussed in Ref.[31]. The determination of plasma rotational velocities in connection with the ABEL-inversion problem is treated in Ref.[32]. The deconvolution of the measurements can be avoided when the emission originates from well defined and physically small spatial regions.

5. REABSORPTION AND SCATTERING

Tokamak and Stellarator plasmas have relatively low particle density. They are generally optically thin, i.e. the emitted photons are not (or only very slightly) reabsorbed. In high-density plasmas (e.g. in laser-compressed plasmas) reabsorption and scattering can be important processes. The observed photon flux must therefore be corrected with the aid of the equation of radiative transfer in order to obtain the locally emitted fluxes. For a review see e.g. Ref.[33].

6. SPATIAL AND TEMPORAL RESOLUTION

Finally, in the different types of experiments, the length and time scales of the measurements are very different. Even for the same machine, the time scales for performing similar types of measurements can differ considerably from each other and are a function of the information to be extracted from a measurement. For laser-produced plasmas having diameters of 1000 μ to 100 μ or less a temporal resolution in the range of 10^{-9} to 10^{-11} seconds is desirable, for "quiet" Tokamak discharges 10^{-3} seconds or longer are often sufficient for big machines with diameters now exceeding one meter.

All these facts should be borne in mind in real situations, since they affect directly the reliability of the diagnostic results and conclusions.

7. QUANTITIES OBTAINABLE BY PHOTON SPECTROSCOPY

The following quantities can, in principle, be determined from plasma spectroscopic measurements:

- *Heavy-particle temperature* T_s^z of species "s" in the ion charge state z, from the Doppler profile of their spectral lines[10-13], or from Thomson scattering in the forward direction[37,38].

- *Directional velocity* $\vec{w}_s^{\,z}$ from the line shift of a spectral line[32,39-41] or from the motional Stark effect originating from the Lorentz electric field of strength $\vec{E}^z_{s\perp} = \vec{w}_s^{\,z} \times \vec{B}$ felt by the particles (s,z) moving perpendicular (⊥) to a magnetic field of induction $\vec{B}$, [42], or from the spectral shift of the absorbed or scattered signal of a laser beam[43-45].

- *Electron temperature* T_e from Thomson scattering[37,38] of a laser beam, from the ratio of two or more suitably chosen spectral lines[1-14,46] or from the free-free and/or free-bound emission continuum[1-14].

- *Deviation from Maxwellian velocity distribution,* from the ratio of several suitably chosen spectral lines[47] or from the free-free and/or free-bound emission continuum[48,49]. And Thomson scattering could also give information, at least in principle, about deviations from a Maxwellian distribution.

- *Electron density* n_e, from Thomson scattering[1-4,37,38], from laser interferometry[1,2,50], from absolute intensity measurements of the free-free and/or free-bound continuum[1-14], from the intensity ratio of two or more suitably chosen spectral lines[51-55], from Stark-broadening of a spectral line[1-4,56-57], from the Stark and plasma polarization shift of a spectral line[4,56-60], and from the so-called "plasma satellites" accompanying certain spectral lines[56,57,61,62].

- *Particle density of neutral hydrogen atoms* n_{H°, from an absolute measurement of the intensity of a hydrogen spectral line[1-14] or from resonance fluorescence of laser beam excited hydrogen atoms[63,64].

• *Particle density of ionized impurity atoms* n_s^z, from an absolute measurement of the intensity of corresponding spectral lines[1-14], or from resonance fluorescence of laser beam-excited particles[43,67], or from an absolute measurement of the intensity of one or several spectral lines emitted due to charge exchange with injected neutral atoms[13,68], or from an absolute measurement of the free-free and/or free-bound continuum[1-14,69].

• *Local magnetic field* $\vec{B}$, from the Zeeman effect[70] or from the motional Stark effect[42] of specially injected neutral particles (e.g. lithium). Simultaneous laser excitation of the injected particles will permit the determination of the local values of $|\vec{B}|$ and the direction of $\vec{B}$. One can thus determine the poloidal magnetic field and, hence, the local electric current density in a Tokamak plasma. For special magnetic configurations it should be possible to obtain the local current distribution simply by measuring the plasma density and temperature distributions[71]. It should finally be mentioned that laser light scattering has been applied for measuring magnetic fields in a Tokamak plasma[72] and that it has been proposed to apply the Faraday effect of submillimeter waves[73-75] to the determination of $\vec{B}$.

8. WAVELENGTH REGIONS AND SPECTROSCOPIC INSTRUMENTS

Plasmas can emit radiation over the whole electromagnetic spectrum. At fusion temperatures, the emission of the highly ionized atoms is preferentially in the vacuum-ultraviolet (VUV) and in the X-ray region. The infrared region is generally of little interest for atomic spectroscopy, some special conditions excepted (e.g. study of cleaning discharges).

The following spectroscopic instruments are (preferentially) employed in the different spectral regions (λ is the wavelength in Angström units, 1 Å = 10^{-8} cm) :

• *Microwave and infrared region* : $\lambda > 8000$

- grating spectrographs (- meters) and others.

• *Visible and near ultraviolet* : $2000 \leqslant \lambda$ [Å] $\leqslant 8000$

- grating spectrographs, normal incidence.
- resolving power $\lambda/\Delta\lambda \approx 2.10^5$.
- for special purposes : Fabry-Perot interferometers.

• *Far ultraviolet region* : $100 \leqslant \lambda \leqslant 2000$

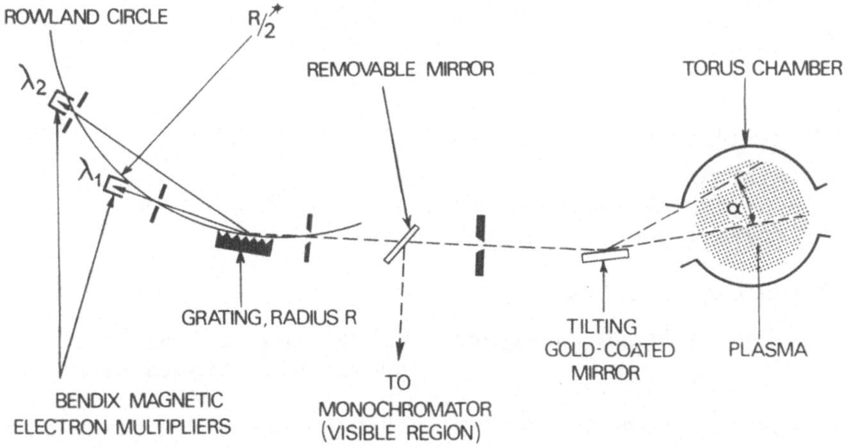

Fig. 1. Schematic drawing of a VUV spectrometer equipment for space- and time-resolved measurements on Tokamak discharges by means of a tilting mirror ; adapted from Ref.[76]. See also Ref.[86].

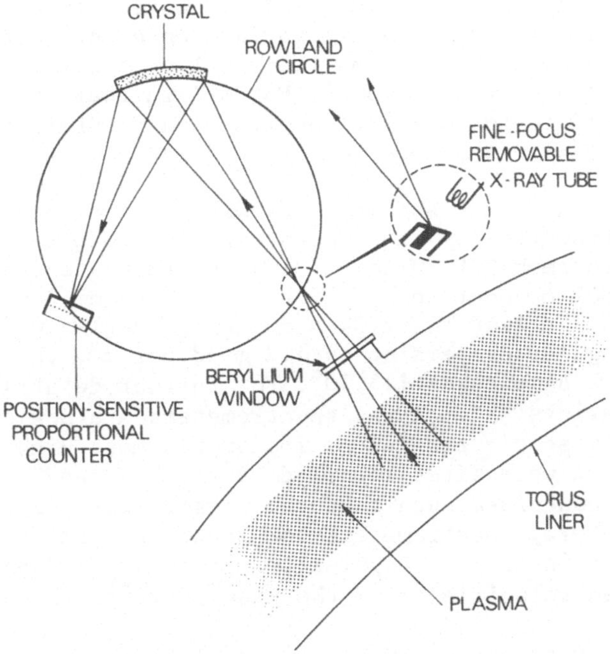

Fig. 2. Quartz crystal spectrometer for measurements in the X-ray region ; adapted from Ref.[79].

- grating spectrographs (- meter)
  in normal incidence : $400 \leq \lambda \leq 3000$
  in grazing incidence ($\sim 10°$) : $100 \leq \lambda \leq 2000$

• *Extreme ultraviolet and soft X-ray region* : $1 \leq \lambda \leq 100$

- grating spectrographs (- meters)
  in grazing incidence ($\sim 1°$) : $5 \leq \lambda \leq 100$
- crystal spectrographs (- meters) : $1 \leq \lambda \leq 25$

• *Hard X-ray region* : $\lambda < 1$ Å

- pulse height analyzers (Li-drifted silicon-detectors, cooled with liquid nitrogen).

Figure 1 shows schematically how space- and time-resolved spectroscopic measurements are performed on plasma discharges in the Tokamak TFR. The whole system, the tilting gold-coated mirror included, is under vacuum. The distance between the tilting mirror and the plasma is 1.8 m. By rotating the mirror by a total angle of $\alpha = 4°$ one can scan one-half of the plasma. Relative calibration is done with spectral lines in the visible and near UV-region using the "branching ratio method"[77,78]. For absolute calibration a carbon ribbon lamp or any other absolute calibration standard can be used. The apparatus is equiped with a holographically ruled diffraction grating with 1200 grooves per mm, a radius of curvature of 1 m and operates at an incidence angle of 83° with a maximum efficiency at $\lambda \approx 300$ Å. The total spectral range covered is from $\lambda \approx 100$ Å to $\lambda \simeq 1800$ Å. For all further details the reader is referred to the original publication[76]. Similar devices are used in all fusion laboratories.

Figure 2 shows the schematic drawing of a curved-crystal reflection spectrometer for measurements in the soft X-ray region, as used on the Tokamak TFR. With a (310) quartz plate (inter-planar spacing d = 1.1802 Å) bent to a radius of R = 288 cm one covers the spectral region $\lambda = 2.18 \ldots 2.24$ Å for Bragg angles θ from 67.5 to 72.5°. The angular dispersion is 1.2 mrad per mÅ, the spectral resolving power $\lambda/\Delta\lambda$ exceeds $1.5.10^4$. For further details, see the original publications[79,46]. X-ray spectrometers are now currently applied in spectroscopic studies of fusion plasmas. High-resolution X-ray spectra were first reported by the PLT-tea[34] whose spectrometer is conceived such that the plasma lies inside the Rowland circle. X-ray spectrometers now reach spectral resolutions which allow determination of Doppler temperatures[46,80] and of possible directed velocities from the line shift[81].

The spectroscopic equipment is completed by instruments serving special purposes, for instance for the repetitive scanning of the whole line profile of an impurity spectral line[82,83] or for a spatially resolved measurement of the soft X-ray emission[84].

9. DETERMINATION OF PARTICLE DENSITIES

An important problem is the determination of the particle density $n^z_{s,1}$ of z-times ionized impurity species of chemical element "s" in the ground state i = 1. The most widely applied method is via the measurement of the radiance of characteristic spectral lines emitted in a transition $|j\rangle \rightarrow |i\rangle$. How this density is obtained from optically thin lines is shown schematically in Fig.3.

The photon or energy flux is measured in the form of a current which is converted into the radiance as a function of time t and of height of observation h. For a given time $t = t_x$, the measured radiances as a function of h are converted into volume emission coefficients $\varepsilon^z_{ij}(\vec{r},t)$ as a function of $t = t_x$ and of space point $\vec{r}$. Applying the equation (see also Eq. (123) in the lecture on Atomic and Molecular Processes ..., Ref.[85])

$$\varepsilon^z_{ij} = \frac{1}{4\pi} A^z_{ij}\, n^z_j\, h\nu^z_{ij} \tag{1}$$

yields the particle density $n^z_j(\vec{r},t)$ of the excited species. At this stage the plasma model intervenes. When we assume that the corona assumption applies, the ground state density n^z_1 is obtained from an equation of type (124) in Ref.[85]. The proportionality factor depends on the particular atomic model adopted.

The particle densities of impurity atoms thus determined vary considerably from machine to machine and depend strongly on the materials employed in constructing the (plasma containing) vacuum vessel and the current-limiting diaphragm(s). There are nevertheless some features commun to all machines when the same chemical elements are employed and when no divertor is used : the oxygen densities can be of the order of several per cent of the electron density n_e (nearly equal to the proton density), heavy species such as titanium, iron, nickel, molybdenum and tungsten have generally densities of the order of 10^{-4} to 10^{-3} times n_e, see e.g. Refs.[86-88].

Figure 4a,b shows as an example the radial distributions of electron density n_e and impurity species n^z_s in the PLT Tokamak. Figure 4a refers to the quasi-stationary state of the discharge just before injection of a hydrogen gas puff, and Fig. 4b represents the distributions appr. 100 ms after the injection of a hydrogen gas puff which results in a 3.5 times higher electron density. Due to a change in the plasma parameters and in the plasma-wall interaction, the impurity densities have dropped by a factor of 5 to 6. Quite similar results have been obtained on other Tokamak machines.

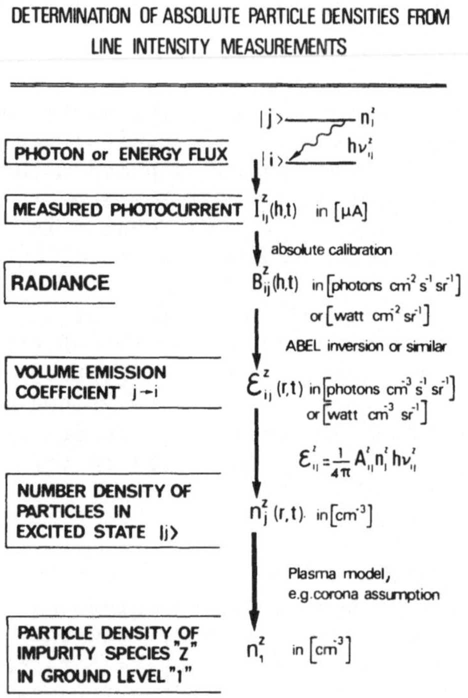

Fig. 3 Determination of absolute particle density n_1^z in the ground level i = 1 from the photon or energy flux emitted in an optically thin (i.e. reabsorption-free) spectral line. The subscript "s" for chemical element has been dropped.

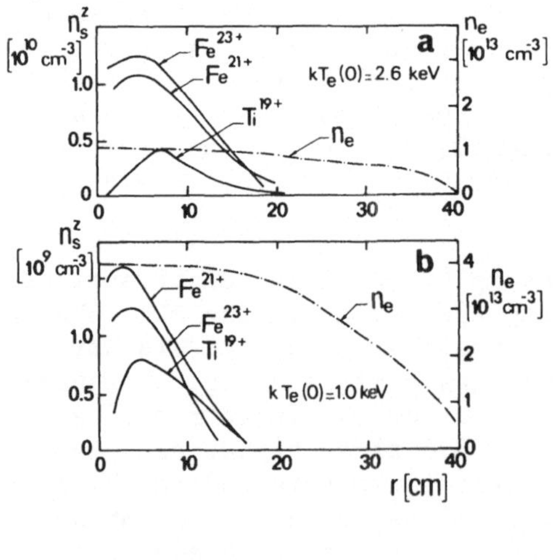

Fig. 4 Radial distributions of electron density n_e and impurity density n_s^z in the PLT Tokamak ; adapted from Ref.[88]. Toroidal magnetic field B_T = 2.5 Tesla ; total plasma current 400 kA ; $T_e(0)$ refers to the axis r = 0.

10. ELECTRON AND ION TEMPERATURES

The temperatures of a plasma depend sensitively on the power input and the energy loss rates. With ohmic heating alone one reaches in Tokamaks axial electron temperatures $T_e(o)$ of the order of 10^7 K, but the ion temperatures $T_{ion}(o)$ stay below this value, except in the peripheral region where $T_{ion} > T_e$, see e.g. Ref.[86,90].

Temperatures higher than 10^7 K are obtained by additional heating, either by the injection of energetic neutral particles or/and by ion-cyclotron-resonance-frequency (ICRF) heating.

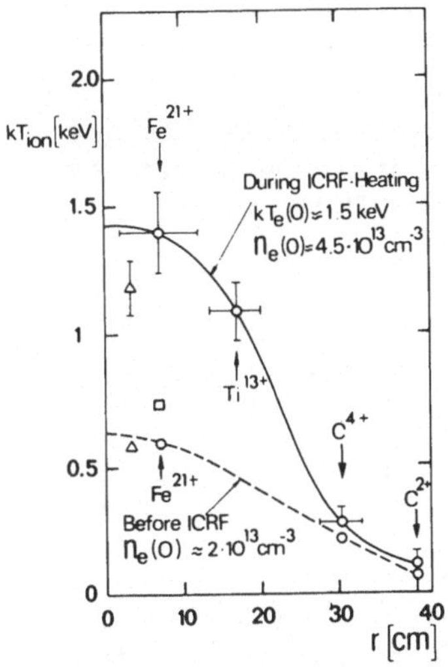

Fig. 5 Ion temperature T_{ion} in the PLT Tokamak before and during ICRF heating, after Ref.[89].

○ from Doppler broadening ;
□ from neutrons ;
Δ from charge exchange neutrals.
$T_e(o)$, $n_e(o)$ are axial electron temperatures and densities.

11. DETERMINATION OF CONFINEMENT TIMES

One distinguishes between particle confinement time τ_p and energy confinement time τ_E, and the latter is subdivided further into the electron and ion energy confinement time (τ_{Ee}, τ_{Eion}). The confinement times are a consequence of the fact that a non-homogeneous plasma transports particles and energy.

Warning : One has introduced the concept of the confinement times τ in order to describe qualitatively and rather phenomenologically transport properties which are still an enigma. The knowledge of τ deviates from the essential physical processes which are responsible for the values of τ. Low (high) values of τ_p and τ_E only mean that particles and energy are transported with high (low) velocity.

The confinement performances of plasma machines are widely characterised by the confinement times. The ideal case would be a machine with $\tau \to \infty$. In present Tokamak and Stellerator machines one has roughly

$$\tau_p \approx (2-3) \cdot \tau_E \tag{2}$$

with τ_E lying between 10 ms and 60 ms for a great number of machines.

11.1 - Particle confinement time τ_p

The determination of τ_p is based on measurements of spectral line intensities. Consider a volume V containing a pure hydrogen plasma. We have particle fluxes $\vec{\Gamma}^{in}$ and $\vec{\Gamma}^{out}$, with

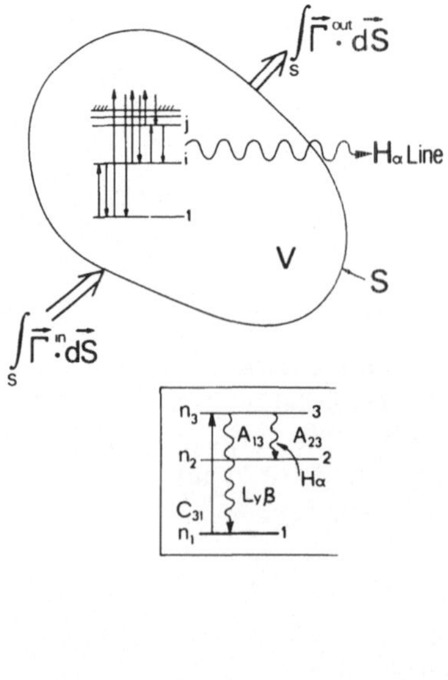

Fig. 6 Particle diffusion fluxes $\vec{\Gamma}^{in}$ and $\vec{\Gamma}^{out}$ cross the surface S enclosing the plasma volume V. Hydrogen atoms are excited and emit radiation, e.g. the Hα line.

Insertion : Corona excitation model. A_{ij} = transition probability, C_{31} = excitation coefficient.

$\vec{\Gamma} = \vec{\Gamma}^{in} + \vec{\Gamma}^{out}$. Charge neutralisation forces electrons and protons to diffuse together with equal flux densities, i.e. $\vec{\Gamma}_e = \vec{\Gamma}_+$. In the stationary state, the constancy of the plasma pressure requires the following relation to be fulfilled for the neutral and charged flux densities, $\vec{\Gamma}_o = -(1/2)(\vec{\Gamma}_+ + \vec{\Gamma}_e) = -\vec{\Gamma}_e$.

The rate equations for the neutral hydrogen atoms and the electrons are

$$\frac{\partial n_o}{\partial t} + \vec{\nabla} \cdot \vec{\Gamma}_o = - n_o n_e S_o + n_+ n_e \alpha, \tag{3a}$$

$$\frac{\partial n_e}{\partial t} + \vec{\nabla} \cdot \vec{\Gamma}_e = + n_o n_e S_o - n_+ n_e \alpha_+ \tag{3b}$$

where S_o and α_+ are the ionization and recombination coefficients, respectively. In the stationary state, $\partial n_o/\partial t = 0$ and $\partial n_e/\partial t = 0$ hold. We now make the following definitions

$$\frac{n_o}{\tau_{po}} = \vec{\nabla} \cdot \vec{\Gamma}_o \quad , \quad \frac{n_e}{\tau_{pe}} = \vec{\nabla} \cdot \vec{\Gamma}_e \tag{4}$$

of the *local* particle confinement time for the neutrals and the electrons (protons). Equations (3a,b) thus yield in the stationary state

$$\frac{n_o}{\tau_{po}} = - n_o n_e S + n_+ n_e \alpha_+ \tag{5a}$$

$$\frac{n_e}{\tau_{pe}} = + n_o n_e S - n_+ n_e \alpha_+ \tag{5b}$$

hence, owing to $\vec{\Gamma}_o = -\vec{\Gamma}_e = -\vec{\Gamma}_+$, it follows that

$$\tau_{po} = \tau_{pe} = \tau_{p+} = \frac{1}{n_o S_o - n_+ \alpha_+} \equiv \tau_p . \tag{6}$$

It is thus possible to determine τ_p when the local values of n_o, $n_+ = n_e$, S_o and α_+ are known. In the general case, S_o and α_+ are given by collisional-radiative models. In the special case of hydrogen, S_o and α_+ as a function of T_e and n_e are available in the literature, see e.g. Refs.[91-93].

The crucial point is the determination of the particle density $n_o \equiv n_{o,1} \equiv n_1$ of the hydrogen atoms in the ground state ($i = 1$). n_o is determined by measuring the emission coefficient ε

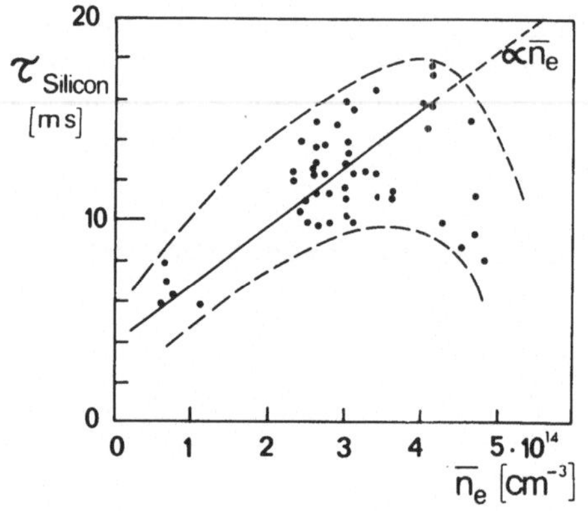

Fig. 7 Confinement time of silicon ions, $\tau_{silicon}$, in the Alcator. A Tokamak, adapted from Ref.[94]. The abscissa gives the line average electron density $\bar{n}_e$.

of a hydrogen spectral line. We have

$$\varepsilon_{23}^{H\alpha} = \frac{1}{4\pi} A_{23}\, n_3\, h\nu_{23} \tag{7}$$

When we assume corona excitation conditions, the particle density n_3 is determined by

$$n_e\, C_{31}\, n_1 = [A_{13} + A_{23}]\, n_3 \tag{8}$$

Equations (7) and (8) yield

$$n_1 \equiv n_o = \frac{4\pi}{n_e\, C_{31}} \; \frac{A_{13} + A_{23}}{A_{23}} \; \frac{\varepsilon_{23}^{H\alpha}}{h\nu_{23}} \tag{9}$$

where $C_{31}(T_e)$ is the coefficient for electronic excitation of level j = 3, from the ground level i = 1.

A measurement of $T_e(r)$, $n_e(r)$ and $\varepsilon_{23}^{H\alpha}(r)$ thus permits the determination of the local values of $\tau_{po} = \tau_{pe}$. Quite similar measurements can be performed on spectral lines originating from impurities. One has only to replace the equations for hydrogen by the corresponding equations for the impurities.

In the transient state the rate equation (3b) yields

$$\tau_{pe} = \frac{1}{n_o S_o - n_+ \alpha - (1/n_e)\, \partial n_e/\partial t} \tag{10}$$

Mean particle confinement times $\bar{\tau}_p$ are defined by the relation

$$\frac{1}{\bar{\tau}_p} \int_V n dV = \int_V \vec{\nabla} \cdot \vec{\Gamma} \, dV \qquad (11)$$

and are determined in connection with the space integrated rate equation for the corresponding species :

$$\frac{\partial}{\partial t} \int_V n dV + \int_V \vec{\nabla} \cdot \vec{\Gamma} = \int_V \left[\frac{\partial n}{\partial t}\right]_{CR} dV \qquad (12)$$

The determination of the confinement time of specially injected impurity species requires the integration of a set of coupled rate equations[95] of type (6) of Ref.[85]. Impurities are injected by gas puffing or by a laser beam evaporation of nonrecycling metals. Results of such measurements are shown in Fig. 7. They were obtained from measurement of the following silicon lines :

Si^{3+} 458 Å ; Si^{10+} 303 Å ; Si^{11+} 499 Å ;
Si^{12+} 6.65 Å ; Si^{13+} 6.18 Å .

A behavior similar to that of Fig. 7 has been found for the Wendelstein W VII-A Stellarator[96]. A possible explanation for the deterioration of the plasma confinement at high densities is the increase of tearing modes[96].

11.2 - Confinement time τ_p and diffusion coefficient $\mathcal{D}$

Using the relation (Fick's law, see Eq. (66) of Ref.[85])

$$\vec{\Gamma} = - \mathcal{D} \, \vec{\nabla} n \qquad (13)$$

we obtain from Eq. (11)

$$\frac{1}{\bar{\tau}_p} \int_V n dV = - \int_V \vec{\nabla} \cdot \mathcal{D} \, \vec{\nabla} n \, dV = - \int_S \mathcal{D} \, (\vec{\nabla} n) \cdot d\vec{S} \qquad (14)$$

For toroidal geometry, it follows from Eq. (14) with the torus length $L = 2\pi R$, the plasma radius a, the mean particle density $\bar{n}$ defined by $\int n dV = \bar{n} L \pi a^2$ and the assumed radial distribution $n(r) = n(o)\,(1 - r^2/a^2)$ that

$$\bar{\tau}_p = \frac{1}{4} \frac{\bar{n}}{n(o)} \frac{a^2}{\bar{\mathcal{D}}} \approx \frac{1}{8} \frac{a^2}{\bar{\mathcal{D}}} \qquad (15)$$

where $\bar{\mathcal{D}}$ represents a space-averaged diffusion coefficient and where we have assumed that $\bar{n} \approx (1/2)\, n(o)$.

11.3 - Energy confinement time

The rate equation for the total energy density $\bar{E}$ is given by (see Eq. (42) of Ref.[85])

$$\frac{\partial \bar{E}}{\partial t} + \vec{v}_0 \cdot \vec{\nabla} \bar{E} + (\bar{E} + p)\vec{\nabla} \cdot \vec{v}_0 + \vec{\nabla} \cdot \vec{Q} + \dot{R} = \dot{\Omega} + \dot{I} + \dot{N} \qquad (16)$$

We now define the *local* energy confinement time τ_E of the plasma with regard to the total energy density $\bar{E}$ by the relation

$$\frac{\bar{E}}{\tau_E} = \vec{\nabla} \cdot \vec{Q} + \dot{R} \qquad (17)$$

where $\vec{Q}$ is the total heat flow vector comprising heat conduction and heat convection. When the internal energy density is neglected we have $E \approx (3/2)\, n_e k T_e + (3/2) n_+ k T_+$. The energy confinement times τ_{Ee} and τ_{Eion} are defined in a similar way. Under all actual conditions, $\tau_E \approx \tau_{Ee}$.

It follows for the stationary state and with the assumption of $\vec{v}_0 = 0$, $\vec{\nabla} \cdot \vec{v}_0 = 0$ that

$$\tau_E = \frac{(3/2)\ (n_e k T_e + N_+ k T_+)}{\dot{\Omega} + \dot{I} + \dot{N}} \qquad (18)$$

The space-averaged value is given by the relation

$$\bar{\tau}_E = \frac{\int_0^a (3/2)\ (n_e k T_e + n_+ k T_+) r dr}{\int_0^a (\dot{\Omega} + \dot{I} + \dot{N}) r dr} \qquad (19)$$

Since local measurements of the ohmic power density $\dot{\Omega}$ do not yield very reliable values one generally applies Eq. (19) with the total ohmic power given by

$$L \int_0^a \dot{\Omega}\ 2\pi r\ dr = JU \qquad (20)$$

(J = plasma current, U = loop voltage).

We can also define an energy confinement time by the relation

$$\frac{\bar{E}}{\tau_E} = \vec{\nabla} \cdot \vec{Q} \qquad (21)$$

In this particular case the radiation losses are excluded, i.e. τ_E refers to the energy losses due to heat conduction and convection only. In the denominator of Eqs. (18) and (19) will then appear the additional term $- \dot{R}$.

The energy confinement time can also be determined from the temperature relaxation after a weak additional heating pulse[87]. The method is analyzed in Ref.[98]. The precision of the measurements depend in a sensitive manner on the experimental conditions.

12. DIELECTRONIC SATELLITE SPECTRA

The dielectronic satellite lines play an important role in the diagnostic of high-temperature, low-density plasmas. They are formed by the radiation-less capture of an electron leading to a doubly excited state which decays by spontaneous de-excitation. The emitted photon yields the dielectronic satellite line accompanying "normal" spectral lines. The physical processes and reactions have in detail been treated in the lectures on the Atomic and Molecular Processes, see Ref.[85].

The evaluation of a sequence of dielectronic satellite and "normal" resonance lines and lines originating from inner-shell excitation permit a determination of both the electron temperature and the relative abundances of impurity species in successive stages of ionization. In connection with the rate equations for the particle densities of the successive stages of ionization, the method should, at least in principle, lead to the divergence of the local flux density.

The practical evaluation of dielectronic satellite spectra requires complicated atomic structure calculations. Further, atomic data such as auto-ionization probabilities and excitation and ionization cross-sections are necessary. And finally, all these data must be incorporated into a plasma model. For all further details, the reader is referred to the review articles Refs.[99,100]. A practical application is described in Ref.[46] for the chromium spectra of Cr XXIII to Cr XIX emitted in the wavelength region $\lambda\lambda$ = 2.175 Å ... 2.245 Å, and in Ref.[34] for iron lines emitted between 1.85 Å and 1.87 Å.

REFERENCES

1. R.H. Huddlestone and S.L. Leonard (Eds.), Plasma Diagnostic Techniques, (Academic Press, New York 1980).
2. W. Lochte-Holtgreven (Ed.), Plasma Diagnostics, (North-Holland, Amsterdam 1968).
3. M. Venugopalan (Ed.), Reaction Under Plasma Conditions, Vol. 1,2, (Wiley-Interscience, New York, 1970).

4. H.R. Griem, Plasma Spectroscopy, (Mc Graw-Hill, New York, 1964).
5. G. Bekefi, Radiation Processes in Plasmas, (Wiley, New York, 1966).
6. W. Neumann, Spectroscopic Methods of Plasma Diagnostics, Article in Progress in Plasmas and Gas Electronics, Vol. 1, edited by R. Rompe and M. Steenbeck, (Akademie Verlag, Berlin 1975).
7. H.R. Griem and R.H. Lovberg (Eds.), Methods of Experimental Physics, Vol. 9-Part A : Plasma Physics, (Academic Press, New York, London, 1970).
8. R.H. Lovberg (Ed.), Methods of Experimental Physics, Vol. 9-Part B : Plasma Physics, (Academic Press, New York, London, 1971).
9. M.H. Key and R.J. Hutcheon, Spectroscopy of Laser-Produced Plasmas, Article in Advances in Atomic and Molecular Physics, Vol. 16, pp. 201-280, edited by D.R. Bates and B. Bederson, (Academic Press, New York, London, 1980).
10. Equipe TFR, Tokamak Plasma Diagnostics (review paper), Nucl. Fusion 18, 647 - 731 (1978).
11. C. De Michelis and M. Mattioli, Soft-X-ray Spectroscopic Diagnostics of Laboratory Plasmas (review paper), Nucl. Fusion 21, 677- 754 (1981).
12. E. Hinnov, Spectroscopy of Highly Ionized Atoms in the Interior of Tokamak Plasma, Article in Atomic and Molecular Processes in Controlled Thermonuclear Fusion edited by M.R.C. Mc Dowell and A.M. Ferendeci, pp. 449-470, (Plenum Press, New York, 1980).
13. S. Suckewer, Spectroscopic Diagnostics of Tokamak Plasmas (a review), Physica Scripta 23, 72-86 (1981).
14. H.W. Drawin, Spectroscopic Measurement of High Temperatures (a review) High Temperatures - High Pressures 2, 359-409 (1970).
15. H.W. Drawin, Validity Conditions for Local Thermodynamic Equilibrium, Article in Progress in Plasmas and Gas Electronics, Vol. 1, pp. 591-660, edited by R. Rompe and M. Steenb ck, (Akademie Verlag, Berlin, 1975).
16. H.W. Drawin, Plasma Impurities and Cooling (a review), Physics Reports 37, 125-163 (1978).
17. R.W.P. Mc Whirter, (Atomic) Data Needs, Priorities and Accuracies for Plasma Spectroscopy (a review), Physics Reports 37, 165-209 (1978).
18. O.H. Nestor and H.N. Olsen, SIAM Review 2, 204 (1960).
19. M.P. Freeman and S. Katz, J. Opt. Soc. Am. 50, 826 (1960).
20. K. Bockasten, J. Opt. Soc. Am. 51,943 (1961).
21. W. Frie, Ann. der Physik (Leipzig) 10, 332 (1963).
22. C.D. Maldonado, A.P. Caron and H.N. Olsen, J. Opt. Soc. Am. 55, 1247 (1965).
23. C.D. Maldonado and H.N. Olsen, J. Opt. Soc. Am. 56, 1305 (1966).

24. H.N. Olsen, C.D. Maldonado and G.D. Duckworth, J. Quant. Spectr. Radiative Transfer 8, 1419 (1968).
25. A. Albers and G. Fussmann, Z. Physik 268, 97 (1974).
26. H. Brunner, J. Computat. Physics 12, 412 (1973).
27. J. Glasser, J. Chapelle and J.C. Boettner, Appl. Optics 17, 3750 (1978).
28. S.J. Young, J. Quant. Spectr. Radiative Transfer 25, 479 (1981).
29. L. Becker and H.W. Drawin, Z. Instrumentenkunde 72, 251 (1964).
30. R. Gorenflo and Y. Kovetz, Numerische Mathematik 8, 392 (1966).
31. V.V. Pikalov, N.G. Preobrazhenskii, E.F. Gippius and V.N. Kolesnikov, Sov. J. Plasma Phys. 4, 518 (1978).
32. H.W. Drawin, Z. Physik 174, 489 (1963).
33. H. Zwicker, Evaluation of Plasma Parameters in Optically Thick Plasmas, Article in Ref. 2.
34. M. Bitter, S. von Goeler, R. Horton, M. Goldman, K.W. Hill, N.R. Sauthoff and W. Stodiek, Phys. Rev. Letts. 42, 304 (1979) ; 43, 129 (1979).
35. T. Sugie, J. Phys. Soc. Jap. 46, 250 (1979).
36. S. Suckewer, J. Cecchi, S. Cohen, R. Fonk and E. Hinnov, Physics Letts. 80A, 259 (1980).
37. D.E. Evans and J. Katzenstein, Laser Light Scattering in Laboratory Plasmas, Repts. Progr. Phys. 32, 207-271 (1969).
38. J. Sheffield, Plasma Scattering of Electromagnetic Radiation (Academic Press, New York, London, 1975).
39. H.W. Drawin and M. Fumelli, Proc. Phys. Soc. 85, 987 (1965), see also Z. Naturforsch. 20a, 859 (1965).
40. S. Suckewer, H.P. Eubank, R.J. Goldston, E. Hinnov and N.R. Sauthoff, Phys. Rev. Letts. 43, 207 (1979).
41. S. Suckewer, H.P. Eubank, R.J. Goldston, J. Mc Enerney, N.R. Sauthoff and H.H. Towner, Nucl. Fusion 21, 1301 (1981).
42. N.J. Wiegart, U. Rebhahn and H.-J. Kunze, to be publ.
43. P. Bogen and E. Hintz, Comments on Plasma Phys. Cont. Fusion 4, 115 (1978).
44. W. Husinsky, R. Bruckmüller and P. Blum, Nucl. Instr. and Methods 70, 287 (1980).
45. B. Schweer, D. Rusbüldt, E. Hintz, J.B. Roberto and W.R. Husinsky, J. Nucl. Materials 93 & 94, 357 (1980).
46. TFR Group, J. Dubau and M. Loulergue, J. Phys. B : At. Mol. Phys. 15, 1007 (1981).
47. H.A. Gabriel and K.J.M. Phillips, Month. Not. Roy. Astr. Soc. 189, 319 (1979).
48. C. Chin-Fatt, Phys. Fluids 17, 1410 (1974).
49. C. Chin-Fatt, Anomalous Bremsstrahlung from a Turbulent Toroidal Plasma, Report PL 82-048, Laboratory for Plasma and Fusion Energy Studies, University of Maryland (USA) 1981.

50. D. Veron, Submillimeter Interferometry of High-Density Plasmas, Article in Vol. 2 of Infrared and Millimeter Waves, pp. 67-135, edited by K.J. Button, (Academic Press, London, New York 1979),appeared also as Report EUR-CEA-FC-980, Fontenay-aux-Roses 1978.
51. G.A. Doschek and U. Feldman, J. Appl. Phys. 47, 3083 (1976).
52. A.V. Vinogradov, I. Yu. Skobelev and E.A. Yukov, Sov. J. Plasma Phys. 3, 389 (1977).
53. V.P. Shevelko, I. Yu. Skobelev and A.V. Vinogradov, Physica Scripta 16, 123 (1977).
54. J.F. Seely, R.H. Dixon and R.C. Elton, Phys. Rev. A 23, 1437 (1981).
55. H.-J. Kunze, Phys. Rev. A 24, 1096 (1981).
56. H.R. Griem, Spectral Line Broadening by Plasmas, (Academic Press, New York, London, 1974).
57. B. Wende (Ed.), Spectral Line Shapes, (W. de Gruyter, Berlin, New York, 1981).
58. S. Skupsky, Phys. Rev. A 21, 1316 (1980).
59. T.L. Pittman, P. Voigt and D. Kelleher, Phys. Rev. Letts. 45, 723 (1980).
60. R. Cauble, J. Quant. Spectr. Radiative Transfer 28, 41 (1982).
61. Truong-Bach and H.W. Drawin, J. Quant. Spectr. Radiative Transfer 27, 627 (1982).
62. H.W. Drawin and Truong-Bach, J. Phys. B : At. Mol. Phys. (in press).
63. K.-H. Steuer and W.-G. Wrobel, Vorschlag zur Neutralteilchendichtemessung in Tokamaks durch Resonanzfluoreszenz mit einem Lyman-Alpha-Exzimerlaser, Report IPP 1/168, IPP III/48, Max-Planck- Institut für Plasmaphysik, Garching (1978).
64. G. Reinhold, J. Hackmann and J. Uhlenbusch, Investigation of Plasma-Wall Interaction by H_α Resonant Scattering in the Tokamak UNITOR, Report 81-01-87, Sonderforschungsbereich Plasmaphysik Bochum/Jülich, 1981.
65. D.W. Koopman, Th. J. Mc Ilrath and V.P. Myerscough, J. Quant. Spectr. Radiative Transfer 19, 555 (1978).
66. V.D. Kotsubanov, A.N. Letucchii and O.S. Pavlichenko, Sov. J. Plasma Phys. 2, 556 (1976).
67. C.H. Muller, III, and K.H. Burrell, Phys. Rev. Letts. 47, 330 (1981).
68. R.C. Isler, L.E. Murray, S. Kasai, J.L. Dunlap, S.C. Bates, P.H. Edmonds, E.A. Lazarus, C.H. Ma and M. Murakami, Phys. Rev. A 24, 2701 (1981).
69. K. Brau, S. von Goeler, M. Bitter, R.D. Cowan, D. Eames, K. Hill, N. Sauthoff, E. Silver and W. Stodiek, Phys. Rev. A 22, 2769 (1980).
70. K. Mc Cormick and J. Olivain, Revue de Phys. Appliquée 13, 85 (1978).
71. J.P. Christiansen and J.B. Taylor, Nucl. Fusion 22, 111 (1982).

72. M.J. Forrest, P.G. Carolan and N.J. Peacock, Measurement of Magnetic Fields in a Tokamak Using Laser Light Scattering, Report CLM-P-499, Culham Laboratory, Abingdon 1977. Nature 271, 718 (1978).
73. S.E. Segrè, Plasma Physics 20, 295 (1978).
74. F. de Marco and S.E. Segrè, Plasma Physics 14, 245 (1972).
75. W. Kunz and G. Dodel, Plasma Physics 20, 171 (1978).
76. C. Breton, C. De Michelis, M. Finkenthal and M. Mattioli, J. Phys. E : Sci. Instrum. 12, 894 (1979).
77. E. Hinnov and F.W. Hofmann, J. Opt. Soc. Am. 53, 1259 (1963).
78. F.E. Irons and N.J. Peacock, J. Phys. E : Sci. Instrum. 6, 857 (1973).
79. P. Platz, J. Ramette, E. Belin, C. Bonnelle and A. Grabriel, J. Phys. E : Sci. Instrum. 14, 448 (1981).
80. M. Bitter, S. von Goeler, K.W. Hill, R. Horton, D. Johnson, W. Roney, N. Sauthoff, E. Silver and W. Stodiek, Phys. Rev. Letts. 47, 921 (1981).
81. M. Bitter, private communication.
82. C. Breton, C. De Michelis, M. Finkenthal and M. Mattioli, J. Phys. E : Sci. Instrum. 13, 216 (1980).
83. S. Suckewer, H.P. Eubank, R.J. Goldston, E. Hinnov and N.R. Sauthoff, Phys. Rev. Letts. 43, 207 (1979).
84. V.A. Vershkov and S.V. Mirnov, Sov. J. Plasma Phys. 3, 109 (1977).
85. H.W. Drawin, this volume.
86. TFR-Group, Plasma Phys. 20, 207 (1978).
87. S. Suckewer and E. Hinnov, Phys. Rev. A 20, 578 (1979).
88. E. Hinnov, J. Hosea, H. Husuan, F. Jobes, E. Meservey, G. Schmidt and S. Suckewer, Nucl. Fusion 22, 325 (1982).
89. S. Suckewer, E. Hinnov, D. Hwang, J. Schiveli, G.L. Schmidt, K. Bol, N. Bretz, P.L. Colestock, H. Dimock, H.P. Eubank, R.J. Goldston, R.J. Hawryluk, J.C. Hosea, H. Hsuan, D.W. Johnson, E. Meservey and D. Mc Neill, Nucl. Fusion 21, 981 (1981).
90. R.J. Hawryluk, S. Suckewer and S.P. Hirshman, Nucl. Fusion 19, 607 (1979).
91. T. Fujimoto, I. Sigiyama and K. Fukuda, Memoirs Faculty of Engineering, Kyoto Univ. (Japan) 34, part 2, 249 (1972).
92. L.C. Johnson and E. Hinnov, J. Quant. Spectr. Radiat. Transfer 13, 333 (1973).
93. H.W. Drawin and F. Emard, Physica 85 C, 333 (1977).
94. E.S. Marmar, J.E. Rice and S.L. Allen, Phys. Rev. Letts. 45, 2025 (1980).
95. P. Gratreau, Plasma Physics 22, 949 (1980).
96. Wendelstein W VII-A team, presented by H. Ringler, in Proceedings 9th Europ. Conf. Contr. Fusion and Plasma Phys., paper A.2.4, Oxford 17-21 Sept. 1979, edited by UKAEA, Culham Laboratories, Abingdon, U.K.
97. A.S. Fisher and G. Bekefi, Phys. Letts. 63 A, 281 (1977).

98. S.W. Simpton, R.M.P. Drozak and R.M.O. Galvao, Phys. Letts. 78 A, 68 (1980).
99. J. Dubau and S. Volonté, Di-electronic Recombination and its Application in Astronomy, Repts. Progr. Phys. 43, 199-251 (1980).
100. J. Dubau and M. Loulergue, Physica Scripta 23, 136-142 (1981).

PARTICLE DIAGNOSTICS FOR MAGNETIC FUSION EXPERIMENTS

Douglass E. Post

Plasma Physics Laboratory, Princeton University
Princeton, New Jersey 08544

I. INTRODUCTION

The statement has been made [1] that the most impressive progress in the past twenty years in research in magnetic fusion has been in the area of diagnostics. Compared to the situation in fusion work in the 1950's and early 1960's, measured plasma parameters for present fusion experiments can be quoted with confidence. Often each parameter, such as temperature or density, is measured by a variety of techniques. Spatially and temporally resolved measurements are quite common (Fig. 1). While the progress in our theoretical understanding of fusion experiments has been impressive, the major thrust of fusion research continues to be the construction of larger experiments, and the measurement of the properties of the plasmas produced in the experiments. Diagnostics play a crucial role in this effort.

The chapter by Dr. J. Hogan has outlined the general nature of the plasmas whose properties must be measured. A typical density and temperature profile for the PLT tokamak with ohmic heating is given in Fig. 1. With auxiliary heating in this size plasma, temperatures as high as 7 keV have been obtained [3,4]. Densities as high as 10^{15} particles/cm^3 have been obtained in the Alcator A tokamak [3,4]. Mirror experiments have achieved electron temperatures in the 100 eV range, ion temperatures in the 10 keV range, and densities in the 10^{13} particles/cm^3 range [5]. Other magnetic fusion experiments have similar characteristics, except that most have lower temperatures and somewhat worse confinement than tokamaks. A good review of tokamak plasmas has recently been done by J. Rawls [4] and H. Furth [3].

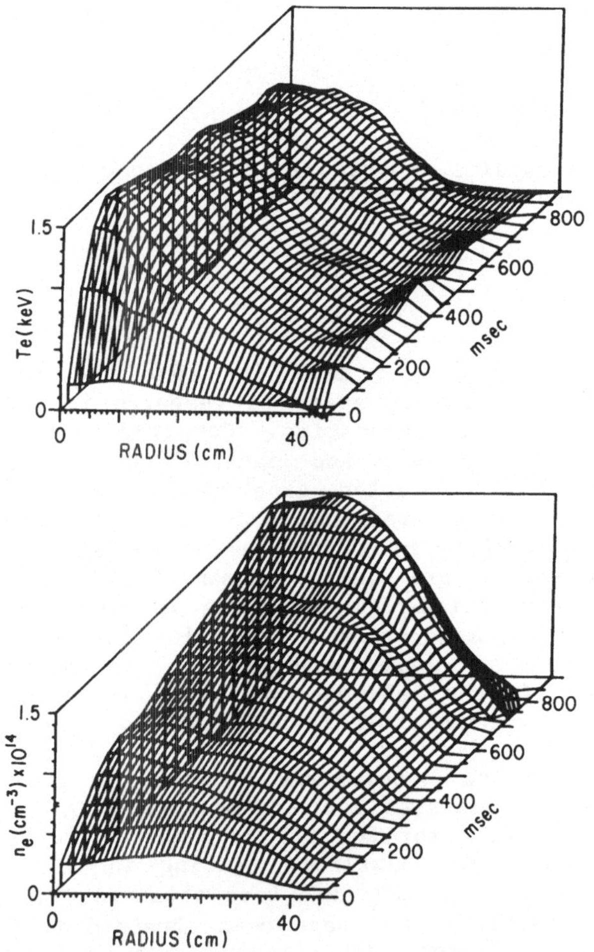

Figure 1. Typical Ohmic Heating electron temperature and density profile in a helium PLT discharge (taken from [2]).

The main parameters of interest for measurement and the commonly used measurement techniques are listed in Table I drawn from several reviews [6,7,8]. Several things apparent from Table I are the large number of quantities to be measured and the wide diversity of techniques employed. Most of these diagnostics represent difficult measurements, each usually requiring several physicists and engineers, and each a challenging experiment in itself. In fact, most of the physicists employed in the experimental portion of fusion research work almost entirely on diagnostics; designing, developing, building, and operating them.

Diagnostic data in the form of time dependent density, temperature, and radiation profiles are being used to unfold the detailed confinement properties of fusion experiments. The new level of understanding possible with these improvements will hopefully make extrapolations to future machines less uncertain. The data will certainly serve as a stimulus for plasma theorists.

I will present a summary of the subset of diagnostics that rely primarily on the use of particles and leave the diagnostics that rely primarily on photon spectroscopy to Dr. Drawin. An excellent review of the general field of tokamak diagnostics has been given recently by the TFR group [6]. Particle diagnostics have been discussed by Goldston [8] and Kislyakov and Krupnik [7]. I will attempt to point out where atomic and molecular data play a role in these diagnostics. Also, I will not discuss neutron and related diagnostics as they do not, in general, involve atomic and molecular physics.

Particle diagnostics can be divided into active and passive systems. Passive systems are those in which emitted particles are measured without the diagnostician doing anything to the plasma. Active systems use a beam to produce the emitted particles.

II. PASSIVE CHARGE-EXCHANGE ION TEMPERATURE MEASUREMENTS

The most common passive systems are neutral particle spectrum measurements and fusion product measurements. Neutral particles are useful because they can freely cross the confining magnetic fields. An outflux of energetic neutral atoms from the plasma is produced by the charge-exchange of energetic confined ions and neutral hydrogen atoms. The energy spectrum of the "charge-exchange" neutrals can be used to calculate the ion temperature of the region where the neutrals were produced. The measurement of neutrons produced from thermonuclear fusion reactions can also be used to measure the ion temperature as well.

The particle balance in a tokamak is illustrated in Fig. 2. Plasma flows out across the field lines until it reaches the plasma edge. There it flows along the field lines, strikes the limiter or wall, and recombines. The neutral atoms formed in this way then

TABLE I

Summary of Plasma Properties and Diagnostic Techniques

PROPERTY	DIAGNOSTIC TECHNIQUE
Electric and Magnetic Fields	- Magnetic Flux Loops - Voltage loops - Thomson scattering - Zeeman splitting - Plasma conductivity - High energy neutral beams - Xray Tomography
Electron Density	- Microwave and far infrared interferometry - Thomson scattering - Beam attenuation
Ion Density and Composition	- H_α and H_β spectroscopy - Neutral particle spectroscopy - Vacuum ultraviolet spectroscopy - Visible spectroscopy - Hard and soft Xray spectroscopy - Angular scattering of fast ions - Radiometry - Collective laser scattering - Charge-exchange recombination - Visible bremsstrahlung

PROPERTY	DIAGNOSTIC TECHNIQUE
Electron temperature and Velocity Distribution	- Thomson scattering - Electron cyclotron emission - Soft and hard Xray spectroscopy - Beam attenuation
Ion Temperature	- Fast neutral analysis - Impurity line broadening - Thermonuclear neutrons - Far infrared laser scattering
Instabilities and Turbulence	- Magnetic flux loops - Soft Xray fluctuations - Microwave and infrared scattering - Heavy ion beam - Electron cyclotron emission
Heating Diagnostics	- Neutral particle analysis - Charged fusion product analysis - Neutrons

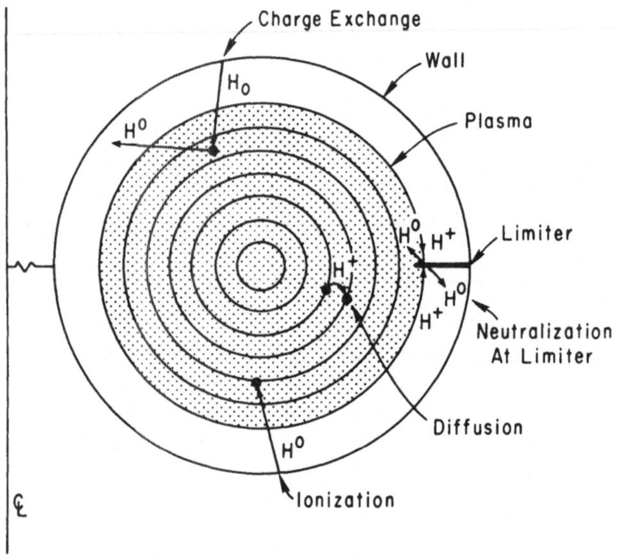

Figure 2. Typical particle balance in a tokamak [9].

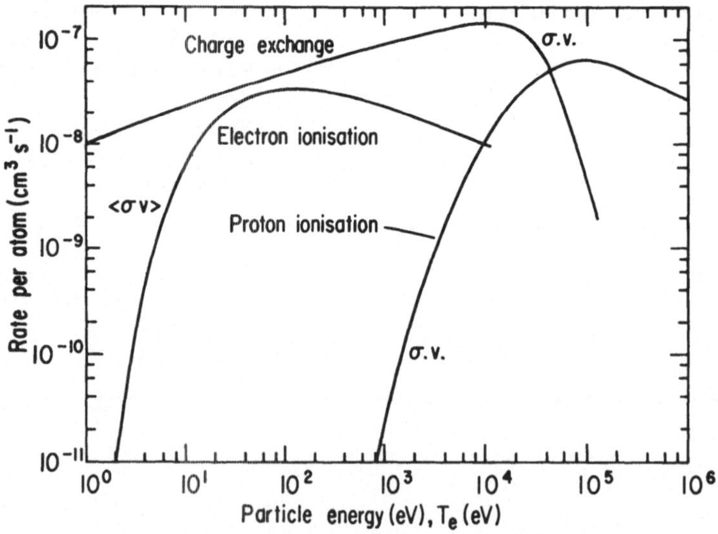

Figure 3. Ionization and charge exchange rates for atomic hydrogen [10].

enter the plasma on straight line trajectories since the magnetic field has no effect on them. In the plasma, the atoms collide with the plasma electrons and ions. The atoms can either be ionized by electron and ion impact, or charge exchange with the plasma ions. The rates for charge exchange are typically a factor of two or more higher than the ionization rates (Fig. 3) so an average neutral atom undergoes several charge-exchange collisions before it is ionized. The bound electron thus changes nuclei several times before it is ionized. The atom's energy, and thus its ionization mean free path due to electron impact ionization, is increased when a "cold" atom charge exchanges with a "hot" ion. This effect of having many generations of charge-exchange neutrals can lead to much higher neutral densities at the plasma center than would be expected from the ionization of the recycling cold gas [11].

At a given neutral energy, the outgoing flux, seen by a detector, is proportional to

$$\int_0^{\ell} (\sigma_{cx} v_{rel}) \left\{\exp[-(E_i/T_i(r))]/T_i^{3/2}\right\} \exp[-(s/\lambda)]\ n_0(s)\ \sqrt{E}\ \Delta\Omega\Delta E\ ds,$$

where $\sigma_{cx} v_{rel}$ is the reaction rate, n_0 the local neutral density, the factor $\exp[-(E_i/T_i)]/T_i^{3/2}\ \sqrt{E}\ \Delta E$ is the portion of the ion Maxwellian that can be detected, $\exp[-(s/\lambda)]$ is the attenuation of the outgoing neutrals in the plasma, n_0 is the local neutral density, and $\Delta\Omega$ is the angular acceptance of the detector. The integration over s for attenuation is taken along the line of sight through the plasma. The $\exp[-(E_i/T_i)]$ factor means that, at high energies ($E_i \sim 5$ or $10 \times T_i$), the neutral flux is dominated by neutrals born with maximum temperature along the chord. At lower energies, the spectrum has contributions from slower neutrals along the whole chord. Thus, if one defines $f = (F(E)/\sigma_{cx} E)$, f should be proportional to $\exp[-(E/T_i)]$ at large energies (providing that n_0 is reasonably large) (Fig. 4).

The energy spectrum of the escaping charge-exchange neutrals is measured by first ionizing the atoms by collisions with a gas and then using magnetic and electrostatic fields to separate the ions of different energies (Fig. 5). Systems are now being built for TFTR which are mass selective and use position sensitive detectors which have the advantage of measuring the energy of essentially all the particles that are stripped in the detector [14]. The stripping cross section of hydrogen in gases is small below 500 eV, making low energy measurements difficult. Several techniques have been used at these energies. One is to use a metal vapor and produce negative ions by charge exchange $H^0 + A^0 \rightarrow H^-$ [15]. Another method using time of flight discrimination with rotating chopping disks has been used on PLT [16]. It has been difficult, even with these techniques, to go much below 100 eV.

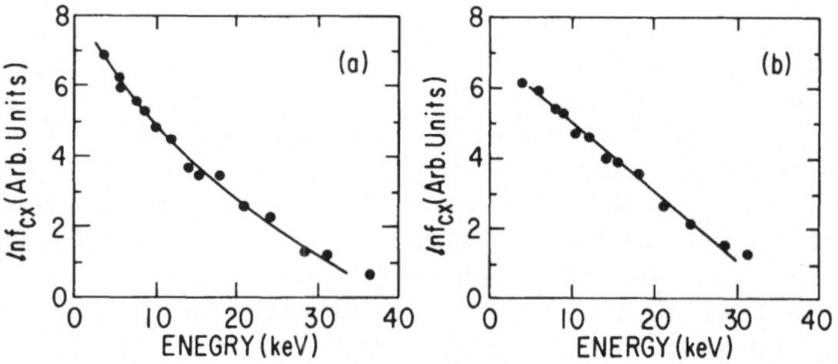

Figure 4. a) "Passive charge exchange spectrum measured during neutral beam injection on PLT. $T_i(o) \approx 5$ keV. The curved spectrum reflects the ion temperature profile" (taken from [8,12]). b) "Active charge exchange spectrum from the same set of discharges" (taken from [8,12]).

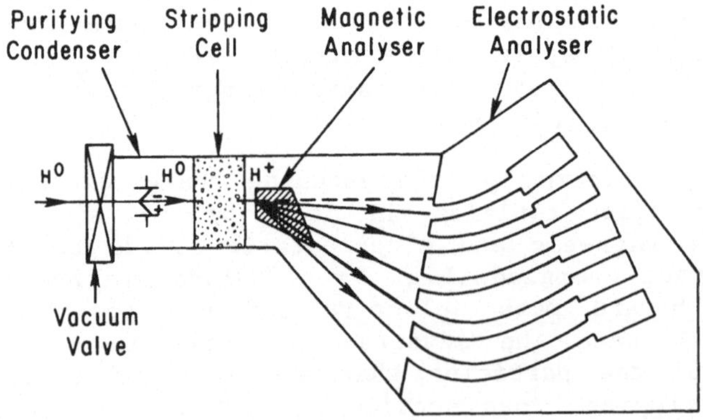

Figure 5. Multichannel analyzer with mass and energy resolution for the measurement of charge exchange spectra [13].

Stripping at higher energies can be accomplished with thin foils (see [8]) which greatly simplifies the stripping system by avoiding pumping problems.

Neutral particle analysis is also used to study the fast ion spectrum of both neutral beam heated and radio-frequency heated plasmas [8]. It is the close agreement between the measured slowing down fast ion distribution of neutral beam ions and the calculations of the spectra based on classical theory that give us the confidence to design large power neutral injection systems for the next generation of tokamaks, such as TFTR, JET, and JT-60. The spectra of hydrogen ions heated by ion cyclotron waves have been measured on PLT [17], and are used to guide the theoretical work and to plan for future heating systems.

III. HYDROGEN BEAMS FOR DENSITY, ION TEMPERATURE, q AND ZEFF MEASUREMENTS

For very large dense plasmas, the central neutral density is too low to produce a measurable number of high energy neutral particles through charge exchange. The neutral particles that come from the edge are ionized before they penetrate. Some neutrals can be formed from radiative recombination [18], but for very large plasmas almost all of the neutrals from the center are ionized before they leave the plasma. The local neutral density can be enhanced by the injection of an energetic neutral hydrogen beam if the plasma is not too dense, making the system an "active" charge-exchange technique. The use of such beams often provides a more accurate measurement than passive systems of the central ion temperature (Fig. 4b).

Injection with neutral beams opens up a whole new range of diagnostics at the cost of doubling the system complexity. The requirements are that reasonable current and energy beams can be made, that the beams can penetrate into the plasma, that they can interact with some plasma constituent in a useful way, and that the reaction products can escape to be detected. Table II [7] lists the penetration limits ($\bar{n}a$) of diagnostic atomic particle beams.

As we have seen, diagnostic hydrogen neutral beams have been used to extend the density range for which charge-exchange analyzer systems can be used to measure the plasma ion temperature or fast ion distribution. Well collimated neutral hydrogen beams have also been used to measure the plasma density, current profile, Zeff, and impurity density.

The plasma density profile in relatively small scale experiments has been measured by the attenuation of energetic neutral beams. By using different species beams, either the proton or the electron density, can be measured. The formula for the beam intensity is [7]

TABLE II.

Maximum Integral Plasma Density $\bar{n}L(cm^{-2})$ at which Diagnostics with Atomic Particle Beams is Still Possible [7]

	Energy range of the probing particles	
PROBING PARTICLES	1-10 keV	50-100 keV
H^o	$(2-5) \times 10^{15}$	$(1.5-2.5) \times 10^{16}$
He^o	$(0.5-1.5) \times 10^{16}$	$(1.5-2) \times 10^{16}$
Ar^o	$(2-6) \times 10^{14}$	$(1-1.5) \times 10^{15}$
Li^o	$(0.5-1) \times 10^{15}$	$(1-2.5) \times 10^{15}$
K^o	$(0.5-2) \times 10^{14}$	$(3-4) \times 10^{14}$
Cs^+	$(3-4) \times 10^{14}$	$(2-4) \times 10^{15}$*
Tl^+	$(1-2) \times 10^{14}$	$(1-2) \times 10^{15}$*

*For energies up to 200 keV.

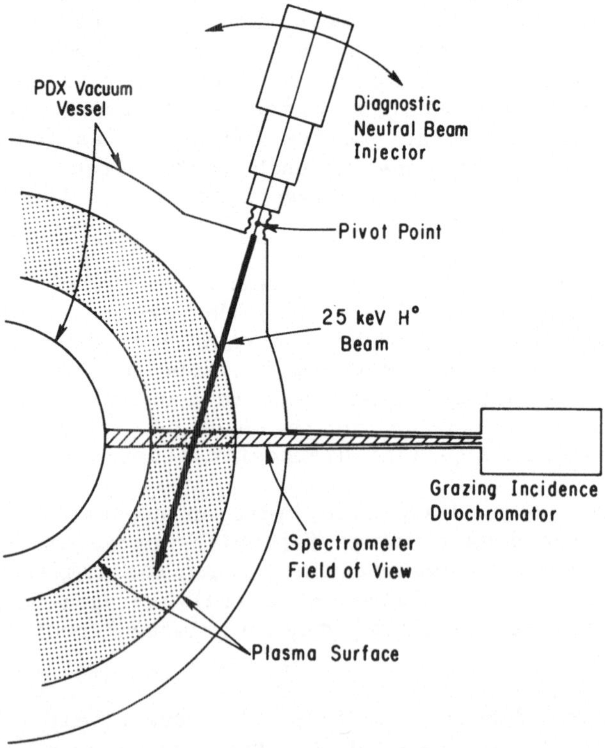

Figure 6. "Schematic of the Experimental Apparatus for Low Z Impurity Diagnostic on PDX" (taken from [25]). 82X0247

$$I_L = I_o \exp[-(1/v_i) \int \int n\sigma f(v) \, |\vec{v}_i - \vec{v}| \, d^3v \, d\ell].$$

The main interaction of a hydrogen beam ($\lesssim$ 20 keV) with a proton-electron plasma is charge exchange and proton ionization, so the $I_L = I_o \exp[-(\bar{n}_p \sigma_p L)]$, and $\bar{n}_p = (1/\sigma_p L) \ln(I_o/I_L)$, where σ_p is the total cross section for proton ionization and charge exchange, I is the unattenuated beam intensity, L is the chord length of the beam in the plasma, $\bar{n}_p$ is the chordal averaged density and I_L is the measured attenuated beam intensity.

If an inert gas beam (He, Ar) is used, charge exchange is not resonant, so electron ionization is the dominant interaction, and the electron density and temperature can be measured

$$\bar{n}_e \bar{\sigma}(T_e) = \frac{\ln(I/I_L)}{L} .$$

This is a limited technique for measuring T_e, as $\sigma(T_e)$ is only a strong function of T_e below about 100 eV.

Beam attenuation techniques have largely been replaced by microwave and far-infrared interferometry for density measurements. The interferometry techniques offer the advantages of convenience, reliability, simplicity, and, at least, for far-infrared, the ability to be used at high density. However, since the techniques involve interferometry (counting fringes), great care must be used in building the density up whereas beam attentuation is independent of the previous plasma history. Interferometry techniques are also more subject to noise problems than beam attenuation.

Well collimated energetic neutral hydrogen beams have been used to measure the plasma current profile and the plasma Zeff [19,20]. The current profile measurement is done by using a tangentially injected beam to deposit particles on a particular flux surface. The center of the orbits of these particles are shifted by an amount

$$\Delta = q \frac{v}{\Omega_i} = qc \frac{mv}{eB} ,$$

where q is the safety factor, v is the particle velocity and Ω_i is the ion cyclotron frequency. One measures this shift as a function of ion momentum and extrapolates to zero momentum to obtain the flux surface. The safety factor q is a function of the plasma current profile. These measurements gave q's generally higher than the q's determined by the plasma conductivity based on Thomson scattering data.

The Zeff in ATC [20] was measured by matching the measured pitch angle scattering rate of the injected beam ions, and the value of Zeff required to reproduce that scattering theoretically. While the measurement is dependent on the close correspondence of the Fokker-Planck theory and the experimental beam slowing down behavior, it gave Zeff's reasonably consistant with the global Zeff measured by other techniques. The beam technique has the advantage of being a local measurement. The technique is being pursued on PDX and TFTR. One of its main drawbacks is the necessity of a movable, well collimated beam, and a movable charge-exchange analysis system which can operate from the high energies of the injected beam ions to thermal energies.

IV. IMPURITY DIAGNOSTICS USING CHARGE-EXCHANGE RECOMBINATION

One of the most promising uses of high energy hydrogen beams is the measurement of the local fully-ionized impurity density. In this technique, one injects a neutral beam into the plasma. The neutrals in the beam charge exchange with the impurity ions

$$C^{+6} + H^{o} \rightarrow H^{+} + \left(C^{+5}\right)^{*}$$

$$\left(C^{+5}\right)^{*} \rightarrow C^{+5} + h\nu .$$

The charge exchange is generally into an excited state of the one electron hydrogen-like ion (generally, $n \sim Z^{3/4}$, where n is the principal quantum number of the excited state [21]). One can measure the radiative decay cascade and from a knowledge of the beam attenuation and impurity charge-exchange cross section compute the impurity density. The techniques were originally applied both to the ORMAK tokamak and the ISX tokamak [22] where the doping beam was the neutral heating beam, and to T-4 and T-10 tokamaks where a well collimated doping beam was used to measure the central concentration of C^{+6} in the T-4 tokamak [23] and the radial distribution of O^{+8} in the T-10 tokamak [24].

This technique has been further developed on the PDX tokamak with the use of a well collimated 4 to 11 kilowatt 25 keV neutral doping beam and a grazing incidence vacuum ultraviolet spectrometer [25]. The experimental layout is sketched in Fig. 6. The beam is modulated and the spectrometer signal is cross-correlated with the beam. A typical signal is shown in Fig. 7. The types of impurity profiles that can be measured are shown in Fig. 8. The measured profile is for a 30 cm radius plasma in PDX with a peak electron temperature of 850 eV and a peak density of $3 \times 10^{13} cm^{-3}$. The

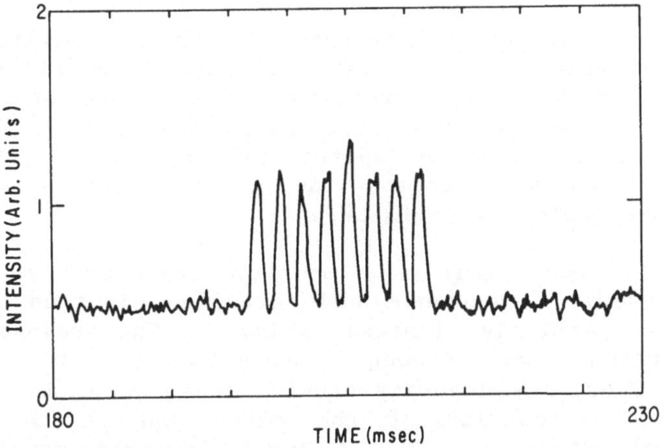

Figure 7. "Excitation of the OVIII 102Å (n=3 to n=2) emission by charge exchange with the pulsed diagnostic neutral beam," (taken from [25]).

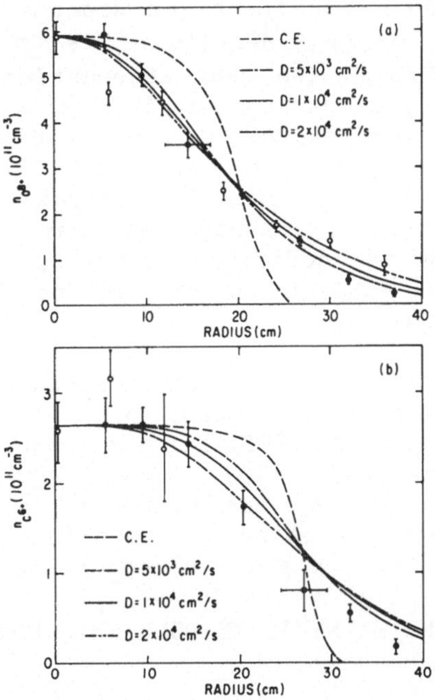

Figure 8. "Radial profiles of fully ionized oxygen and carbon during the steady-state phase of the discharge. Limiter radius is at 30 cm. Closed circles: $R > R_{p\ell}$ = 147 cm. The solid lines are radial profiles of C^{6+} and O^{8+} calculated from an impurity transport code. C. E. = distribution expected from coronal equilibrium assuming a constant impurity density. D = constant impurity diffusion coefficient." The limiter is at 30 cm (taken from [25]).

carbon and oxygen profiles show several striking features. The first is that they are not consistent with coronal equilibrium, but show transport features that require diffusion coefficients of ~ 10,000 cm^2/sec to match the profiles. This value of D is much larger than neoclassical transport would indicate. The second feature is that there is an appreciable density of the fully-stripped ions at radii beyond the limiter radius where the temperatures are quite low (~ 10 eV).

Ordinarily, the study of impurity transport in tokamaks involves the study of resonance and forbidden electron excitation lines of the partially ionized atoms. The resonance line excitation rates are strong functions of the electron temperature. Thus, the impurity density measurement is subject to errors in the calculation of the rates and in the electron temperature measurements. Ten to twenty percent errors in the electron temperature can change some rates by an order of magnitude or more. In addition, fully-stripped ions cannot be studied at all by electron excitation since there are no bound electrons. The charge-exchange recombination technique depends only on the rate $R = n_0 n_1 \sigma_{cx} v_{beam}$. The neutral density n_0 is only weakly dependent on the temperature through the beam attenuation calculation. The electron temperature does not enter into the rest of the rate. Analysis of the data does depend on the calculated cross sections for populating the excited states $(\sigma(n,\ell))$. Thus, there is great interest in improvements to existing calculations [21], and in experimental benchmarks for these calculations. The charge-exchange technique has the additional advantage that it is a local technique and is not subject to the errors introduced by Abel inversion of chordal data.

The major drawback of the charge-exchange technique is the need for a complicated beam system, but such beam systems are already under development for TFTR for ion temperature measurements by charge-exchange neutrals [26]. Very little is understood quantitatively about impurity and hydrogen transport at present, and the charge-exchange recombination technique offers the promise of a new and powerful tool to study plasma transport.

V. PLASMA ELECTRIC AND MAGNETIC MEASUREMENTS USING BEAMS HEAVIER THAN HYDROGEN

Beams of elements heavier than hydrogen such as Rb^+, $T\ell^+$, and Cs^+ have been used to measure the plasma electric potential distribution in tokamaks, mirrors, and bumpy toruses [27,28]. In these measurements, the singly and doubly ionized ion beams must have energies large enough so that their Larmor orbits are of the order of the plasma size. The energy required is thus (Fig. 9)

$$E(eV) = 48\ (B\rho)^2\ \frac{Z^2}{A}\ ,$$

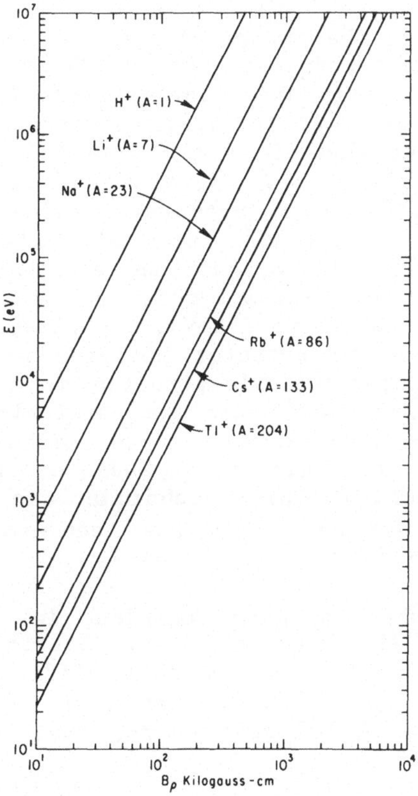

Figure 9. Minimum energy required for a beam of Z = 1 so that the particles in hte beam have a gyro-orbit equal to ρ. The dependent coordinate is the product of the magnetic field in kilogauss and system size ρ in centimeters (after [7]).

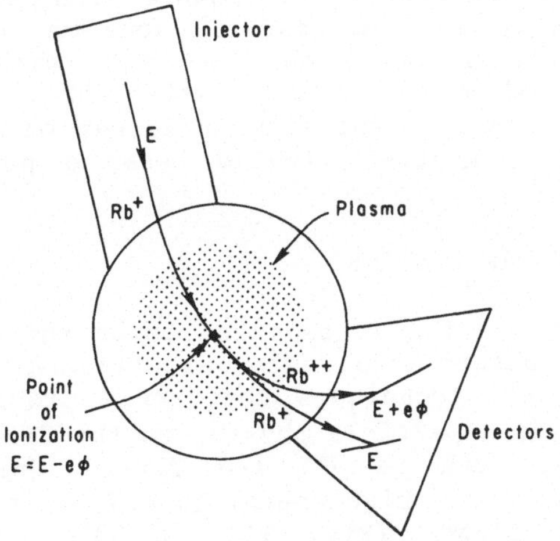

Figure 10. Schematic illustration of a Rubidium ion beam used to measure the electric potential in a plasma (after [27]). The beam is ionized again at x where it has energy E-eφ.

where B is the magnetic field in kilogauss, ρ is the required gyroradius in cm, Z is the charge of the ion, and A is the atomic number of the ion. In one application, a 100 keV Rb^+ beam was used [27] (Fig. 10). The beam was electrostatically positioned and injected into the plasma. At the point of interest, the energy of an ion would be $E - e\phi$ where E is the beam injection energy and ϕ is the local plasma potential. Ions which are ionized again by electron impact to Rb^{++} have a gyro-orbit radius which is one-half of the radius of Rb^+. The Rb^{++} ions will also pick up $2e\phi$ of energy as they leave the plasma, and thus will have energies of $E - e\phi + 2e\phi = E + e\phi$ at the plasma edge. By measuring this energy difference (typically 100 eV to 1 keV), the plasma potential can be measured. By varying the detector position and acceptance, and beam position, the potential distribution in the plasma can be mapped out [27]. Such measurements are particularly important for mirror experiments, such as tandem mirrors and bumpy toruses, where much of the plasma confinement depends on the electrostatic potential distribution. This technique can have good time resolution (10^{-5} - 10^{-6} sec), and has been used to measure the potential fluctuations in various plasmas.

One serious drawback of this technique is that it requires a precise knowledge of the magnetic field distribution so that the orbits can be mapped out. It is this difficulty and the high fields and densities in tokamaks that have limited its use in present day tokamaks. A 100 keV neutral thallium beam was tried unsuccessfully on PLT.

As pointed out by Dr. Hogan in a previous lecture, many of the stability properties of a tokamak depend on the current profile. One technique employed to do this measurement was to inject a neutral lithium beam into a plasma and measure the Zeeman splitting of the collisionally excited Li^0 resonance line (λ = 6708Å , 2s → 2p transition). This was successfully done on Pulsator [29,30]. The technique is being tried on the TEXT tokamak using laser fluorescence to produce the light rather than electron impact excitations [31]. Its applicability to larger machines is limited by the ability of a neutral lithium beam to penetrate a large plasma.

VI. ALPHA PARTICLE DIAGNOSTICS

Alpha particle heating is expected to be the dominant heating mechanism on all reactor-sized fusion experiments. Thus, there is interest in diagnostic techniques that could provide quantitative information about alpha particle physics in the first generation of fusion experiments, such as JET and TFTR. One such proposed diagnostic is the use of a high energy (6 MeV) neutral lithium beam injected into the plasma (Fig. 11) [32,33]. The fast alpha particles would be neutralized and escape the plasma by double charge exchange with the neutral lithium atoms in the beams (Li^0 +

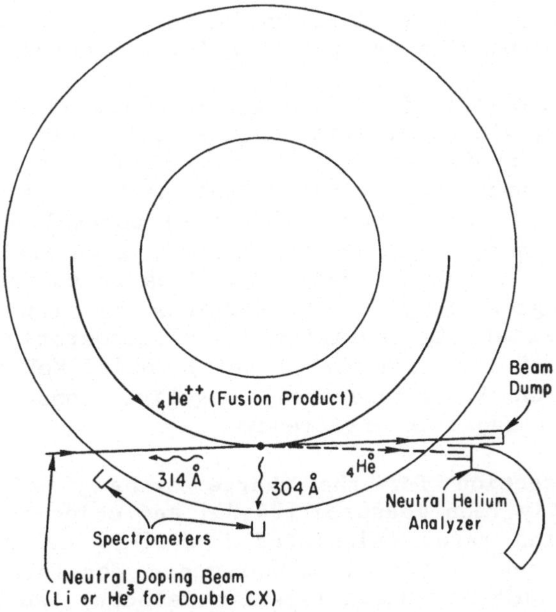

Figure 11. Schematic of alpha particle diagnostic (taken from [32]).

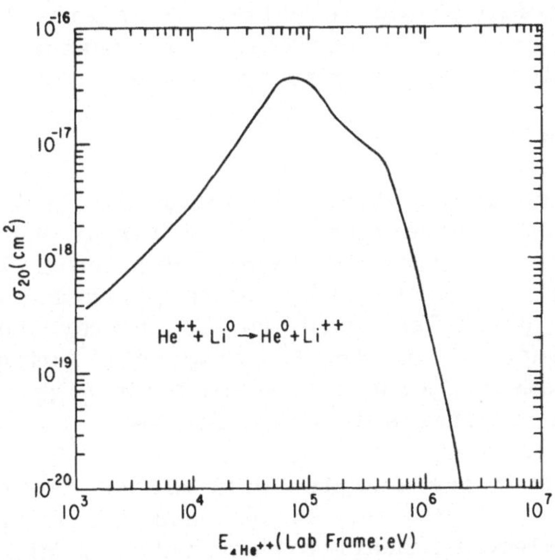

Figure 12. Double charge-exchange cross section for $He^{++} + Li^0 \rightarrow He^0 + Li^{++}$ [39, 35]. The curve has been extrapolated below 28 keV and above 800 keV (taken from [36]). 81X1327

$He^{++} \rightarrow He^{0}$). The fast neutral helium atoms would be ionized in the detector and analyzed with energy sensitive detectrons.

The major features of this diagnostic are the production of the fast Li^{0} beam, the penetration of the fast neutral lithium beam into the plasma, the double charge exchange of neutral lithium and alpha particles, and the escape of the neutral helium from the plasma. Almost all low energy (< 200 keV) neutral beams are made by forming positive ions in an ĩon source, accelerating the ions electrostatically, and neutralizing the ions by charge exchange in a gas. As charge-exchange cross sections for the reactions of interest are small at high energies and electrostatic acceleration difficult above 500 keV, the most feasible multi-MeV neutral beams would be based on RF accelerated, negative ions which can be neutralized with reasonable efficiency.

The cross section for the charge exchange of lithium and alphas has recently been measured [34,35] and calculated [36] (Fig. 12). The reaction rates calculated from this cross section are large enough to make detailed measurements possible. The cross section is large only for small relative velocity, so that only the alphas almost parallel to the beam will charge exchange and be available to be measured. In a sense, it is a "merged beam" experiment with alphas as one merged beam, and the lithium beam as the other.

Given the expense and complications of the high energy lithium beam, simpler techniques would be extremely interesting. However, no other methods have been proposed for measuring the alpha particles as they slow down and heat the plasma.

VII. FURTHER COMMENTS

Particle diagnostics play an important role in fusion experiments. As fusion experiments become larger and hotter, most traditional particle diagnostics become difficult because large plasmas are difficult for neutral atoms to penetrate and the gyro-orbits of charged particles need to be larger than typically obtained with present beams to be comparable with the plasma size. These requirements point to the need for high energy probing beams, with energies considerably above 100 keV.

Neutral beams of these energies will almost certainly be based on negative ions. Fortunately, as we have heard from Dr. Pyle, both hydrogen and elements heavier than hydrogen are suitable for reasonable current negative ion based beam systems. Electrostatic acceleration of the beams will only be feasible up to ~ 500 keV, so that Z > 1 beams will probably require RF accelerators. All of this will probably increase the cost and complexity of diagnostic beams.

Charge transfer cross sections drop rapidly above 20-30 keV/amu, so high energy beams will only measure the portion of the particle distribution parallel to the beam.

Charge-exchange recombination of impurity ions and probing hydrogen beams will be an extremely important technique for studying impurities. Its lack of sensitivity to the electron temperature makes it very powerful. Accurate cross sections for the (n,ℓ) distribution will be essential to exploiting the full potential of this diagnostic technique.

If new, more convenient, techniques can be developed for measuring the plasma current or magnetic field, they will be extremely important. Not only does the current profile affect the plasma stability, but there is a growing opinion that any serious fusion reactor will have to be steady state. This will require some type of steady-state current drive for tokamaks, such as neutral beams or RF. Thus any technique for measuring the current will be extremely important.

ACKNOWLEDGMENTS

The author is grateful to Drs. L. Grisham, R. Fonck, D. Mikkelsen, R. Hulse, and R. Goldston for useful discussions and comments. The author thanks Dr. Goldston for the use of his manuscript [8] prior to publication.

This work was supported by the United States Department of Energy Contract No. DE-AC02-76-CHO-3073.

REFERENCES

[1] Harold Grad, private communication.
[2] N. Bretz et al., Appl. Opt. 17, 192 (1978).
[3] H. P. Furth, "The Tokamak," in Fusion, Vol. 1, E. Teller, Ed., Academic Press (1981).
[4] J. M. Rawls et al., Status of Tokamak Research, DOE/ER-0034, U.S. Department of Energy (1979).
[5] F. H. Coensgen et al., Phys. Rev. Lett. 44, 1132 (1980).
[6] C. DeMichelis, Ed., Nuclear Fusion 18, 647 (1978).
[7] A. Kislyakov and L. Krupnik, Sov. J. Plasma Phys. 7, 478 (1981).
[8] R. J. Goldston, "Diagnostic Techniques for Magnetically Confined High Temperature Plasmas II," (to be published in the Handbook of Plasma Physics; Rosenbluth, Sudan, Sagdeev, Galeev, Ed., North Holland (1982).
[9] D. E. Post, XII International Conference on the Physics of Electronic and Atomic Collisions, Sheldon Datz, Ed., North Holland, New York (1981).

[10] R. Freeman and E. Jones, CLM-R-137, Culham Laboratory, Oxfordshire, (1974).
[11] D. F. Duchs, D. E. Post, P. H. Rutherford, Nuclear Fusion 17, 565 (1977).
[12] D. Mueller, S. L. Davis, C. Keane (to be published).
[13] V. V. Afrosimov, E. L. Berezovskii, I. P. Gladkovskii, A. I. Kislyakov, M. P. Petrov, V. A. Sadovnikov, Sov. Tech. Phys. 20, 33 (1975).
[14] R. Kaita and S. S. Medley, "A Study of the Mass and Energy Resolution of the $E_{||}B$ Charge Exchange Analyzer for TFTR," Princeton University, Plasma Physics Laboratory, PPPL-1582 (1979).
[15] D. Brisson, F. N. Baily, B. H. Quon, J. A. Ray, C. F. Barnett, Rev. Sci. Instr. 51, 511 (1980).
[16] D. Voss and S. Cohen, J. of Nuc. Mat 93 and 94, 405 (1980).
[17] J. Hosea et al., "Fast Wave Ion Cyclotron Heating in the Princeton Large Torus," IAEA-CN-38-D-5-1, Plasma Physics and Controlled Nuclear Fusion Research, Vol. II, IAEA, Vienna, (1980), p. 95.
[18] Yu. S. Gordeev, A. N. Zinov'ev, M. P. Petrov, JETP Letters 25, 204 (1977).
[19] R. J. Goldston, Phys. Fluids 21, 2346 (1978).
[20] R. J. Goldston, E. Mazzucato, R. E. Slusher, and C. M. Surko, Proceedings of the Sixth Conference on Plasma Physics and Controlled Nuclear Fusion Research, Berchtedyaden (1976), p. 371.
[21] R. E. Olson, Phys. Rev. A. 24, 1726 (1981), see also T. A. Green, E. J. Shipsey, and J. C. Browne, Phys. Rev. A 23, 546 (1981); A. Salop, J. Phys. B. 12, 919 (1979); R. K. Janev and I. S. Belic, "Final State Distribution in the Low-Energy Electron Capture Reactions of Hydrogen Atoms with Fully-Stripped Ions," (submitted to Phys. Rev. A); H. Ryufuku, JAERI-M-82-081, April 1982.
[22] R. C. Isler, Phys. Rev. Lett. 38, 1359 (1977); R. C. Isler, L. E. Murray, S. Ksai, J. L. Dunlap, S. C. Batso, P. H. Edmonds, E.A. Lazarus, C. H. Ma, and M. Murakami, Phys. Rev. A 24, 2701 (1981).
[23] V. V. Afrosimov, Yu. S. Gordeev, A. N. Zimov'ev, and A. A. Korotov, Sov. J. Plasma Phys. 5, 551 (1979).
[24] A. N. Zimov'ev, A. A. Korotko, E. R. Krzhizhanovskii, V. V. Afrosimov and Yu. S. Gordeev, JETP Lett. 32, 539 (1980).
[25] R. Fonck, M. Finkenthal, R. Goldston, D. Herdon, R. Hulse, R. Kaita, and D. Meyerhofer, "Spatially Resolved Measurements of Fully-Ionized Low-Z Impurities in the PDX Tokamak," (submitted for publication), PPPL-1906 (1982), Princeton University Plasma Physics Laboratory.
[26] S. S. Medley, R. J. Goldston, and H. H. Towner, "Performance Study of the TFTR Diagnostic Neutral Beam for Active Charge Exchange Measurements," Princeton University, Plasma Physics Laboratory Report, PPPL-1673 (1980).

[27] R. L. Hickok, "Heavy Ion Beam Probing," Rensselear Polytechnic Institute, Troy, New York (1980), RPDL Report 80-14.

[28] J. C. Hosea, F. C. Jobes, R. L. Hickok, A. N. Dellis, Phys. Rev. Lett. 30, 839 (1973).

[29] F. Fujita and K. McCormick, Proceedings of the Sixth European Conference on Controlled Fusion and Plasma Physics, Moscow 1, 191 (1973).

[30] K. McCormick, M. Kick, and J. Olivarin, Proceedings of the Eighth European Conference on Controlled Fusion and Plasma Physics, Prague 1, 140 (1977).

[31] K. Burrell, private communication.

[32] D. Post, D. Mikkelsen, R. Hulse, L. Stewart, and J. Weisheit, Journal of Fusion Energy 1, 129 (1981).

[33] L. Grisham, D. Post, and D. Mikkelsen, "A Multi-MeV Li^{0} Beam as a Diagnostic For Fast Confined Alpha Particles," PPPL-1886 (to appear in Nuclear Technology/Fusion, October 1982).

[34] R. McCullough, T. Goffe, M. Shah, M. Lennon, and H. Gilbody, J. Phys. B 15, 111 (1982).

[35] G. Murray, J. Stone, M. Mazo, and F. Morgan, "Single and Double Electron Transfer in He^{2+} + Li^{0} Collisions," (to appear in Phys. Rev. A).

[36] R. Olson, J. Phys. B 15, L163 (1982).

LECTURERS

B.H. Bransden
Department of Physics, University of Durham, Science Laboratories, South Road, Durham DH1 3LE, England.

F. Brouillard
Institut de Physique, Université Catholique de Louvain, B-1348 Louvain-la-Neuve, Belgium.

H.W. Drawin
Association EURATOM-CEA sur la Fusion, F-92260 Fontenay-aux-Roses, France.

F.J. de Heer
FOM Institute for Atomic and Molecular Physics, Kruislaan 407, 1098 SJ Amsterdam, The Netherlands.

K.T. Dolder
School of Physics, The University, Newcastle upon Tyne, NE1 7RU, England.

G. Ferrante
Istituto di Fisica dell'Università degli Studi, Via Archirafi, 36, 90123 Palermo, Italy.

G. Grieger
Max Planck-Institut für Plasmaphysik, EURATOM Association D-8046 Garching, Federal Republic of Germany.

R. Haas
University of California, Lawrence Livermore National Laboratory, Livermore, California 94550, U.S.A.

M.F.A. Harrison
UKAEA/Euratom Fusion Association, Culham Laboratory, Abingdon, Oxon OX14 3DB, England.

J.T. Hogan
Oak Ridge National Laboratory, P.O. Box Y, Oak Ridge, Tennessee, 37830, U.S.A.

C.J. Joachain
Physique Théorique, Faculté des Sciences, Université Libre de Bruxelles, Campus Plaine, C.P. 227, boulevard du Triomphe, 1050 Bruxelles, Belgium.

M.R.C. McDowell
Department of Mathematics, Royal Holloway College (University of London), Egham Hill, Egham, Surrey TW20 OEX, England.

R.M. More
University of California, Lawrence Livermore National Laboratory, Livermore, California, 94550, U.S.A.

D.E. Post
Plasma Physics Laboratory, Princeton University, Princeton, New Jersey, 08544, U.S.A.

R. Pyle
University of California, Lawrence Berkeley Laboratory, Berkeley, California, 94720, U.S.A.

PARTICIPANTS

R.J. Allan	Department of Physics, The University Newcastle upon Tyne, NE1 7RU England
M. Bacal	Laboratoire de Physique des Milieux Ionisés Ecole Polytechnique 91128 Palaiseau France
J. Bailey	287, 16th Place H5 Costa Mesa, California 92627 U.S.A.
F. Begay	Los Alamos National Laboratory, Mail stop 554 Los Alamos, New Mexico 87545 U.S.A.
S. Bienstock	Harvard Smithsonian Center for Astrophysics Cambridge, Mass. 02138 U.S.A.
M. Bitter	Princeton Plasma Physics Laboratory James Forrestal Campus P.O. Box 451 Princeton, N.J. 08544 U.S.A.
S. Bivona	Universita di Palermo, Istituto di Fisica, Facolta di Ingegneria Viale delle Scienze 90128 Palermo Italy
S. Bliman	S.I.G. Centre d'Etudes Nucléaires de Grenoble 85X - 38041 Grenoble France

Ch. Bouquegneau	Faculté Polytechnique de Mons Département de Physique, rue de Houdain 7000 Mons Belgium
A.M. Bruneteau	Laboratoire de Physique des Milieux Ionisés Ecole Polytechnique 91128 Palaiseau France
R. Burlon	Universita di Palermo, Istituto di Fisica Facolta di Ingegneria, Viale delle Scienze 90128 Palermo Italy
R. Camilloni	Istituto di Methodologia Avanzate Inorganiche Aero Ricerca Romà, C.N.R. CP 10 00016 Monterotondo Italy
P. Cavaliere	Universita di Palermo, Istituto di Fisica Facolta di Ingegneria, Viale delle Scienze 90128 Palermo Italy
S. Cisneros	Universidad Nacional Autonoma Mexico Instituto di Fisica, Apart. post. 20-364 Delegacion Alvaro Obregon 01000 Mexico D.F. Mexico
C. Conde	University of Coïmbra, Departamento de Fisica 3000 Coïmbra Portugal
T.B. Cook	Oak Ridge National Lab., Solid State Division Oak Ridge, Tn. 37830 U.S.A.
R. Daniele	Universita di Palermo, Istituto di Fisica Via Archirafi, 36 90123 Palermo Italy
P. Defrance	Université Catholique de Louvain Institut de Physique, chemin du Cyclotron, 2 1348 Louvain-la-Neuve Belgium

T. Dias	University of Coīmbra, Departamento de Fisica da Universidade 3000 Coīmbra Portugal
R. Dubois	Battelle - Pacific Northwest Lab. P.O.Box 999 Richland, Washington 99352 U.S.A.
E. Fiordilino	Universita di Palermo, Istituto di Fisica Facolta di Ingegneria, Vialle delle Scienze 90128 Palermo Italy
W. Fritsch	Bereich Kern-und Strahlenphysik Hahn Meitner Instituut für Kernforschung Glienicker str. 100 1000 Berlin 39 West Germany
M. Gargaud	Department of Mathematics Royal Holloway College, University of London Egham Hill, Egham, Surrey TW20 0EX England
R. Geller	C.E.N. - G. - DRFC. 85X 98041 Grenoble France
G. Gillespie	Physical Dynamics Inc. P.O.Box 1883 La Jolla, California 92038 U.S.A.
P.T. Greenland	Theoretical Physics Division AERE Harwell, Oxfordshire OX11 0RA England
Y. Hahn	Physics Department, University of Connecticut Storrs, Connecticut 06268 U.S.A.
M. Hendrickx	Université Libre de Bruxelles, Physique Théorique Campus Plaine CP 227, bd du Triomphe 1050 Bruxelles Belgium
R. Hewitt	University of Durham, Department of Physics South Road, Durham City, DH1 3LE England

J. Hiskes	University of California, Lawrence Livermore National Laboratory, L-630 P.O. Box 5511 Livermore, California 94550 U.S.A.
R. Hulse	Princeton University, Plasma Physics Laboratory Princeton, N.J. 08544 U.S.A.
M.S. Huq	Department of Physics, College of William & Mary Williamsburg, Virginia 23185 U.S.A.
B.M. Johnson	Brookhaven National Laboratory Department of Physics, Building 901A Upton, Long Island, New York 11973 U.S.A.
G. Kerlet	Department of Physics, MIT Cambridge, Massachusetts U.S.A.
M. Kilic	Department of Physics, Stevens Institute of Technology Hoboken, N.J. 07030 U.S.A.
G.P. Lafyatis	Harvard University, Center for Astrophysics Harvard College Observatory 60 Garden Str. Cambridge, Massachusetts 02138 U.S.A.
K. Lagattuta	University of Connecticut, Physics Dept. U-46 Storrs, Connecticut 06268 U.S.A.
J.P. Lanquart	Université Libre de Bruxelles Laboratoire de Physique des Plasmas, CP 165 50, av. F.D. Roosevelt 1050 Bruxelles Belgium
L. Lathouwers	Rijksuniversitair Centrum Antwerpen Dienst Theoretische en Wiskunde Natuurkunde Groenenborgerlaan, 171 2020 Antwerpen Belgium

M.A. Lennon	Queen's University of Belfast Department of Pure and Applied Physics Belfast BT7 1NN Northern Ireland
C. Leone	Universita di Palermo, Istituto di Fisica Facolta di Ingegneria, Viale delle Scienze 90128 Palermo Italy
L. Lo Cascio	Universita di Palermo, Istituto di Fisica via Archirafi, 36 90123 Palermo Italy
M. Lieber	University of Arkansas, Physics Department Fayetteville, Ar. 72701 U.S.A.
H. Makowitz	University of Texas at Austin Institute for Fusion Studies Austin, Texas 78712 U.S.A.
G. Mastrocinque	Universita di Napoli Istituto di Fisica, Facolta di Ingegneria P. le Tecchio 80125 Napoli Italy
D.J. McLaughlin	University of Connecticut, Physics Dept. U-46 Storrs, Connecticut 06268 U.S.A.
F. Morales	Universita di Palermo, Istituto di Fisica Facolta di Ingegneria, Viale delle Scienze 90128 Palermo Italy
T. Morgan	Wesleyan University, Physics Department Middletown, Connecticut 06457 U.S.A.
W.L. Morgan	University of California Lawrence Livermore National Lab. Box 808, L-18 Livermore, California 94550 U.S.A.

E. Mund	Université Libre de Bruxelles Service de Metrologie Nucléaire Faculté des Sciences Appliquées 50, av. F.D. Roosevelt, 1050 Bruxelles Belgium
S. Nuzzo	Department of Mathematics Royal Holloway College, University of London Egham Hill, Egham, Surrey TW20 OEX England
S. Oss	Libera Universita degli Studi di Trento Departimento di Fisica 38050 Povo (Trento) Italy
G. Peach	Department of Physics and Astronomy University College London, Gower Street London WC1E 6BT England
B. Piraux	Université Catholique de Louvain Institut de Physique, chemin du Cyclotron, 2 1348 Louvain-la-Neuve Belgium
S. Smith	University of Kentucky College of Arts and Science Department of Physics and Astronomy Lexington, Kentucky 40506 U.S.A.
M. Snyckers	Centre d'Etudes de l'Energie Nucléaire C.E.N./S.C.K., Boeretang, 200 2400 Mol Belgium
B. Spagnolo	Universita di Palermo, Istituto di Fisica Facolta di Ingegneria, Viale delle Scienze 90128 Palermo Italy
A.D. Stauffer	University of York, Faculty of Science Keele str. 4700 Downsview, Toronto M3J 1P3 Canada
G. Stefani	Istituto di Methodologia Avanzate Inorganiche Aereo Ricerea Roma, C.N.R. CP 10 00016 Monterotondo Italy

M. Terao
Université Catholique de Louvain
Institut de Physique, chemin du Cyclotron, 2
1348 Louvain-la-Neuve
Belgium

J. Trebes
Yale University, Applied Physics Mason Lab.
P.O. Box 2159 Yale Station
New Haven, Ct 06520
U.S.A.

F. Trombetta
Universita di Palermo, Istituto di Fisica
Facolta di Ingegneria, Viale delle Scienze
90128 Palermo
Italy

H. Ujc
York University, Faculty of Science
4700 Keelse str.
Downsview, Toronto M3J 1P3
Canada

D. Valenza
Universita di Palermo, Istituto di Fisica
Facolta di Ingegneria, Viale delle Scienze
90128 Palermo
Italy

R. Walling
University of California
Lawrence Livermore National Lab. P.O.Box 808,L-71
Livermore, Ca 94550
U.S.A.

C. Whelan
Depart. of Applied Mathematics & Theo. Physics
University of Cambridge
Silver Street
Cambridge CB3 9EW
England

L. Whitten
University of California
Lawrence Livermore National Lab. P.O.Box 808
Livermore, Ca 94550
U.S.A.

I.D. Williams
Queen's University of Belfast
Department of Pure and Applied Physics
Belfast BT7 1NN
Northern Ireland

H. Winter
Technische Universität Wien
Institut für Allgemeine Physik
Karlsplatz, 13
1040 Wien
Austria

F. Zadvorny Centre d'Etudes Nucléaires de Grenoble
C.E.A. - 85X
38041 Grenoble
France

R. Zangara Universita di Palermo, Istituto di Fisica
Facolta di Ingegneria, Viale delle Scienze
90128 Palermo
Italy

M. Zarcone Department of Mathematics
Royal Holloway College, University of London
Egham Hill, Egham, Surrey TW20 0EX
England

G.B. Zimmerman University of California
Lawrence Livermore National Lab. P.O. Box 808
Livermore, Ca 94550
U.S.A.

INDEX